Environmental Information Management and Analysis: Ecosystem to Global Scales

Environmental Information Management and Analysis: Ecosystem to Global Scales

Series Editors:

Donna J. Peuquet, The Pennsylvania State University
Duane F. Marble, The Ohio State University

Also in this series:

Environmental Information Management and Analysis:
Ecosystem to Global Scales

Edited by

William K. Michener,[1] **James W. Brunt,**[2]
and Susan G. Stafford[3]

[1]Joseph W. Jones Ecological Research Center
Route 2, Box 2324
Newton, GA 31770

[2]Department of Biology
University of New Mexico
Albuquerque, NM 87131-1091

[3]Department of Forest Science
Oregon State University
Corvallis, OR 97331-7501

Taylor & Francis
Publishers since 1798

UK   Taylor & Francis Ltd, 4 John St, London WC1N 2ET
USA Taylor & Francis Inc., 1900 Frost Road, Suite 101, Bristol PA 19007

British Library Cataloguing in Publication Data

A catalogue record for this book is available from the British Library

ISBN 0 7484 0123 7 (cased)

Library of Congress Cataloging in Publication Data are available

Cover design by Amanda Barragry

Typeset by RGM, Southport

Printed in Great Britain by Burgess Science Press, Basingstoke on paper which has a specified pH value on final paper manufacture of not less than 7.5 and is therefore 'acid free'.

Contents

List of figures

List of tables

Preface

Most environmental studies are based on data collected at fine spatial scales (plots, sediment cores, etc.). Furthermore, temporal scales of these studies have been relatively short (days, weeks, months), and few studies have exceeded three years in duration (the typical funding cycle). Despite this history, environmental scientists are now being called upon to extrapolate findings from 'plot-level' studies to broader spatial scales and from short-term studies to longer temporal scales (up to decades for questions related to long-term processes such as global warming and the rise in sea level). In addition, new research and funding initiatives are providing scientists with the opportunity to design and conduct studies that address long-term, broad-scale questions. The complex questions being addressed internationally require that scientists take advantage of new technologies including remote sensing, geographic information systems (GIS), and powerful climatic and environmental simulation models. As more environmental scientists begin to work at these broader spatial and temporal scales, and to utilize many of the newer technologies, they are recognizing a whole new class of problems. For example:

- 'Scale' has become a critical issue. Environmental scientists are now attempting to relate observations made at various scales (e.g. SPOT, 10 m resolution; Landsat, 30 m resolution; AVHRR, 1 km resolution; etc.) to ecological properties occurring in watersheds, regions and the biosphere. Emerging questions are related to the loss or gain of information as one changes scales, the temporal and spatial resolution necessary to identify patterns and change, and how best to incorporate 'the human dimension' into ecosystem, landscape, regional and biosphere models.
- Remote sensing and GIS technologies have been primarily utilized for obtaining 'snap-shots in time' of landscape patterns. Environmental scientists may, however, be more interested in utilizing these tools for examining 'change' at broad spatial scales, linking identifiable patterns to ecological processes, and linking multiple 'snap-shots' to simulation models for both data input and model validation.
- Those scientific database management and information system software packages that are commercially available have not been developed with broad-scale and long-term environmental research in mind. They are especially weak when massive volumes of complex data from multiple sources are required and when spatial 'change' is being examined.
- Ecologists working at scales ranging from individual sites up to the

biosphere are being overwhelmed by the sheer volume of data being generated. Many are seeking advice on how best to design scientific database management and information systems so that data quality is assured, efficiency is optimized and data are protected yet readily available.

- For many of the broad-scale and long-term questions being addressed, it will no longer be feasible for a single scientist to oversee all data collection and processing efforts. Thus, data collected to satisfy a specific objective may be repackaged and utilized in additional studies by numerous scientists who were not associated with the original study. This repackaging will raise many questions for the environmental sciences: how to design data collection efforts for maximizing use; how to facilitate data sharing among investigators, institutions and nations; how to develop appropriate standards for metadata (documentation about the data) so that the data collected by one grouup of scientists may be effectively and appropriately utilized by other scientists, etc.

We recognize that no single book could comprehensively address all of the problems outlined above. We have, however, chosen to include 31 chapters that address several pertinent issues. The book includes comprehensive reviews of selected topics, case studies and theoretical discussions, and is divided into seven sections, each of which is preceded by a brief introduction.

Section I (A research perspective) includes four chapters that identify specific research challenges related to broadening environmental research from ecosystem to global scales and the ramifications of these challenges for the management and analysis of scientific information. The five chapters in Section II (Scientific databases and information systems) include applied examples and theoretical discussions related to the design, implementation and operation of the scientific databases and information systems that can facilitate long-term and broad-scale research. In Section III (Quality assurance/quality control), three chapters provide relevant examples of mechanisms and standards for assessing and increasing data quality. The three chapters in Section IV (Data sharing issues) focus on many of the institutional and technical impediments to data sharing.

The final three sections of the book emphasize many of the analytical aspects associated with addressing broad-scale and long-term environmental issues. Section V (Databases for broad-scale research) includes five chapters that examine many of the issues related to design, development and utilization of environmental databases for broad-scale research. The five chapters in Section VI (Environmental modelling and geographic information systems) include a review of the status of GIS as a tool to support environmental modelling, as well as discussions of technical impediments and some potential solutions that can facilitate environmental modelling and its integration with GIS. The six case studies presented in Section VII (New analytical approaches) provide insights into how some of the latest technological developments are being used to address broad-scale environmental issues.

The foundation for this book was established at an international symposium held during May 1993 in Albuquerque, New Mexico. Support for the symposium and for technical editing was provided by the National Science Foundation (grant no. BIR 9120237). The symposium attracted a large and diverse group of scientists representing the computational, environmental, geographical and statistical sciences. Presentations at this symposium were made by the authors in this book.

The editors would like to thank the authors for their contributions to this book and also for meeting stringent deadlines. The success of the symposium and the realization of this book were made possible through the efforts of many individuals to whom we express our sincere thanks. The contributions of Anne Miller (technical editor) and Ellie Trotter (symposium co-ordinator) are especially appreciated. Discussions with Peter Arzberger, Caroline Bledsoe, Tom Callahan and Jim Edwards provided the impetus for the symposium and governed the selection of topics to be addressed. Approximately 100 scientists helped review the various chapters. Their timely and critical suggestions significantly enhanced the content and organization of each chapter.

Reference to hardware and software, many items of which carry registered trademarks, does not imply endorsement by the authors, editors or sponsoring organizations.

Finally, the editors hope that this volume will prove useful to scientists in the academic community, government agencies and private sector who are in the process of developing environmental databases, managing environmental data and information, and addressing important questions at scales ranging from the ecosystem to the biosphere. We recognize that this volume does not provide a comprehensive treatment of all relevant issues, but we hope that it will inspire future meetings and publications in which scientists from all pertinent disciplines tackle the difficult problems involved.

Contributors

John D. Aber
Complex Systems Center, Institute for the Study of Earth, Oceans, and Space, University of New Hampshire, Durham, NH 03824, USA.

Richard J. Aspinall
GIS and Remote Sensing Unit, Macaulay Land Use Research Institute, Aberdeen, Scotland, AB9 1QJ, United Kingdom.

Lawrence E. Band
Department of Geography & Institute for Land Information Management, University of Toronto, Toronto, Ontario, Canada M5S 1A1.

Tamara L. Bearly
Department of Forest Sciences and Natural Resource Ecology Laboratory, Colorado State University, Fort Collins, CO 80523, USA.

John A. Bognar
Department of Natural Resources, Cook College – Rutgers University, New Brunswick, NJ 08903-0231, USA.

G. A. Bradshaw
Deparment of Forest Science, Oregon State University, Corvallis, OR 97331, USA.

John M. Briggs
Division of Biology, Kansas State University, Manhattan, KS 66506-4901, USA.

Daniel G. Brown
Department of Geography, Michigan State University, East Lansing, MI 48824-1115, USA.

James H. Brown
Department of Biology, University of New Mexico, Albuquerque, NM 87131-1091, USA.

Jesslyn F. Brown
Center for Advanced Land Management Information Technologies, Conservation and Survey Division, University of Nebraska-Lincoln, Lincoln, NB 68588-0517, USA.

James W. Brunt
Department of Biology, University of New Mexico, Albuquerque, NM 87131-1091, USA.

Ingrid C. Burke
Department of Forest Sciences and Natural Resource Ecology Laboratory, Colorado State University, Fort Collins, CO 80523, USA.

David R. Butler
Department of Geography, University of North Carolina, Chapel Hill, NC 27599-3220, USA.

David M. Cairns
Department of Geography, University of Iowa, Iowa City, IA 52242, USA.

James T. Callahan
Long-Term Projects in Environmental Biology, Division of Environmental Biology, National Science Foundation, 4201 Wilson Boulevard, Arlington, VA 22230, USA.

Stephane Casset
Department of Natural Resources, Cook College – Rutgers University, New Brunswick, NJ 08903-0231, USA.

Scott E. Chapal
Joseph W. Jones Ecological Research Center, Route 2, Box 2324, Newton, GA 31770, USA.

Warren B. Cohen
Forestry Sciences Laboratory, 3200 SW Jefferson, Corvallis OR 97331, USA.

Martha B. Coleman
Department of Forest Sciences and Natural Resource Ecology Laboratory, Colorado State University, Fort Collins, CO 80523, USA.

Nicholas R. Chrisman
Department of Geography DP-10, University of Washington, Seattle, WA 98195, USA.

Paul J. Curran
Department of Geography, University of Southampton, Highfield, Southampton S09 5NH, United Kingdom.

D. Richard Cutler
Department of Mathematics and Statistics, Utah State University, Logan, UT 84322-3900, USA.

Jennifer L. Dungan
MS 242-4, NASA Ames Research Center, Moffett Field, CA 94035-1000, USA.

Don Edwards
Department of Statistics, University of South Carolina, Columbia, SC 29208, USA.

Thomas C. Edwards, Jr
U.S. Fish and Wildlife Service, Utah Cooperative Fish and Wildlife Research Unit, Utah State University, Logan, UT 84322-5210, USA.

Jennifer M. Ellis
Complex Systems Center, Institute for the Study of Earth, Oceans, and Space, University of New Hampshire, Durham, NH 03824, USA.

John Evans
Massachusetts Institute of Technology, School of Architecture and Planning, Computer Resource Laboratory, 77 Massachusetts Avenue, Cambridge, MA 02139, USA.

William K. Ferrell
Department of Forest Science, Oregon State University, Corvallis, OR 97331, USA.

Maria Fiorella
Forestry Sciences Laboratory, 3200 SW Jefferson, Corvallis, OR 97331, USA.

James C. French
Institute for Parallel Computation, Department of Computer Science, University of Virginia, Charlottesville, VA 22903, USA.

James Frew
Sequoia 2000 Project, Computer Science Division, University of California, Berkeley, CA 94720, USA.

Steven L. Garman
Department of Forest Science, Oregon State University, Corvallis, OR 97331, USA.

Michael F. Goodchild
National Center for Geographic Information and Analysis, and Department of Geography, University of California, Santa Barbara, CA 93106-4060, USA.

James R. Gosz
Division Director, Division of Environmental Biology, National Science Foundation, 4201 Wilson Boulevard, Arlington, VA 22230, USA.

Mark E. Harmon
Department of Forest Science, Oregon State University, Corvallis, OR 97331, USA.

Bruce P. Hayden
Department of Environmental Science, University of Virginia, Charlottesville, VA 22903, USA.

Karl A. Hermann
ManTech Environmental Technology, Inc., 2 Triangle Drive, Research Triangle Park, NC 27709.

Peter Homann
Department of Forest Science, Oregon State University, Corvallis, OR 97331, USA.

Dennis E. Jelinski
Department of Forestry, Fisheries and Wildlife, Institute of Agriculture and Natural Resources, University of Nebraska-Lincoln, Lincoln, NE 68583-0814, USA.

John J. Kineman
NOAA National Geophysical Data Center, Boulder, CO 80303, USA.

Thomas B. Kirchner
Natural Resource Ecology Laboratory, Colorado State University, Fort Collins, CO 80523, USA.

David P. Lanter
Department of Geography, University of California, Santa Barbara, CA 93105, USA.

Richard G. Lathrop, Jr
Department of Natural Resources, Cook College – Rutgers University, New Brunswick, NJ 08903-0231, USA.

William K. Lauenroth
Department of Range Science and Natural Resource Ecology Laboratory, Colorado State University, Fort Collins, CO 80523, USA.

Charles I. Liff
Bureau of Land Management, USEPA Environmental Monitoring Systems Laboratory, P.O. Box 93478, Las Vegas, NV 89193-3478, USA.

Thomas R. Loveland
US Geological Survey, EROS Data Center, Sioux Falls, SD 57198, USA.

D. Scott Mackay
Institute for Land Information Management, University of Toronto, Erindale Campus, Mississauga, Ontario L5L 1C6, USA.

George P. Malanson
Department of Geography, University of Iowa, Iowa City, IA 52242, USA.

Pamela A. Matson
ESPM Soil Science, University of California, Berkeley, CA 94720, USA.

Blanche W. Meeson
NASA Goddard Space Flight Center, Code 935, Greenbelt, MD 20771, USA.

James W. Merchant
Center for Advanced Land Management Information Technologies, Conservation and Survey Division, University of Nebraska-Lincoln, Lincoln, NB 68588-0517, USA.

William K. Michener
Joseph W. Jones Ecological Research Center, Route 2, Box 2324, Newton, GA 31770, USA.

David Miller
GIS and Remote Sensing Unit, Macaulay Land Use Research Institute, Aberdeen, Scotland, AB9 2QJ, United Kingdom.

Gretchen G. Moisen
USDA Forest Service, Intermountain Research Station, Ogden, UT 84401, USA.

Alan K. Nelson
Universities Space Research Association, NASA Goddard Space Flight Center, Code 610.3, Greenbelt, MD 20771, USA.

Donald O. Ohlen
Hughes STX Corporation, EROS Data Center, Sioux Falls, SD 57198, USA.

Scott V. Ollinger
Complex Systems Center, Institute for the Study of Earth, Oceans, and Space, University of New Hampshire, Durham, NH 03824, USA.

David L. Peterson
MS 242-4, NASA Ames Research Center, Moffett Field, CA 94035-1000, USA.

John L. Pfaltz
Institute for Parallel Computation, Department of Computer Science, University of Virginia, Charlottesville, VA 22903, USA.

Donald L. Phillips
EPA Environmental Research Laboratory, Corvallis, OR, USA.

John H. Porter
Department of Environmental Science, University of Virginia, Charlottesville, VA 22903, USA.

Christopher S. Potter
Johnson Controls, Inc., Mail Stop 239-20, NASA Ames Research Center, Moffett Field, CA 94035, USA.

Bradley C. Reed
Hughes STX Corporation, EROS Data Center, Sioux Falls, SD 57198, USA.

Kurt H. Riiters
Tennesee Valley Authority, Ridgeway Road, Norris, TN 37828, USA.

Vincent B. Robinson
Institute for Land Information Management, University of Toronto, Erindale Campus, Mississauga, Ontario, Canada, L5L 1C6.

Rodney L. Slagle
Lockheed, Environmental Systems and Technologies, 980 Kelly Johnson Drive, Las Vegas, NV 89119, USA.

Phillip Sollins
Department of Forest Science, Oregon State University, Corvallis, OR 97331-7501, USA.

Susan G. Stafford
Department of Forest Science, Oregon State University, Corvallis, OR 97331-7501, USA.

Louis T. Steyaert
521 National Center, United States Geological Survey, Reston, VA 22092.

Donald E. Strebel
Versar, Inc., 9200 Rumsey Road, Columbia, MD 21045-1934, USA.

Haiping Su
Division of Biology, Kansas State University, Manhattan, KS 66506-4901, USA.

Peter M. Vitousek
Department of Biological Sciences, Stanford University, Stanford, CA 94035, USA.

David O. Wallin
Department of Forest Science, Oregon State University, Corvallis, OR 97331, USA.

Stephen J. Walsh
Department of Geography, University of North Carolina, Chapel Hill, NC 27599-3220, USA.

SECTION I

A research perspective

The scope of environmental research is expanding as issues related to scale, spatial variability, global change and the human dimension receive increased attention. In the first chapter, Stafford, Brunt and Michener discuss the ramifications of these new research directions for managing scientific information, specifically, the need for rapid and easier data access; timely, broad-scale, high-resolution data; new analytical approaches; better sampling resolution; and a shift in focus from data to information to knowledge. They discuss advances in several relevant technologies that will affect the management and analysis of environmental information: distributed analytical environments, database management systems, integrated GIS and remote sensing in the time domain, user interfaces, visualization software and knowledge discovery, improved spatial sampling resolution and standardization.

In the remaining three chapters of this section, specific challenges related to broadening research from ecosystem to regional to global scales are identified. Brown concentrates on general scientific challenges including the development of non-experimental methods, new statistical tools and improved methods for ensuring data quality; incorporating new types of data from other disciplines into ecological studies, especially data relating to humans and their activities; and better integrating, synthesizing and modelling ecological knowledge. Gosz discusses many of the technological and cultural changes that will be needed to address issues related to global change, biodiversity and sustainability. Furthermore, he highlights the need for making significant changes in how research projects are funded and carried out, and emphasizes the need for sharing and managing data for long-term access and utility. In the final chapter of the section, Jelinski, Goodchild and Steyaert identify several initiatives needed to facilitate the better use of GIS for research on ecosystem, landscape, regional and global change: better integration of GIS and environmental modelling; new mechanisms for assessing, increasing and communicating data quality; new spatial techniques for analysing data at multiple scales; and the linkage of human and physical databases.

1

Integration of scientific information management and environmental research

Susan G. Stafford, James W. Brunt and William K. Michener

A new dynamic between science and technology is changing the management and analysis of environmental information. As a result of the expanding scope of environmental research, developers of environmental databases must now address difficult and diverse issues: the wide variety of data being considered for inclusion; the size and complexity of databases being collected; their design, development and utilization; the different scales of data collection; increasingly sophisticated analytical requirements; and the integration of data from various sources and disciplines. Increased attention to spatial variability, scale and the integration of basic and applied research to address societal issues has several implications for the management of scientific information. These new research directions have created a need for rapid and easier data analysis; timely, broad-scale, high-resolution data; new analytical approaches; better sampling resolution in space; and a shift in focus from data to information to knowledge. This will require the use of remote sensing, geographic information systems (GIS), spatial statistics and other methods. Several challenges are discussed for future environmental information management and analysis systems: distributed analytical environments, database management systems, integrated GIS and remote sensing in the time domain, user interfaces, visualization software and knowledge discovery, improved spatial sampling resolution and standardization.

Introduction

Historically, environmental research has been conducted as small-scale studies involving one or a few investigators in a single discipline and funded for relatively short periods. Recently, increased understanding of small-scale environmental patterns and processes, coupled with burgeoning interest in landscape, regional and global patterns and processes, has led to the development and funding of studies addressing broad-scale, long-term questions. For example, the Long-Term Ecological Research Program (LTER), initiated by the National Science Foundation in the early 1980s, supports long-term investigations that could not be addressed effectively in the normal 1–3-year funding cycle of most US granting agencies (Franklin *et al.*, 1990). These and related programmes generally involve experiments and monitoring lasting years, decades or

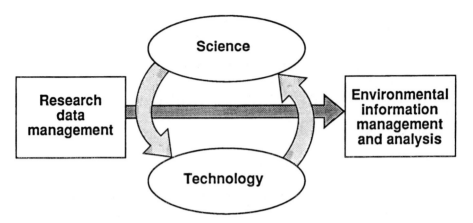

Figure 1.1 A new dynamic between science and technology has forced evolution of the way in which environmental information is now managed and analyzed.

centuries (e.g., Harmon's (1992) 200-year study of log decomposition), and they frequently include many scientists from several disciplines.

As a result of the expanding scope of environmental research, developers of environmental databases must now address difficult and diverse issues: the wide variety of data, both spatial and non-spatial, being considered for inclusion; the size and complexity of databases being created; their design, development and utilization; the different spatial and temporal scales of data collection; increasingly sophisticated analytical requirements; new systems (e.g., Gophers) to provide information about and access to databases; and the integration of data from various sources and disciplines. Concomitantly, attention to management of environmental data within scientific organizations has increased. The intensified focus on effective management of environmental data is based on the premise that data, like people, machines and capital, are a significant organizational resource and product (Michener, 1986).

Coupled with new research directions in the environmental sciences and with improved technologies, data management will evolve far beyond its traditional scope (Figure 1.1). Data management within scientific organizations historically has been driven by the need to support specific tasks (e.g., data entry, archiving, security and quality assurance), many of which may be viewed simplistically as scientific custodial services. The importance of fully integrating data management into the research process – beginning with the design of data format, collection and documentation; following through in data collection, quality assurance, analysis and interpretation; and concluding with synthesis, review and publication – has been emphasized in previous workshops (Stafford *et al.*, 1986; Gorentz, 1992). We see data management within scientific organizations necessarily expanding and evolving to emphasize the timely and effective transformation of data into information and provision of that information to scientists, managers, policy makers and the public. We anticipate that this evolution will lead to the emergence of 'scientific information management' as a discipline. The focus on 'information' will ensure that both management and research are fundamental components of this activity.

In the following discussion, we examine how both the environmental sciences and computational technologies are changing and the implication of these changes for scientific information management. Finally, we describe challenges and research needs in scientific information management.

New Directions in Environmental Research

Increased attention to spatial variability and scale (Levin, 1992) and the emergence of co-directed research (integrated basic and applied research to address societal issues) in environmental science (Sharitz *et al.*, 1992; Swanson and Franklin, 1992) have several implications for the management of scientific information.

Spatial Variability and Issues of Scale

Characteristic scales define the space and time over which processes can be detected, as well as the characteristic dimensions of geographic phenomena. A phenomenon may therefore appear homogeneous at one spatial scale, but heterogeneous at another (Nellis and Briggs, 1989). Many natural systems show hierarchical organization, with nested patterns and processes occurring over a wide range of characteristic spatial and temporal scales (Allen and Starr, 1982; O'Neill *et al.*, 1988; Bradshaw, 1991). Concepts embedded in hierarchy theory highlight the importance of identifying appropriate scales for investigation, and recognizing that new patterns and processes may emerge as the scale of observation is varied (Allen and Starr, 1982). For example, climatic fluctuations over relatively long periods (decades) and large areas (106 m^2) are associated with changes in phenology and population dynamics that produce broad-scale patterns in plant communities. At smaller scales, wind and lightning associated with localized thunderstorms may be responsible for many of the patterns observed in forest stands and ecosystems (Gosz *et al.*, unpublished data).

To address phenomena that depend on spatial pattern, we must find ways to better quantify patterns of variability in space and time and to understand the causes and consequences of pattern and how patterns vary with scale (Levin, 1992). Achieving this will require the use of remote sensing, spatial statistics and other methods.

Co-directed Research

Co-directed research may be defined as the integration of basic and applied research to better address issues with management, policy and societal implications. Foci of co-directed research include natural resource management and sustainable development, environmental monitoring, environmental risk analysis and restoration ecology.

Natural resource management has received much attention recently. In the Pacific Northwest, for example, a new paradigm for forestry is being defined that focuses on forest management primarily for ecosystem structure, diversity and function, rather than for timber and commodity production (Forest Ecosystem Management Assessment Team, 1993). The International Geosphere–Biosphere Programme (Rosswall *et al.*, 1988) and the Sustainable Biosphere Initiative of the Ecological Society of America (Lubchenco *et al.*, 1991; Gosz, this volume) are focusing the efforts of the environmental research community on developing the knowledge and technologies for supporting more sustainable development.

Better management of our natural resources requires that we develop environmental monitoring programmes that allow us to identify and track the condition of specific resources at appropriate spatial and temporal scales. For example, the United States Environmental Protection Agency is launching a multi-year, multi-hundred-million dollar programme to assess status and trends of the earth's natural resources. This programme, the Environmental Monitoring and Assessment Program (EMAP), will focus initially at a national level and eventually expand to global proportions (Environmental Protection Agency, 1991). The United States Geological Survey has begun a more modest programme, the National Water Quality and Assessment Program (NAWQA), to monitor the nation's water resources (Hirsch *et al.*, 1988). Similarly, many new strategies for management of forest resources in the Pacific Northwest may be viewed as broad-scale experiments, the outcome of which can be followed through long-term monitoring.

The success of new resource management strategies and their effects on ecosystems and industrial development cannot be assessed without precise methods of analysing environmental risk. The relatively new field of environmental risk analysis seeks to develop a robust capability for predicting ecosystem risk that can be used for management, policy development and validation of ecosystem models (Whyte and Burton, 1980).

Research in restoration ecology (Jordan *et al.*, 1987; National Research Council, 1992) focuses on applying fundamental ecological principles to the restoration of natural ecological patterns and processes in degraded habitats.

Implications of New Research Directions

These new research directions have created a need for rapid and easier data analysis; timely, broad-scale, high-resolution data; new analytical approaches; better spatial sampling resolution; and a shift in focus from data to information to knowledge. Recent events like the Forest Summit in Portland, Oregon (Forest Ecosystem Management Assessment Team, 1993) and the

Four-Corners virus outbreak in New Mexico (Parmenter *et al.*, 1993) are good examples of the need to alter our trajectory from a data-rich to an information-rich science. The Forest Summit, for example, required panels of experts to integrate and synthesize vast amounts of data into meaningful information within 60 days in order to produce scientifically credible options for policies guiding management of forest and aquatic resources in the Pacific Northwest. Rapid analysis of data from disparate sources was critical in relating the Four Corners virus outbreak to the increasing population densities of rodents.

In order to address issues of scale and spatial variability, data are being collected over a broader spectral range, at finer spatial sampling resolution, and at shorter sampling intervals (Wickland, 1991). Most data used in analysing the spatial dependence of surface and atmospheric phenomena are obtained by satellite- and aircraft-mounted, multispectral sensors. Environmental scientists are increasingly employing geographic information systems (GIS) and remotely sensed data to examine questions at multiple spatial and temporal scales. Data density, which depends jointly on spatial and temporal resolution, is a technical consideration. Continuous increases in computing and storage capacities have shifted studies toward smaller temporal and spatial scales. As a result, processes operating at relatively fine spatial and temporal scales can now be investigated with finer spectral resolution over increasingly large areas (Davis *et al.*, 1991).

According to one estimate, the amount of data in the world doubles every 20 months (Frawley *et al.*, 1992). Earth-observing satellites launched in this decade will generate one terabyte of data every day – more than all previous missions combined. The Earth Observing System (EOS) will create roughly the equivalent of the entire 17-year Landsat archive every two weeks (Marshall, 1989). Similarly, the federally funded Human Genome project will store thousands of bytes for each of the several billion genetic nucleotide bases (National Research Council, 1988). Moreover, scientists are loath to discard any data for fear that a scientist in the future may need exactly that data set to test a specific hypothesis (French *et al.*, 1990; Silberschatz *et al.*, 1991). As a result, the gap between data generation and data understanding is growing (Frawley *et al.*, 1992), and our ability to gather and store data is much more advanced than our ability to manage, analyse and interpret them.

The new analyses needed for examining spatially correlated data require abandoning many of the traditional experimental designs developed primarily for agricultural experimentation (e.g., focusing on Type I and Type II errors). New experimental designs, broad-scale, spatially explicit sampling programs, geostatistical tools, and other methods that will allow us to better identify patterns and link them with processes must be developed. The extensive databases generated by EOS and other programs highlight the need for better tools that will support time-series analyses of spatial data.

Evolving Technologies Applicable to New Directions in Environmental Research

Several technologies that have been developed and improved during the past decade are now being applied increasingly to environmental research. Major advances include greatly increased desktop computing power, new analytical approaches and software, new database designs and the proliferation of remote sensing and GIS.

Advances in Computer Technology

The 1980s saw a dramatic convergence in capabilities of mini-, micro- and mainframe computers. The introduction in 1981 of the IBM personal computer (PC), based on the Intel 8086 computer chip, propelled microcomputers into a high profile. IBM extended PC capabilities initially with the Intel 80286, a 16-bit processor, and then the Intel 80386, a 32-bit processor, providing faster data access and combining multiple instruction executions in clock time (Faust *et al.*, 1991). Minicomputers and workstations have achieved mainframe speeds, and distributed processing strategies via networked PCs have fundamentally altered computing environments. In the short time between the development of the VAX 11/780 minicomputer in the late 1970s and the present, we have gone from a realizable speed of 1 million instructions per second (MIPS) to about 50 MIPS on current-generation single-CPU workstations.

The rate of development of new computer hardware tends to be under-estimated, whereas that of software improvements usually is overestimated. Nevertheless, according to recent projections for hardware development (Frank *et al.*, 1991), processing speed will approximately double each year; prices for main memory will decrease about 50 per cent biannually; hard-disk access times will increase very slowly; the capacity of hard disks will increase as prices decrease; and communication networks for the exchange of large volumes of data will proliferate.

In the near future, we anticipate faster communication networks with speeds of 100 megabits/second (Mb s^{-1}); faster CPUs, with 500–1000 MIPS; more efficient devices that can store terabytes; higher resolution monitors; and higher level programming languages, both visual and object-oriented. Networks will allow complete transparency among machines world-wide, and desktop access to supercomputers will become common. By the late 1990s, the specifications for a personal GIS workstation may be a CPU with 500–1000 MIPS; 500 megabytes of main memory; 5 gigabytes of storage space on hard disks and an additional 50 gigabytes on optical disk; a workstation screen 2000 × 2000 pixels; and a communication device with a transfer rate of 100 Mb s^{-1}.

Development of software for these hardware platforms is expensive and time-consuming and lags significantly behind hardware improvements. Popular 'new' programming languages such as C and Pascal are almost 20 years old. Current operating systems (VMS, MS-DOS and UNIX), also were developed 10–20 years ago to circumvent obstacles made obsolete by rapidly changing hardware technologies. UNIX, for example, with its cryptic short commands, was invented to reduce typing when input was commonly by teletype (Winkler, 1986).

Although software development lags far behind, extensible software systems promise relief. Visual programming environments like Khoros (Rasure *et al.*, 1990), and database management systems (DBMS) and analytical systems like S+ (Becker *et al.*, 1988) allow application building and customization for specific projects.

Analytical Approaches

New emphasis on spatial scale and structure has increased characterization and modelling by spatial statistics (Robertson, 1987). Spatial variation can be modelled in a variety of ways by covariance, variograms or fractal analysis. Meteorologists and oceanographers are able to use the covariance or the power-spectrum approach (the Fourier transform of the covariance), since they can overcome some of the inherent limitations of these techniques by assuming stationarity of their data (Davis *et al.*, 1991). Semi-variograms have been used to characterize spatial variability in topography (Dubayah *et al.*, 1989), solar radiation (Dubayah *et al.*, 1990), surface albedo (Webster *et al.*, 1989), spectral vegetation indices (Davis *et al.*, 1989), soil properties (Robertson *et al.*, 1988) and vegetation (Pastor and Broschart, 1990). Precipitation (Lovejoy and Schertzer, 1985), soil properties (Burrough, 1983a, 1983b) and land use (Milne, 1991) have recently been analysed with fractals. Explicit modelling of spatial statistical structure should improve the interpolation of point data, optimize sampling designs for field collections and contribute to our understanding of underlying processes (Davis *et al.*, 1991).

New research directions have also created the need for better resolution in spatial sampling in order to understand phenomena at relevant spatial scales. For example, in order to predict spotted owl habitat, Ripple *et al.* (1991) decomposed landscape pattern by scale (coarse to fine), using the wavelet transform. Phenomena were matched to their appropriate scale of investigation: nest tree to the tree level; thermal cover to the stand level; prey base and old-growth patch connectivity to the landscape level; and conifer cover to the subregion level. This technique has numerous potential applications in land-scape ecology, where analysis may focus on several scales of pattern simultaneously.

Remote Sensing and GIS

Remote sensing and GIS technology, which support the computerized capture, management, display and analysis of spatial data (Thapa and Bossler, 1992),

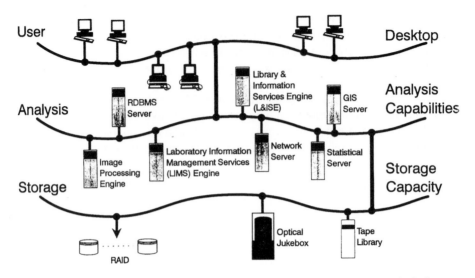

Figure 1.2 Analytical and storage capabilities accessible from the future user's desktop. Note: RAID: Redundant Array of Inexpensive Disks.

have opened new frontiers in our ability to monitor and manage natural resources – from mapping soil mycorrhizal mats (Griffiths *et al.*, 1993) to analysing regional phenomena (Burke *et al.*, 1991). GIS technology has migrated to smaller machines with ever-increasing processing power and ever-decreasing costs (Juhl, 1989). Unfortunately, although many view GIS as a panacea solving data volume and management problems, more important problems still underlie GIS, including the lack of adequate DBMS software to manage scientific data effectively and the lack of sophisticated statistical software and other analytical tools within the GIS environment.

Challenges for Future Environmental Information Management and Analysis Systems

Distributed Analytical Environments

A primary challenge in this area is creation of a highly interactive metacomputer composed of workstations with graphics accelerators, high-speed networks, terabyte data vaults and supercomputers that can address multidisciplinary 'grand challenge' problems. In the future, scientific organizations will likely use multiple file and computer servers with ease from their desktop computers. File servers may be dedicated to one or more specific tasks (e.g., statistics, a DBMS, library support, or image processing; Figure 1.2). Hierarchical storage servers will provide transparent access to huge amounts of optical disk and tape storage previously maintained off-line.

Organizations may have heterogeneous DBMSs differing in their capabilities and structure and dispersed over several sites. The heterogeneity, distribution and sheer number of databases in the public and private domains make it imperative to share access to these databases while maintaining autonomy of individual systems. Thus, new techniques are needed that support the interoperability of autonomous databases without requiring their global integration (Scheuermann *et al.*, 1990).

In the system envisioned for the future, a user submits a database query, completely unaware of and unconcerned with the location and format of the data. The software automatically locates the data across any number of DBMS, processes them and returns the answer. Security, error correction and rollback are independent of user, DBMS, vendor and location. Multimedia, hypertext and hypermedia technologies will be increasingly used for creating, organizing and disseminating on-line information (Mohageg, 1992) and for combining text and graphics with working models and imagery.

The complex nature of spatial, as compared with record-oriented, data complicates data distribution and communication for distributed GIS. Network technology ultimately will develop to the point where intermachine communication is no longer an issue. The challenge will be in providing software to utilize this type of network efficiently. Thus, scientists in Albuquerque, New Mexico, will be able to interactively access soils data stored on the United States Geological Survey server in Denver, Colorado, and climate data stored on the National Climate Data Center server in Asheville, North Carolina.

Database Management Systems

The promise of a DBMS lies in its ability to provide timely and consistent data, to enable users to access those data directly without technical assistance and to evolve easily to meet changing user requirements (Curtice, 1986). Environmental research at the project level requires integrated computational services that provide access to sophisticated database, statistical and numeric operations and to technical text support (Conley *et al.*, 1986). These operations could be supported collectively by a scientific DBMS; however, most commercial DBMS software is record-oriented and directed specifically at business solutions.

If the next-generation database applications are to meet the needs of environmental scientists, they are unlikely to resemble current business-oriented DBMSs (Silberschatz *et al.*, 1991). Recently, considerable research has focused on developing general-purpose DBMS prototypes that present some advantages over relational DBMS (RDBMS) in representing complex (i.e., spatial) data, which do not lend themselves easily to representation or efficient management within a record-oriented system. One such prototype is POSTGRES (for post Ingres) (Pollock and McLaughlin, 1991). Additional research in this area focuses on extending current RDBMS technology and developing object-

Table 1.1 Advanced database management system (DBMS) strategies with potential for scientific information management.

Extended relational:	incorporates new features in the relational model, such as inheritance, user-defined functions and data types, and procedures for data types and nested relations.
Object-oriented:	accesses a complex object as a single concept and performs operations on that object in the DBMS, rather than in an external application program (Dittrich, 1986); supports inclusion of 'unconventional' data types, such as text, voice, graphics or CAD (computer-assisted drafting) (Scheuermann *et al.*, 1990).
Extensible:	utilizes a 'toolkit' approach to tailor DBMS internals (e.g., storage structures, query optimizers) to meet application requirements that are inadequately served by existing DBMSs; supports easier customization of DBMSs for scientific applications.
Logic-based languages:	support access to relational RDBMS by providing a consistent interface between the tools and the RDBMS (Hessinger, 1987); e.g., Structured Query Language (SQL).

oriented and extensible DBMSs and logic-based languages (Table 1.1) (Shoshani, 1990). The overall challenge will be integrating the best of each of these techniques to manage scientific data.

Integrated GIS and Remote Sensing in the Time Domain

Overcoming the technological obstacles associated with jointly processing image and GIS data, such as converting between vectors and rasters and jointly displaying and updating digital imagery and maps, has already entailed much work. Continued efforts in this area should lead to the evolution of integrated geographic information systems (IGIS) (Davis *et al.*, 1991). Since not all physical processes can be mapped directly from remotely sensed data or existing maps, an important role of IGIS will be the generation of such information by appropriate modelling based on more than one information source (Risser, 1986). The coupling of satellite measurements with other spatial data has tremendous potential for describing earth surfaces, predicting future conditions and validating physical representations produced solely through remote sensing or GIS. One key impediment to total integration of GIS and remote sensing is that current GIS technology does not adequately represent the time domain, whereas remote sensing samples in both space and time. Causal relationships among events or entities are embedded in temporal information, and data therefore should be interpreted in a temporal context (Snodgrass and Ahn, 1986).

Errors associated with remotely sensed and GIS data can significantly impact the confidence with which decisions are made (Lunetta *et al.*, 1991). GIS provides new information without simultaneously establishing its reliability. Uncertainty is propagated and transformed each time a conceptual or physical model is constructed in GIS processing (Bedard, 1987). Lack of concern for error in spatial databases reflects the lack of a standard framework for analysing the propagation of modelling error (Lanter and Veregin, 1992).

As these technologies continue to improve, the potential of IGIS analysis will come to be limited only by our conceptual understanding of the phenomena under investigation and their representation in spatial databases (Davis *et al.*, 1991).

User Interfaces

The improvement of interfaces within and among image processing, GIS, DBMS, expert systems, models and statistical software will facilitate analyses significantly. To meet the needs of environmental scientists, future systems will emphasize sharing data and tools among disparate software environments, rather than computing power. More powerful and more efficient languages and user interfaces are needed that can lead less computer-literate users systematically through common data management and information processing without requiring them to master the intricacies of the DBMS and the computer operating system. Ideally, the further development of higher level languages will allow the investigator to focus on the problem at hand (Peled, 1987).

There is also a strong need for better management of critical information associated with models and model input and output – that is, model metadata. This includes keeping good records of model evolution and of the lineage of data as they are used to produce publications or as input to other models or analyses that may cascade to further models. The bottom line is clearly linking the data with the models. Intuitively designed user interfaces can help expedite this.

Visualization Software and Knowledge Discovery

Visualization tools have been developed and primarily utilized by physical scientists. One of the challenges for environmental researchers will be the use of such tools for extracting information from the massive volumes of data being collected. Visualization software will be a required part of the environmental scientist's toolkit.

Knowledge discovery, the use of computers to extract implicit and potentially useful information from data, carries potential for tackling some of the enormous data analysis challenges of the next generation of environmental scientists. Knowledge discovery is expensive even with today's computers and requires additional knowledge regarding the form and content of data.

Visualization software, knowledge discovery algorithms and expert systems will necessarily be used increasingly to facilitate the processing of vast amounts of data into the information needed by scientists, managers and policy makers.

Improved Spatial Sampling Resolution and Standardization

In most environmental research, data are collected in plots of $1 \, m^2$ or less (Kareiva and Anderson, 1988; Brown and Roughgarden, 1990). Few studies entail collection of pattern- and process-oriented data on the ground that can be readily related to the scale of data sampled by Landsat Thematic Mapper (30 m × 30 m) or SPOT (10 m × 10 m, panchromatic; 20 m × 20 m, multispectral). Thus, there is currently a wide discrepancy between the spatial scales of data collected by remote sensing and those collected on the ground that can be used for verification and validation. This gap likely will decrease as scientists take advantage of new airborne remote sensing systems that sample data at the submetre scale.

Nevertheless, several impediments to scaling up from the plot to broader scales likely will remain. For example, several networks (including the LTER Network, Global Change Sites funded by the National Park Service, and National Oceanic Atmospheric Administration Marine Sanctuaries) have been established to examine ecological patterns and processes at the scale of the continental USA. Certain attributes of primary production are examined at most or all of the sites; however, specific parameters being measured, temporal and spatial resolution, field and laboratory protocols, and data management and analysis generally differ significantly among sites. In addition, numerous biomes within any given network are generally not sampled, and spatial replication within biomes is rare or non-existent. Expanding individual networks to encompass more of the inter- and intra-biome variability, as well as integrating individual networks into a 'network of networks', could facilitate broad-scale research by ensuring better spatial coverage and encouraging data comparability and information exchange (Bledsoe and Barber, 1993). We need to focus attention for intersite comparison at the information stage, however, in part because the big differences among biomes and ecosystems may make standardized data collection much less useful than standardized approaches to information assembly and management. Spatial and temporal variability, standardization of monitoring programmes, laboratory and analytical protocols, and data and information management practices likely will continue to challenge us.

Conclusion

The virtual explosion in computing power and accessibility has significantly revolutionized the scientific environment. Managing scientific information has gone far beyond a mere 'custodial service'. Users of scientific databases work in a great variety of environments, from a single personal computer or workstation to supercomputers to heterogeneous networked systems. Network technology ultimately will develop to the point where intermachine communication is no longer an issue, and the potential for analysis will be limited only by our conceptual understanding and our ability to frame the questions.

We face challenges in monitoring and modelling global change. Of these, two interrelated problems are key: (1) combining remote sensing and GIS, and (2) determining the nesting of observations required to bridge scales between short-term, fine-scale and long-term, broad-scale measurements and process models (Rosswall *et al.*, 1988). Tight linkages between research and application will be critical to monitoring the environmental and social consequences of global change (Davis *et al.*, 1991).

Because large and diverse data sets, documents and images are becoming common in both business and science, the gap between business and scientific software is closing. DBMS software that will be useful to both business and scientific applications should result. A properly instituted and supported DBMS can increase the productivity of environmental researchers and enhance the quality of environmental science (Risser and Treworgy, 1986), reducing the time scientists devote to learning details of hardware and software. This level of DBMS for a large project must be supported with a great deal of forethought and ongoing commitment.

Technologically advanced societies frequently cannot transform new ideas into successful products quickly (Strome and Lauer, 1977). Acceptance of technology in any field is complex, and decades often elapse before practical applications of research are achieved. Barriers to communication among engineers, environmental scientists, computer scientists, policy makers and others further delay product development.

A sociology of information management is developing that is as critical as the hardware and software issues. Because bridges must be built between researchers and data managers, training a new breed of scientists and scientific information managers will be a prerequisite to the conscientious efforts required to close these gaps. We anticipate the emergence of 'scientific information management' as a discipline with both a management and a research component, emphasizing the timely and effective transformation of data into information and knowledge for scientists, managers, policy makers and the public.

Acknowledgement

We acknowledge National Science Foundation support from DEB No. 90-11663 to the H. J. Andrews LTER project and BSR No. 88-110946 support to the Sevilleta LTER. This chapter benefited from discussions with Fred Swanson. This is paper 2974, Forest Research Laboratory, Oregon State University and contribution 48 to Sevilleta LTER, University of New Mexico.

References

Allen, T. F. H. and Starr, T. B., 1982, *Hierarchy, Perspectives for Ecological Complexity*, Chicago: University of Chicago Press.

Becker, R. A., Chambers, J. M. and Wilks, A. R., 1988, *The New S Language: A Programming Environment for Data Analysis and Graphics*, Pacific Grove, California: Wadsworth and Brooks/Cole Advanced Books and Software.

Bedard, Y., 1987, Uncertainties in land information systems databases, *Proceedings, Auto-Carto*, **8**, 175–84.

Bledsoe, C. and Barber, M., 1993, *Ecological Network of Networks*, Washington, DC: US MAB Secretariat, US Department of State.

Bradshaw, G. A., 1991, 'Hierarchical analysis of pattern and processes of Douglas-fir forests', unpublished PhD dissertation, Corvallis, Oregon: Department of Forest Science, Oregon State University.

Brown, J. H. and Roughgarden, J., 1990, Ecology for a changing earth, *Bulletin of the Ecological Society of America*, **71**, 173–88.

Burke, I. C., Kittel, T. G. F., Lauenroth, W. K., Snook, P., Yonker, C. M. and Parton, W. J., 1991, Regional analysis of the central Great Plains, *BioScience*, **14** (10), 685–92.

Burrough, P. A., 1983a, Multiscale sources of spatial variation in soil, I: The application of fractal concepts to nested levels of soil variation, *Journal of Soil Science*, **34**, 577–97.

Burrough, P. A., 1983b, Multiscale sources of spatial variation in soil, II: A non-Brownian fractal model and its application in soil survey, *Journal of Soil Science*, **34**, 599–620.

Conley, W., Slator, B. M., Anderson, M. P. and Sitze, R. A., 1986, Designing and prototyping a scientific problem solving environment: the NMSU science workbench, in Michener, W. K. (Ed.) *Research Data Management in the Ecological Sciences*, Belle W. Baruch Library in Marine Science, No. 16, pp. 383–409, Columbia, South Carolina: University of South Carolina Press.

Curtice, R. M., 1986, Getting the database right, *Datamation*, 1 October, 99–104.

Davis, F., Dubayah, R., Dozier, J. and Hall, F., 1989, Covariance of greenness and terrain variables over the Konza Prairie, *Proceedings IGARSS '89*, 1322–5.

Davis, F. W., Quattrochi, D. A., Ridd, M. K., Lam, N. S.-N., Walsh, S. J., Michaelson, J. C., Franklin, J., Stow, D. A., Johannsen, C. J. and Johnson, C. A., 1991, Environmental analysis using integrated GIS and remotely sensed data: Some research needs and priorities, *Photogrammetric Engineering and Remote Sensing*, **57** (6), 689–97.

Dittrich, K., 1986, Object-oriented database systems: The notion and the issues, in *Proceedings of the International Workshop on Object-oriented Database Systems*, pp. 2–4, Washington, DC: IEEE. Computer Society Press.

Dubayah, R., Dozier, J. and Davis, F. W., 1989, The distribution of clear-sky radiation over varying terrain, *Proceedings IGARSS '89*, 885–8.

Dubayah, R., Dozier, J. and Davis, F. W., 1990, Topographic distribution of clear-sky radiation over the Konza Prairie, Kansas, *Water Resources Research*, **267**, 679–90.

Environmental Protection Agency, 1991, *An Overview of the Environmental Monitoring and Assessment Program, EMAP Monitor, January 1991*, EPA-6001 M-90/022, Washington, DC: US EPA, Office of Research and Development.

Faust, N. L., Anderson, W. H. and Star, J. L., 1991, Geographic information systems and remote sensing future computing environment, *Photogrammetric Engineering and Remote Sensing*, **57** (6), 655–68.

Forest Ecosystem Management Assessment Team, 1993, *Forest Ecosystem Management: An Ecological, Economic, and Social Assessment, Report of the Forest*

Ecosystem Management Assessment Team, 1993-793-071, Washington, DC: US Government Printing Office.

Frank, A. U., Egenhofer, M. J. and Kuhn, W., 1991, A perspective on GIS technology in the nineties, *Photogrammetric Engineering and Remote Sensing*, **57** (11), 1431 6.

Franklin, J. F., Bledsoe, C. S. and Callahan, J. T., 1990, Contributions of the long-term ecological research program, *BioScience*, **40** (7), 509 23.

Frawley, W. J., Piatetsky-Shapiro, G. and Mathews, C. J., 1992, Knowledge discovery in databases: an overview, *AI Magazine*, fall, 57 70.

French, J. C., Jones, A. K. and Pfaltz, J. L., 1990, Summary of the final report of the NSF Workshop on Scientific Database Management, *Sigmod Record*, **19** (4), 32 40.

Gorentz, J. B. (Ed.), 1992, *Data Management at Biological Field Stations and Coastal Marine Laboratories*, Report of an invitational workshop, W. K. Kellogg Biological Station, Michigan: Michigan State University April 1990.

Griffiths, R. P., Bradshaw, G. A. and Caldwell, B. A., 1993, 'Distribution of ectomycorrhizal mats in coniferous forests in the Pacific Northwest', presentation at an international symposium: The Functional Significance and Regulation of Soil Biodiversity, East Lansing, Michigan, May.

Harmon, M. E., 1992, *Long-term Experiments on Log Decomposition at the H. J. Andrews Experimental Forest*, USDA Forest Service General Technical Report PNW-GTR-280.

Hessinger, P. R., 1987, DBMS: Adding value to vanilla, *Datamation*, 1 March, 50 4.

Hirsch, R. M., Alley, W. M. and Wilbur, W. G., 1988, *Concepts for a National Water-Quality Assessment Program*, US Geological Survey Circular 1021.

Jordan, W. R., III, Gilpin, M. E. and Aber, J. D. (Eds), 1987, *Restoration Ecology*, Cambridge, UK: Cambridge University Press.

Juhl, G., 1989, GIS technology coming of age, *American City and County*, April, 50 4.

Kareiva, P. and Anderson, M., 1988, Spatial aspects of species interactions: the wedding of models and experiments, in Hastings, A. (Ed.) *Community Ecology*, New York: Springer Verlag.

Lanter, D. P. and Veregin, H., 1992, A research paradigm for propagating error in layer-based GIS, *Photogrammetric Engineering and Remote Sensing*, **58** (6), 825 33.

Levin, S. A., 1992, The problem of pattern and scale in ecology, *Ecology*, **73** (6), 1943 67.

Lovejoy, S. and Schertzer, D., 1985, Generalized scale invariance in the atmosphere and fractal models of rain, *Water Resources Research*, **21**, 1233 50.

Lubchenco, J., Olson, A. M., Brubaker, L. B., Carpenter, S. R., Holland, M. M., Hubbell, S. P., Levin, S. A., MacMahon, J. A., Matson, P. A., Melillo, J. M., Mooney, H. A., Peterson, C. H., Pulliam, H. R., Real, L. A., Regal, P. J. and Risser, P. G., 1991, The sustainable biosphere initiative: an ecological research agenda, *Ecology*, **72** (2), 371-412.

Lunetta, R. S., Congalton, R. G., Fenstermaker, L. K., Jensen, J. R., McGuire, K. C. and Tinney, L. R., 1991, Remote sensing and geographic information system data integration: error sources and research issues, *Photogrammetric Engineering and Remote Sensing*, **57** (6), 677-87.

Marshall, E., 1989, Bringing NASA down to earth, *Science*, **244** (16 June), 1248 51.

Michener, W. K., 1986, Data management and long-term ecological research, in Michener, W. K. (Ed.) *Research Data Management in the Ecological Sciences*, Belle W. Baruch Library in Marine Science, No. 16, pp. 1 8, Columbia, South Carolina: University of South Carolina Press.

Milne, B. T., 1991, Lessons from applying fractal models to landscape patterns, in Turner, M. G. and Gardner, R. H. (Eds) *Quantitative Methods in Landscape Ecology*, pp. 199 235, New York: Springer Verlag.

Mohageg, M. F., 1992, The influence of hypertext linking structures on the efficiency of information retrieval, *Human Factors*, **34** (3), 351-67.

National Research Council, 1988, *Mapping and Sequencing the Human Genome*, Report of the Committee on Mapping and Sequencing the Human Genome, Washington, DC: National Research Council.

National Research Council, 1992, *Restoration of Aquatic Ecosystems: Science, Technology and Public Policy*, Washington, DC: National Academy Press.

Nellis, M. D. and Briggs, J. M., 1989, The effect of spatial scale on Konza landscape classification using textural analysis, *Landscape Ecology*, 2, 93-100.

O'Neill, R. V., Krummel, J. R., Gardner, R. H., Sugihara, G., Jackson, B., DeAngelis, D. L., Milne, B. T., Turner, M. G., Zygmunt, B., Christensen, S. W., Dale, V. H. and Graham, R. L., 1988, Indices of landscape pattern, *Landscape Ecology*, 1, 153-62.

Parmenter, R. R., Brunt, J. W., Moore, D. I. and Ernest, S., 1993, *The Hantavirus Epidemic in the Southwest: Rodent Population Dynamics and the Implications for Transmission of Hantavirus-associated Adult Respiratory Distress Syndrome (HARDS) in the Four Corners Region*, a special report to the Federal Centers for Disease Control and Prevention and the New Mexico Department of Health, Publication No. 41, Albuquerque, New Mexico: Sevilleta Long-Term Ecological Research Program.

Pastor, J. and Broschart, M., 1990, The spatial pattern of a northern conifer-hardwood landscape, *Landscape Ecology*, 4, 55-68.

Peled, A., 1987, The next computer revolution, *Scientific American*, 257 (4), 57-64.

Pollock, R. J. and McLaughlin, J. D., 1991, Data-based management system technology and geographic information systems, *Journal of Surveying Engineering*, 117 (1), 9-26.

Rasure, J., Williams, C., Argiro, D. and Sauer, T., 1990, A visual language and software development environment for image processing, *International Journal of Imaging Systems and Technology*, 2, 183-99.

Ripple, W. J., Bradshaw, G. A. and Spies, T. A., 1991, Measuring forest landscape patterns in the Cascade Range of Oregon, USA, *Biological Conservation*, 57, 73-88.

Risser, P. G., 1986, *Spatial and Temporal Variability of Biosphere and Geosphere Processes: Research Needed to Determine Interactions with Global Environmental Change*, Paris: ICSU Press.

Risser, P. G. and Treworgy, C. G., 1986, Overview of research data management, in Michener, W. K. (Ed.) *Research Data Management in the Ecological Sciences*, Belle W. Baruch Library in Marine Science, No. 16, pp. 9-22, Columbia, South Carolina: University of South Carolina Press.

Robertson, G. P., 1987, Geostatistics in ecology: Interpolating with known variance, *Ecology*, 68, 744-8.

Robertson, G. P., Huston, M. A., Evans, F. C. and Tiedge, J. M., 1988, Spatial variability in a successional plant community: Patterns of nitrogen availability, *Ecology*, 69, 1517-24.

Rosswall, T., Woodmansee, R. G. and Risser, P. G. (Eds), 1988, *Scales and Global Change: Spatial and Temporal Variability in Biospheric and Geospheric Processes, Scientific Committee on Problems of the Environment* (SCOPE) Report 35, New York: Wiley.

Scheuermann, P., Yu, C., Elmagarmid, A., Manola, F., Rosenthal, A., Garcia-Molina, H., McLeod, D. and Templeton, M., 1990, Report on the workshop on heterogenous database systems held at Northwestern University, Evanston, Illinois, December, 1989, *Sigmod Record*, 19 (4), 23-31.

Sharitz, R. R., Boring, L. R., Van Lear, D. H. and Pinder, J. E., III, 1992, Integrating ecological concepts with natural resource management of southern forests, *Ecological Applications*, 2 (3), 226-37.

Shoshani, A., 1990, Emerging and new technologies, in French, J. C., Jones, A. K. and Pfaltz, J. L. (Eds) *Scientific Database Management, Report of the Invitational NSF*

Workshop on Scientific Database Management, Computer Science Report No. TR-90-22, pp. 13–21, Charlottesville, Virginia: University of Virginia.

Silberschatz, A., Stonebraker, M. and Ullman, J. (Eds), 1991, Database systems: Achievements and opportunities, *Communications of the ACM*, **34** (10), 110–20.

Snodgrass, R. and Ahn, I., 1986, Temporal databases, *Computer*, September, 35–42.

Stafford, S. G., Alaback, P. B., Waddell, K. L. and Slagle, R. L., 1986, Data management procedures in ecological research, in Michener, W. K. (Ed.) *Research Data Management in the Ecological Sciences*, Belle W. Baruch Library in Marine Science, No. 16, pp. 93–114, Columbia, South Carolina: University of South Carolina Press.

Strome, W. M. and Lauer, D. T., 1977, An overview of remote sensing technology transfer in the US and Canada, *Proceedings, 11th International Symposium on Remote Sensing of Environment*, Vol. 1, pp. 325–31, Ann Arbor, Michigan: Environmental Research Institute of Michigan.

Swanson, F. J. and Franklin, J. F., 1992, New forestry principles from ecosystem analysis of Pacific Northwest forests, *Ecological Applications*, **2** (3), 262–74.

Thapa, K. and Bossler, J., 1992, Accuracy of spatial data used in geographic information systems, *Photogrammetric Engineering and Remote Sensing*, **58** (6), 835–41.

Webster, R., Curran, P. J. and Munden, J. W., 1989, Spatial correlation in reflected radiation from the ground and its implications for sampling and mapping by ground-based radiometry, *Remote Sensing of Environment*, **29**, 67–78.

Whyte, A. V. T. and Burton, I. (Eds), 1980, *Environmental Risk Assessment*, prepared by the Scientific Committee on Problems of the Environment (SCOPE) of the International Council of Scientific Unions (ICSU), New York: Wiley.

Wickland, D. E., 1991, Mission to planet earth: The ecological perspective, *Ecology*, **72** (6), 1923–45.

Winkler, C., 1986, Getting from here to there: No way, *Datamation*, 1 June, 202–5.

2

Grand challenges in scaling up environmental research

James H. Brown

Ecologists must confront six challenges in their efforts to address environmental questions at regional to global scales: (1) developing non-experimental methods to conduct large-scale research; (2) incorporating information from new data sources and other disciplines; (3) standardizing and controlling the quality of data; (4) developing new statistical tools; (5) integrating, synthesizing and modelling knowledge about ecological systems; and (6) incorporating humans and their activities explicitly into ecological studies.

Introduction

Formidable challenges face the environmental sciences in the coming decade. It is necessary to greatly increase the scale of inquiry, from the sample plot, habitat patch and small watershed of traditional ecology, to the landscape, geographic region, continent, ocean and entire earth. This must be done in all areas of ecology: population, community and ecosystem. It must be done so as to understand contemporary patterns and processes, and then to use them to interpret the past and to predict the future.

The reasons to increase the scale of environmental research are increasingly apparent. Our own species is transforming the environment of the earth on a scale and with a magnitude that has only begun to be appreciated. These changes are exemplified by the rapid decrease in global biological diversity owing to extinction of species and the alteration of the air, water and soil by the products of modern technology. There is a pressing need to understand these changes so as to manage the environmental impacts of humans and approach a sustainable relationship between our species and the biosphere (Lubchenco *et al.*, 1991).

There are also basic science reasons to increase the scale of environmental research. Ecologists, biogeographers and earth scientists have long been aware of important phenomena that occur on regional, continental and global scales, but until recently the lack of adequate data and analytical tools has severely limited the kinds of research that could be conducted. Now these barriers are rapidly diminishing with the availability of new data, technology and analytical/

modelling tools (e.g., large databases, information from remote sensing, reconstructions of the climate and vegetation of the Pleistocene, and increasingly fast, powerful computers to store, manage, analyse and simulate; see also Stafford *et al.*, this volume).

In this chapter six challenges are identified. These are by no means the only challenges that face the next generation of ecologists, but they illustrate both the problems and the opportunities that lie ahead.

Develop Non-experimental Approaches to Increase the Scale of Ecological Research

The ecology of the last few decades has been largely experimental. In order to manipulate and control the environment, ecologists have worked on necessarily small scales. Reviews of these experimental studies reveal just how small: 50 per cent of them have used plots measuring less than 1 m^2 in area, and 40 per cent of them have been completed in less than one year (Kareiva and Anderson, 1988; Tilman, 1989; Brown and Roughgarden, 1990).

Now ecologists are being asked to characterize patterns and processes that occur on scales from landscapes to the entire globe. Not only is it impractical or impossible to use the experimental approaches of the last few decades at these scales, but it is also hazardous to extrapolate from the results of small-scale experimental studies. Additional processes, neither detected nor important at small scales, often dominate the structure and dynamics of ecological systems at large spatial and long temporal scales.

An excellent example is provided by the work of Roughgarden *et al.* (1987, 1988) on the processes that determine the abundance, distribution and diversity of species on the rocky intertidal shores of Monterey Bay, California. The interactions of the species present at the start of the season with their abiotic and biotic environment determine the composition of the community at the end of the season. These interactions can be demonstrated by traditional experimental methods: controlled manipulations of small, replicated plots. But what determines the composition of species at the start of the season? Roughgarden and colleagues have shown that it is influenced strongly by the settling of the planktonic larvae on the rocky shore, and this varies annually in response to changes in the ocean currents along the California coast. These ocean current and larval settlement patterns can be demonstrated by analysis of satellite images that show sea surface temperature.

In their efforts to work at larger scales, ecologists have more in common with astronomers, geologists, climatologists and oceanographers than with most biologists, chemists and physicists. The success of the earth sciences in developing the theory of plate tectonics and in evaluating the hypothesis of a Cretaceous–Tertiary asteroid impact offer inspirational models of how to infer process from pattern, to build and evaluate models, and to develop and test hypotheses without the benefits of experimental manipulation and replication.

Develop New Statistical Tools

The vast majority of today's statistical techniques were developed to analyse data from experiments in plant and animal breeding and psychology. These techniques rely on the principles of experimental design to examine the effects of one or a small number of manipulated variables while holding constant the influence of many other variables. They are concerned primarily with estimating Type I errors (i.e., using differences among means and variances to calculate confidence levels for rejecting null hypotheses when they are true).

To conduct large-scale ecological research without experimental manipulation and yet still be able to test hypotheses rigorously will require new statistical tools to deal with the complex patterns of variation over both space and time. Applications of multivariate techniques, time series analysis and geostatistics, and development of geographic information systems (GIS) are promising beginnings. Additional methods are needed to characterize the complexities of spatial and temporal patterns and processes.

Establish Linkages With Other Disciplines

Ecologists should not and cannot tackle the large-scale problems in isolation. The traditional ecological research unit of a senior scientist, a few graduate students and perhaps a post-doc, collecting their own data and working largely independently of other research groups, is not well suited to tackle the scale and complexity of regional to global problems. There is both the need and opportunity for interdisciplinary research. Other disciplines can contribute invaluable data and insights. Three examples will illustrate some of the many possibilities, a fourth — linkages with the social sciences — will be addressed later.

The interaction between organisms and their environment is governed ultimately by the physical template of the earth: its climate, geology, soil, physical oceanography and limnology. Recent advances in the earth sciences provide not only data on the structure and dynamics of this physical template but also understanding of the mechanistic processes responsible for these patterns. These interpretations come from data sources ranging from remote sensing to general climate models (see Hayden, this volume), and they need to be integrated with the distributions of organisms, and their interactions, over this physical template in both space and time.

Data on the distribution, abundance and diversity of organisms on regional to global scales are increasingly available. Some can be obtained from large databases, such as the North American Breeding Bird Survey (BBS), the

Audubon Society's Christmas Bird Counts, the Nature Conservancy's Data Bank, and comparable data sets for Europe and Australia. Invaluable data documenting historical distributions of organisms is available from museum collections, which provide records of many species that go back for decades to centuries.

Longer term data on the past are also available and can offer invaluable insights. For example, data from tree rings, pollen deposits and pack rat middens provide a record of climate and vegetation change for the last several thousand years (e.g., Davis, 1986; Delcourt and Delcourt, 1987; Betancourt *et al.*, 1990). These changes in species distribution in response to past climatic changes are of obvious relevance in assessing possible impacts of future global climate change.

Standardize and Control the Quality of Data

This scale of research means that most researchers will no longer be able to collect their own data. They will have to rely on data collected by other investigators and by technological means (from automated meteorological stations to thematic mapper satellite images). Often data will have been collected originally for a purpose other than the one for which they are to be used. Not all data will be of equal quality and precision, and all will contain errors — sometimes serious errors of omission and commission.

This variation in the quality of information makes the need for standards for data collection, management and analysis critical. Data do not have to be perfect. In fact, the standards of accuracy and precision required vary with the problem being addressed. What is critical, however, is that the quality of data be known so as to ensure that it is sufficient for the application. This requires attention to documentation (i.e., metadata) and standardization at all stages of data processing, from initial collection through management to final analysis.

Emphasize Synthesis, Integration and Modelling

Scaling up ecological research necessarily involves processing large quantities of information about the structure and dynamics of complex ecological systems. It is easy to become overwhelmed by the sheer magnitude of the problems and the quantities of data. Imagine the problem of trying to understand not only the pattern but also the causes of the variation in the abundance of a single bird species at 600 BBS census sites distributed over eastern North America during each of the last 20 years.

To deal with such complexity it is equally necessary to produce models that simplify systems, that identify their most important characteristics, and that can be generalized to make predictions and to understand other systems. Models are required both to synthesize results of basic science and for application of these results to management and policy. Ultimately, models should incorporate and integrate both space and time; they should map temporal dynamics on to spatially explicit landscapes. The single unidirectional dimension of time has been exploited to great advantage in modelling the dynamics of systems that can be considered to occupy just one point in space, but the problems become much greater when dealing with two (i.e., area) or three dimensions (i.e., volume).

Incorporate Humans Explicitly Into the Systems

Traditionally, ecologists have treated human beings as something separate from nature. They have tried to work on nature preserves and on 'natural' systems minimally influenced by any human activities. They have left human-impacted and -managed systems to applied scientists. This has left ecologists woefully unprepared to address the environmental impacts of humans. Furthermore, while it might be possible to find small local experimental study sites that are still relatively pristine, it is impossible to disregard the effects of humans on ecological processes at regional to global scales.

The only way to address this problem is to consider the human species and all relevant human activities as explicit, integral parts of the system. For most ecologists, this will not only require a fundamental change in research philosophy, it will also require incorporating additional information and levels of complexity from the applied environmental sciences and social sciences. These changes are essential if ecologists want to understand the earth as it is today, rather than the pristine world that they would like it to be.

Conclusion

Six critical challenges that face ecologists in their efforts to scale up their research from small-size, short-term studies of relatively pristine local areas to investigations of basic and applied environmental problems at regional to global scales have been outlined. All of these challenges require new ways of studying the relationships between organisms — including humans — and their environments on unprecedented scales of space, time and complexity. They require new ways of acquiring, managing, analysing, interpreting and synthesizing information. This book represents an important advance in the effort to identify and supply some of these informational needs.

Acknowledgement

I thank A. Kodric-Brown and four anonymous reviewers for suggestions that improved this chapter. My research has been generously funded by the National Science Foundation, most recently by grant DEB-9221238.

References

Betancourt, J. L., Van Devender, T. R. and Martin, P. S., 1990, *Packrat Middens: The Last 40 000 Years of Biotic Change*, Tucson: University of Arizona Press.

Brown, J. H. and Roughgarden, J., 1990, Ecology for a changing earth, *Bulletin of the Ecological Society of America*, **71**, 173–88.

Davis, M. B., 1986, Climatic instability, time lags, and community disequilibrium, in Diamond, J. and Case, T. J. (Eds) *Community Ecology*, pp. 269–84, New York: Harper & Row.

Delcourt, P. A. and Delcourt, H. R., 1987, *Long-term Forest Dynamics of the Temperate Zone*, Berlin: Springer.

Kareiva, P. and Anderson, M., 1988, Spatial aspects of species interactions: The wedding of models and experiments, in Hastings, A. (Ed.) *Community Ecology*, pp. 38–54, New York: Springer.

Lubchenco, J., Olson, A. O., Brubaker, L. B., Carpenter, S. R., Holland, M. M., Hubbell, S. P., Levin, S. A., MacMahon, J. A., Matson, P. A., Melillo, J. M., Mooney, H. A., Peterson, C. H., Pulliam, H. R., Real, L. A., Regal, P. J. and Risser, P. G., 1991, The sustainable biosphere initiative: An ecological research agenda, *Ecology*, **72**, 371–412.

Roughgarden, J., Gaines, S. D. and Pacala, S. W., 1987, Supply side ecology: The role of physical transport processes, in Gee, J. H. R. and Giller, P. S. (Eds) *Organization of Communities: Past and Present*, pp. 491–518, Oxford: Blackwell.

Roughgarden, J., Gaines, S. D. and Possingham, H., 1988, Recruitment dynamics in complex life cycles, *Science*, **241**, 1460–6.

Tilman, D., 1989, Ecological experimentation: Strengths and conceptual problems, in Likens, G. E. (Ed.) *Long-term Studies in Ecology*, pp. 136–57, New York: Springer.

3

Sustainable Biosphere Initiative: Data management challenges

James R. Gosz

Implementation of the Sustainable Biosphere Initiative will require a significant increase in interdisciplinary interactions and unprecedented data management challenges and opportunities. Major challenges presented by issues such as global change, biodiversity and sustainability will be met only if disciplines in the physical, biological, social and economic sciences become more integrated as well as develop strong associations with policy and management efforts. In addition to massive changes in federal research programmes, the university-based research community will become more critical to the national effort, providing that community can make significant changes in the management of research data and demonstrate an increased willingness to share these data with the scientific community. Technological advances and cultural changes are needed.

Introduction

The Sustainable Biosphere Initiative (SBI) resulted from Ecological Society of America (ESA) activities focused on the necessary role of ecological science in the wise management of the earth's resources and the maintenance of the earth's life support systems (Lubchenco *et al.*, 1991). Citizens, policy makers, resource managers and leaders of business and industry all need to make decisions concerning the earth's resources, but such decisions cannot be made effectively without a fundamental understanding of the ways in which the natural systems are affected by human activities. Thus, the SBI was developed as a framework for the acquisition, dissemination and utilization of ecological knowledge that supports efforts to ensure the sustainability of the biosphere. The SBI calls for basic research for the acquisition of ecological knowledge, communication of that knowledge to citizens, and incorporation of that knoweldge into policy and management decisions (Lubchenco *et al.*, 1991). Based on the potential to contribute to fundamental ecological knowledge and respond to major human concerns about the sustainability of the biosphere, three research priorities were identified:

- *Global change*, including the ecological causes and consequences of changes in climate; in atmospheric, soil, and water chemistry; and in land- and water-use patterns;

- *Biological diversity*, including natural and anthropogenic changes in patterns of genetic, species and habitat diversity; ecological determinants and consequences of diversity; the conservation of rare and declining species; and the effects of global and regional change on biological diversity;
- *Sustainable ecological systems*, including the definition and detection of stress in natural and managed ecological systems; the restoration of damaged systems; the management of sustainable ecological systems; the role of pests, pathogens and disease; and the interface between ecological processes and human social systems.

It was recognized that the implementation of the SBI would require a significant increase in interdisciplinary interactions linking biological, physical, social and economic scientists with mass media and educational organizations, policy makers, and resource managers in all sectors of society. Linking, interpreting and communicating that research will require unprecedented data management challenges and opportunities. The global change and biological diversity (biodiversity) research priorities will illustrate these challenges.

Global Change

Global change, with its ramifications on biodiversity and sustainability, can be used as an example because it alone presents one of the grandest challenges to computational and data management needs (see Hayden, this volume, and Jelinski *et al.*, this volume). Modelling global climate dynamics will be minor compared with modelling the effects of climate change on all scales of biological, ecological, social and economic factors. Our knowledge deficiencies about patterns and processes of individuals, populations, communities and landscapes from local to regional scales are extreme, especially dealing with long-term changes. We are challenged to predict how such global changes will be reflected in the genetic structure of organisms, biodiversity and community behaviour. The measurement of these impacts will underlie the design of strategies to mitigate problems associated with global change and to support sustainable development. A primary goal is the development of approaches for dealing with and relating phenomena across disparate scales of space, time and organizational complexity (Kingsbury, 1993).

Perhaps the major challenge is the variety of disciplinary data sets that must be used in, for example, regional modelling of global change. Problem solving and decision making require effective and efficient use of the substantial existing knowledge bases associated with these disciplines. Typically, domain-specific information such as ecology, geography, forestry, economics, psychology and sociology form the knowledge base for a given issue. Our track record is not good in identifying and demonstrating the tools and techniques needed to integrate the various elements of the knowledge base so that it can be used in an effective manner. To cite one approach, the Knowledge Engineering Laboratory of Texas A & M University is emphasizing object-oriented simulation, expert systems, knowledge-based systems, intelligent geographic information systems and the knowledge system environment (R. Coulson, personal communication). These techniques permit use of spatially referenced data, tabular information and heuristic knowledge of technical experts for applied problems in environmental management and for basic studies in environmental science. This has been easy to say in a few words; in practice it is very difficult and uncommon.

In another effort, the Washington Office for the Sustainable Biosphere Initiative has developed an interagency working group that has developed plans for two regional demonstration projects; south Florida and the Rio Grande Basin of the Southwest USA and Mexico. The purpose of these projects is to demonstrate how agencies (regulatory, scientific and management; federal, state and local) can work together with the scientific community and other stakeholders to meet a common objective: sustainability of critical systems — ecological, cultural and economic — on a regional scale. The need for this approach is based on:

- Issues of scale — the appropriate scale of focus may be regional because of the attempts of global research to scale down and ecological research to scale up. Methods should address multiple scales and integration across scales.
- Representation of social and economic effects — the area of focus must be large enough to include representative social, economic and cultural values but small enough to allow a sense of ownership and belonging among the inhabitants.
- Management techniques addressing national and local needs — management may fail if it does not account for local variations in culture, ecology and economics. Local perception is important if the management is designed by outsiders.
- Mixed ownership issues — there are many variations in management techniques, goals, land classification, information quality and quantity, etc., across the government, commercial and private onwerships in a region.
- Complexity of the landscape mosaic — management plans must address the reality of the complexity caused by physical, ecological and political boundaries.
- Availability of advanced technology — there is improved access to su_ipercomputers, GIS, remote sensing information and modelling.
- Ecological theory and approaches — there have been advances in many areas of science, including landscape ecology, ecosystem ecology, boundary/ecotone dynamics, island biogeography, population dynamics, percolation theory, hierarchy theory, nonlinear mathematics and statistics. Developments in these areas allow management to be based on sound scientific principles.

Clearly, these examples demonstrate that scientists recognize the challenge ahead.

Data Management for Global Change Research Policy Statements

The US Global Change Research Program was conceived and developed to be policy-relevant. Hence, it supports the needs of the USA and other nations to address significant uncertainties in knowledge concerning the natural and human-induced changes now occurring in the earth's life-sustaining environmental envelope. The scientific goal is to gain a predictive understanding of the interactive physical, geological, chemical, biological, economic and social processes that regulate the total earth system. The scientific programme addresses earth system processes that vary on time scales ranging from seasons to many decades to several centuries. There are a number of policy statements, quoted below (US Global Change Research Program, 1993), that help introduce the challenges to data management in this endeavour. Key phrases have been bracketed.

> The Global Change Research Program requires an early and continuing [commitment to the establishment, maintenance, validation, description, accessibility, and distribution] of high-quality, long-term data sets.

> [Full and open sharing of the full suite] of global data sets for all global change researchers is a fundamental objective.

> [Preservation of all data needed] for long-term global change research is required. For each and every global change data parameter, there should be at least one explicitly designated archive. Procedures and criteria for setting priorities for data acquisition, retention, and purging should be developed by participating agencies, both nationally and internationally. A clearinghouse process should be established to prevent the purging and loss of important data sets.

> Data archives must include [easily accessible information] about the data holdings, including quality assessment, supporting ancillary information, and guidance and aids for locating and obtaining the data.

> [National and international standards] should be used to the greatest extent possible for media and for processing and communication of global data sets. Data should be provided at the lowest possible cost to global change researchers in the interest of full and open access to data. This cost should, as a first principle, be no more than the marginal cost of filling a specific user request. Agencies should act to streamline administrative arrangements for exchanging data among researchers. For those programs in which selected principal investigators have initial periods of exclusive data use, [data should be made openly available] as soon as they become widely useful. In each case, the [funding agency should explicitly define the duration of any exclusive use period].

The key phrases bracketed above will be difficult to address, especially in the university-based research community. The culture in much of the academic community is one of the investigator-initiated research where a premium is placed on the quality and diversity of ideas and approaches; a system that is extremely valuable in the progress of science. However, the negative aspect of this approach is that it generates data systems that are at odds with many of the policies stated above (see Porter and Callahan, this volume). There is a large array of 'conditions' of data in the university research community, most of which would not allow their easy use in the Global Change Research Program as envisioned above. The emphasis in much of the research community is on the publication of interpretations of the primary data, resulting in few published primary data sets. The best examples of data called for by the US Global Change Research Program are held by agency monitoring programmes and a few university-operated programmes. These include the Long-Term Ecological Research and Land Margin Ecosystem Research programmes funded by the National Science Foundation.

Of the preceding seven policy statements, the last is the most controversial. This policy suggests that decisions about when the data should become widely useful are the responsibility of the funding agency. The agency also should explicitly define the periods of restricted access, if any. Major changes in policy may be needed by some agencies (e.g., National Science Foundation) to reverse the practice of allowing principal investigators to retain their data for indefinite periods, restricting or inhibiting their widespread use. Issues of intellectual property will not be easy to resolve. A major challenge may be to change the social, philosophical and psychological attitudes of researchers regarding 'their' research data. There will be increasing pressure for scientists to 'contribute' their data sets to common data archives such as is occurring in the National Institute of Health. There are a number of 'databank' activities demonstrating how research communities are beginning to change attitudes about sharing data. These result in investigators entering data into systems that are accessible to the entire community. Examples are:

- GenBank — this is a gene sequence databank administered through the National Library of Medicine (connected with Los Alamos National Laboratory) that includes nucleotide sequences of DNA and RNA from all kinds of organisms including viruses and also organelles (chloroplast and mitochondrial DNAs). This databank is widely used and many journals require GenBank accession numbers on all sequences announced in the submitted papers.
- Protein Data Bank — this databank is administered from Brookhaven National Laboratory, and contains atomic co-ordinates and related structural and bibliographic information for virtually all proteins and most oligonucleotides whose three-dimensional structures have been determined by the use of X-ray crystallography.

- The University of Illinois School of Pharmacy administers a database on natural products chemistry, chemical structures, organismal source, publication references and therapeutic reports.

The Nature of Environmental Data

There is a consensus in the biological disciplines that there is an overwhelming amount of data. However, what sets biology apart from other data-rich fields is the complexity of data and the heterogeneous data environment rather than the sheer volume of data produced (Kingsbury, 1993). Environmental science requirements encompass a broad range of levels of biological organization, from genetics and physiological aspects of the individual through populations, communities and entire landscapes. Large-scale research and analysis, by necessity, must combine data from a variety of sources to model natural processes. Extremely large data streams are available from satellite remote sensing and these include the added difficulties of obtaining the appropriate amounts and types of information. It may be analogous to attempting to obtain a drink of water from the jetstream of an eight-inch fire hose or equivalent to one Library of Congress per day! In addition, the wide array of ground-based data on topography, soils, climate, vegetation, land-use dynamics, human demography, economic pressures, factors influencing societal perceptions and values, etc. have made, and will continue to make, modelling broad-scale dynamics a serious challenge. The complexity of data also is compounded by the heterogeneous environment in which data are archived.

A further difficulty is that each organism, species, population and community represents the potential for a unique solution to the constraints provided by a dynamic environment. The true complexity of ecological systems often is inadequately sampled in our research programmes, making interpretation and extrapolation from data collected a difficult challenge. Concerns about biodiversity illustrate this point. We have recognised only ~1.5 million species globally but estimate that there actually may be 10 million to 100 million species. Clearly, we cannot appreciate the roles that those species play in the array of functions involved in global change dynamics if we do not even know what they are! The sampling that we have performed also has generated information that is highly skewed toward fine spatial scales and short-term data sets (see Brown, this volume).

There are a number of issues dealing with the traditional approach to generating, protecting and archiving data as well as making them available to other users. Unfortunately, many data are generated to be analysed once, published, and often not visited again; these can be thought of as 'throw-away' data. An appropriate database that documents the many conditions and associated data sets (metadata) is necessary to make the primary data set valuable for subsequent measurements (e.g., trend analysis). Data become more valuable for subsequent studies if the appropriate ancillary data (i.e., documentation or metadata) are archived. The attitude that a given study is a 'one-time' project, suitable only for the testing of a hypothesis and subsequent publication, significantly limits future data utility. In addition this attitude also effectively prohibits future repetition of the experiment to quantify change or to validate interpretation. It would not be surprising to see federal agencies increasingly demand a certain measure of database management to guard against such 'throw-away' data. Furthermore, data generation is done in a highly distributed mode, often with little attention to standard format or syntax. The needs of global change, biodiversity and sustainability issues include extrapolation and interpretation at regional and larger scales. The state of many of the biological data, collected at fine scales and poorly managed, makes this extremely difficult, hence the need to redo studies at larger scales and across a number of scales in hopes of generating relationships that will allow incorporation of much of the existing data. Data developed in an *ad hoc* manner often do not conform to modelling needs. Thus, even if an attempt to incorporate more and broader scale data into models is made, the modelling environment becomes increasingly difficult to adapt to new data sources, to modify for purposes of experimentation, or to merge with other complementary modelling efforts. It soon becomes obvious that there is a dramatic increase in the number of data sources that must be supported, the number of data sets that must be managed and the number of versions of process-level modelling components that must be maintained. Each new application adds more complexity to this *ad hoc* collection of information, causing scientists to be faced with mazes of computer programs, data representations and data sources, with little overall system structure.

Biological data also must be captured in some type of machine-readable form in order to be usable. It is likely that many biological data reside in handwritten notes, proprietary software formats, and in physical media such as gels, journal texts and printed graphs. Additional efforts are needed to generate machine-readable forms using technologies such as pattern recognition, document and diagram understanding, and intelligent instrumentation to reduce the dependence on human processing (i.e., keyboarding, manual data logging). Data acquisition systems need to go directly from the laboratory or field measurement systems into the database management systems of all relevant users.

The proliferation of information from remote sensing introduces the need for geographical information systems that provide a framework for classifying information, spatial statistics for analysing patterns, and dynamic simulation models that allow the integration of information across multiple spatial, temporal and organizational scales. Today, the application software is mostly non-existent except in a few special cases such as image processing and remote sensing. The situation will only become more complex as new satellites with the capability for finer spatial resolution and spectral resolution and new instrument types increase the volume of data from space.

The complexity inherent in biological systems requires increasingly sophisticated methods of analysis, and these need to be readily accessible to

biological scientists untrained in computer technology. Visualization tools will
be especially needed, not only in actual analysis procedures but for data
exploratory analysis, much in the way that a statistician evaluates data to guide
subsequent analyses. It is often said that 70 per cent of the human brain's
function is directed to the interpretation of three-dimensional images. Visualization will greatly speed the understanding of complex data and the exploration of
experimental simulations.

Biodiversity

There are some additional unique aspects of biodiversity that warrant further
discussion. The proposed creation of the National Biological Survey in the
Department of Interior (DOI) will be an immense undertaking and involve
consolidation of information about the nation's ecosystems from various
agencies in DOI and other departments; integration of information from other
non-governmental agencies like The Nature Conservancy and university collections; and the collection of new data (Yoon, 1993). A primary difficulty is that
'There is no place on earth — none — where the entire biota is known. We are
talking about doing a new kind of biology' (Yoon, 1993). An example of that
new kind of biology was identified as an All Taxa Biodiversity Inventory
(ATBI), an exhaustive survey of all the various kinds of organisms in a single
area that may be as large as 200 square miles. In addition to myriad vertebrates,
vascular plants and large insects that we may know relatively well, there will be
thousands of unknowns. Some studies have indicated that there may be as many
as 10 000 species of bacteria in a single gram of temperate forest soil, yet only
about 3000 species of bacteria have actually been named (Yoon, 1993).
Sampling soil to a depth of 15 cm can result in finding 2 million to 6 million
nematodes per square metre with the potential for 200 species being present.
Most of them are unnamed and unknown. In addition to the immense number
of species, associated data on location, dynamics, ecological interactions and
associations (e.g. associated parasites, viruses, symbionts), genetics and chemical
information also will be needed. The need to archive and access such data is
obvious. Unfortunately, in the taxonomic fields where researchers historically
have tended to work independently, there is no standard for a large, interactive
network of biodiversity data. In order to develop such a network, researchers in
biodiversity will have to learn to handle the compromises and conflicts inherent
in massive collaborations (Yoon, 1993).

The Need for Other Disciplines

Programmes designed to address issues such as Global Change and Biodiversity
often identify the physical, social and biological disciplines as areas that collect

data on relevant variables. Other disciplines will become increasingly important in this research. For example, the mathematical sciences have long played a fundamental role in the study of complex systems. Mathematical models of fluid mechanics continue to be among the main vehicles to analyse and predict climate system behaviour. The analytical and computational methods for deterministic and stochastic partial differential equations are the main tools for these models. By extension of time series and stochastic processes, statistical techniques that encompass the full range of temporal and spatial scales have been developed to handle the substantial global climate database in prediction and climate impact studies. For continued advances, more sophisticated deterministic and stochastic models and better ways to manage, analyse and visualize data sets of increasingly higher resolution are needed. These require mathematical, statistical and computational methods neither currently available nor accessible to the scientists, such as dynamical systems, waves and reaction-diffusion systems arising in chemistry, biology, geosciences and associated computational methods; computer visualization of evolving systems or of high-dimensional data; seismic tomography, inverse problems and control theory; interpolation, smoothing, filtering and prediction; multiple time series, spatial statistics and multivariate analysis; inference for stochastic processes; parametric and non-parametric function estimation and wavelets; and extreme value theory.

Interactions with Computer Sciences

Collaboration between computer engineers, computer scientists and environmental scientists will be necessary to design information platforms to accommodate the various representations of biological data. Distributed data acquisition systems pose significant challenges for data integration as do the myriad demands placed on different data sets and algorithms for data comparison. Typical users traditionally depend on inexpensive commercial technology for database access on inexpensive personal computers. A major challenge is the development of new data management systems as well as specific funding to speed up commercialization. Several novel approaches involve research support programmes that encourage biologists, computer scientists and database designers to be co-investigators on projects. Database designers need to be brought into research programmes in the same way that statisticians are included in the initial phases of an experimental research programme. The Khoros development programme at the University of New Mexico (Rasure *et al.*, 1990) is an example of such an interaction. Khoros is a broad software environment for data processing and visualization that contains a User Interface Development System, libraries that range in function from numerical analysis to X Windows display utilities, distributed computing and file services, and collaboration technology. The new version (Khoros 2.0) creates a highly interactive meta-computer composed of workstations with graphics accelerators, high-speed networks, terabyte data vaults and supercomputers. This unique, high performance computing environment provides the collaboration technology needed to address multidisciplinary challenge problems. The Sevilleta LTER Program at the University of New Mexico is using Khoros for near-ground-level image classification, lightning strike data analysis and animation, raster-based geographical information analysis, remote sensing analysis and ecosystem modelling. Combining this capability with the Sevilleta Information Management System (SIMS) provides a powerful approach to complex, broad-scale analysis.

Imaging technology is central to almost all of the biological sciences, and data representation through image construction may be the most powerful tool in our arsenal. For example, applications of computer image processing have become necessary for nuclear magnetic resonance imaging in the medical profession. Similarly, computer imaging of plant biology, landscapes, regions, etc., will become an integral component of the sciences. Recent developments in video photography provide opportunities that range from microvideo recording of root systems and associated rhizosphere dynamics to broader scale landscape dynamics. These video records can serve as comparative standards for other scientific disciplines. Visual information systems integrated with additional sophisticated computer algorithms that compress images will significantly improve the usefulness of these techniques by the scientific community.

There is general agreement that current database systems are inadequate for managing large heterogeneous sets of scientific data (Box *et al.*, 1989). The full utilization of modern image processing tools in environmental computation will require intense software and hardware development. Much of our understanding of complex environmental phenomena comes from comparison studies, whether they be comparing gene sequences or landscape change. In the same way that statistical techniques were developed to address complexity in the environment, visualization hardware and software must be developed to improve our understanding of the complexity of huge data sets of many types. Images provide a great deal of spatial and geometric information about many real-world objects and, although all sciences interpret and extract measurements from images, image processing presents a dilemma. Human interpretation is usually fast and accurate but a large sequence of images is difficult to deal with. In contrast, computer processing is precise and recording is easy, allowing proper managing of a large sequence of images; however, development of robust algorithms for recognizing important three-dimensional image features and detecting and quantifying change continues to be a significant research problem. The manipulation and analysis of 3-D data will require additional development of data management platforms integrated with techniques for edge detection, and volume and density analysis. The current topic of how to search large parameter spaces in biological models is a growing and important analysis issue (Kingsbury, 1993).

The Role of Synthesis

The Sustainable Biosphere Initiative called for a framework for the acquisition, dissemination and utilization of ecological knowledge (Lubchenco *et al.*, 1991). Success in this initiative will depend on an ability to support small and large research groups with a genuine interdisciplinary focus on synthesis. The recent workshop on the need for a National Center for Ecological Synthesis and Analysis (Brown and Carpenter, 1993) identified the need to integrate ecological knowledge across temporal, spatial and biological scales; across ecological systems; and across interrelated disciplines of the biological, geological, hydrological, climatological and social sciences. Currently, there is no mechanism or institution dedicated to the advancement of synthesis in the ecological sciences and, without such studies, it will be impossible for many sciences to have the predictive capability required by current and future environmental problems. Acceleration of synthesis efforts will lead to more cost-effective use of the large array of accumulating data. Currently, many high-quality data, collected for one specific purpose, are not used for a variety of other purposes for which they are suitable. Additionally, many national programmes that include synthetic efforts as part of their mandate could benefit from a scientific programme that makes synthesis its primary goal rather than merely a desirable by-product.

A Synthesis Centre, along with synthesis activities in the other programmes of federal agencies, university centres, and among individual scientists, will be completely dependent on the developing technologies and capabilities of data management systems. The present generation of biological databases will fail in the next decade because they were not designed to deal with the volume, complexity and diversity of the data that will need to be accessible for environmental research (Kingsbury, 1993). Kingsbury further suggests that the term 'database' itself tends to trivialize the problem and is misleading. Current database technology and theory represent an inadequate paradigm for what is needed now and in the future to represent and organize biological information. Advances in synthesis and data management activities are highly co-dependent. Human abilities to synthesize information will be a determinant in developments in database management in much the same way as the availability of data will determine synthesis activities. Unfortunately, the two activities are somewhat independent and occurring in different communities. Successful data management and knowledge (synthesis) generation can only be guaranteed if these activities are merged.

Conclusion

This introductory chapter identifies a number of primary concerns that must be addressed by scientists in order to successfully meet the challenges posed by issues such as global change, biodiversity and sustainability. The scientific community must find ways of integrating disciplines, and the funding agencies must find ways of providing resources for those efforts and maintaining high science standards. Technological advancements are needed to establish the capability for integrating many, large data sets of variable forms, and new procedures are needed for the analysis and synthesis of the information. Cultural changes in the way data are perceived, collected, managed and shared may be necessary for some disciplines. The opportunities are present for a revolution in the way science can be performed.

Acknowledgement

Challenges for data management arise from many issues. I would like to acknowledge the many discussions with science colleagues as well as direct input from programme officers in the Division of Environmental Biology of the National Science Foundation. A number of anonymous reviewers provided valuable input to the manuscript. I also thank Peter Arzberger of the National Science Foundation for providing draft versions of the NSF report on Computational Biology compiled by David Kingsbury. A number of individuals at the University of New Mexico aided this effort, in particular, James Brunt for increasing my appreciation and understanding of the role of data management in science and Greg Shore who helped with visuals for the presentation and programming support to allow someone in Washington, DC to interact with the conference organizers in Albuquerque, New Mexico. This is contribution number 43 of the Sevilleta Long-Term Ecological Research Program supported by the National Science Foundation (BSR8811906).

References

Box, J. E., Jr., Smucker, A. J. M. and Ritchie, J. T., 1989, Minirhizotron installation techniques for investigating root responses to drought and oxygen stress, *Soil Science Society of America Journal*, **53**, 115–18.
Brown, J. and Carpenter, S., 1993, National Center for Ecological Synthesis: Scientific Objectives, Structure, and Implementation. A report from a joint committee of the Ecological Society of America and the Association of Ecosystem Research Centers. 8 February 1993, Albuquerque, New Mexico, USA.
Kingsbury, D. T., 1933, 'Research opportunities in computational biology: Results from a series of invitational workshops', draft report for the National Science Foundation.
Lubchenco, J., Olson, A. M., Brubaker, L. B., Carpenter, S. R., Holland, M. M., Hubbell, S. P., Levin, S. A., MacMahon, J. A., Matson, P. A., Melillo, J. M., Mooney, H. A., Peterson, C. H., Pulliam, H. R., Real, L. A., Regal, P. J. and Risser, P. G., 1991, The sustainable biosphere initiative: An ecological research agenda, *Ecology*, **72**, 371–412.

Rasure, J., Argiro, D., Sauer, T. and Williams, C., 1990, Visual language and software development environment for image processing, *International Journal of Imaging Systems and Technology*, **2**, 183–99.

US Global Change Research Program, 1993, *Policy Statements on Data Management for Global Change Research*, DOE/EP-0001P, Washington, DC: National Science Foundation.

Yoon, C. K., 1993, Counting creatures great and small, *Science*, **260**, 620–2.

4

Multiple roles for GIS in global change research: Towards a research agenda

Dennis E. Jelinski, Michael F. Goodchild and Louis T. Steyaert

The National Center for Geographic Information and Analysis (NCGIA) is a National Science Foundation-sponsored, three-university consortium whose mandate is to facilitate geographic research using geographic information systems (GIS). An upcoming research initiative of the centre, Multiple Roles for GIS in US Global Change Research (I-15), concerns the ways in which GIS could better support ecosystem, landscape, regional and global change research. For this initiative, we propose a research agenda with five principal objectives. First, this initiative will support the increased integration of GIS and environmental modelling using techniques for managing, manipulating, analysing and displaying spatial data. Second, we will critically assess the quality of existing global data in terms of spatially varying accuracy, sampling methodologies and completeness of coverage, and develop means of managing and visually communicating components of environmental data quality (e.g., error). Our third objective is to develop spatial database techniques that can manage hierarchical structures across multiple scales of spatial patterning. Fourth, we propose to develop methods for dynamically linking human and biophysical databases within GIS and for exploring the regional impacts of global change. Last, means of using GIS for detecting, characterizing and modelling changes in ecotones will be explored.

Introduction

During the last decade, questions about potential changes in climate, deterioration of the stratospheric ozone layer and decreasing biodiversity have raised concerns among both scientists and the general public. At the same time, new political and economic transformations and structures are emerging. These phenomena are described as 'global change' (Botkin, 1989; Price, 1989; Turner *et al.*, 1990; Committee on the Human Dimensions of Global Change, 1992), and can be classified into two basic types (Botkin, 1989; Turner *et al.*, 1990). In one sense the term applies wherever actions are *systemic* (i.e., global in extent; on a spatial scale at which perturbations in the system have consequences anywhere or reverberate throughout the system). For example, there is concern over increasing concentrations of greenhouse gases and other climate forcing agents that are manifested globally. The second type is *cumulative* global change. The loss of biological diversity at so many locations throughout the world is global in scale because its implications are global, even though the causes of extinctions are localized.

The US Global Change Research Program (IGBP, 1990; NRC, 1990; USGCRP) has grown out of the need for scientific assessments of both types of global change, and is ultimately intended to aid in policy decisions. Emphasis has been placed largely on the study of interactions between the earth's biosphere, oceans, ice and atmosphere. The research strategies that help to provide this scientific foundation were developed in the mid-1980s (ESSC, 1986, 1988; ICSU, 1986; NRC, 1986) and feature an advanced systems approach to scientific research based on data observation, collection and documentation; focused studies on the underlying processes; and the development of quantitative, earth system models for diagnostic and prognostic analyses. Concepts such as earth system science (ESSC, 1986), global geosphere–biosphere modelling (IGBP, 1990), and integrated systems modelling at multiple scales (NRC, 1990) have emerged, focusing broadly on the earth system but including subsystems such as atmosphere–ocean coupling. The USGCRP is one example of a national-level effort to implement this three-component strategy for scientific research (CES, 1989, 1990; CEES, 1991) with the more recent addition of scientific and economic assessment components to global change research (CEES, 1992a, 1992b).

In addition to the geophysical, biogeochemical and other natural processes that drive the earth system, changes in human land use, energy use, industrial processes, social values and economic conditions are also increasingly being recognized as major forces in global change (Task Group 6, 1991; Committee on the Human Dimensions of Global Change, 1992). The relationship of these activities and behaviours to global change is critical because they may affect the biophysical systems that sustain the geosphere–biosphere. The National Research Council's (NRC) Committee on Global Change has stated that the development of a coherent and systematic assessment and understanding of global change phenomena requires the establishment of better linkages between environmental and human dimensions (e.g., social, economic) (Committee on the Human Dimensions of Global Change, 1992). However, there are difficulties in collecting requisite socio-economic and demographic data. Those data that do exist often span a range of time and space scales, lack appropriate intercalibration, have incomplete coverages, are inadequately checked for error and have unsuitable archiving-retrieval formats (Task Group 6, 1991; Committee on the Human Dimensions of Global Change, 1992).

Both the human and physical dimensions of global change research require large amounts of spatial data and place heavy demands on systems used to handle those data. The calibration and verification of models of the earth system, the investigation of the relationships that exist between various aspects of biogeochemical components of this system, and the examination of human populations that both cause change and experience global change present daunting problems for data management and spatial analysis. It is widely believed that GIS and related technologies (e.g., remote sensing, global position-ing systems (GPS), image processing, visualization, high bandwidth commu-nications) will play an increasingly important role in global change research (Mounsey, 1988; Townshend, 1991; Goodchild *et al.*, 1993). In particular, GIS is seen as an effective vehicle for collecting, manipulating and preprocessing data for models; for integrating data from disparate sources with potentially different data models, spatial and temporal resolutions and definitions; for monitoring global change at a range of scales; and for visual presentation of modelling results in a policy-supportive, decision-making environment. For example, two major areas in global change research for which GIS could play an important role concern: (1) enhancement of models of earth system phenomena operating at a variety of spatial and temporal scales across local, regional and global landscapes, and (2) improvements in the capacity to assess the effects of global change on biophysical (ecological) systems over a range of spatial and temporal scales. This paper proposes a research agenda whose primary objective is to contribute to the development of a quantitative understanding of global change through enhanced use of GIS and spatial data analysis.

The National Center for Geographic Information and Analysis (NCGIA)

The NCGIA is a National Science Foundation-sponsored consortium of the University of California (Santa Barbara campus), the State University of New York (Buffalo campus) and the University of Maine. Its primary mission is to facilitate basic geographic research of lasting and fundamental significance using GIS. More specifically, the centre is concerned with advancing the theory, methods and techniques of geographic analysis based on GIS in the many disciplines and professions involved in geographic research; augmenting the nation's supply of experts in GIS and geographic information analysis (GIA); and promoting the dissemination of analysis techniques based on GIS and GIA. Research at the NCGIA is undertaken primarily by means of Research Initiatives, or projects designed to investigate fully impediments to the more widespread implementation of GIS and spatial analysis methods. These initiatives are interdisciplinary and span a range of topics, such as improved user interfaces for GIS products, better understanding of the nature of error in GIS databases and its propagation, enhanced languages for GIS support of model-ling, and more complete integration of spatial statistics and GIS.

The initiative 'Multiple Roles for GIS in US Global Change Research' (I-15) will bring together a group of experts in a 'Specialist Meeting' to design a comprehensive research agenda for GIS in the context of global change research. This project will focus for a period of two years on appropriate aspects of the research agenda. In the remainder of this chapter major foci for this research initiative based on the multiple roles GIS could play in global change research are identified, existing hurdles that prevent GIS from being used more broadly are analysed, and a research agenda designed to enhance the role of GIS in ecosystem-to-global-scale environmental analyses is presented.

Use of GIS to Support Integrative Modelling and Spatial Analysis

GIS can play a major role in the development, calibration, initialization, validation and application of complex global change models that inherently emphasize strong spatial components of environmental processes. A GIS includes a spectrum of tools designed for a range of applications including environmental modelling. However, although GIS support a wide range of data models, many of the fundamental primitives needed to support environmental modelling are missing and must be added by the user (Goodchild, 1991). At present, environmental simulations must be carried out by a separate package linked to the GIS. The ability to write the environmental model directly in the command language of the GIS is still some distance away (Steyaert and Goodchild, this volume). Despite the potential, a number of technical impediments stand in the way of more complete integration of GIS and global environmental modelling (Nyerges, 1993). Moreover, an understanding of the modelling environment, such as outlined by Steyaert (1993), is essential to developing the needed links between the models and GIS.

Scientifically based mathematical models for computer analysis are fundamental to the development of reliable quantitative assessment tools. One major purpose of these computer-based models is to realistically simulate spatially distributed, time-dependent environmental processes. Environmental simulation models are, at best, simplifications and inexact representations of real-world environmental processes. The models are limited because basic physical processes are not well understood, and because complex feedback mechanisms and other interrelationships are not known. The sheer complexity of environmental processes (usually three-dimensional, dynamic, nonlinear and stochastic, involving feedback loops operating across multiple time and space scales) necessarily leads to simplifying assumptions and approximations (Steyaert, 1993).

Frequently, further simplifications are needed to permit numerical simulations on digital computers. For example, the conversion of mathematical equations for numerical analysis on a grid (discretization) can necessitate the parametrization of small-scale complex processes that cannot be represented explicitly in the model because they operate at subgrid scales. In addition, there may be significant qualitative understanding of a particular process, but quantitative understanding may be limited. The ability to express the physical process as a set of detailed mathematical equations may not exist, or the equations may be too complicated to solve without simplifications (Steyaert, 1993).

In addition to incomplete knowledge, simplifications and inadequate parametrizations of real-world processes, more difficulties arise when modelling crosses disciplinary boundaries. This can be illustrated by the concept of modelling water and energy exchange processes within the soil–plant–atmosphere system (e.g., Hall *et al.*, 1988). Another example is ecosystem dynamics modelling with the environmentally and physiologically structured Forest-BGC model (Running and Coughlan, 1988), which includes models from the disciplines of atmospheric science, hydrology, plant physiology and soil science. Clearly, not all processes can be modelled within a GIS, but we need to define those processes that are amenable to such treatment, and develop ways of visualizing them.

Effective use of GIS requires attention to several generic issues, many of which are also of concern to environmental modellers. First, the discretization of space that is inherent in both fields forces the user to approximate true geographical distributions; the impacts of such approximations on the results of modelling are often unknown or not evaluated. Space can be discretized in numerous ways — finite differences and finite elements are two examples well known to environmental modellers and each has its own set of impacts on the results. Second, effective display of the results of modelling, particularly for use in policy formulation, requires attention to principles of cartographic design. Work is also needed on enhanced data conversion algorithms that intelligently move data between the GIS and simulation model environments. Finally, spatial databases tend to be large, and effective environmental modelling necessitates careful attention to the efficiency of algorithms and storage techniques. Many of these generic issues are identified in the larger NCGIA research agenda (NCGIA, 1992), and are subjects of active research within the GIS community.

GIS-linked Models and Conceptual Frameworks for Hierarchical and Aggregated Data Structures

The requirements of global change research place significant emphasis on modelling at multiple time scales and across multiple spatial scales. Spatial scaling involves significant research issues, such as how to parametrize and aggregate or integrate water and energy fluxes from the plant leaf level to the regional level. For example, the land surface parametrizations for water and energy exchange between the biosphere and atmosphere within long-term climate change simulation models must eventually account for climate-induced vegetation successional changes, and modified land characteristics that control water and energy fluxes. One approach to spatial and temporal scaling involves the use of nested gridded models such as those illustrated by Hay *et al.* (1993). There is also the problem of extrapolating research results from local to regional scales (Burke *et al.*, 1991; King, 1991; Nemani *et al.*, 1993). Last, there are hurdles in linking vegetation, atmosphere, climate and remote sensing across a range of spatial and temporal scales (Hall *et al.*, 1988).

The use of physically based models of spatially distributed processes in complex terrain and heterogeneous landscapes is also challenging (Steyaert, 1993). Factors such as terrain and landscape variability are important considerations for land–atmosphere interactions (Pielke and Avissar, 1990; Carleton *et al.*, in press). Distributed parameter approaches are increasingly used instead of classic lumped analyses as models become more sophisticated, allowing for more realistic, physically based parametrizations of a wide variety (e.g., leaf area index, evapotranspiration) of land surface characteristics (Running *et al.*, 1989; King, 1991).

Environmental simulation modelling may be greatly enhanced by the results of field experiments such as the First International Satellite Land Surface Climatology Project (ISLSCP) Field Experiment (FIFE) (Sellers *et al.*, 1988; Strebel *et al.*, this volume), an intensive study of interactions between land surface vegetation and the atmosphere in the prairie grasslands biome of central Kansas, and the planned Boreal Ecosystem–Atmosphere Study (BOREAS) for the boreal forest biome of central Canada (NASA, 1991). Such experiments are integral to the development and testing of models based on direct measurements and remote sensing data gathered by various ground-based, aircraft and satellite systems. Focused research to understand processes and to develop remote-sensing-drive algorithms for regional extrapolations will be supplemented by a range of simulation models in BOREAS (NASA, 1991).

In addition to the issue of spatial processes operating at multiple time and space scales, GIS and environmental simulation models share similar spatial data requirements. GIS, by definition, is a technology designed to capture, store, manipulate, analyse and visualize diverse sets of spatially referenced data. Advanced simulation models also require a rich variety of land-surface-characteristics data of many types in order to investigate environmental processes that are functions of complex terrain and heterogeneous landscapes.

Spatial data are available from an increasing number of sources, including land cover characteristics based on multitemporal satellite data. Land-surface-characteristics data required by scientific research include land cover, land use, ecoregions, topography, soils and other properties of the land surface; these help to enhance the understanding of environmental processes and to develop environmental simulation models (Loveland *et al.*, 1991). Advanced land-surface-process models also require data on many other types of land surface characteristics, such as albedo, slope, aspect, leaf area index, potential solar insolation, canopy resistance, surface roughness, soils and the morphological and physiological characteristics of vegetation.

In addition to the problems of spatial and temporal scales, parametrization is confounded by structuring processes that take place at different hierarchical levels (e.g., physiological, autecological, competitive, landscape). The interactions between levels are often asymmetrical, such that larger, slower levels maintain constraints within which faster levels operate (Allen and Starr, 1982). Because GIS are, by definition, models, there may be misconceptions about the dynamics of, say the physical or ecological, systems they represent due to their emphasis on a particular 'observation set' (O'Neill *et al.*, 1986), which is a reflection of the way in which one views the world. Since the layers in GIS result in a form of one-to-one 'mapping', it is easy to overlook the complexity of a system that is represented in a GIS because of structural and functional constraints.

Extensions of the more commonly used GIS techniques are required to create an integrated modelling environment. Further, the themes of multiple space and time scales are basic to coupled systems modelling and demand a highly cross-disciplinary modelling approach. GIS could help meet these requirements and provide the flexibility for the development, validation, testing and evaluation of innovative data sets that have distinct temporal components. For example, detailed consideration of landscape properties and spatially distributed processes on the land surface is fundamental to global and mesoscale climate models. There is also the need to derive data sets from existing data; GIS tools may be used for flexible scaling; parametrization and reclassification; creating variable grid cell resolutions; and aggregation and integration of spatial data. At the same time, it is necessary to preserve information across a range of scales and quantify the loss of information with changing scales. This overall environment seems suited for GIS to support integrative modelling, to conduct interactive spatial analysis across multiple scales, and to derive complex land surface properties for model input based on data sets created from innovative thematic mapping of primary land surface characteristics.

GIS is a tool that can play many roles in global change research; however, there is a need to identify those roles more clearly, and also to identify impediments that prevent GIS from being used more broadly. We need also to address the generic needs in global change research for spatial data handling tools, and for hierarchical concepts and data structures. Addressing these fundamental research issues will require a research programme in global change that includes several objectives, including, but not limited to, the following:

- To identify technical impediments and problems that obstruct the use of GIS in global change research.
- To assess critically the quality of existing global data in terms of spatially varying accuracy, sampling methodologies and completeness of coverage, and develop improved methods for analysis and visualization of such data.
- To develop theoretical and computational structures capable of integrating knowledge at finer spatial scales and lower levels of aggregation, within the context of global change.
- To develop methods for dynamically linking human and physical databases within a GIS, exploring the regional impacts of global change, and understanding the interactions between human systems and regional and global environmental systems.
- To develop methods for detecting, characterizing and modelling change in ecological transition zones.

These objectives form the scientific core of the initiative, and will provide the focus of the discussions at the specialist meeting. Although the five objectives together present a massive challenge, our intent is to focus on specific topics within each area to identify the larger research agendas that must be pursued. The following sections address each of these topics in greater detail within the context of the relevant literature.

Technical Impediments

The development of linkages between environmental systems as well as between these biophysical systems and the human dimensions is critical to global change research. We begin with the premise that GIS can help to provide such a linkage, by making the results of environmental modelling more accessible to natural and social scientists, and by providing the means of integrating a wide spectrum of data types as inputs to environmental models. Development of such linkages, however, is impeded by factors such as incompatible data models, inconsistent scales and lack of access to accurate and complete data. We view overcoming these impediments as a two-step process. First, we need to determine what new GIS functionality is needed to meet the needs of environmental modellers for building databases using GIS, linking models with GIS and modelling within a GIS. Second, additional knowledge is needed to determine the nature of environmental processes in the ecological, hydrological and atmospheric sciences, and how these processes, including uncertainties, are modelled and parametrized. GIS can be used to enhance this interdisciplinary effort.

In addition to questions of functionality and linkage, several other technical impediments need to be removed if GIS is to play a more useful role in support of environmental systems modelling. GIS user interfaces are unfriendly, requiring excessive investment of time and effort for effective use. There is a need for streamlined data format conversion algorithms, and for more comprehensive standards for data to promote exchange and sharing. Other impediments include the difficulty of handling time and hierarchical relationships, and the inability to visualize the output of dynamic models. Many of these can be addressed through the concepts and techniques of object-oriented programming and design. Some of these are being dealt with in complementary NCGIA initiatives.

For example, current GIS are best suited for the representation and analysis of relatively static landscape features, such as change detection analysis based on two separate land cover classifications or estimation of soil erosion risk factors associated with the mean environmental conditions. In contrast, environmental simulation models deal with environmental processes involving time steps from seconds to hours for water and energy exchange processes, days to seasons for biological and biogeochemical processes, and annual time steps for the multidecade to century processes of soil development or forest succession (NRC, 1990). New temporal GIS tools and functions are needed to address these types of multitemporal processes, especially for the analysis of multitemporal satellite data indicative of such land surface processes and dynamic land surface characteristics.

Regional and Global Data

Multiscale land cover, vegetation, soils and topographic data are needed in global change studies at the plot, community, ecosystem, landscape, region, continent and global levels. Sources of such data include national-level surveys (e.g., topographic, cartographic, land cover, land use and soils mapping) and focused global change efforts involving data set development in conjunction with Long-Term Ecological Research (LTER) sites, field experiments such as FIFE and BOREAS, global 1-km land cover mapping projects based on multitemporal advanced very high resolution radiometer data (AVHRR), or socio-economic databases for regional assessments. However, there are substantial variations on the availability and quality of these data sets. To illustrate, the prospects for the development of a consistent global 1-km land cover characteristics data set from multitemporal AVHRR are excellent based on the efforts of the IGBP and USGCRP, and others. Notwithstanding, the availability of corresponding, consistently developed land use practice, soils and DEM databases is less clearly determined. This initiative will critically assess the quality of existing regional and global data in terms of accuracy, sampling methodologies and completeness of coverage.

Further, although GIS uses many methods for exploration, analysis and visualization of geographic data, it does so almost exclusively in two dimensions, and relies on classical map projections to reduce the curved surface of the globe to a plane. These projections distort space; for example, the familiar Mercator projection introduces an artificial discontinuity at the 180th meridian, and shows the poles as lines equal in length to the equator. The results of modelling and analysis displayed on the projected 2-D plane can differ substantially from those as displayed on the 3-D globe. Goodchild and Yang (1992) have described a hierarchical data structure for global GIS that avoids these problems of projection distortion. Although several significant contributions have been made (Wahba, 1981; Legates and Willmott, 1986), our set of spatial techniques for the globe is still very limited. Many global data sets, particularly those relevant to the human dimensions of global change, are constructed from multiple sources and have variable quality. Yet most techniques of spatial analysis, modelling and display assume the absence of uncertainty in the data, or statistically stationary models of uncertainty, and have no way of dealing with uncertainty that varies spatially in a known fashion. NCGIA Initiatives 1 (Accuracy of Spatial Data Quality) and 7 (Visualization of Spatial Data Quality) have developed several models and techniques in the area of uncertainty, and their adaptation to the spherical case, and application to global data, will be explored under I-15.

The Problem of Scale

The problem of inference across scales has been recognized for decades, but it has assumed greater importance in environmental sciences with the current focus on global change. In social science, it manifests itself in the so-called 'ecological fallacy' (Robinson, 1950) and the modifiable areal unit problem (Openshaw, 1984; Fotheringham and Wong, 1991). In ecology, it has been a central problem. Levin (1992) notes that 'Applied challenges, such as the prediction of the ecological causes and consequences of global climate change, require the interfacing of phenomena that occur on very different scales of space, time, and ecological organization'. Although the specific problems of cross-scale inference in disciplines such as ecology are not of direct concern in this initiative, successful upward scaling requires appropriate data, and data models and structures that can provide the necessary support. For this reason, there has been a strong call for the development of hierarchical GIS, defined as spatial data handling systems that can provide views of data at multiple scales. Although some GIS support concepts of hierarchy (e.g., quadtree-based systems; Samet, 1990), most systems are orientated to projects and databases with a single, uniform level of spatial resolution.

Many of the issues of hierarchical GIS have already been studied in NCGIA Initiative 3 (Multiple Representations). This initiative will look at hierarchical GIS from the specific perspective of global change research. Software capabilities and published research in this area will be reviewed and examined for their adequacy in aiding research on both environmental and human dimensions of global change. We currently lack tight linkages between the modelling and analytic techniques and the capabilities of GIS. The principal objective of this initiative will be to produce a set of functional requirements and specifications for hierarchical GIS to be used to support global change research.

Human–physical Linkages

Linkage of the human and physical dimensions of environmental change lies at the heart of many concerns of regional and global change research. Although GIS is widely acknowledged as an integrative tool, such linkage is impeded by fundamental incompatibilities between data models. Many global environmental models use finite elements defined as regular tessellations of a latitude–longitude space, while models of global climate frequently abandon geographic space altogether in favour of the spectral domain. However, it is recognized that dynamic flows may be calculated in spectral space but are connected within the model to interact with grid cell physical parametrizations. By contrast, many social and economic data are defined for irregular tessellations such as nation-states, which vary in area by factors of as much as 10^5. Social and economic data often concern interactions that are defined as attributes of pairs of spatial objects, having no analogous equivalent in environmental modelling. Many environmental variables are defined for fields, whereas social variables are often defined for discrete objects. Also, the different approaches to the time dimension used by physical and human modellers present a further source of incompatibility. Physical data are often available at regular time intervals (e.g., the orbital frequency of satellite sensors) while acquisition of social data is not synchronous among nations (e.g., census timing).

Tobler (1992) has recently provided an example of the types of transformations that can help bridge this gap: a world population density data set expressed as a sum of spherical harmonics on a regular spatial tessellation. Databases such as the Digital Chart of the World (DCW) may help by providing environmental modellers with ready access to the irregular tessellations used by social scientists. The initiative will review the data models used for data collection, dissemination and modelling in global change research in both human and physical areas, particularly with respect to time. It will develop models that can facilitate dynamic linkage between the two fields, and examine the problems of implementing them in GIS software.

Transition Zones

The biogeographic effects of impending global change are likely to appear first at ecotones, either because these sensitive areas have intense biotic–abiotic interactions, or because they are situated at physiological thresholds (di Castri *et al.*, 1988; Hansen *et al.*, 1988). Further ecotones are considered important in influencing ecological flows (energy, resources, information) and in detecting change in the global environment (Hansen *et al.*, 1988). It is crucial to understanding the nature of these effects to be able to recognize ecotonal boundaries systematically, to determine their size and, in the context of global change scenarios, to characterize their functional response to large-scale climatic features.

Unfortunately, relatively little is known about the fundamental characteristics of ecotones because of a general absence of techniques for measuring the dynamic processes characteristic of ecotonal areas. Instead, environmental models use characterizations of the earth's surface based on degree of homogeneity and uniformity. Moreover, to date, much research on the response of vegetation to climatic change has been measured over time scales of minutes or hours, rather than the more appropriate ecological time frames of weeks or months (Woodward and Diament, 1991). Similarly, much of our knowledge of population and ecosystem structure comes from the analysis of small areas (less than 1 km^2), represented by even smaller plots (1–30 m^2); however, it is over larger areas (at smaller scales) that global climatic change can best be assessed and understood (Burke *et al.*, 1991).

There has yet to be a comprehensive attempt to understand the sensitivity of ecotones to global change (Gosz and Sharpe, 1989; Holland and Risser, 1991). This initiative will explore the means of detecting, characterizing and modelling change in ecotones, principally through the use of GIS and satellite-derived data. From a GIS perspective, ecotones are areas of continuous spatial change that are incompatible with standard discrete data models such as polygonized area class maps. Under Initiative 1 (Accuracy of Spatial Databases), NCGIA explored the representation of continuous change fields expressed on rasters, and developed models that can be used to represent ecotones in spatial databases and to link the concepts of ecotone gradients with those discrete cartographic objects. Under I-15, a systematic review of existing environmental models will examine the role of homogeneous earth surface characterization, and the problems of modelling based on spatial gradients.

Summary and Conclusion

New directions for improving scientific database development in the context of environmental research and for addressing questions at ecosystem to global scales are discussed throughout this volume. This chapter proposes a research agenda that addresses several specific themes including the development of advanced GIS tools for multiscale analysis and visualization of spatially complex systems, and the use of these tools to overcome technical impediments and accelerate the integration of GIS and remote sensing technologies with environmental simulation models. We have developed an integrated framework for enhancing the role of GIS in facilitating environmental research at ecosystem to global scales, which takes as its starting point the need for increased integration of GIS and modelling. As a visual technology, GIS is a very powerful tool for the communication of scientific results and simulations and the assessment of data quality, but there remain hurdles in the discretization of space and visual portrayal of multiscaled results. Another key topic is the need for increased recognition of the multiple time and space scales at which data for global change research are being collected and analysed. An increased understanding of the feedback mechanisms and interactions that link multiple scales demands additional research on incorporating hierarchical frameworks and data structures into GIS. In view of these challenges, we have recommended a five-objective programme and identified 'targeted' research within each of five broad objectives to contribute to improving the role of GIS in global change research. These include identification of technical impediments, assessment of the quality of global data, tackling problems in making cross-scale inferences, developing methods for human–physical linkages, and monitoring and modelling in ecological transition zones.

We are still in our infancy in terms of understanding the nature of impending global change. Advancements in GIS can help garner a better understanding of environmental problems from landscape to global scales. It is essential that work proceed rapidly in this area as greater challenges await us, for soon we must turn to solutions to reduce the problem of climate change, preserve biological diversity and reverse the decline in the productive capacity and health of the planet.

Acknowledgements

The National Center for Geographic Information and Analysis is supported by the National Science Foundation, grant SES 88-10917. DEJ is grateful for the support of the NCGIA and McIntire-Stennis funds. The authors acknowledge critical reviews by W. Easterling, J. Merchant, M. Palecki and J. Wu.

References

Allen, T. F. H. and Starr, T. B., 1982 *Hierarchy: Perspectives for Ecological Complexity*, Chicago: University of Chicago Press.

Botkin, D. B., 1989, Science and the global environment, in Botkin, D. B., Caswell, M. F., Estes, J. E. and Orio, A. A. (Eds) *Changing the Global Environment: Perspectives on Human Involvement*, pp. 3–14, New York: Academic Press.

Burke, I. C., Kittel, T. G. F., Lauenroth, W. K., Snook, P., Yonker, C. M. and Parton, W. J., 1991, Regional analysis of the Central Great Plains, *BioScience*, **24**, 685–92.

Carleton, A. M., Jelinski, D. E., Travis, D., Arnold, D., Brinegar, R. and Easterling, D. R., in press, Climatic-scale vegetation–cloud interactions during drought using satellite data, *International Journal of Climatology*.

Committee on Earth Sciences (CES), 1989, *Our Changing Planet: A US Strategy for Global Change Research*, Washington, DC: Federal Coordinating Council for Science, Engineering, and Technology, Office of Science and Technology Policy.

Committee on Earth Sciences (CES), 1990, *Our Changing Planet — The FY 1990 Research Plan*, Washington, DC: Federal Coordinating Council for Science, Engineering, and Technology, Office of Science and Technology Policy.

Committee on Earth and Environmental Sciences (CEES), 1991, *Our Changing Planet — The FY 1992 US Global Change Research Program*, Washington, DC: Federal Coordinating Council for Science, Engineering, and Technology, Office of Science and Technology Policy.

Committee on Earth and Environmental Sciences (CEES), 1992a, *Our Changing Planet — The FY 1993 US Global Change Research Program*, Washington, DC: Federal Coordinating Council for Science, Engineering, and Technology, Office of Science and Technology Policy.

Committee on Earth and Environmental Sciences (CEES), 1992b, *Economics and Global Change: The FY 1993 Research Program on the Economics of Global Change*, Washington, DC: Federal Coordinating Council for Science, Engineering, and Technology, Office of Science and Technology Policy.

Committee on the Human Dimensions of Global Change, Commission on the Behavioral and Social Sciences and Education, National Research Council, 1992, *Global Environmental Change: Understanding the Human Dimensions*, Eds Stern, P. C., Young, O. R. and Druckman, D., Washington, DC: National Academy Press.

di Castri, F., Hansen, A. and Holland, M. M. (Eds), 1988, A new look at ecotones: Emerging international projects on landscape boundaries, *Biology International*, Special Issue no. 17, 9–46.

Earth System Sciences Committee (ESSC), 1986, *Earth System Science Overview: A Program for Global Change*, Washington, DC: National Aeronautics and Space Administration.

Earth System Sciences Committee (ESSC), 1988, *Earth System Science: A Closer View*, Washington, DC: National Aeronautics and Space Administration.

Fotheringham, A. S. and Wong, D. W. S., 1991, The modifiable areal unit problem in multivariate statistical analysis, *Environment and Planning*, *A*, **23**, 1025–44.

Goodchild, M. F., 1991, Integrating GIS and environmental modeling at global scales, in *Proceedings, GIS/LIS 91*, Vol. 1, pp. 117–27, Washington, DC: ASPRS/ACSM/AAG/URISA/AMFM.

Goodchild, M. F., Parks, B. O. and Steyaert, L. T., (Eds), 1993, *Environmental Modeling with GIS*, New York: Oxford University Press.

Goodchild, M. F. and Yang, S., 1992, A hierarchical spatial data structure for global geographic information systems, *CVGIP-Graphical Models and Image Processing*, **54**, 31–44.

Gosz, J. R. and Sharp, P. J. H., 1989, Broad-scale concepts for interactions of climate, topography, and biota at biome transitions, *Landscape Ecology*, **3**, 229–43.

Hall, F. G., Strebel, D. E. and Sellers, P. J., 1988, Linking knowledge among spatial and temporal scales: Vegetation, atmosphere, climate, and remote sensing, *Landscape Ecology*, **2**, 3–22.

Hansen, A. J., di Castri, F. and Naiman, R., 1988, Ecotones: What and why?, in di Castri, F., Hansen, A. and Holland, M. M. (Eds) A New Look at Ecotones: Emerging International Projects on Landscape Boundaries, *Biology International*, Special Issue 17, 9–46.

Hay, L. E., Battaglin, W. A., Parker, R. S. and Leavesley, G. H., 1993, Modeling the effects of climate change on water resources in the Gunnison River basin, Colorado, in Goodchild, M. F., Parks, B. O. and Steyaert, L. T. (Eds) *Environmental Modeling with GIS*, pp. 173–81, New York: Oxford University Press.

Holland, M. M. and Risser, P. G., 1991, The role of landscape boundaries in the management and restoration of changing environments: Introduction, in Holland, M. M., Risser, P. G. and Naiman, R. J. (Eds) *Ecotones: The Role of Landscape Boundaries in the Management and Restoration of Changing Environments*, pp. 1–7, New York: Chapman and Hall.

International Council of Scientific Unions (ICSU), 1986, *The International Geosphere-Biosphere Program — A Study of Global Change: Report No. 1*, Final Report of the Ad Hoc Planning Group, ICSU Twenty-first General Assembly, 14–19 September 1986, Bern, Switzerland: International Council of Scientific Unions.

International Geosphere–Biosphere Programme (IGBP), 1990, *The International Geosphere-Biosphere Programme — A Study of Global Change: The Initial Core Projects, Report No. 12*, Stockholm, Sweden: IGBP Secretariat.

King, A. W., 1991, Translating models across scales in the landscape, in Turner, M. G. and Gardner, R. H. (Eds) *Quantitative Methods in Landscape Ecology*, pp. 479–517, New York: Springer.

Legates, D. R. and Willmott, C. J., 1986, Interpolation of point values from isoline maps, *The American Cartographer*, **13**, 308–23.

Levin, S., 1992, The problem of pattern and scale in ecology, *Ecology*, **73**, 1943–67.

Loveland, T. R., Merchant, J. W., Ohlen, D. and Brown, J. F., 1991, Development of a land-cover characteristics database for the conterminous US, *Photogrammetric Engineering and Remote Sensing*, **57**, 1453–63.

Mounsey, H. M., 1988, *Building Databases for Global Science*, London: Taylor & Francis.

National Aeronautics and Space Administration (NASA), 1991, BOREAS (Boreal Ecosystem–Atmosphere Study): Global Change and Biosphere–Atmosphere Interactions in the Boreal Forest Biome, Science Plan.

National Center for Geographic Information and Analysis (NCGIA), 1992, A Research Agenda for Geographic Information and Analysis, Technical Report 92-7, Santa Barbara, California: National Center for Geographic Information and Analysis.

National Research Council (NRC), 1986, *Global Change in the Geosphere–Biosphere, Initial Priorities for an IGBP*, Washington, DC: US Committee for an International Geosphere–Biosphere Program, National Academy of Sciences.

National Research Council (NRC), 1990, *Research Strategies for the US Global Change Research Program*, Washington, DC: Committee on Global Change, US National Committee for the IGBP, National Academy of Sciences.

Nemani, R. R., Running, S. W., Band, L. E. and Peterson, D. L., 1993, Regional hydro-ecological simulation system — an illustration of the integration of ecosystem models in a GIS, in Goodchild, M. F., Parks, B. O. and Steyaert, L. T. (Eds) *Environmental Modeling with GIS*, pp. 291–304, New York: Oxford University Press.

Nyerges, T. L., 1993, Understanding the scope of GIS: Its relationship to environmental modeling, in Goodchild, M. F., Parks, B. O. and Steyaert, L. T. (Eds) *Environmental Modeling with GIS*, pp. 75–93, New York: Oxford University Press.

O'Neill, R. V., DeAngelis, D. L., Waide, J. B. and Allen, T. F. H., 1986, *A Hierarchical Concept of Ecosystems*, Princeton, New Jersey: Princeton University Press.

Openshaw, S., 1984, *The Modifiable Areal Unit Problem, Concepts and Techniques in Modern Geography*, No. 38, Norwich, UK: Geo Books.

Pielke, R. A. and Avissar, R., 1990, Influence of landscape structure on local and regional climate, *Landscape Ecology*, **4**, 133–55.

Price, M. F., 1989, Global change: Defining the ill-defined, *Environment*, **3**, 18–20.

Robinson, W. S., 1950, Ecological correlations and the behavior of individuals, *American Sociological Review*, **15**, 351–7.

Running, S. W. and Coughlan, J. C., 1988, A general model of forest ecosystem processes for regional applications. I: Hydrologic balance, canopy gas exchange, and primary production processes, *Ecological Modelling*, **42**, 125–54.

Running, S. W., Nemani, R. R., Peterson, D. L., Band, L. E., Potts, D. F., Pierce, L. L. and Spanner, M. A., 1989, Mapping regional forest evapotranspiration and photosynthesis by coupling satellite data with ecosystem simulation, *Ecology*, **70**, 1090–1101.

Samet, H., 1990, *The Design and Analysis of Spatial Data Structures*, Reading, Massachusetts: Addison-Wesley.

Sellers, P. J., Hall, F. G., Asrar, G., Strebel, D. E. and Murphy, R. E., 1988, The First ISLSCP Field Experiment (FIFE), *Bulletin of the American Meteorological Society*, **69**, 22–7.

Steyaert, L. T., 1993, A perspective on the state of environmental simulation modeling, in Goodchild, M. F., Parks, B. O. and Steyaert, L. T. (Eds) *Environmental Modeling with GIS*, pp. 16–30, New York: Oxford University Press.

Task Group 6, 1991, Human dimensions of climate change, in Jager, J. L. and Ferguson, H. L. (Eds) *Climate Change: Impacts and Policy*, Proceedings of the Second World Climate Conference, pp. 459–62, Cambridge: Cambridge University Press.

Tobler, W. R., 1992, Preliminary representation of world population by spherical harmonics, *Proceedings of the National Academy of Sciences*, **89**, 6262–4.

Townshend, J. R. G., 1991, Environmental databases and GIS, in Maguire, D. J., Goodchild, M. F. and Rhind, D. W. (Eds) *Geographical Information Systems: Principles and Applications*, pp. 201–16, London: Longman.

Turner, B. L., II, Kasperson, R. E., Meyer, W. B., Dow, K. M., Godling, D., Kasperson, J. X., Mitchell, R. C. and Ratick, S. J., 1990, Two types of global environmental change: Definitional and spatial-scale issues in their human dimensions, *Global Environmental Change*, **1**, 15–22.

Wahba, G., 1981, Spline interpolation and smoothing on the sphere, *SIAM Journal of Scientific and Statistical Computing*, **2**, 5–16.

Woodward, F. I. and Diament, A. D., 1991, Functional approaches to predicting the ecological effects of global change, *Functional Ecology*, **5**, 202–12.

SECTION II

Scientific databases and information systems

Databases and information management systems specifically designed for use by scientists and for environmental data have been slow in development. Environmental data are voluminous, diverse and complex, and frequently cannot be contained in a single database. With few exceptions, commercially available software has been developed for business applications in which all data are logically related and may be represented in a single database. Thus, most theory and software on database management are focused on facilitating the design, development and implementation of systems for business and are seldom optimal for environmental research. Although available 'solutions' may not physically or logically meet the needs of environmental research, relevant data are typically 'shoe-horned' into commercial database management systems. Such systems frequently complicate data processing and may also constrain the science.

The five chapters in the second section are based on extensive experience in designing, developing and managing scientific databases and information systems. The authors discuss many of the problems they have encountered and suggest solutions and alternative strategies. Strebel, Meeson and Nelson present a conceptual framework for the design, implementation and operation of scientific information systems. On the basis of their experiences in developing environmental information systems, they provide a comprehensive discussion of four basic components (management and organization, science requirements, data flow from source to archive and required resources) that underlie successful system design and implementation. Briggs and Su discuss characteristics of the Konza Prairie Long-Term Ecological Research Information Management Program and, importantly, present examples of how the program has evolved in response to changing computer technology (GIS, remote sensing, communication networks), growth of the research programme and personnel changes. Liff, Riiters and Hermann describe how field data at numerous distributed sites are rapidly converted into information that is used in assessing the ecological status of forests.

New strategies for database management will be required to deal with the massive multidisciplinary data sets needed for addressing issues related to global change. Frew discusses the Sequoia 2000 Project, the focus of which is to develop solutions for massive data storage, data access, data analysis and visualization, and wide-area networking. In the final chapter of the section, Pfaltz and French discuss a critical issue pertinent to environmental research at all scales — specifically, how best to represent change in environmental databases.

5

Scientific information systems: A conceptual framework

Donald E. Strebel, Blanche W. Meeson and Alan K. Nelson

A conceptual framework for the design, implementation and operation of scientific information systems has been developed. This framework is based on experience with focused field experiments, long-term data archiving and data publication. The elements of the framework include management and organizational constraints, the requirements of the scientific community, the flow of data from source to archive and resource requirements. Examples of these principles, drawn from successful information systems experience with the FIFE Information System and the Pilot Land Data System, are discussed.

Introduction

The design, implementation and operation of information systems to support scientific research are not well understood. It is already clear, however, that the structured engineering approach often applied in business applications usually fails in the science arena (Lucas, 1975; Preheim, 1992). The dogmatic application of structured methodologies to information systems for exploratory science, in which all relations are seldom known *a priori*, guarantees such failure. Failures have been particularly acute in the environmental sciences, which often deal with data from monitoring programmes with rapidly changing instrumentation or from one-time field experiments with varying degrees of structure. Integrating many such disparate data sets and preserving them for use over decades are major challenges that must be pursued if global change and other earth systems questions are to be successfully studied (see Jelinski *et al.*, this volume).

Our primary assumption in addressing this problem is that the fundamental role of a scientific information system is to publish data in a manner analogous to publication of research results. The detailed argument for this assumption (Meeson and Strebel, unpublished) will not be explored here. On the basis of this paradigm, however, we have established a conceptual framework of four key components of successful scientific information systems. These four components are:

- Management and organization;
- Science requirements;

- Data flow;
- Resources.

In each component, there are basic principles of design, implementation and operations. Ignoring these principles greatly increases the chances of failure of the system.

This chapter represents a first attempt to derive and state the principles in a general way that can be widely applied. The following sections develop principles and examples of each of the four components in some detail. These sections are based on our experience with the FIFE (First International Satellite Land Surface Climatology Project Field Experiment) Information System (FIS) (Strebel *et al.*, 1990; Briggs and Su, this volume) and the Pilot Land Data System (PLDS) (Sellers *et al.*, 1992; Agbu and Meeson, 1993; Meeson *et al.*, 1992) as well as design efforts for several emerging scientific information systems (e.g. the Earth Observing System Data and Information System (EOSDIS) and the Environmental Protection Agency's Environmental Monitoring and Assessment Program (EMAP)). The principles address many aspects of scientific information systems, including data quality, data retrievability and distribution, data system usability, data system continuity and responsiveness of data system management.

Management and Organization

Principles

(a) No rules are absolute. To accommodate the exploratory and evolving nature of scientific research, flexibility is critical.

(b) There must be a partnership between the scientists being served and the scientific information system management.

(c) Because scientific data management is partially a scientific task, the information system staff must have scientists as an integral part of the team.

(d) A scientific information system must be a service organization, with emphasis on delivering services that the science community will use.

Discussion

The management and organization of the information system staff and services are key elements in the success or failure of any information system (Lucas, 1975). Our experience has led us to identify several essential management and organization features.

(a) The flexibility of the management, organization and services associated with a scientific information system is a fundamental component of the framework. All facets of a scientific information system should be easily modified, with little or no adverse effect on the user community. In particular, the services or management structure should not undergo major revisions each time an internal structural change is required. However, the services and organization must respond to the changing and diverse needs of the scientific community, to changes in the composition of the scientific community itself, to changes in available technology and to budgetary fluctuations. Scientific information system managers should design systems and procedures that can cope easily and cost-effectively with this constant and sometimes dramatic change — all without loss of continuity.

(b) A critical element in the success of a scientific information system is developing a truly functional partnership between scientific information system managers and the science community. Through such a partnership, the service providers and their customers establish a daily working relationship in which responsibilities are shared and an environment of mutual respect and trust is developed. Such a working relationship is necessary for effective partnerships in which the information managers and the scientific community have a common set of objectives. Ideally, this results in an information system that becomes integrated into and a vital part of the scientific community.

In our experience, these partnerships have three different origins: (1) agencies or institutions assign information managers to a scientific activity or community, (2) information managers originate from within the scientific community in grass-roots, *ad hoc* activities, or (3) leaders of scientific activities actively seek information managers because their value is recognized. The obstacles and risks associated with establishing a truly functional partnership differ for each. The first case is the most risky since it does not originate from within the scientific community and the community does not select the service providers. The least risky is the last case, since not only is the partnership formed from within the scientific community, but the information partners are deliberately sought out and selected by the scientific community.

In an effective partnership, the scientific partners should hold the responsibility for defining the science objectives for the information managers and for working with the information managers to establish priorities and assign responsibilities within resource realities. The information managers and staff, in turn, hold the responsibility to adopt the management of the scientific information system, its organization and its services to these objectives and priorities in a way that is consistent with and acceptable to the scientific community. In addition, funding agency programme managers who are supporting the services of the information system have certain related responsibilities. They must work to ensure that the information managers and scientific community have the same understanding of the goals and objectives, and that these are consistent with budgetary realities and the agency's priorities.

Organizationally, this partnership is fostered by a management council made up of members of the scientific community, information managers and appropriate agency management (Figure 5.1). The management council should be characterized by frank, open, direct and frequent interactions among the members.

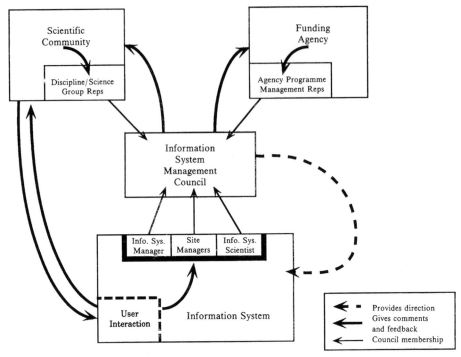

Figure 5.1 The information system management council should bring together representatives of the scientific community, funding agency management and information system staff for frank, open, direct and frequent interactions.

Note: Feedback loops are important to detect critical or changing needs and to adapt the information system to respond to them.

(c) The scientists on this management council should be chartered to represent a well-defined subgroup of the relevant scientific community, to build consensus within their sector of the scientific community, to represent that consensus at the council and to build scientific consensus within the management council. They may also be called upon to arbitrate customer-based conflict within the information system team. These scientists are most effective when they are active researchers with direct experience in using scientific data and providing that data to the user communities that they represent. Direct participation of members of the scientific community in this council provides an active and guiding scientific voice in the direction, priorities and decisions of the information system managers.

The information managers on the council should be chartered to listen to, comprehend and act on the direction of their scientific partners. The information managers are most effective when they have direct experience doing scientific research, because they then have an understanding of the scientific community and an ability to communicate with both the scientists and the technologists.

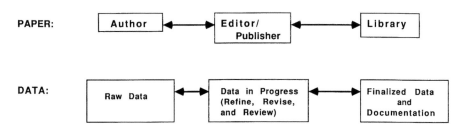

Figure 5.2 The analogy between data publication and publication of scientific research results.

Note: To archive research data sets for public use, an intermediate scientific information systems organization that functions as an editor and publisher is required.

Information managers are responsible for translating the directions of the science partners into pragmatic, realistic, concrete goals and services that can be provided within the necessary scientific timelines and within budget constraints.

The end result of a successful partnership is the integration of the information system managers and staff into the scientific community. The science community then relies on this relationship for increased productivity, and the information managers in turn rely on the relationship to foster improved data flow and accelerated data maturation.

(d) Recognition of the fundamental nature of the scientific information system partnership and the associated responsibilities leads to the conclusion that the information system must assume a service role in the scientific community. Conceptually, an information system's services are analogous to those provided by a publisher (Figure 5.2). The information manager assists in the final preparation of data and its documentation for distribution, taking on the tasks of editor, printer and publisher for the scientist/authors who collected or assembled the data.

It is also critical for information managers to realize that the service role means that building services is subordinate to delivering services. The information system managers, in conjunction with their science partners in the management council, should identify and then concentrate on providing services that the scientific community recognizes as critical to their research objectives. The information managers should then work closely with their scientific partners as active members of the development and design teams to ensure that the services develop in a way consistent with scientific needs and wants, and that the services produced will be used by their colleagues. This co-operative identification and iterative development of services is essential because the evolving requirements of the scientific community often cannot be stated exactly. A service should initially meet the basic need, and expand or be enhanced only when a demand is demonstrated. This approach ensures that the information managers and staff retain the focus on service and do not get lost in developing nice but unnecessary enhancements.

The service role suggests that the information system organize itself around the services that it provides. Within this context, the services fall into five general categories: (1) find and collect data for users, (2) process, integrate and assure the quality of the data, (3) assist with data publication, (4) provide access to staff who educate users and answer technical and scientific questions about the information system and the data it handles, and (5) provide access to data.

(1) Services to acquire data: Data acquisition services assist in obtaining relevant data from external institutional sources and from science project investigators. Staff work with the investigator or research project management to determine which data sets are required for the research activity (e.g., for site selection, site planning, data processing, calibration, modelling, etc.), if those data sets exist, who has them and how to acquire them. The staff then acquire the existing historical data that are needed and schedule future data collections for data that have not been collected. For example, maps and aerial photographs may be ordered from the United States Geological Survey (USGS) for site planning, and acquisitions of commercial data from Landsat and SPOT may be scheduled during the data collection period. Acquisition of data from individual investigators is encouraged by gentle urging and by reiterating the benefits that data integration, common data-quality checking and co-operative data exchange provide to their research objectives.

(2) Services to process, integrate and assure quality of data: The purpose of these services is to ensure that data sets made available to the scientific community are consistent and of high quality. Services to process data, for example, enable science projects with a large number of investigators to process data that many investigators will need (such as image data) using a consistent set of processing procedures and algorithms, and to place the data in a standard format.

Integration services include integration of the scientific data and integration of the corresponding descriptive information (sometimes called documentation or metadata). The integration of data within a research project includes combining similar data from different sources into a common format, standardizing units across data sets, standardizing variable names (e.g., temperatures measured for different entities are indicated in variable names), standardizing reference points (solar angles measured from zenith vs. horizon), and establishing the first cross-links between data sets (e.g., common instruments, methodologies, etc.). The data and metadata also should be integrated into the larger context of data beyond that of the research project that collected the data. This portion of the service strives to ensure that between research projects, and even across archives, similar data and their documentation are integrated.

(3) Data publication services: These services focus on preparing the data first for access and use among the project investigators and then for long-term archiving. Ultimately, these services strive to prepare the data so that they meet the 20-year test (Webster, 1991). That is, the archived data should be usable after 20 years by a scientist unfamiliar with the data or their collection.

The publication services assist scientists in the documentation and preparation of quality data sets, by assisting with the organization and quality review of the data, and with the writing, reviewing or editing of relevant documentation and descriptive information that is required to describe a data set fully. Through this service, scientific information system managers work with the scientific community to establish standards, guidelines and procedures for documenting and preparing data for publication. They also work to improve overall quality assurance by establishing measures for the accuracy of the descriptive information and documentation, and by providing a peer review procedure to ensure the overall quality of the data, documentation and other descriptive information.

It is through these publication activities that co-ordinated data sets such as those from interdisciplinary field experiments, or those that focus on a research topic/problem (e.g., bidirectional reflectance, greenhouse gases, etc.) are combined into integrated data sets and published for the scientific community. These publication services often bring together hard-to-find or difficult-to-use data sets into a common location and structure, which are then readily available for reference and use.

(4) User education and support: User services should concentrate on providing access to scientifically and technically knowledgeable staff who serve to educate users and to support science projects and general community research activities. The services that these staff provide have three main thrusts: (a) providing direct (telephone, email, mail) assistance to the scientific community, (b) general community and science project liaison, and (c) project or data set planning activities. Direct assistance requires individuals with a wide range of expertise because it covers a number of areas including: providing information about data set content, availability, format, errors, quality, reliability, collection and processing methods, and revisions; information about data content within co-ordinated or topical data set collections; status of requests; information about access to the information system services; and ordering of data and documentation.

General community or science project liaison activities include participation in, assistance with and attendance at scientific conferences, workshops and field campaigns. The information management staff may assist with the organization of the activity or they may contribute technically or scientifically to the workshop or field campaign. For example, in field campaigns, information system staff need to participate beforehand in workshops and site selection activities. Then during the campaign they need to be on site to experience the difficulties encountered and to record changes in original plans, including changes in sites, instrument locations and operation, experiment teams, etc. General community liaison activities also include training and communicating publication standards to the scientific community and to data publishers.

Similarly, project and data set planning requires that individuals participate in project planning meetings covering both information management and scientific aspects of the project and its data sets. In these meetings, information managers contribute their expertise in the areas of data set formats, data set compilation, supporting data sets, data set access and support, and information system design.

(5) Services to provide access to data: These services centre on tools to find, view and distribute data. The tools to find data should support simple access to data that are held in very large or very small collections. Very large data collections are similar in many ways to university libraries. Both have large and diverse holdings that are open to anyone; the users often do not know which book or journal they want, and they usually do not know what is available. Institutions responsible for large data collections also have the responsibility for long-term maintenance and care of their data holdings for future generations, just as libraries have these responsibilities for written materials.

Services to access data and to find data sets of interest in these very large collections require the scientific information system equivalent of the library catalogue system and library abstracting and indexing services. To provide these abstracting, indexing, card catalogue and check-out services, scientific information systems should provide data set abstracts or summaries, detailed information about available data sets, browsable data products and inventories of available data elements (e.g., image, flightlines, transects, profiles, etc.). This information must be accessible by a variety of criteria, and supported by an ordering capability. These services mean that information systems that provide access to very large data collections require completed materials, as do libraries, that are fully documented with their descriptive information (abstracts, detailed descriptions and inventory listings) and are ready for distribution.

Methods of providing less formal but similar services for small scientific data collections include placing data in anonymous File Transfer Protocol (FTP) accounts to which anyone can connect, scan the contents and extract data files of interest to them. Other services might include sending lists of available draft or archived data to members of a project. They could then select items from the list and request copies, much as they would do for materials held in small personal or departmental libraries. Projects might also provide small collections of data directly through subscriptions for the data, under which the data collections would be sent to their members automatically as they were produced.

Science Requirements

Principles

(a) The science requirements must drive the development and operation of the information system. Performing the science cannot become hostage to information system development.

(b) Exploratory research communities cannot often specify exact information system requirements *a priori*. The real requirements of the user community start with basic functionality.

(c) Co-operative identification and iterative development of services are essential.

(d) Information management staff should not be put into the position of making the determination as to whether data are scientifically acceptable.

(e) Peer review procedures are necessary for successful data and documentation publication and archiving.

Discussion

(a) The requirements of the scientific community must drive the development and delivery of scientific information system services (Preheim, 1992). A subtle but significant distinction should be made here. Science projects are not formed to feed an information system, they are created to accomplish specific scientific research. The research objectives result in user requirements that then drive the information system. The key point is that the research objectives are primary and the information system supports them. In projects where the information system development becomes more important than the science goals, both information system and science fail.

(b) The real requirements of the scientific community often cannot be stated exactly, although scientists will readily recognize when a service meets (or does not meet) their needs. Many research projects, by their exploratory nature, cannot specify *a priori* their exact information system requirements. Only when such a project is well under way are the requirements clear, at which point it is critical for the information system to reassess its objectives rapidly and respond to meet project deadlines.

(c) The information system managers and staff should work co-operatively with their science partners to identify requirements iteratively and provide highly reliable, basic services that are critical to the scientific research objectives and are structured to adapt easily to the community's rapidly changing needs. If services are initially provided in their basic form, they can then be incrementally developed and easily adapted to address changing needs. Co-operative identification and incremental development of services are essential to ensure that the services produced are needed and will be used.

In general, scientists need and will use a variety of off-line and on-line services that (a) provide a simple way to find and obtain quality, well-documented data, (b) provide data that are easy to use and compatible with common/familiar data analysis tools, (c) help manage the data they have collected and help make those data and the associated information available to their colleagues and the broader community, and (d) provide data at their own location, either on their desktop or bookshelf. Scientists also assume that the services will be adequately supported with hardware and staff that are responsive to their requests, and that the services will require little of their time to learn or to use.

(d) Scientific involvement in scientific data integration is critical. Scientists, especially those contributing the data, should be involved in the integration activities. Integration activities across research projects depend less on the contributing scientists because integration is mostly of metadata. For example, converting different names and units of identical variables in two different data sets to the same name and units requires little involvement of the contributing scientists. Integration across archives requires even less involvement of the contributing scientists since integration at this level is almost exclusively integration of metadata.

Quality assurance (QA) procedures are checks for the accuracy of the scientific data, for the accuracy of the information which describes the scientific data and for ensuring that the standards are met. QA checks for the scientific data are of two types: procedural (range checks, outliers, constant readings, etc.) and scientific (comparison of related data sets for consistent results). Information system staff should not make the determination whether data are acceptable or not — the data contributors make that decision. QA of the metadata and the data dictionary information are primarily procedural (e.g., valid ranges, valid values, required fields, etc.), although some scientific checks (e.g., comparison of collection times for data collected from the same location at the same time, comparison of descriptor definitions for data collected by similar instruments, etc.) may be required. Checks for adherence to standards are procedural but require examining both the metadata (e.g., adherence to naming conventions, use of existing descriptors, complete descriptions, etc.) and the scientific data (e.g., format of data).

(e) The end result of data integration and quality assurance is ideally a stand-alone published product that will stand the test of time. As in any publication process, this result cannot be achieved by investigators (data set authors) or scientific information managers (data set publishers) alone. The best results are achieved by an independent external review of both the data and the documentation.

Data Flow

Principles

   (a) As data mature, they pass through distinct stages that require physically and procedurally separate implementation of the information system.

   (b) As data mature, they require less scientific manipulation and more information management rigour.

   (c) The flow of data from producers to users is more important to the design of an information system than specific subsystems.

   (d) Scientific data integration is facilitated by assembling the data for quality assurance, documentation and integration at a single location.

Discussion

   (a) Data are not a static component of an information system. They are evaluated, refined, documented and augmented in many ways between collection by an instrument or an observer and their publication and archiving. Thus the data *mature*, both in a scientific sense and an information management sense, as they are processed by scientific investigators and the information system staff. Such changes are not independent of the community and the evolution of the information system itself. Often, only when users see the data in the context of an integrated information system do they become aware of and express their true needs for consistent formatting, quality assurance, relational links with other data, and documentation.

   Because the scientific understanding of the data and the requirements for handling them are actively evolving, it is not possible to design and build a scientific information system subsystem-by-subsystem according to predefined specifications, as has been done traditionally. Rather, the design effort must focus on procedures and capabilities that address the process of data maturation. This process has identifiable features and stages through which each data set must go. Within this data flow framework, the treatment of an individual data set may vary considerably.

   There are three general stages in the maturation of a data set. Like a maturing individual, there is first a period of infancy, followed by a transition phase that could be called adolescence and, finally, adulthood. This terminology emphasizes the data set and the care given to it by the information system; the three stages also correspond nicely to the publication analogy developed earlier. The distinct character of the three stages, as discussed below, is best addressed by organizationally distinct entities, rather than a single end-to-end information system.

   (1) The infant data set (stage 1): A data set is created and prepared for its scientific role in this first stage. That is, the data set is acquired or collected,

it is subjected to basic scientific manipulations that make it usable (e.g., calibration, extraction of signal from noise, conversion to physical units), it is recorded on appropriate media, and it is processed into a basic scientific information system to provide for tracking and further manipulation. Using the information system, the data set may be reformatted or processed to standard products, integrated with other data to which it is scientifically related, and associated with documentation that describes its content, pedigree and use.

The integration and formatting provided by the information system at this stage begin to prepare the data set for release to a larger group of investigators. Without this information management component and broader distribution intent, the data set dies an early death. It is likely to decay in a dusty corner in the individual collector's office with at best a brief note in a research paper or two. In general, science is not well served by such early termination of the data maturation process, and it is the responsibility of the scientist–information manager partnership to transition as much useful data as possible to the next stage.

(2) The adolescent data set (stage 2): In the second stage, the data set is better able to stand on its own, but is not completely independent of its scientific origins. Peer review and revision by a limited group of scientists help to prepare the data for adult life as a free-standing, independent, published data set. At this stage, also, preliminary versions of the data set may be released for use in related investigations. Additional integration with related data sets from other investigations or research projects may also occur in the information system.

The release of the data requires a more operational approach with increasing information management involvement and rigour compared with the earlier stage. In the context of a large, co-ordinated field experiment, for example, this stage requires a formal data distribution mechanism. In general, a shift from science-driven manipulations of the data set toward archival formatting, documentation and associated information management activities is evident at this stage.

Because the information management activities of this stage would get in the way of the more science-orientated activities of the first stage, they are often best conducted by an associated but distinct organization with more formal operational policies and procedures. Thus, as the data set matures into this stage, it acquires a certain stamp of quality and reliability, just by having passed the entrance criteria for the second-stage scientific information system. Conversely, the entrance criteria set a standard toward which the data set team must work during the first stage.

(3) The adult data set (stage 3): At the beginning of the third stage, revised, quality assured and documented data are released in standard format to the general scientific community. The fully mature (adult) data set may now be used (and abused) by the scientific community in the same way that a published research work is accessed in the open literature or obtained from a library. It must stand on its own, independent of its scientific progenitors.

In this stage, the data set may be formally placed in the care of the professional information management staff of a data archive. The characteristics of such an archival scientific information system are even more procedural than in the second stage. While the staff must be scientifically knowledgeable, they are unlikely to be conversant with the details of any particular data set. Their activities focus on ensuring that the data set meets archive standards and is suitable for interarchive exchanges, that it is documented for general users, that its availability is appropriately catalogued and advertised, and that it is permanently preserved.

(b) In each of the three stages of maturing data, different services are provided as described above. Table 5.1 illustrates the matrix of five services and three stages. This matrix is the basis for designing, implementing, scheduling and operating scientific information systems that successfully assist and promote the data maturation process. Not all of the services are required for each stage. However, if the services in one of the non-empty cells are ignored, the delivery of the data set to the scientific community will suffer, and the science requirements will not be fully met. In accordance with the 20-year test previously mentioned, if one of the services (rows) or one of the stages (columns) is missing, the system will fail.

Table 5.1 makes it very clear that the services to be provided by information management vary significantly from stage to stage in the life of the data set. Because of the user community size, the duration the data set is in each maturation stage, and ability of the data set to stand on its own, the level of information management formalism in handling the data set increases in each successive stage. Effective organization of these information management activities, including establishing a realistic partnership with the relevant scientific community, argues strongly that distinct organizations should be created to address the user and data interactions in each stage.

The publication analogy, shown in Figure 5.2 and important from the science requirements perspective, would replace the organizational terms just used with author information system, publisher information system and library information system, respectively. No matter what terms are used, the descriptions convey the fact that there is a different span of control for the information management organization in each stage of data maturation. Restricting the span of control in this way not only provides more focused services, but it also makes efficient use of resources. The Resource Requirements section below discusses how an expanded list of the steps in data maturation can assist in assessing scientific information system resource requirements.

Table 5.1 Matrix of data maturation stages and information management services.

Service	Infant data sets	Stage Adolescent data sets	Adult data sets
Acquire data	Identify and acquire supporting data sets (includes commercial data) Cajole data from investigators	Identify and acquire supporting data sets (includes commercial data not acquired by projects)	
Process, QA, integrate data	Process data to standard products Integrate scientific data within project Perform scientific QA	Create value-added products Integrate data across research projects Information management QA Propose data formats	Integrate data for interarchive exchange Archive requirement QA
Publish data	Initial data set documentation Construct and populate data dictionary	Final data set documentation Propose publication standards User guides Data publication Assemble data set collections	Document for general users Set publication standards Advertise data (brochures, user guides, tutorials)
Educate and support users	Assist PI technical users Participate in experiment planning and field campaigns	PI user assistance Science/research project meetings and workshops Observe field campaigns Track (orders, revisions)	Assist general-community users Community outreach (meetings, workshops, campaigns, communicate standards) Track archive usage (orders, activity, etc.)
Provide access to data		Develop tools to find data Provide limited data distribution	Develop and provide tools to find and browse data Distribute data Maintain permanent archive

(c) As stated previously, a successful scientific information system must be science-driven and service-orientated. The focus of the scientific information system must be the movement of data through the system, and not the treatment of the data within the system. The necessary treatment of the data will evolve as the science community's understanding of the data evolves. This fact necessitates delivery of data to the science community before some of the processing steps are even conceived. Complete reliance on advanced planning of compartmentalized subsystems removes the flexibility needed to respond to the evolving demands of a science community discovering new relationships in their data. Maintaining a focus on data flow-through both achieves the science goals of the information systems (by delivering data) and promotes the most rapid refinement of the system requirements (since scientific analysis of the data drives data maturation).

(d) The services listed in Table 5.1 imply that data flow in a scientific information system must be more than a pass-through process. Common data integration and quality assessment make more data available to a wider range of scientific analyses. Thus, at the centre point of the data flow stream, data from multiple sources are assembled, refined and then distributed in many different combinations to diverse users. It is difficult to achieve this focusing effect if the data are not brought together at a single location where an individual or small team can do the hard, creative work of cross-data set quality assurance, documentation and integration.

Resource Requirements

Principles

(a) Data synthesis (integration) is an expensive (time, personnel, intellectual) investment with associated risks and payoffs.
(b) The magnitude and character of the required fiscal and personnel resources change as a data set matures from stage to stage.
(c) Any given stage of data maturation requires an investment of human resources that is largely independent of the data set being processed.

Discussion

(a) The resources required to design, implement and operate scientific information systems are appreciable and often the subject of contention. Good science practice demands that quality data are documented and archived for long-term use by the scientific community. This is particularly true in earth systems studies where each data point is potentially critical in detecting and understanding causes and effects of ongoing environmental changes that may outlast generations of scientists.

Yet, good scientists frequently debate the wisdom of using funds, which could be used to collect and analyse more data now, for information system activities that only promise potential long-term payoffs. This debate is not readily terminated, particularly when the projected costs of scientific information system activities are largely hypothetical.

Table 5.2 *Information system personnel requirements for handling data.*

Step	Activity	Skills	Personnel	Time	Maturity stage
1	Initial data submission/acquisition	Communication, persuasiveness	Data manager	1–2 weeks	Infant
2	Log data set into tracking system	Attention to detail	Data manager, data technician	1 day	Infant
3	Perform acceptance QA (media, format, etc.)	Always alert, an eye for pattern	Data analyst	1 day	Infant
4	Review initial documentation	Science background	Data analyst	1 week	Infant
5	(Return/discuss with source if either of previous two steps unsatisfactory)	Persuasiveness, investigatory abilities	Data manager	1 week	Infant
6	Plan integration into system (logical design; links to other data, inventory procedures, processing algorithm)	Programming, scientific knowledge, database (DB) design, configuration control	Information scientist, programmer, DB analyst, GIS specialist	2 weeks	Infant
7	Implement plan (create programs, procedures, DB tables)	Programming, DB programming	Programmer	1 week	Infant
8	Process data to output product and/or load it into development DB	Computer, DB familiarity	Programmer, data technician	1 week	Infant
9	Load documentation and inventory entries into developmental DB	DB skills, science understanding	Programmer, user support	2 weeks	Infant

Table 5.2 Continued

Step	Activity	Skills	Personnel	Time	Maturity stage
10	Submit copies to back-up archive	Routine processing	Data manager, data technician	1 day	Infant
11	QA of data processing and entry	Making comparisons, investigating discrepancies	Data analyst	3 days	Infant
12	Resolve loading, verification or science questions identified by processing	Communications, investigatory abilities	Data manager, data analyst	1–2 weeks	Infant
13	Source review of processed/loaded data and documentation	Scientific review	Information scientist, data manager	1 week	Infant/ adolescent
14	Prepare automated data management procedures to track data, updates, usage	Organization, programming	Data manager, programmer	2 days	Adolescent
15	Release data set and initiate user support	Personality, scientific knowledge	User support	1–2 weeks	Adolescent
16	Begin secondary QA (PI, other investigators)	Scientific analysis	Investigators, research staff	1–2 weeks	Adolescent
17	Obtain revised/refined data and documentation	Communication	Data manager	1–2 weeks	Adolescent
18	Release updated data/documentation	Scientific review, decisiveness	Information scientist	1 week	Adolescent
19	Expand user group, begin public distribution, initiate intermediate-level user support	Personality, scientific background	User support	1 week	Adolescent

Table 5.2 Continued

Step	Activity	Skills	Personnel	Time	Maturity stage
20	Scientific integration of disparate data sets	Scientific analysis	Information scientist, research staff	2–4 weeks	Adolescent
21	Final documentation preparation and QA	Technical writing	Technical editor, user support	2 weeks	Adolescent
22	Documentation and data set peer review	Scientific review	Information scientist	1 week	Adolescent
23	Archive formatting	Computer	Programmer	1 week	Adolescent
24	Data publication/ release to community	Co-ordination, organization	Information scientist, user support	1 week	Adolescent/ adult
25	Archive requirements QA	Technical review	Data analyst	3 days	Adult
26	Initiate long-term archive user support	Personality, library skills	User support	1–2 weeks	Adult
27	Prepare general user documentation and data set advertising	Education skills, writing	User support	1 week	Adult
28	Data integration for interarchive exchange	Computer, database	Data analyst, programmer	1 week	Adult

(b) Better estimates of the amount and timing of resources required by information systems can be mined from our experience developing and operating such systems. It is natural to apply the concepts developed in the previous section and ask what services are required as data mature, when are they required, what kind of people it takes to do them, and how much time they take. It is then a relatively easy step to convert this information into costs and timelines.

As in the previous section, the focus of our discussion will be on data sets. The basic question is: what resources does it take to create and archive a believable (trusted), adequately documented data set that is usable over a long period by people not directly familiar with the data or the people who collected it? The activities which are required are outlined in Table 5.1; in this section those activities are detailed more explicitly, ordered sequentially, and associated with a personnel type and time investment.

Hardware and software resources are fundamental to this process, but they are essentially single-time start-up costs that must be provided if there is going to be any serious information system effort. Under today's conditions, the bulk of the cost of any information system is in the cost of the people who process, document and support the use of a data set through time. The magnitude and character of the required personnel resources change as each data set matures. The number and timing of data sets handled by the system then determine the instantaneous and cumulative fiscal resources required.

The first step in estimating information system costs is to describe the personnel resources required at each stage of data set maturation. Table 5.2 lists the steps required to handle a data set, the skills and type of personnel required for each step, an average length of time for the step, and the general data maturity stage in which the step occurs. Recall that we have argued that in general there should be distinct scientific information system organizations, one for each maturity stage. The steps in Table 5.2 do not include all of the activities of the three information management organizations — only those which directly involve handling a data set. The other activities (community outreach, setting publication standards, developing browse and access tools, participation in field campaign activities, etc.) are part of the overhead costs of the system.

The range of skills required is best provided by a team of generalists with multiple skills. In practice, the information manager assembles a small task force, or data set team, for each data set. This data set team then works the data set through the steps in Table 5.2, from data receipt to publication. The team members include people with computer and scientific skills appropriate to the data set. They also include people from each of the three information management organizations, as required, to ensure smooth transitions as the data set moves between stages. Each member of the team, in addition to specific activities, has a general responsibility for quality assurance. In addition to providing flexibility, this team approach instils a sense of ownership and an in-depth knowledge of the data set, thus helping to integrate the information system staff with the scientific community and to ensure that high-quality products and services are delivered.

Information Management Resources

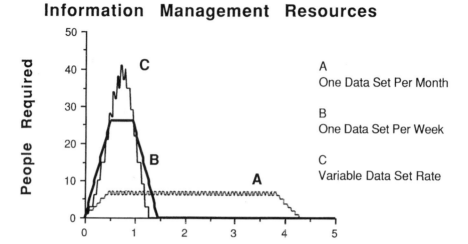

Figure 5.3 Three personnel scenarios for handling 50 data sets, each of which requires 26 weeks of human effort to achieve maturity.

Note: Scenario C has a data set submission distribution like that encountered in FIFE; scenario A is closer to the actual resources available to handle the FIFE data.

(c) The actual size of the data set team varies with the complexity of the data the information system has to handle, the priorities and commitments of the funding agencies, and other practical constraints. Based on Table 5.2, though, each data set requires between 26 weeks and 33 weeks of collective effort to make it self-sufficient, ready for a long-term archive, and usable for 20 years or more.

Some who have not pushed a data set through the entire process outlined in Table 5.2 may think that this is an overestimate. However, in our experience with approximately 100 diverse data sets it is a realistic average. Some particularly simple data sets go through some steps a little more quickly, some complex or high-volume data sets may take a little longer. Volume is not as critical as the number of parameters. Data sets with more or less volume are handled more or less automatically, using fixed-cost computing resources, so that the net human resource investment is about the same for each data set. The operational costs may become significant for a single type of data collected continuously for long intervals. Such monitoring activities usually support focused scientific analyses but do not dominate them. They can be accommodated in resource estimates by breaking the collection interval into segments and treating each as if it were a separate data set.

It is worth considering how this estimate of the overall personnel cost of a data set interacts with the number and timing of the data sets that an information management effort expects to handle. For simplicity, let us assume

Penalties of Delayed Data Maturation

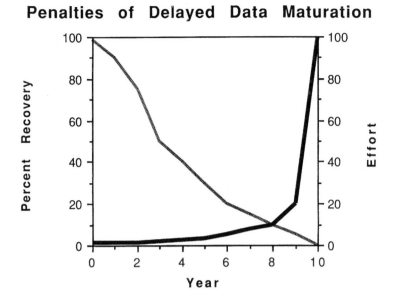

Figure 5.4 Two factors that can increase the cost of handling data sets long after they are collected or submitted.

Note: The recovery of data and information about the data (stippled curve) becomes increasingly difficult. At the same time, the effort required to achieve that level of recovery (solid curve, relative to 1 unit of effort at time 0) increases and eventually becomes prohibitive.

that each data set requires 26 weeks to handle and that there are 50 data sets from a particular project. Figure 5.3 shows the personnel required, as a function of time, under three scenarios: (a) one data set received each month, (b) one data set received each week, (c) a hypothetical distribution of 1, 2, 3, 4, 5, 6, 7, 7, 6, 6, and 3 data sets per month, respectively.

The last scenario, which is similar to our experience in the FIFE Information System (Strebel *et al.*, 1990), suggests a peak requirement for over 40 people staffing the data set teams. In reality, only 6–10 people were available, reducing the rate at which data sets were handled to something much more like scenario (a). The penalty for this delay is not only reduced availability of the data to the scientific community, but an overall reduction in quality because of the loss of information about data sets that were not handled immediately (see Figure 5.4). The increase in the effort required to document and recover such decaying data sets, also shown in Figure 5.4, is not reflected in either Table 5.2 or Figure 5.3. It will, though, increase the overall cost.

The implications of these illustrations for data set submission management, tracking and processing, and for personnel recruitment, training and management are explored further in the following section. It is critical to recognize, however, that the dominant operational delays in scientific information management are driven by the large personnel investment required to obtain full data set maturation.

Examples and Discussion

Management and Organization

An organization similar to the management council described above was created for NASA's Pilot Land Data System (PLDS) in 1988. This organization comprised five individuals from the scientific community (the PLDS Science Working Group, or SWG), the PLDS project manager, the PLDS project scientist, and two NASA programme managers, one each from the Earth Science and the Information Systems offices. This group was significantly different from pre-1988 advisory bodies in several ways: (1) the experience and training of the individual members, (2) the charter and composition of the SWG, and (3) the ability and willingness to develop common goals and objectives.

The experience and education of the members of this group was a critical element. The PLDS project manager had direct experience doing scientific research in the earth sciences, had a strong background in scientific information systems, and was committed to providing services that met the needs of the scientific community. Similarly, the five SWG members represented a broad cross-section of the land science community. They were selected not only because they were active researchers in a particular earth science discipline but because they could build consensus among their colleagues, had experience providing information management services to their colleagues, and represented the wide range of computer sophistication that existed within the community.

The charter of the SWG was significant in that it was charged to build consensus among the community and to participate as active partners in the information system development team. The members of the SWG were often in daily contact with the development staff and information system project management. They worked as partners to define, design and build the information system. As this management council worked together, the goals and objectives of all parties coalesced to a well-defined focus: providing basic services that met the immediate needs of the scientific community.

The end result of this approach was an executive/management council that was a truly functional partnership between the scientific community, the agency programme managers, and the information system managers; that had a singular objective and focus accurately reflecting the needs of the scientific community; and that comprised individuals who were able to work well together to develop common goals and objectives. These elements were sufficient to turn a previously failing project around and create an information system that served its community effectively.

Science Requirements

The PLDS, in response to the direction of its management council, developed services to assist scientists and science projects acquire and develop mature data sets, provided investigators with restricted access to these data during their adolescent stage, and provided broad scientific access to these data once they had matured. Basic services were developed initially, and then expanded as demand increased. The PLDS staff assisted individual investigators and science projects to find and acquire data that they needed for experiment planning and as precursor data sets for their primary research. The staff also served as a focal point for acquisition of commercial data for individual investigators. Library services were provided through an on-line cataloguing and data ordering system. The cataloguing system allowed scientists to find a data set and to determine its quality, collection procedures, processing methods and structure. The ordering system then allowed scientists to order individual data items that met their criteria.

Trained user-assistance staff, who focused on educating and training users, were a key element of the information system's services. They were charged with assisting users to find the data they wanted within the information system, and in the use of the services. They also served as a feedback channel for determining the evolving needs of the science community for information system services.

PLDS also provided services to assist with the publication of maturing data sets for general distribution. Scientifically trained personnel advised and assisted science projects or individual investigators (i.e., the data set authors) with the publication of their data. For final publication of complex data set collections, PLDS assisted in designing and implementing peer review workshops.

Data Flow

Figure 5.5 illustrates how data matured in the FIFE Information System (FIS). Data were exchanged between sources, science investigators, information system staff and the user community before reaching the long-term archive form on CD-ROM (Landis *et al.*, 1992). Some of the data (e.g., digital elevation maps) were acquired from institutional data sources. Some data were collected as part of a project staff core data collection effort (e.g., soil moisture and plant biomass time series), and the science investigators, of course, contributed their data. The science members of the FIS staff performed the bulk of the data integration and scientific QA effort. In areas where sufficient expertise did not reside within the FIS staff, an exchange of data between the scientific information management staff and investigators was needed to complete the integration and scientific QA tasks. Such interactions sometimes led to 'derived' data sets: new submissions of data with different computational methods, or new data derived from the combination of two or more previously submitted data sets. Such a derived data set effectively moved backward along the maturity axis in Figure 5.5 because it re-entered the system as a new data set that was subjected to preliminary QA

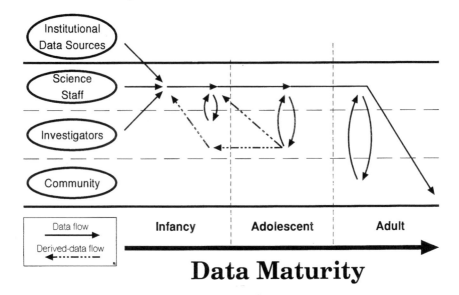

Data Maturity

Figure 5.5 Data flow diagram illustrating how data matured in the FIFE information system.

Note: As the data moved forward along the maturity axis, they were reviewed and used by increasingly wider audiences.

and then took on a life of its own (complete with new documentation). After the scientific integration and QA was completed, and the contributing investigator(s) cleared the data, the data set was ready to progress to the adolescent stage.

In FIS, the adolescent stage was most clearly delineated by on-line access to the data (or inventory tables for image data) by all the FIFE investigators. This allowed a peer-review level of data comparison and further opportunities for derived data submissions. The FIFE experience also illustrated a deeper level of QA available in an integrated project that may not be possible in other scenarios. For example, comparisons of surface flux data by the surface flux science working group revealed previously undetected differences between two different designs of net radiometers.

After a period of data exchange and revision in the adolescent stage, the data were ready for the adult stage. An intensive review and documentation effort was applied to each data set, within the constraints of available time and financial resources, to prepare it for publication. A series of five CD-ROMs (Strebel *et al.*, 1992a,b,c, 1993, 1994a,b) containing documentation, PC interface software, image display software, project data, investigator data and selected image data was then published and distributed to some 300 members of the land science community. In addition, the entire data set collection was submitted to NASA's Global Change Data Center for permanent archiving.

Resource Requirements

The illustrations in Figure 5.3 are somewhat theoretical. It is not realistic almost instantaneously (i.e., within a few months) to create a staff of 40 people to handle a pulse of data from a remote-sensing field experiment, and then disband this team equally rapidly. Neither is it desirable to draw out the experiment for years as a small group ploddingly processes one data set per month.

Nonetheless, the net amount of resources required is realistic based on our experience with FIFE. Approximately 30 person-years of FIS staff effort were used in information management activities to support the experiment and bring 70 plus data sets from field data collection to publication on CD-ROM The preparation of documentation, back-up archiving, computer operations and some data processing were handled with resources outside this total.

Conclusion

Scientific research into environmental problems at a variety of scales is generating demands for better data access, integration and long-term preservation. We have presented a conceptual framework for the design, implementation and operation of scientific information systems to support such scientific research. This framework is based on management, science requirements, data flow and resource principles drawn from our experience in implementing successful systems for complex environmental data set collections.

The principles led us to develop a 15-cell matrix of scientific information system activities classified by services that meet the science requirements and stages in the data maturation process. This matrix focuses on both organization issues and resource issues, and therefore can be a guide to the design and implementation of a scientific information system.

Successful systems with which we are familiar all have characteristics of the three-stage organization implicit in the matrix. One convenient way (among many) to describe this organization is that it is equivalent to the Author–Publisher–Library distinction familiar to scientists from their research publication activities. A key point is that the scientific information system must provide services that the science community will utilize.

The management and organization of a scientific information system can help or hinder the services that it provides. It is critical that the system be viewed and organized as a scientifically driven partnership between the scientific communities that produce and use the data and the information management staff that handle it. Only such a partnership can create the highly flexible system necessary to cope with the inherent exploratory and evolutionary nature of scientific research. Such a partnership also leads to an appreciation of the processes and resources required to achieve high-quality data archives and provides a mechanism for setting priorities for the use of available resources.

The resources required to support a maturing data set from field data collection to final archival publication are appreciable. An estimate of the personnel costs based on our experience with FIFE and PLDS is at least 26 weeks (one-half year) of full-time effort. This intensity of effort, combined with the pattern of data set submissions, can greatly affect the distribution of resources required as a function of time for a given information management effort. Peak requirements for an intensive data collection effort may exceed the available personnel, delaying data maturation and archiving. Such delays may result in the loss of vital information and render a data set far less valuable than it might have been.

In the long run, the value and usability of the data sets that are successfully archived and preserved will be the basis on which a scientific information management effort is judged. We believe that the risks of failure for future environmental data management efforts can be reduced by examining each such effort in the context of the framework presented in this chapter.

References

Agbu, P. A. and Meeson, B. W., 1993, Field experiment data management: NASA's Pilot Land Data System's experience, *ACSM/ASPERS*, **3**, 1–11.

Landis, D. R., Strebel, D. E., Newcomer, J. A. and Meeson, B. W., 1992, Archiving the FIFE data on CD-ROM, *Proceedings of IGARRS'92, Houston*, May 1992, pp. 65–67, Pistacaway, New Jersey: IEEE.

Lucas, H. C., 1975, *Why Information Systems Fail*, New York: Columbia University Press.

Meeson, B. W., Strebel, D. E. and Paylor, E. D., 1993, Earth science information systems: A perspective from the Pilot Land Data System, in Ziegelbaum, A. (Ed.) *Earth and Space Science Information Systems*, Pasadena, California 1992. AIP Conference Proceedings 283, American Institute of Physics, New York.

Preheim, L. (Ed.), 1992, *Proceedings of the Second Discipline Data System Forum, November 1992*, McLean, Virginia: NASA/Jet Propulsion Laboratory.

Sellers, P. J., Hall, F. G., Asrar, G., Strebel, D. E. and Murphy, R. E., 1992, An overview of the First International Satellite Land Surface Climatology Project Field Experiment (FIFE), *Journal of Geophysical Research*, **97** (D17), 18345–71.

Strebel, D. E., Landis, D. R., Newcomer, J. A., van Elburg-Obler, D., Meeson, B. W. and Agbu, P. A., 1992a, *Collected Data of the First ISLSCP Field Experiment*, Vol. 2: Satellite Imagery 1987–1989, published on CD-ROM by NASA.

Strebel, D. E., Landis, D. R., Newcomer, J. A., Goetz, S. J., Meeson, B. W. Agbu, P. A., Huemmrich, K.F., 1992b, *Collected Data of the First ISLSCP Field Experiment*, Vol. 3: NS001 Imagery, published on CD-ROM by NASA.

Strebel, D. E., Landis, D. R., Newcomer, J. A., Meeson, B. W. Agbu, P. A., 1992c, *Collected Data of the First ISLSCP Field Experiment*, Vol. 4: ASAS & DBMR Imagery, published on CD-ROM by NASA.

Strebel, D. E. and McManus, J. M. P., Landis, D. R., Nickson, J. E., Goetz, S. J., Meeson, B. W. Agbu, P. A., 1994, *Collected Data of the First ISLSCP Field Experiment*, Vol. 1: Point Data, published on CD-ROM by NASA.

Strebel, D. E., Landis, D. R., Newcomer, J. A., van Elburg-Obler, D., Meeson, B. W. Agbu, P. A. and McManus, J. M. P., 1993b, *Collected Data of the First ISLSCP Field Experiment*, Vol. 5: Derived Image Products, published on CD-ROM by NASA.

Strebel, D. E., Newcomer, J. A., Ormsby, J. P., Hall, F. G. and Sellers, P. J., 1990, FIFE information system, *IEEE Transactions in Geosciences and Remote Sensing*, **28**, 703–10.

Webster, F., 1991, *Solving the Global Change Puzzle: A US Strategy for Managing Data and Information*, Report by Committee on Geophysical Data Commission on Geosciences, Environment and Resources, Washington, DC: National Research Council, National Academy Press.

6

Development and refinement of the Konza Prairie LTER Research Information Management Program

John M. Briggs and Haiping Su

The primary goal of the long-term ecological research programme at Konza Prairie is to understand how grazing influences biotic and ecosystem processes and patterns imposed by fire frequency over the landscape mosaic, all of which are subjected to a variable (and possibly directional) climatic regime. A vital component of the research programme has been the development of the Konza Prairie LTER Research Information Management Program with the overall objectives of assuring data integrity, providing security for the database, and facilitating use of data by the original investigator(s) as well as by future investigators. This programme has evolved considerably from its original version in 1981, which served only a localized research group, to its present capabilities of responding to requests for data from investigators across the globe. Programme development has benefited from experiences gained through various administrative changes associated with Konza Prairie, growth of the research programme on Konza Prairie (especially in the fields of remote sensing and geographical information systems (GIS)), knowledge gained from other multidisciplinary research efforts (i.e., other LTER sites and First ISLSCP (International Satellite Land Surface Climatology Project) Field Experiment (FIFE), and with changing computer technology.

Introduction

The Konza Prairie Long-Term Ecological Research (LTER) Program has developed a research information management programme (KPRIMP) that supports long-term studies on the Konza Prairie Research Natural Area (KPRNA). Objectives of the KPRIMP are to (1) assure data integrity (correctness, at all times, of all items in the database), (2) provide security (protection against loss of data), and (3) facilitate use of data by the original investigator(s) as well as by future researchers. The overall objective is to develop a research database that will allow us to address scientific questions ranging from local to global scales.

The information management programme has evolved from its inception in 1981, when it served only a localized research group at Kansas State University (most in the Division of Biology), to its present version, which works with a multidisciplinary team of investigators from a variety of universities and government agencies. During the early 1980s, considerable effort was made by the Konza LTER staff to implement a base-level research information management plan, with its primary goal being to have all interested researchers locate, interpret and utilize data. This plan was designed using guidelines established by Gorentz *et al.* (1983) and is documented in Gurtz (1986).

In this chapter we briefly describe the KPRNA and the Konza Prairie LTER Program, and discuss how the KPRIMP has developed and evolved in response to administrative changes, growth of the Konza Prairie research programme, and association with broad-scale, multidisciplinary research efforts.

Site Description

Konza Prairie Research Natural Area (KPRNA) was established as a research facility in 1972, primarily as a result of the efforts of Dr Lloyd C. Hulbert. Initially KPRNA included only 371 ha, but additional purchases in the early 1980s expanded the site to its present size of 3487 ha. The area is owned by The Nature Conservancy and is leased to the Division of Biology, Kansas State University for long-term research purposes. A watershed-level (catchment unit) fire frequency experimental design that includes replicated, long-term unburned (20 yr) and prescribed spring burns (1-, 2-, 4-, and 10-yr frequencies) was devised and initiated by Hulbert. Overlaid on this design is a grazing experiment with blocks of watersheds designated as ungrazed, grazed by native ungulates (*Bos bison*), and grazed by domestic cattle (*Bos taurus;* (Figure 6.1) (Hulbert, 1985; Marzolf, 1988). Fire treatments have been maintained in the southernmost watersheds since 1971 and throughout the rest of Konza Prairie since 1979. Bison have been on Konza since 1987 and experimental cattle herds were added in 1992. Thus, considering the experimental design of fire and grazing, coupled with the unique ecosystem, it is vital to have an intensive information management programme to manage adequately the multifaceted ecological data from this research site.

The Konza Prairie LTER Program

The Long-Term Ecological Research (LTER) Program of the National Science Foundation (NSF) began funding research in 1981 (Callahan, 1984; Franklin *et al.,* 1990). The justification for long-term, well-documented ecological studies has previously been recognized (Likens, 1989; Krebs, 1991). LTER sites conduct research in five 'core' areas: (1) patterns and controls of primary production (to study the energy limits of system function), (2) spatial and temporal distribution

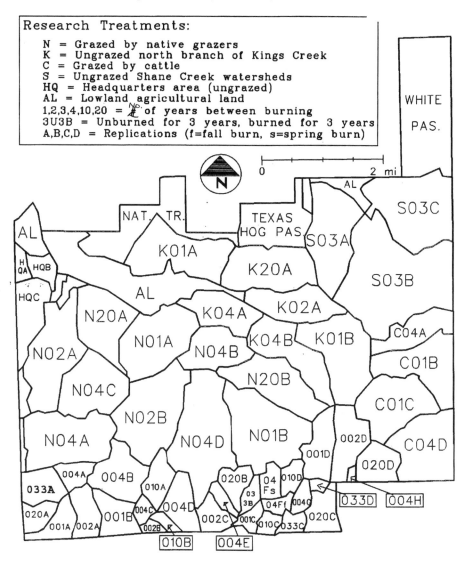

Research Treatments:

N = Grazed by native grazers
K = Ungrazed north branch of Kings Creek
C = Grazed by cattle
S = Ungrazed Shane Creek watersheds
HQ = Headquarters area (ungrazed)
AL = Lowland agricultural land
1,2,3,4,10,20 = No. of years between burning
3U3B = Unburned for 3 years, burned for 3 years
A,B,C,D = Replications (f=fall burn, s=spring burn)

Figure 6.1 Konza Prairie research experimental design.

of populations (selected to represent the trophic structure and thereby identify major pathways of mass and energy transfer and the structural organization of the ecosystem), (3) patterns and controls of organic matter accumulation in surface layers and sediments (to elucidate storage and processing of biological materials), (4) patterns of inorganic input and movement through soils, groundwater and surface water (to assess interactions among geochemical and biological processes), and (5) patterns of frequency of disturbances to the system (to compare nature and humans as perturbers of natural systems; Callahan, 1984).

Konza Prairie, one of the six original LTER sites selected by the NSF in 1981, is now in its third funding cycle (1991–6). Since it is a relatively young site in terms of ecological research (20 yrs), the research goals at Konza Prairie have been redirected and expanded several times, but the five core areas have been and continue to be a baseline research effort.

LTER I (1981–6)

The initial research programme was focused on comparative investigations of biotic responses to fire and climatic variability. Long-term research sites and sampling protocols were established during this period with an emphasis on studies of the extremes of annually burned vs. unburned watersheds and upland vs. lowland sites.

LTER II (1986–90)

During LTER II, the research efforts expanded to include a wide range of fire frequencies (specifically 4-yr fire cycles). As a result of the collaborative NASA FIFE (First ISLSCP (International Satellite Land Surface Climatology Project) Field Experiment) Sellers *et al.*, 1988; Strebel *et al.*, this volume) program from 1987 to 1989, Konza LTER investigators began to address more complex questions of scale and to utilize remotely sensed satellite data to explore landscape-level issues (summarized by Nellis *et al.*, 1992).

LTER III (1991–6)

Research under way with the current Konza LTER programme represents the most significant expansion to date. The primary goal of the long-term ecological research programme at Konza Prairie is to understand how grazing influences biotic and ecosystem processes and patterns imposed by fire frequency over the landscape mosaic, all of which are subjected to a variable (and possibly directional) climatic regime. The documentation and archival storage of data collected at this site are considered two of the most important tasks that each investigator performs as part of their effort in the Konza LTER programme. In fact, most LTER investigators routinely include in their published manuscripts the name and data set code of the data set(s) associated with their publication(s). This should inform any interested researcher(s) that the data set that was used in a manuscript is fully documented and archived.

Konza Prairie Information Management

The KPRIMP includes archived LTER data with an electronic data catalogue that allows any investigator connected through the Novell network to browse necessary documentation such as data format, site location, etc. Specialized data entry programs and data checking programs have been developed to aid us in reducing errors while entering data, reducing the time spent entering data, and maintaining data integrity. In addition, a user-interface to the LTER database was developed to facilitate proper and prompt data documentation by the current LTER researchers, which facilitates database development. Most quality control checks are managed by the principal investigator(s) associated with a data set. All archived files (data files that have been entered and verified as correct by the investigator(s)) are stored on a variety of electronic media: ½″ magnetic tapes, 8-mm tapes, hard disks, and re-writable optical disks, with the goal being to have at least three copies of the database stored in different physical locations. A recent water pipe break in our building has reinforced the importance of maintaining multiple back-up copies in different buildings. System back-ups are executed on a weekly basis on 8-mm tape cartridges and twice a year on optical disks. In addition, back-ups of data sets also exist on ½″ magnetic tapes that are stored at the Kansas State University (KSU) mainframe facilities.

The personal microcomputers of most LTER researchers (located at KSU) are linked directly to each other and the Konza Prairie data bank through a Novell network. This allows data, reports and manuscripts to be transferred among researchers. The design of the current Konza Prairie LTER database is straightforward. All data sets (Appendix 1) are in ASCII format (with the exception of GIS coverages, satellite images and WordPerfect files). The entire database resides on a Novell network (Version 3.14), running on an IBM-compatible 80486 computer (33 Mhz with a 1.6 gigabyte hard disk). The database is divided into subdirectories that represent the data set directly, correspond to the research groups that have developed on Konza, or represent the name of the data sets. The subdirectories are: ABIOTIC, CONSUMER, NUTRIENT, ORGANIC, OTHER, WP5 and WOODY. In most instances, the first letter of the data set code indicates the subdirectory location for the file. The extension of the file name represents the year of the data set. For example, the data set associated with prairie precipitation for 1986 (data set code APT01), is found in the subdirectory ABIOTIC under the file name of apt011.86. The subdirectory WP5 contains the WordPerfect files (Version 5.1) that are mentioned (i.e., KPL01, KFH01) in Appendix 1. The subdirectory WOODY contains the files with the data set code PWV01 (the location of trees and woody shrubs on various watersheds on Konza Prairie; these files are separated owing to the size of these files). The subdirectory OTHER is reserved for data sets that do not conform to the naming procedures (generally data sets that come from non-LTER sources such as United States Geological Survey flow data or National Atmospheric Deposition Program data) or future data sets (for now, data sets from a recently initiated water supplementation experiment (WAT01) are here).

To maintain a consistent format for LTER data files, most data sets have the first 16 columns of each line organized as: data code (columns 1–5, A5 format); record type (column 6, I1 format); year (columns 7–8, I2 format); month (columns 9–10, I2 format); day (columns 11–12, I2 format); and watershed (columns 13–16, A4 format). Thus, each line has the data set code associated with it. Although this format dates from the days of 'computer cards' when it was important to have the data set code on each card, it is still useful to have this information in case of hardware failures (disk crashes, bad tapes, etc.)

The most difficult and time-consuming component of our (any?) research information management plan has been the proper documentation of data sets such that they can be used by investigators who did not originally collect the data (Tate and Jones, 1991). This is especially true considering the turnover of personnel and the changes that occur in data collection procedures because of technology enhancements (i.e., new laboratory equipment, etc.). During the period 1981 to 1992, the site director (Dr Lloyd Hulbert) passed away and two LTER principal investigators left Kansas State University. These personnel losses, coupled with the usual 'leavings and goings' of post-docs and graduate students that occur at any major research unit, has strengthened our need for proper documentation. One of the more important and useful products of our information management system, in this regard, has been a 'Methods Manual', maintained since 1981, that details procedures for ongoing and prior studies. The current version is a 172-page document that details how each LTER data set is collected. It includes items such as precise maps of the vegetation survey, sample data sheets, and very detailed procedures on instrument installation and use. This manual provides the necessary details to interpret the more extensive data documentation files maintained for each data set. This document is updated yearly and a complete revised manual is produced every five years; 25 requests for this document were received in 1992.

One of the goals of the Konza Prairie LTER programme is to support, as for as possible, collaborative as well as independent research efforts on Konza Prairie. Thus, we developed a protocol, implemented in 1983, that allows outside (non-Konza LTER) investigators access to LTER data. We have three levels of access: unrestricted, limited restriction and full restriction. Briefly, *unrestricted* refers to archived data sets accessible to all interested researchers on notification to the data manager. This is read-only access: any errors discovered (or suspected) in an unrestricted data file must be brought to the immediate attention of the data manager, who will confer with the data set investigator. *Limited restriction* refers to archived data sets with read-only access available to current Konza LTER researchers or to outside researchers with the written permission of the current LTER PI(s) and the data set investigator. The PI(s) may deem that the investigator's approval is not necessary if he/she has waived that privilege, is deceased or cannot be reached within a reasonable amount of time. *Restricted* data are accessible only to the investigator or persons designated by the investigator. These may be raw data files or other data files that are considered incomplete, unverified or otherwise uncertain as to their correctness. Our goal is to have all LTER data archived and in the unrestricted category within one year after the last datum is collected. These restrictions apply to data sets only after they are archived; before that, the data are entirely the responsibility of the investigator(s). Regardless of access restriction, no researcher outside Konza Prairie LTER will be given access to LTER data without the written approval of the LTER principal investigator.

Several guidelines have been adopted for the release and citation of data collected as part of the Konza Prairie LTER project to individuals not directly associated with the Konza Prairie LTER project. Specifically, data request guidelines include: (1) submission of a formal written request and statement of intended use; (2) approval of the investigator and/or the Konza Prairie LTER Principal Investigator; (3) request must be filed with the Konza Prairie LTER data manager; (4) release of data (following approval) should include a cover letter specifying that: 'These data are released for your use only and for the purposes outlined in your request'; (5) manuscripts using the data are to be provided to the Principal Investigator, LTER, Division of Biology, Ackert Hall, Kansas State University, Manhattan, KS 66506 so that he/she may notify the appropriate investigators; (6) publication of these data is allowed on the express permission of Konza Prairie LTER investigators (named), who have primary responsibility for the data sets; and (7) acknowledgement of the contribution of data by Konza Prairie LTER should be made. Data set citations should use the following format:

Data from the Konza Prairie Research Natural Area were collected as part of the Konza Prairie LTER program (NSF grants DEB-8012166 and BSR-8514327, BSR-9011662), Division of Biology, Kansas State University, Manhattan, KS. Data and supporting documentation are stored (Data Set Code(s)= _____) in the Konza Prairie Research Natural Area LTER Data Bank.

Additionally, specific investigators should be cited for their contributions to the paper.

How successful has this data sharing policy been? From 1984 to 1992, requests for data from the Konza Prairie LTER programme have grown from virtually no requests to an average of > 2 requests a week (Figure 6.2). In fact, handling data requests now occupies approximately 20 per cent of our time, and we expect this aspect of our job to continue to increase.

In 1992, 115 data requests from outside investigators (i.e., non-Konza Prairie LTER scientists) were handled by the Konza LTER information management staff. The data requests can be partitioned as: 76 per cent Konza Publications List or general information; 34 per cent weather data; 3 per cent above-ground production data; 3 per cent fire history; and 1 per cent plan species composition. Other requests included soil moisture and soil temperature data.

To alleviate the major time constraints that these requests impose, we have developed and installed an Interactive Data Access system (IDA; called the Konza LTER Information System (KIMS)) using the Oracle database system, which uses structured query language (SQL). Our IDA is modelled after the FIFE Information System (FIS) (Strebel *et al.*, 1989; Strebel *et al.*, this volume). Investigators can log on to the system, look at the Data Summary

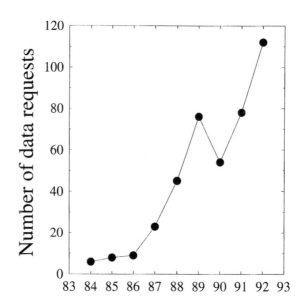

Figure 6.2 Data requests to Konza Prairie LTER data manager from 1984 to 1993 from non-Konza LTER investigators.

Table, and download certain Konza LTER data. Currently available data include the Konza Publication List, weather data, stream flow records, soil moisture and list of satellite images. A common characteristic of most data requests is that outside investigators typically do not want raw data that are collected at fine temporal or spatial scales. For example, our weather station is set up to collect temperatures every minute, store the hourly mean of these readings, and compute a daily mean temperature. The weather data set is the most requested data set, and since 1984 only three of 345 requests have been for hourly readings. Over 98 per cent of all requests have wanted monthly (not daily!) values. We are hoping that investigators using the IDA via Oracle will be able to obtain the data they need at the appropriate scale.

Since its creation in November 1992, the IDA system has had an average of two users (non-Konza LTER investigators) per week. The estimate of time savings for staff is approximately 2–3 hours a week; in addition, more importantly, outside investigators can retrieve the data when needed, instead of relying on the Konza LTER information management staff. In the future, KIMS will handle most data requests.

In addition to the KIMS, a Gopher server has recently been implemented. Gopher is an interactive, network navigation tool. A simple, text-based, menu-driven interface is used to retrieve files and, when used with a Wide-Area Information Server (WAIS), it can be used for simple searches.

Importance of Konza Prairie Information Management

Having a database at Konza Prairie complete with the necessary documentation has proved extremely useful over the relatively short time that research has been conducted at this site. A key example of this has been the coupling of more 'traditional' ecological data (i.e., vegetative species composition, estimates of above-ground biomass) with data generated by high-technology tools (i.e., digital satellite data and GIS). Although a watershed-level focus has been implicit at Konza Prairie since its design by Hulbert (Figure 6.1), our experiences with the NASA FIFE program (Sellers *et al.*, 1988) on Konza Prairie and its accompanying database have allowed us to address many new questions at multiple scales.

We routinely acquire and archive digital satellite data ranging from National Oceanic and Atmospheric Administration (NOAA) Advanced Very High Resolution Radiometer (AVHRR) with a pixel size of 1.1 km to SPOT (System pour l'Observation de la Terre) with a 10-m pixel size in the panchromatic spectrum and a 20-m pixel size with its multispectral mode. A special effort has been made to acquire at least one Thematic Mapper (TM) and/or Multispectral scanner (MSS) image of Konza Prairie during the mid-to-late growing season (July to early September) each year. These satellite data have been used in a variety of studies such as estimating above-ground biomass (Briggs and Nellis, 1989), studying the impact of drought on native tallgrass prairie (Nellis and Briggs, 1992), quantifying the impact of spatial and temporal scale on landscape heterogeneity (Nellis and Briggs, 1989; Briggs and Nellis, 1991), detecting changes in grasslands (Henebry, in press), and mapping different soil types (Su *et al.*, 1989).

Shortcomings

With the addition of remote sensing, GIS and modelling efforts to the Konza Prairie Research Program, we are struggling with the proper development of procedures to document fully these very large data sets. Presently, all satellite data are stored on 8-mm tapes, ½" magnetic tapes, or on optical disks using band sequential format. Each image has an associated ASCII file that provides information such as date, platform (i.e., SPOT, TM, etc.), ground control points and the method of geometric correction. Because many of the GIS coverages are derived in various ways, they are much more difficult to document fully, and procedures to document these files adequately are still being explored.

The focus of the Konza Prairie LTER information management staff is documenting and maintaining the long-term data sets on Konza Prairie; however, some of the most interesting data sets collected on Konza Prairie involve short-term studies. Although most of these studies are published, publications typically do not include enough detail to be considered proper documentation of the data sets. Tracking, properly documenting and providing access to these data sets for the entire research community is beyond the scope of our present Konza Prairie LTER information management staff. Data from short-term studies, if properly documented, could be examined in the future and study sites possibly re-sampled to address new ecological questions. In this way, any data set that is properly documented and archived in reality becomes a long-term data set over time.

Conclusion

The value of having long-term, well-documented data sets cannot be overemphasized. For example, we are examining landscape heterogeneity across a broad geographical area by using satellite-derived digital data that are coupled with our extensive ground-base measurements. When these ground-based measurements were set up, they were designed to describe small plots, not to be scaled up to landscapes. However, by having a complete information management system, these data are being used in ways that the initial investigator did not foresee. With very few exceptions, databases have not outlived the investigators who collected them (Strayer *et al.*, 1986). Those that have survived are now ecological treasures (Seastedt and Briggs, 1991). Scientists cannot measure decade-to-century phenomena without having a serious commitment to ecological information management.

Acknowledgement

The authors would like to thank Alan K. Knapp for his constructive comments on an earlier draft of this manuscript. Comments from three anonymous reviewers improved this manuscript. Many of our ideas and concepts for ecological information management have resulted from our interactions with scientists involved at Konza Prairie, especially T.R. Seastedt. Konza Prairie Research Natural Area, a preserve of The Nature Conservancy, is managed by the Division of Biology, Kansas State University. Support was provided by the National Science Foundation (grant no. BSR-9011662) for Long-Term Ecological Research to Kansas State University.

References

Briggs, J. M. and Nellis, M. D., 1989, Landsat thematic mapper digital data for predicting aboveground biomass in a tallgrass prairie ecosystem, in Bragg, T.B. and Stubbendieck, J. (Eds) *Proceedings of the Eleventh North American Prairie Conference, Prairie Pioneers: Ecology, History and Culture,* pp. 53-5, Lincoln, Nebraska: University of Nebraska Press.

Briggs, J. M. and Nellis, M. D., 1991, Seasonal variation of heterogeneity in tallgrass prairie: A quantitative measure using remote sensing, *Photogrammetric Engineering and Remote Sensing,* **57**, 407–11.

Callahan, J. T., 1984, Long-term ecological research, *BioScience,* **34**, 363–7.

Franklin, J. E., Bledsoe, C. S. and Callahan, J. T., 1990, Contributions of the Long-Term Ecological Research Program, *BioScience,* **40**, 509–23.

Gorentz, J., Koerper, G., Marozas, M., Weiss, S., Alaback, P., Farrell, M., Dyer, M. and Marzolf, G. R., 1983, 'Data Management at Biological Field Stations', report of a workshop at W.K. Kellogg Biological Station, Michigan State University, 17–20 May 1982, prepared for the National Science Foundation.

Gurtz, M. E., 1986, Development of a research data management system, in Michener, W.K. (Ed.) *Research Data Management in the Ecological Sciences,* The Belle W. Baruch Library in Marine Science No. 16, pp. 23–38, Columbia, South Carolina: University of South Carolina Press.

Henebry, G. M., 1993, Detecting change in grasslands using measures of spatial dependence with Landsat TM data, *Remote Sensing of the Environment.* 46: 223–34.

Hulbert, L. C., 1985, History and use of Konza Prairie Research Natural Area, *The Prairie Scout,* **5**, 63–93.

Krebs, C. J., 1991, The experimental paradigm and long-term population studies, *Ibis,* **133** (suppl. 1), 3–8.

Likens, G. E., 1989, *Long-term Studies in Ecology: Approaches and Alternatives,* New York: Springer.

Marzolf, R., 1988, Konza Prairie Research Natural Area of Kansas State University, *Transactions of the Kansas Academy of Science,* **91**, 24–9.

Nellis, M. D. and Briggs, J. M., 1989, The impact of spatial scale on Konza landscape classification using textural analysis, *Landscape Ecology,* **2**, 93–100.

Nellis, M. D. and Briggs, J. M., 1992, Transformed vegetation index for measuring spatial variation in drought impacted biomass on Konza Prairie, Kansas, *Transactions of the Kansas Academy of Science,* **95**, 93–9.

Nellis, M. D., Briggs, J. M. and Seyler, H. L., 1992, Growth and transition: Remote sensing and geographic information systems at Kansas State University, *Photogrammetric Engineering and Remote Sensing,* **58**, 1159–61.

Seastedt, T. R. and Briggs, J. M., 1991, Long-term ecological questions and considerations for taking long-term measurements: Lessons from the LTER and FIFE programs on tallgrass prairie, in Risser, P. J. (Ed.) *Long-term Ecological Research: An International Perspective,* SCOPE 47, pp. 153–72, Chichester: Wiley.

Sellers, P. J., Hall, F.G., Asrar, G., Strebel, D. E. and Murphy, R. E., 1988, The first ISLSCP field experiment (FIFE), *Bulletin of the American Meteorological Society,* **69**, 22–7.

Strayer, D., Glitzenstein, Jones, J. S., Kolasa, C. G., Likens, J., McDonnell, G. E., Parker, M. J. and Pickett, S. T. A., 1986, *Long-term Ecological Studies: An Illustrated Account of Their Design, Operation, and Importance to Ecology,* Occasional Publications of the Institute of Ecosystem Studies No. 2, Millbrook, New York: Institute of Ecosystem Studies, New York Botanical Garden.

Strebel, D. E., Newcomer, J. A., Ormsby, J. P., Hall, F. G. and Sellers, P. J., 1989, Data management in the FIFE information system, in *Proceedings of IGARSS'89 Symposium,* pp. 42–5, Vancouver, British Columbia.

Su, H., Ransom, M. D. and Kanemasu, E. T., 1989, Detecting soil information on a native prairie using Landsat TM and SPOT satellite data, *Soil Science Society of America Journal,* **53**, 1479–83.

Tate, C. M. and Jones, C. G., 1991, Improving use of existing data, in Cole, J., Lovett, G. and Findlay, S. (Eds) *Comparative Analyses of Ecosystems: Patterns, Mechanisms and Theories,* pp. 348–50, New York: Springer.

Appendix I

Version 1.0 of the Konza-LTER Information System can now be accessed via internet. This interactive data system allows investigators to browse the Konza LTER data catalog, the Konza Prairie Publication List, and to download data from the Konza LTER database. Please feel free to use the system and direct any/all comments to John M. Briggs. Instructions to use the system are:

(1) Connect to the machine (machine is bison.konza.ksu.edu, ip = 129.130.110.2).
(2) Log in as user 'lterknz' (no quotes and no upper case!)
(3) Password is 'infosys' (again no quotes and no upper case!)
(4) Follow instructions on the screen.
(5) Problems? Contact jmb@andro.konza.ksu.edu

Konza Prairie Database

The following is a list of data sets which are being maintained by the KPLDM staff. Those data sets without an ending date are presently ongoing. Contact the Konza LTER data manager concerning the availability of these data sets. In addition, up-to-date lists are kept on the Konza LTER Information System.

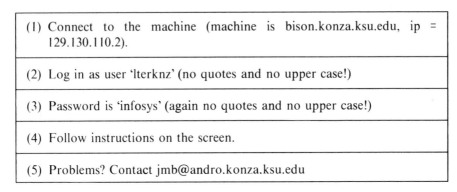

CODE	TITLE	Start	End
AGW01	Belowground water chemistry from wells on N04D	06/01/90	/ /
ANA01	National Atmospheric Deposition Program	08/17/82	/ /
APT01	Prairie Precipitation	06/01/82	/ /
APT02	Monthly total precipitation and mean temperature of Manhattan	1895	
ASD01	Stream Discharge in Kings Creek at USGS Site	04/01/79	/ /
ADS04	Stream Discharge at the Flume on Watersheds N04D, N00B, N01B and N04D	06/14/85	/ /
ASM01	Soil Moisture	05/01/83	/ /
AST01	Soil Temperature during Growing Season	04/23/87	/ /
AWE01	Meteorological Data	04/22/82	/ /
AWT01	Water Temperature – discontinuous measurements in prairie streams	04/24/85	/ /
AWT02	Water Temperature continuous measurements in prairie streams	04/10/86	/ /
BIS01	Bison Locations	01/18/88	
BMS01	Belowground Plots: Mycorrhizae	06/11/87	/ /
CAA01	Sweep samples for Selected Aboveground Arthropods (81)	06/01/81	09/01/81
CAA02	D-Vac Samples for Aboveground Arthropods (81)	08/03/81	08/07/81

CAA03	D-Vac Sampling for Aboveground Arthropods (82)	06/21/82	08/03/82
CAA04	Sweep Samples for Aboveground Homoptera, Hemiptera, Tettigoniidae Coleoptera, Ants, and Spiders (1983)	06/01/83	/ /
CBC01	Bird Check-List	01/01/71	/ /
CBD01	Bird Dates	01/01/71	/ /
CBN01	Bird Nest	01/01/71	/ /
CBP01	Bird Populations	06/01/81	/ /
CGP01	Gall-Plant Interactions	05/01/86	
CGR01	Sweep Samples for Grasshoppers on LTER Watersheds (1981)	04/01/81	12/01/81
CGR02	Sweep Samples for Grasshoppers on LTER Watersheds	04/01/82	/ /
CGR03	Effects of Spring Burning on Grasshopper Nymphs	06/01/82	09/02/82
CGR04	Sampling grasshoppers in tallgrass prairie: night trapping vs. sweeping (1982)	06/01/82	09/30/81
CGR05	Effects of fire frequency on composition of grasshopper assemblages (1983)	08/05/83	08/10/83
CPC01	Census of greater prairie chicken on leks	03/01/81	/ /
CSA01	Soil Microarthropods	04/01/81	06/01/82
CSA02	Soil Macroarthropod Densities and Biomass	11/22/81	04/01/83
CSM04	Seasonal Summary of relative density of small mammals on the LTER traplines	10/01/81	/ /
KPH01	Konza Prairie Publication List	1972	/ /
KFH01	Konza Prairie Fire History	1972	/ /
GIS01	GIS coverages for Konza Prairie (All are in ERDAS or ARC/INFO format)		
NBC01	Belowground Studies: Soil Chemistry	06/01/86	/ /.
NPL01	Prairie Litterfall	07/01/81	/ /
NSC01	Soil Chemistry and Bulk Density	10/01/81	
NSW01	Soil Water Chemistry	03/01/82	/ /
NTF01	Throughfall	03/19/82	/ /
NWC01	Stream Water Chemistry	04/01/83	/ /
OGD01	Gallery Forest Foliage Decomposition	10/31/81	10/26/83
OMB01	Microbial Biomass	04/23/89	/ /
OPD01	Prairie standing dead and litter decomposition	10/31/81	10/26/83
PAB01	Aboveground primary production	04/01/84	/ /
PAB03	Aboveground primary production prior to LTER	04/01/75	10/01/83
PBB01	Aboveground biomass on belowground plots	11/15/86	/ /
PBB02	Belowground Plant Biomass and Nitrogen Content on Belowground plots	07/01/87	/ /
PFS01	Reproductive effort of Big Bluestem, Indiangrass, and Little Bluestem on Belowground Plots	01/10/86	/ /
PGL01	Gallery Forest Litterfall	10/01/81	/ /
PHOTO	Aerial Photographs of Konza Prairie	10/01/34	/ /
PPH01	Plant Phenology	06/13/81	10/31/88
PRE02	Reproductive effort of Big Bluestem, Indiangrass and Little Bluestem	09/01/82	/ /
PRW01	Root Windows	02/01/84	/ /
PTN01	Transect Net Primary Productivity — Soil Moisture	06/01/89	
PVC02	Vegetation species composition	04/01/82	/ /

PVC03	Vegetation species composition of elevation plots on N04D	06/13/84	08/31/84
PWV01	Woody vegetation mapping	05/01/81	/ /
PWV02	Importance values of gallery forest vegetation	05/15/83	07/15/83
SAT01	Satellite derived digital images of Konza	08/23/83	/ /
WAT01	Water Supplementation Transect	06/01/91	/ /
XMS01	Mycorrhizal fungi species survey of Hulbert's Plots	07/01/86	/ /
XNS01	Soil Nematode survey on belowground plots	05/01/87	/ /

7

Forest health monitoring case study

Charles I. Liff, Kurt H. Riitters and Karl A. Hermann

Forest Health Monitoring (FHM) is a national, interagency programme to determine the ecological status and trends of forest condition. The scale of data collection, the variety of data sources and users, and the scope of assessments present a unique challenge; this case study describes how FHM information is managed. The goals of FHM are to produce statistical reports nine months after field data collection, and to produce periodic interpretive reports after accessing a variety of auxiliary databases. In support of these goals, field data collection and quality assurance procedures are automated using portable data recorders, laptop computers and modem links to a central data-processing and distribution centre. The computer system also allows nearly real-time communication with the field crews. Specialists within the FHM programme analyse FHM and auxiliary data, and supply data summaries and standard algorithms to the assessment team.

Introduction

Forest Health Monitoring (FHM) is a co-operative effort of federal and state agencies to monitor and assess the ecological status and trends of the forests in the USA and is part of the US Environmental Protection Agency's Environmental Monitoring and Assessment Program (EMAP). FHM data are intended to be used by federal agencies, decision makers and the academic community. The variety of stakeholders, the inherent complexity of national-scale field data collection and assessment, and the FHM commitments to quality assurance and timely data delivery set the stage for an interesting case study of information management.

Support for both the operational and the research functions of FHM is a key element of the information management system design. FHM draws a distinction between data collected for research purposes and data collected as part of an operational monitoring system. The data collected for research purposes refine field procedures, while the operational data produce statistical summaries and assessment documents. In the summer of 1992, the monitoring system was operational in the 12 eastern states of Maine, New Hampshire, Vermont, Massachusetts, Rhode Island, Connecticut, New Jersey, Delaware, Maryland, Virginia, Georgia and Alabama (Figure 7.1). FHM field personnel collected data on over 900 detection monitoring plots in 1992. Research measurements were made in two demonstration regions (the Southern Appalachian Man and the Biosphere (SAMAB) and the Southeastern loblolly/shortleaf pine demonstrations), in two pilot test states (Colorado and California) (Figure 7.2), and in selected 'off-frame' areas. In all, over 200 field plot locations were visited for research measurements.

To satisfy the many stakeholders, the FHM-information management system (FHM-IM) is designed to interface with a variety of computing systems around the nation. The majority of information management tasks are co-ordinated on a Digital Equipment Corporation VAX minicomputer cluster at the Environmental Protection Agency Laboratory in Las Vegas, Nevada. The other major component of the FHM-IM system is the Forest Service network of Data General systems. FHM data also are managed and analysed on a variety of workstations and microcomputers. For example, the FHM geographic information system is based on workstations running ARC/INFO[1] under UNIX.[2]

Creating an FHM information management system was not simply a matter of using an existing template. Using an existing forest inventory system (Hansen *et al.*, 1991) was contemplated, but such systems were not consistent for the entire nation; and they lacked sufficient documentation for FHM purposes.

The major categories of FHM measurements are provided in Table 7.1. For an overview of FHM field measurements see Palmer *et al.* (1991), Conkling and Byers (1992), and Riitters *et al.* (1992). A core data management and analysis system has been designed around the detection measurements. The research measurements are different in terms of consistency over time and the data user group. Auxiliary data from other sources also enter some FHM analyses, but they are not directly managed by the FHM system.

Numerous problems are related to quality assurance (QA) requirements and the reporting schedule. QA is a routine part of field operations, and after each field season, a rigorous quality assurance procedure is applied to all FHM data to ensure that the data are of known quality for the intended uses. Only verified and validated data are released for analysis and assessment purposes. Timeliness is important because the FHM programme is committed to producing a final statistical report nine months after the end of the field season.

FHM-Information Management System (FHM-IM)

Information management supports and facilitates many aspects of the inter-agency FHM programme. IM personnel work with the technical directors, project scientists, quality assurance/quality control (QA/QC) personnel and logistics staff throughout the project. FHM-IM has a continuing commitment to developing a system that is responsive to overall project needs. During operational phases of data collection and transfer, software systems support the timely incorporation of data into the IM system. After data collection, IM supports the scientists working on verification, validation, assessment and reporting of the data. Figure 7.3 provides an overview of the FHM-IM system, which is detailed in the following sections.

Figure 7.1 Location of 1992 FHM detection monitoring activities.

FHM Field System

The FHM field system is used by field crews to collect and transmit data to processing nodes. The design, development and implementation of FHM field

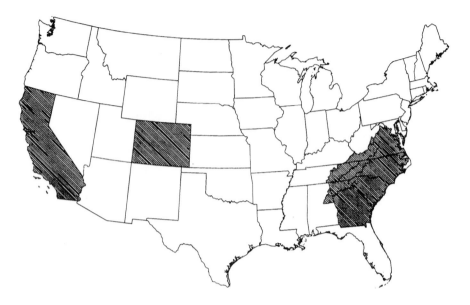

Figure 7.2 Location of 1992 FHM pilot and demonstration activities.

systems are guided by the following goals: to assure that the data collected are of the highest quality possible, and to provide quick data turnaround. To meet those goals in the national EMAP programme (over 4000 field plots when fully implemented) requires electronic data collection and transmittal.

The field crew employs one or more Portable Data Recorders (PDRs) to record field measurements. Global Positioning Systems (GPS), quantum sensors and ceptometers are also used for field data collection. In the mobile base station (i.e., the motel) the PDR and electronic instruments are uploaded to a laptop computer. Sample and shipment tracking data are entered on the laptop using bar code labels and laser scanners. The data and electronic mail messages are sent to the FHM information centre, located in Las Vegas, via a modem connection to a VAX minicomputer. Once they arrive at the VAX, the data are either parsed into databases, or sent via an electronic network to off-site FHM scientists (Figure 7.3). A portable printer enables local printing of data. Components of the system are detailed in the following sections.

PDR Hardware

The PDR is a rugged, hand-held computer used by the field crew to enter data. Before the start of the 1990 field season, FHM personnel evaluated PDRs from many vendors for the following criteria: ruggedness; MS-DOS compatibility; screen size; battery life; vendor-supplied PDR to PC communication software; and non-volatile data storage media. Since 1990, FHM has been using the

Table 7.1 *Detection monitoring and research measurements made by FHM in 1992.*

Detection monitoring:
  Plot-level

Plot number	State
County	Type of measurement
Plot status	(Aerial) photo year
Date of measurement	Elevation
Terrain position	Crew identification

  Subplot-level information

Land use class	Forest type class
Stand origin class	Stand size class
Past disturbances	Slope correction
Slope per cent	Aspect
Microrelief	Subplot map

  Condition class information

Land use class	Forest type class
Stand origin class	Stand size class
Past disturbances	

  Tree-level information

Number	Location
History	Species
Stem diameter	Crown class
Crown ratio	Crown diameter
Crown density	Foliage transparency

Research measurements (examples):
  Soil profile descriptions and soil chemistry
  Foliar chemistry
  Light interception by the forest canopy
  Root diseases
  Vertical vegetation structure
  Lichen community and chemistry

Paravant RHC-44. For the most part, the machines have functioned very well, and crew acceptance is high.

Field Software

While the selection of field computer hardware is important, it is critical to have the correct software running on the PDR. It is the software that allows the crew to enter the data, and will determine whether the data collection activities function as planned. The criteria for field software are easy operation by crew members, who generally do not have a computer background, and the means to make QA checks at the point of data collection.

To meet those criteria, FHM field software has the following characteristics:

- Knowledge-based system. The software guides the user through the

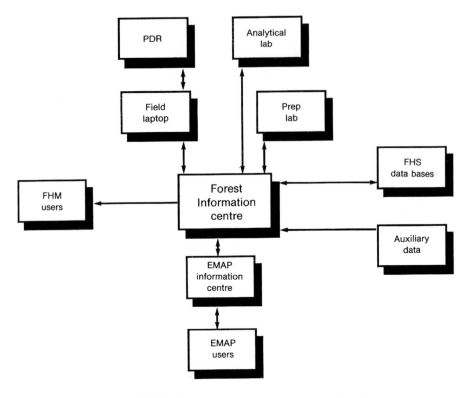

Figure 7.3 FHM information management system data flow.

program. For example, after the user enters the tree status (dead or alive) specific data entry fields are presented as a function of the status (e.g., mortality year is prompted for a dead tree, not a live tree).

- On-line help. FHM uses two types of help information: (1) help on program operations, and (2) context-sensitive help that provides a list of valid values and/or a brief synopsis of data collection methodology.
- Completion checks. Before the crew exits a plot file, completion checks are performed, and the user is warned if data items are missing.
- Historic data. Data from previous surveys can be included in the PDR. By including the species, distance and azimuth to a tree, the crew can easily relocate that individual tree for remeasurement.
- Real-time QA checks. The inclusion of QA checks at the point of data entry is one of the best features of using a PDR. FHM employs two types of QA checks: range checks and logic checks. Each type of check can return either a warning or an error message. A range check tests the value of a single variable (e.g., tree diameter), while a logic check tests the values of two or more variables (e.g., tree species versus region; for example, it is unlikely that a Douglas fir would be found in Maine).

Field software can be developed in-house for a specific project, existing programs can be modified to meet the needs of a new project, or commercial software can be used. FHM relies on software specifically developed for the programme and software developed by the Forest Inventory and Analysis (FIA) programme that has been modified for FHM. To date, no commercial products have been identified to meet the needs of FHM. The in-house software, written in C, has performed very well.

Field Base System

The laptop is the centre of the base system, performing the following functions: electronic data transfer from PDR, GPS and PAR (Photosynthetically Active Radiation) equipment; storage and archiving data files; viewing and printing data files; and sample and shipment tracking using bar code labels and laser scanners. Electronic communications with the operations centre is also supported. A script file initiates communications via an 800 number to the EPA's Las Vegas VAX system. Data are transferred via the modem link. To save time, data are compressed before transmission. The communication link allows for two-way email between crew members and other FHM personnel.

Data Processing at Central Site

Data are automatically processed once they reach the Las Vegas VAX system. A batch job runs daily at midnight to process all files received from the crews in the past 24 hours. SAS[3] programs update tracking databases, including sample and shipment databases, and crew tracking databases. The data that are used by off-site scientists are automatically sent via Internet or DECNet.

Data that are processed in Las Vegas are loaded into SAS tables for further processing. QA programs are run against those data sets to perform checks that are too sophisticated for inclusion on the PDR. A list of discrepancies is produced for review by project scientists. The discrepancy list is sent back electronically to the field crew. The next time the crew communicates with the VAX, the discrepancy information is downloaded to the laptop computer. The field crew member then reviews and resolves the discrepancies, sometimes the day after the data were collected in the field. The crew member's comments are then emailed back to the operation centre. This QA feedback loop has proved to be very successful in maintaining high quality standards.

Testing, Documentation, Training, Support and Debriefing

Most crew members do not want to wade through reams of software documentation. The best strategy is to keep the documentation short, simple and fun. FHM has had great success using self-guided manuals that are sent out before training sessions. The crew member can go through the tutorial at her/his own pace, and come to training sessions with questions or problems. The production of high-quality guides is time consuming, so ample time must be scheduled for this activity. Software development must also allow adequate time for documentation.

Training and support for the national FHM programme has been regionalized. Since 1992, a training and support team composed of regional FHM participants has been co-ordinated by the national information management staff. The team provides regional training and is available to answer crew questions during the field season.

At the conclusion of the field season, regional debriefing sessions are held. Questionnaires are distributed to crew members evaluating the performance of the field systems and the quality of training, documentation and support. The results are summarized, and are used in the planning of the next season's activities.

FHM Database Systems

FHM makes a distinction between data collected for research purposes, and monitoring data. Research data are managed by the national information management group at the EPA's Environmental Monitoring and Systems Laboratory in Las Vegas (EMSL-LV). Detection monitoring data are managed by regional Forest Service offices: Radnor, Pennsylvania; Asheville, North Carolina; Starkville, Mississippi; and Ogden, Utah.

Currently, FHM is using ASCII files and SAS data sets to maintain non-spatial data. The FHM information centre will ultimately be linked with the EMAP Information Center (EIC). There is an ongoing EMAP effort to develop a distributed database management system for EMAP data using Oracle as a relational database management system. FHM data will be included in this system.

The history behind the status of the current database carries some important lessons. FHM is composed of a geographically dispersed group of users with heterogeneous computing facilities. The FHM Programmers Workgroup was established in 1991 to co-ordinate interagency information management issues between FHM participants. The Bureau of Land Management, Environmental Protection Agency, Forest Service, Tennessee Valley Authority and state forestry agencies are represented in the FHM Programmers Workgroup. Topics for workgroup sessions have included file formats, data processing standards and system requirements.

The first meeting of the FHM Programmers Workgroup took place in August 1991. The purpose of the meeting was to determine data processing and transfer standards for the detection monitoring data, which are the responsibility of regional Forest Service units. The target date for a completed data set was 31

October 1991, three months from the programmers' meeting. Due to the short time frame and the heterogeneity of computer systems in the programme, it was decided that an ASCII file transfer format would be adopted for that year. In 1993, FHM was still using an ASCII file format for data exchange, although there are numerous problems: updates to the ASCII file are difficult to make; programming skills are necessary to extract data from the file; and, since the file is physically distributed to users, there are concurrency problems when an update is made to the master copy of the file and not all users receive the update.

Data collected for FHM demonstration projects (research data) are managed by the information management staff at the EPA laboratory in Las Vegas. Those data are stored in SAS tables. The decision to use SAS as a data management tool at the EPA laboratory was predicated by the fact that the EPA did not have an RDBMS at the inception of FHM. There are several advantages for using SAS over ASCII files: the data sets were designed to be third normal form (3NF) to ease the conversion to a true RDBMS; project scientists can access the data directly through SAS to perform their analyses; SAS has several data manipulation tools (including Structured Query Language) that simplify data management. However, SAS lacks many features of an RDBMS, such as transaction logging, rollback capability and security features.

Recognizing the problems with the ASCII and SAS systems, the Programmers Workgroup decided in September 1992 to initiate transition to an Oracle-based system. Although the EPA and the FS both have Oracle, the transition is slow. The large base of computer programs designed to work with the current system require some modification to work in an RDBMS framework. Insufficient resources for staffing impede the transition.

Convincing FHM scientists to turn over their data management functions to the FHM information management staff has been problematic. Most scientists are accustomed to both managing and analysing their data. They view giving up control of the data management aspects with great apprehension. Most scientists are not trained data managers but, that fact notwithstanding, they feel qualified to design, implement and maintain databases. A paradigm shift is in order. The scientist and database manager must form a partnership, with a clearly defined division of roles (Strebel *et al.,* this volume). Scientists need to express clearly the design of their research and requirements for data management and information processing. Database professionals need to take the input from the scientist, and deliver a system that will meet the user's needs. Ultimately, unburdening the scientist from the data management load will free her/him for more creative pursuits.

FHM-Geographic Information System (FHM-GIS)

GIS is employed within the planning, logistical, assessment and reporting areas of FHM, to manage, integrate and analyse spatial information. FHM-GIS is a part of the overall FHM-IM system and provides a spatial framework, visualization, spatial display capabilities, and analysis and integration tools. The objective of the FHM-GIS is to provide a system of hardware, software and staff to input, store, manage, manipulate, display, analyse, interpret and output spatial information of interest to FHM. In fulfilling the objective, the FHM-GIS facilitates FHM planning and logistical tasks, provides analysis and integration tools for scientific and assessment tasks, and provides output (graphics, maps, statistics, digital files and printed reports) capabilities for FHM reporting needs.

FHM-GIS Roles

FHM-GIS has an important role in estimating status, trends and changes in ecological indicators with known statistical confidence. GIS is used to: determine the sampling frame, which is the statistical basis for population estimates; estimate the geographic coverage and extent of forest resources; and seek associations between selected indicators of natural and anthropogenic stresses and indicators of the condition of ecological resources.

FHM-GIS Approach

FHM-GIS is based on UNIX, ARC/INFO and Internet, and adheres to EPA and EMAP GIS standards, such as the locational data accuracy policy and the Federal Geographic Data Committee (FGDC) Spatial Data Transfer Standard (SDTS). FHM-GIS nodes will be established at FHM sites throughout the country to provide FHM scientists and managers access to GIS technology, FHM plot data and auxiliary information.

Another important element of the FHM-GIS approach is to provide novice users access to GIS technology and appropriate data sets. The approach includes development of specific project modules (project databases) with a user-friendly interface. The EMAP GIS Interface (EGI), developed by EMAP personnel at the Environmental Monitoring Systems Laboratory in Las Vegas, Nevada, is a major step toward that goal. The EGI provides both generic and EMAP-specific analysis and display capabilities with an easy module construction process. The interface gives FHM-GIS users the ability to integrate quickly FHM sample data and auxiliary data for exploratory analysis.

FHM-GIS Progress

The FHM-GIS was sufficiently developed and utilized in both the preparation tasks and the data analysis tasks of the 1992 FHM field season. GIS was employed in the planning, design and reconnaissance material preparation tasks for the 1992 FHM demonstration projects. The employment of GIS proved to be a facilitating factor in all of these efforts. In particular, the plot determination tasks accomplished with GIS techniques provided significant cost savings to the programme.

Several FHM-GIS modules were developed for use with the first release of the EGI. FHM-collected field data and auxiliary information were included in the EGI. The auxiliary information included base mapping information, such as topography, roads, streams and political boundaries, as well as information on climate, air quality and forest pests. These FHM GIS modules were provided to a number of FHM-GIS sites.

FHM Assessment Process

Assessment is a process by which data are converted into useful information (National Research Council, 1990). The FHM assessment process is designed to enable knowledgeable analysts to organize, synthesize and interpret data to test for differences and change, to generate hypotheses, predict future states and to evaluate the uncertainty of conclusions. Many environmental problems are complicated, and scientists have limited understanding. There are many public, regulatory and management concerns about forests. Data and models will be used in different ways to interpret ecological condition for various purposes. Thus, the challenge is to create a flexible system with long-term value that functions within the multitiered, multiregional, multi-agency framework of FHM.

The FHM assessment strategy has the following foundations:

- Both statistical and biological models are utilized to address a variety of environmental issues.
- A distinction is drawn between routine statistical reports and special interpretive assessments. This permits rapid turnaround of statistical field data, and allows the incorporation of auxiliary information into the more in-depth studies.
- There is no separate assessment staff. Individual specialists from through-out the FHM community analyse and report on the data with which they are intimately familiar.
- The FHM assessment process consists of the FHM-IM system, the FHM-GIS system, standardized statistical procedures and links to other databases outside FHM.

The most recent FHM statistical summary report (Forest Health Monitoring, 1992) illustrates the type of information contained in routine reports. By restricting the contents of these reports to the results of pre-reviewed and standardized analyses of FHM plot data, FHM can produce useful products each year within nine months of the end of the season. Future interpretive reports will include information from other data sources and will use innovative statistical and biological models to explore and assess particular aspects of forest condition.

Conclusion

In conclusion, the FHM information system is a complicated effort that succeeds through the co-operation of all participants. Field data are collected, managed and distributed electronically to ensure rapid turnaround and high quality. FHM-GIS is designed to support both logistical and assessment needs, and provides links to external databases. These systems are the core of an assessment process that analysts use to produce statistical and interpretive reports about forest condition in the USA. As FHM evolves, the information management system will be improved in response to new challenges.

Notes

1 ARC/INFO is a registered trademark of Environmental Systems Research Institute, Inc., Redlands, California.
2 UNIX is a registered trademark of AT&T Bell Labs, Cambridge, Massachusetts.
3 SAS is a registered trademark of SAS Institute, Inc., Cary, North Carolina.

References

Conkling, B. L. and Byers, G. E. (Eds), 1992, *Forest Health Monitoring Field Methods Guide,* EPA/600/X-92/073, Las Vegas, Nevada: US Environmental Protection Agency.

Hansen, M. H., Waddell, K. L. and Liff, C. I., 1991, Nation-wide forest inventory databases in the United States, in *Proceedings of the Symposium on Integrated Forest Management Information Systems,* pp. 398–406, Japan Society of Forest Planning Press.

National Research Council (NRC), 1990, *Managing Troubled Waters: The Role of Marine Environmental Monitoring,* Committee on a Systems Assessment of Marine Environmental Monitoring, Marine Board, Commission on Engineering and Technical Systems, National Research Council, Washington, DC: National Academy Press.

Palmer, C. J., Riitters, K. H., Strictland, T., Cassell, D. L., Byers, G. E., Papp, M. L. and Liff, C. I., 1991, *Monitoring and Research Strategy for Forests — Environmental Monitoring and Assessment Program (EMAP),* EPA/600/4-91/012, Washington, DC: US Environmental Protection Agency.

Riitters, K., Law, B., Kucera, R., Gallant, A., DeVelice, R. and Palmer, C., 1992, A selection of forest condition indicators for monitoring, *Environmental Monitoring and Assessment,* **20**, 21–33.

8

Bigfoot: An earth science computing environment for the Sequoia 2000 Project

James Frew

The Sequoia 2000 Project is a large-scale collaboration between the Digital Equipment Corporation, the University of California, and several industrial partners and government agencies, for developing new computing environments for global change research. The primary focus of the project is to develop solutions for massive data storage, data access, data analysis and visualization, and wide-area networking. These solutions are embodied in 'Bigfoot', a computing environment the project is building at the University of California's Berkeley campus. Bigfoot comprises 10 terabytes of tertiary storage supporting commercial and experimental file systems, managed by an extended relational database management system supporting earth science data types and operations. Bigfoot is linked to a complex of private high-speed local and wide area networks, running both standard and experimental protocols. Research projects are extending Bigfoot to couple directly to general circulation models, and to provide visualization, full-text retrieval and geographic information system capabilities.

Introduction

This chapter discusses 'Bigfoot', the computing environment developed for the Sequoia 2000 Project at the University of California (UC). It begins with a very brief overview of the Sequoia 2000 Project, followed by a discussion of the computing problems that Bigfoot was designed to address. The architecture of the current implementation of Bigfoot is then described in some detail.

What is the Sequoia 2000 Project?

The Sequoia 2000 Project is a three-year collaboration between computer and earth scientists at UC; the Digital Equipment Corporation (DEC); other industrial sponsors; and several state and federal agencies. The overall goal of Sequoia 2000 is to further the state of the art in data management for global change research. The project has developed several interpretations of this goal, owing largely to the diversity of the Sequoia 2000 community.

For example, Sequoia 2000 computer scientists view the project as an engineering effort to develop a data system capable of storing, retrieving and manipulating several terabytes of heterogeneous earth science data (one terabyte = one trillion bytes). The earth scientists are viewed as 'clients', providing system specifications and testing the various prototypes.

Sequoia 2000 earth scientists view the project as an opportunity to apply massive data management resources to previously intractable problems. Since they are geographically distributed throughout California, they also use the Sequoia 2000 networked computing environment as a 'collaboratory', furthering the kinds of interdisciplinary research essential to global change studies (National Research Council Committee on Global Change, 1990).

Finally, the Sequoia 2000 industrial and governmental partners view the project as an opportunity to share new technologies. Sequoia 2000 is a source of new ideas for products and services, as well as a sophisticated community of beta-testers for products and services that have reached the development stage.

Who is the Sequoia 2000 Project?

The Sequoia 2000 Project is organized around a partnership between DEC and UC. DEC is Sequoia 2000's primary sponsor, contributing about $5 million per year in direct funding and equipment credits, plus four full-time staff positions in engineering and management. Sequoia 2000 is DEC's 'flagship' external research project, the successor to Project Athena (Champine, 1991), but with a much different focus: Athena concentrated on workstations and local area networks, whereas Sequoia 2000 concentrates on data management and wide-area networks.

Five UC campuses are involved in Sequoia 2000. The project is headquartered in the Computer Science Division at UC Berkeley. Most of the computer science and engineering activities are concentrated at Berkeley, with significant contributions from the Computer Science Department at UC San Diego and from the San Diego Supercomputer Center. Earth scientists participating in Sequoia 2000 are affiliated with the Scripps Institution of Oceanography at UC San Diego; the Atmospheric Sciences Department at UC Los Angeles (UCLA); the Center for Remote Sensing and Environmental Optics at UC Santa Barbara; and the Department of Land, Air, and Water Resources at UC Davis.

Several government agencies sponsor Sequoia 2000. Most of them (California State Resources Agency, National Aeronautics and Space Administration, National Oceanic and Atmospheric Administration, US Geological Survey) face data management challenges that could possibly be addressed by early adoption of some Sequoia 2000 technologies. Others (US Army Corps of Engineers) are integrating their own data management technologies into Bigfoot.

Besides DEC, there are several sponsors of Sequoia 2000. They currently include Epoch Systems Inc., Hewlett-Packard Co. (HP), Hughes Aircraft Co., MCI, Metrum Information Storage, North Carolina Supercomputer Center,

Table 8.1 Examples of earth science data sets.

Gigabytes	Data Set
4	normalized-difference vegetation index (NDVI) from Advanced Very High Resolution Radiometer (AVHRR) (Eidenshink, 1992) (one year of bi-weekly composite images; contiguous US)
23	UCLA hybrid coupled general circulation model output (Neelin, 1990) (one simulated year; global)
675	all Coastal Zone Color Scanner raw data ever collected (Feldman, 1989)
3000	AVHRR level 1B 'Pathfinder' data set (Wiscombe, personal communication) (daily 1981–present; global)

PictureTel Corp., Research Systems Inc. (RSI), Science Applications International Corp. (SAIC), Siemens Corporate Research Inc. and TRW. Besides contributing direct financial support or products, the industrial partners provide invaluable feedback on the viability of the technologies being developed by Sequoia 2000.

Shortcomings of Current Computing Environments

The development of Bigfoot has been driven by the inability of current computing environments to accommodate the demands of Sequoia 2000 earth scientists, particularly in the following areas (Stonebraker *et al.*, 1993):

- storing huge data sets;
- accessing huge data sets;
- analysing and visualizing complex data;
- connecting remote investigators and data.

Illustration of these shortcomings and some of their implications follows.

Storing Huge Data Sets

Sequoia 2000 earth scientists must routinely deal with data sets far too large to keep on-line in their local computing environments. To understand the magnitude of this problem, consider a computing environment consisting of a file server with 10 gigabytes of disk storage, on a network of 10 workstations, each with one gigabyte of local storage (one gigabyte = one billion bytes). In 1993, this was a reasonable 'off-the-shelf' scientific computing system. Now consider some data sets, presented in Table 8.1, that are typical of those used by Sequoia 2000 investigators.

Only the smallest of data sets may be kept on-line in the typical computing environment specified; the others must be stored off-line and processed in (relatively) small pieces. The larger the data set, the more physical volumes it will

Table 8.2 Data types used in the Sequoia 2000 Project.

Type	Examples
Raster	Satellite image; digital elevation grid
Vector	Drainage basin boundary; river network
Point	Weather station data; river discharge
Text	Algorithm description; instrument manual

consume (e.g., 3000 gigabytes = 600 8-mm tapes), and thus the more human intervention (fetching, mounting and unmounting volumes) that will be required to process the data set. The delays introduced by requiring a human operator in the middle of a processing sequence can profoundly discourage attempts to process huge data sets.

Accessing Huge Data Sets

Not only are earth science data sets often huge, but they are of many different types. Some of the data types used by Sequoia 2000 earth scientists are presented in Table 8.2.

A typical analysis may involve combining data of several types (Davis *et al.,* 1992). Furthermore, the criteria by which the data are selected may involve both metadata (static attributes of the entire data set; e.g., date and time of acquisition) and the data values themselves. For example, consider the variety of data types and manipulations necessary to satisfy a request for 'all ocean colour values for the melting margin of the ice pack beneath the Antarctic ozone hole for the dates nearest 15 October 1991'.

To extract a portion of a large data set using current computing environments requires mounting off-line physical volumes and then scanning them to extract a region of interest. If the organization of the desired subset is a transposition of the data set (e.g., multispectral values extracted from a sequence of single-band images) then the entire data set may have to be mounted and scanned. If several data sets are involved, the search time multiplies, with additional overheads if the results have to be merged.

Of course, locating and gaining access to large data sets in the first place can be a considerable challenge. Data sets are usually organized logically into files and physically into volumes. The file and volume divisions may, but need not, correspond to some logical properties of the data set (e.g., single image). The point is that file and volume names describe only how the data are stored, not any attributes relevant to the data themselves. Yet in most current environments these are the only metadata available for large data sets.

Collectively, these data access restrictions make it hard to find data of interest and difficult to extract them from the data set. Many analyses that are conceptually simple are never begun because it is so difficult to obtain the relevant portions of huge data sets.

Analysis and Visualization

Earth science data analysis in current computing environments typically involves the successive application of several barely compatible tools, each of which may use a different data format, different units, etc. Users must spend much time both converting data as they flow from one tool to the next, and tracking the sequence of tools applied to each data set. For many earth scientists, this bookkeeping is a major share of their data analysis activities.

Visualization tools are becoming a critical part of the earth scientist's analytical toolkit. Yet current visualization tools have two distinct shortcomings. First, they are large, slow, cumbersome software systems. This is partly because visualization is an inherently complex activity, but also because visualization packages must perform substantial data management activities (e.g., keeping track of files). In any event, the complexity of current visualization software makes it difficult to use effectively. Many earth scientists must employ specially trained programmers to operate their visualization software.

A second shortcoming of many current visualization tools is that they are output-only: the screen, film recorder or whatever is viewed as a data sink. This makes these tools suitable for preparing publication graphics (certainly an important activity!) but less useful for the kind of interactive processing necessary to refine, or even direct, an analysis sequence. For example, if a visualization tool shows that a general circulation model (GCM) is predicting boiling-water surface temperatures over North America, then it would be desirable to use the visualization tool to probe the executing GCM and discover which parameter was causing the (apparent) error.

The current state of data analysis and visualization causes much time to be wasted 'gluing' incompatible stand-alone tools together. There are no automatically maintained audit trails to keep track of how each data set is modified by specific tools. Furthermore, since most visualization is done in what amounts to a 'batch' mode, problems or opportunities arising early in the analysis sequence cannot be detected visually until the entire analysis is complete.

Connecting Remote Investigators and Data

Sequoia 2000 earth scientists are currently based at four UC campuses, the furthest distant of which are separated by several hundred miles. Internet connections between these sites, while adequate for electronic mail and remote log-ins, are too slow and erratic to be used for sustained high-volume data transfers (Pasquale *et al.,* 1991). Insufficient network capacity significantly impedes the scientists' ability to share both computing resources and data. The latter is more important; while a constant unit of computing power continues to

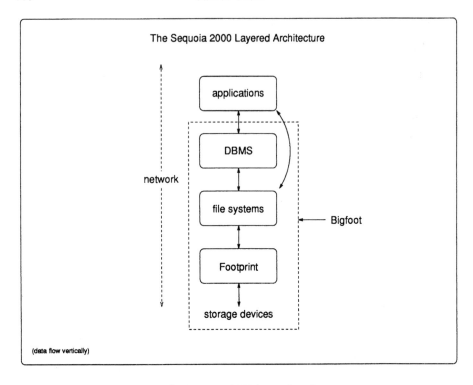

Figure 8.1 The Sequoia 2000 layered architecture.

plummet in price, large storage systems remain relatively expensive, and thus less likely to be part of one's local computing environment. A reliable, high-throughput, wide-area network (WAN) is thus critical to Sequoia 2000.

Bigfoot Architecture and Implementation

In response to the concerns outlined above, the Sequoia 2000 Project has implemented the Bigfoot computing environment. This section describes the current implementation of Bigfoot. Figure 8.1 gives the overall system architecture.

The Storage Layer

Current computing environments all have at least a two-level storage hierarchy, with primary memory (RAM) at the 'top' and magnetic disk 'underneath'. Bigfoot adds a third layer (tertiary memory) to the bottom of this hierarchy, comprising various persistent storage media (tapes, optical disks, etc.) configured as multi-volume 'jukeboxes' with robotic media manipulation. These

Table 8.3 Bigfoot storage devices.

Device	Capacity (gigabytes)
HP magneto-optical disk jukebox	100
Sony write-once optical disk (WORM) jukebox	360
Exabyte 8-mm tape jukebox	580
Metrum VHS tape jukebox	9000

robotic jukeboxes are called 'near-line' storage (Katz, 1992), since any volume may be brought on-line subject only to the latency of the robotics (e.g., the time for a robot arm to fetch and load a tape), as opposed to the much greater and more variable latency of a human exchanging tapes in a tape drive.

Bigfoot currently supports about 40 gigabytes of magnetic disk storage (the number fluctuates as workstations are added to or removed from the network) and 10 terabytes of tertiary storage, on the devices listed in Table 8.3.

The Footprint Layer

The Sony and Exabyte jukeboxes are managed by the project-developed 'Footprint' software. Footprint is a generic program interface for robotic storage devices. As such, Footprint provides a uniform interface to devices with often wildly varying physical characteristics (robot commands, tape sizes, etc.). These device-specific details are normally concealed by Footprint but may be accessed if needed (for example, some devices provide a local 'duplicate media' command that is useful for regenerating worn tapes).

The File System Layer

The project supports four independent file system interfaces. The HP and Metrum jukeboxes are managed by the commercial packages EpochServ and UniTree, respectively. Both of these products are 'hierarchical storage managers', which means that they invisibly migrate files between the jukeboxes and magnetic disk caches; in effect, they make each jukebox look like a single huge disk drive.

The Exabyte jukebox is managed by the locally developed 'Jaquith' software package, which makes the jukebox appear as a single huge tape drive. The Sony jukebox is managed by the 'Inversion' file system, described in the next section.

The Database Management System (DBMS) Layer

Any data stored on Bigfoot may, at the owner's discretion, be managed by the POSTGRES database management system, an extended relational DBMS developed at UC Berkeley (Stonebraker and Kemnitz, 1991). At a minimum, data set owners associate metadata with their data sets. POSTGRES can then use these metadata to do attribute-based retrieval of portions of huge data sets. A POSTGRES 'retrieve' operation may reference multiple data sets. If portions of more than one data set are returned, a new data structure ('class') is automatically created to hold them.

In addition to the usual relational DBMS capabilities, POSTGRES provides some major functionality of particular importance to the management of earth science data:

- large objects: POSTGRES supports binary large objects of unlimited size, and the 'Inversion' file system (Olson, 1993) to access their contents. The name Inversion is derived from the fact that this is an 'upside-down' file system — the file system is implemented on top of the database, rather than vice versa. Inversion offers the standard UNIX file system access functions (read, write, lseek, etc.), and supports both File Transfer Protocol (FTP) and Network File System (NFS) servers. Inversion combines the familiar, efficient file-level access model with all the benefits of a DBMS: transaction protection, time travel (recovery of old versions), audit trails, plus access to the file and its contents by POSTGRES queries and functions.

- Footprint: POSTGRES uses Footprint to access tertiary storage directly, bypassing any file systems. The POSTGRES query optimizer can then take into account the peculiar latency and throughput characteristics of tertiary storage devices.

- User-specified types: in addition to the usual scalar types supported by any DBMS (characters, integers, text strings, floating-point numbers, etc.), POSTGRES allows users to define their own types and associated functions. These types may be implemented as large objects, if necessary. Multidimensional arrays, polygons and physical quantities such as temperature are some of the types added to POSTGRES for Sequoia 2000. These types may also have multiple representations. For example, the type 'temperature' might be defined with a 'units' component that indicates the units of measure (Fahrenheit, Celsius, etc.). Functions defined on temperature may then consult the units to determine whether a conversion is necessary for the current operation and, if so, do the conversion transparently.

- User-specified functions: users may define functions that operate on either built-in or user-specified types. These functions may be written in the POSTQUEL query language, in which case they are stored in POST-GRES, or they may be written in C and compiled, in which case they are dynamically loaded when they are referenced. POSTGRES functions have access to the internals of large objects, thereby enabling data-intensive operations to be done entirely within the DBMS. The return value of a function (built-in or user-defined) may be used as a selection criterion in a query; this is how a large object may be queried by content as well as by metadata. Indices may also be built on function values to accelerate such queries dramatically.

By using POSTGRES for data access, Bigfoot users are freed from having to remember arbitrary file and volume names and how they map into (portions of) data sets. Collectively, the POSTGRES extensions support a new model of scientific computing in which the DBMS is treated less as a data server, and more as a procedure server (i.e., in addition to asking the DBMS for data, the user may ask it for results).

A simple example of image data retrieval will illustrate some of the capabilities of POSTGRES on Bigfoot. Assume a query of the form:

I'm studying seasonal vegetation patterns in southern California. I want AVHRR NDVI data for a certain date, within a geographic rectangle corresponding to my study area.

The following POSTQUEL query would extract the AVHRR sub-images from the relevant images and leave the new sub-images in Inversion files. The return value of the query is a list of the new Inversion file names.

```
retrieve(
    clip_avhrr(
        AVHRR_table, output_format,
        northwest corner, southeast corner
    )
) where AVHRR_table.date = . . .
```

The implementation of the query is straightforward. The AVHRR images themselves are stored in POSTGRES large objects as Inversion files. The metadata (in this simple case, the data format, date of acquisition and geographic bounding box) are stored in the class 'AVHRR_table', which also contains a reference to the appropriate Inversion file. The function 'clip-avhrr' is a user-defined function that returns instances of class AVHRR_table.

Data analysis and visualization on Bigfoot exploit several features of POSTGRES. Many of the functions normally invoked as tools are built into POSTGRES, so much of a typical analysis is done inside the DBMS. To perform functions that are not built in, POSTGRES can invoke an external tool. In such cases, POSTGRES is told what data formats the tool reads and writes, and then does the necessary data conversions invisibly. All function invocations, internal and external, are tracked using standard DBMS logging mechanisms, which permit data recovery at any stage of the processing.

POSTGRES facilitates visualization by assigning a 'renderable form' to each data type that is likely to be visualized. The renderable form is a specific kind of metadata that an external rendering tool can use to determine how the data ought to be displayed. The 'Tioga' effort, described in the next section, is building a renderer that incorporates feedback to POSTGRES, allowing manipulators (keyboard, mouse, etc.) in the renderer to provide input to functions executing in POSTGRES.

POSTGRES communicates with existing tools in two ways. The first, already described, involves POSTGRES invoking the tool with data formats that the tool expects. The second involves modifying the tool to talk to POSTGRES, typically by issuing queries to obtain data instead of opening files. Major existing tools already modified in this fashion include the commercial visualization packages AVS and IDL, the 'S-plus' commercial statistics package, and the public-domain GRASS geographic information system (GIS) (Westervelt, 1988).

Data analysis on Bigfoot is still complicated, but POSTGRES does many of the more tedious tasks, freeing the scientist to concentrate on science instead of data reformatting. POSTGRES internal functions minimize data transfer outside the DBMS. The use of renderable forms as data attributes allows almost any data in the DBMS to be visualized at any point during an analysis sequence.

The Network Layer

All Bigfoot components are networked: the DBMS, the file systems and tertiary storage devices, and the display systems. Standard protocols like NFS and the X window system are used where applicable. The Bigfoot networking strategy is to make all services available to all users. The network should not be the transfer bottleneck; remote users should receive the same level of service as local users.

The Sequoia 2000 Project maintains a private WAN, comprising T1 connections (1.54 Mbit s^{-1}, donated by MCI) linking the University of California's San Diego, Los Angeles, Santa Barbara, Berkeley and Davis campuses, and the California Department of Water Resources. The WAN was upgraded to T3 (45 Mbit s^{-1}) service in April 1994. Each site has a gateway between the WAN and a local FDDI (100 Mbit s^{-1}) fibre-optic ring, to which the investigators' workstations are connected.

The Sequoia 2000 network is being used as a testbed for a new suite of protocols called RTIP (Ferrari, 1990), developed at UC Berkeley. The RTIP protocols enable 'guaranteed delivery', whereby a client program may reserve a fixed portion of network bandwidth. This is important for continuous-media programs (sound, video) where the delay between successive transfers is critical to the application's performance. Sequoia 2000 researchers are using RTIP, together with prototype video compression hardware, to develop a desktop teleconferencing system that can be installed in any Sequoia 2000 investigator's workstation.

The Application Layer

This section discusses some of the more ambitious applications being developed on Bigfoot.

Tioga:

Tioga (Stonebraker *et al.*, 1992) is an attempt to build an integrated programming and visualization system within and on top of POSTGRES. Tioga has three major components:

- *Recipes:* Tioga allows users to build programs or 'recipes' out of POSTGRES functions and to store those recipes in POSTGRES.
- Graphical programming and query language: Tioga includes a graphical programming environment analogous to AVS, in which functions are depicted as boxes and data flows by directed lines. Programs built graphically are stored as recipes. A skeletal program, with some or all boxes replaced by regular expressions, may be used to query POSTGRES and retrieve any recipes whose structure matches the skeleton.
- Smart renderer: Tioga will send renderable forms of recipe outputs over a network connection to a 'smart renderer' being built at the San Diego Supercomputer Center. In addition to the usual mechanisms for visualizing geometry and images, the smart renderer will incorporate a knowledge base allowing it to suggest appropriate representations for specific types of objects (e.g., use of a thermometer as an icon for temperature).

The Big Lift: The Big Lift is a project to connect the UCLA GCM directly to POSTGRES (i.e., the output from the GCM will be deposited directly into the DBMS, without being saved in intermediate files). This involves modifying both the GCM (replacing its output routines) and POSTGRES (enabling it to accept data at the GCM's output rate).

There are two expected benefits from the Big Lift. The first and most immediate is the ability to browse the intermediate output of the GCM while it is running. This will allow the user to detect errors and adjust parameters or restart the model, using POSTGRES to extract subsets from the GCM's voluminous output. The second benefit, the ability visually to control the GCM's execution, is a longer term goal that depends on the completion of Tioga. If successful, this should radically change the way GCMs are used.

Post-GRASS: With the co-operation of the US Army's Construction Engineering Research Laboratory (the originators of the GRASS GIS), the Sequoia 2000 Project is working on subsuming the functions of GRASS into POSTGRES. The first phase of this effort, already complete, has replaced the input/output subsystem of GRASS with calls to POSTGRES. The relevent GRASS external data structures have been replaced by a custom POSTGRES schema. This 'GRASS-on-top-of-POSTGRES' has been dubbed 'Post-GRASS'.

The second phase of Post-GRASS, currently in progress, will migrate specific GRASS commands into POSTGRES as internal functions. The GRASS user interface will then be modified to issue POSTQUEL queries instead of UNIX commands. In the third phase, Tioga will replace the GRASS user interface, and POSTGRES will be able to provide the equivalent functions to GRASS, but with access to all the data in Bigfoot.

Lassen: Lassen is a full-text document retrieval system built on to and on top of POSTGRES. Text pages are stored as large objects, with weighted keyword indices (Larson, 1991) built automatically by POSTGRES as the text is entered. Lassen includes a separate natural-language query tool that interfaces to POSTGRES, allowing text retrieval in a manner familiar to users of automated library catalogues.

Lassen will be used to access a growing on-line collection of text on Bigfoot. The Computer Science Division at UC Berkeley is currently scanning all their technical reports into Bigfoot, saving the text as images and using optical character recognition (OCR) to extract keywords. Eventually all printed material associated with Sequoia 2000 will be accessible via Lassen.

Conclusion

Bigfoot addresses several pressing needs in earth science data management, and does so in novel ways. Robotic tertiary storage, POSTGRES and a fast project-wide network combine to both speed and simplify access to previously unmanageable quantities of data. POSTGRES manages both metadata and data, thus freeing users from worrying about low-level file management, and allowing them instead to concentrate on the data's information content.

Bigfoot is also well positioned to address some of the major future challenges in earth science data management. Bigfoot's seamless integration of storage, database and networks is a prototype of the transparent, distributed environments that will be commonplace in the next century. The tight coupling of data and functions provided by POSTGRES enables new modes of analysis that the project believes will become increasingly important.

Sources of Further Information

Further information about Bigfoot and the Sequoia 2000 Project may be obtained via anonymous FTP to the Internet host *toe.cs.berkeley.edu*. A Sequoia 2000 software distribution, planned for late 1993, will include POST-GRES, Inversion, Footprint, RTIP and any other project-developed software. Email correspondence regarding the project may be addressed to *frew@crseo.ucsb.edu*.

References

Champine, G. A., 1991, *MIT Project Athena: A Model for Distributed Campus Computing,* Bedford, Massachusetts: Digital Press.

Davis, F., Schimel, D., Friedl, M., Michaelsen, J., Kittel, T., Dubayah, R. and Dozier, J., 1992, Covariance of biophysical data with digital topographic and land-use maps over the FIFE site, *Journal of Geophysical Research,* **97**(D17), 19009–21.

Eidenshink, J., 1992, The 1990 conterminous US AVHRR data set, *Photogrammetric Engineering and Remote Sensing,* **58**(6), 809–13.

Feldman, G., 1989, Ocean color: Availability of the global data set, *Eos,* **70**(23), 634–40.

Ferrari, D., 1990, Client requirements for real-time communication services, *IEEE Communications Magazine,* **28**, 11.

Katz, R., 1992, High-performance network and channel based storage, *Proceedings of the IEEE,* **80**(8), 1238–61.

Larson, R., 1991, Classification, clustering, probabilistic information retrieval and the online catalog, *Library Quarterly,* **61**(2), 137–73.

National Research Council Committee on Global Change, 1990, *Research Strategies for the US Global Change Research Program,* Washington, DC: National Academy Press.

Neelin, J. D., 1990, A hybrid coupled general circulation model for El-Nino studies, *Journal of the Atmospheric Sciences,* **47**(5), 674–93.

Olson, M., 1993, The design and implementation of the Inversion File System, *USENIX Association Winter 1993 Conference Proceedings,* El Cerrito, California: USENIX Association.

Pasquale, J., Polyzos, G., Fall, K. and Kompella, V., 1991, *Internet Throughput and Delay Measurements between Sequoia 2000 Sites,* Sequoia 2000 Technical Report 91/7, Berkeley, California: University of California.

Stonebraker, M. and Kemnitz, G., 1991, The POSTGRES next-generation database management system, *Communications of the ACM,* **34**(10), 78–92.

Stonebraker, M., Chen, J., Nathan, N. and Paxson, C., 1992, *Tioga: Providing Data Management Support for Scientific Visualization Applications,* Sequoia 2000 Technical Report 92/20, Berkeley, California: University of California.

Stonebraker, A. M., Frew, J. and Dozier, J., 1993, *The Sequoia 2000 Architecture and Implementation Plan, Sequoia 2000,* Technical Report 93/23, Berkeley, California: University of California.

Westervelt, J., 1988, 'An introduction to GRASS', program documentation, Champaign, Illinois: US Army Corps of Engineers, US Army Construction Engineering Research Laboratory.

9

Representing spatial change in environmental databases

John L. Pfaltz and James C. French

Many scientific investigations centre around the study of the change of state associated with some phenomenon. Traditional database design is concerned with identifying and representing those attributes that characterize the system state itself, not its change. This chapter examines some of the issues raised when we consider the explicit representation of change in a database. Among them are: precisely what do we mean by change and what is required to implement efficient queries involving change?

Statement of the Problem

Scientific inquiry often involves the study of the dynamics of physical systems. Change in global temperature, in atmospheric ozone, in wetlands and in coastal barrier islands are just a few examples of changing phenomena that have been brought to public attention by the media. To study change, it must first be detected. This may be accomplished by comparing sequences of recorded data that have been stored in a database. To the extent that such changes in system state are important in a model of the underlying phenomena, it may be useful to represent explicitly the observed change for later study. We are not concerned here with the detection of change, but rather focus on issues raised when we seek an explicit representation of the change itself in a database.

Databases are designed to represent state, that is, the properties or attributes of some system of interest. Traditional commercial database applications are primarily concerned with the representation of the 'current' state of sales or inventory or financial accounts. The extensive literature on transaction processing (Gray and Reuter, 1993) is devoted to ensuring that processes using the database recognize system attributes in a consistent manner. The concept of serializability in database systems similarly seeks to ensure that concurrent execution cannot result in a process seeing or creating an inconsistent current state. There are many issues concerned with the effective representation of system state, for example:

- What attributes capture the important properties of the system state?
- How should the relationships between components of the system be configured?
- Should a relational model or an object-oriented model be used?

However, none of these issues, which form the core of traditional databases, addresses the issue of representing change.

One approach to this issue has been the introduction of versions into databases (Klahold *et al.*, 1986). This notion was later generalized in temporal databases (Snodgrass and Ahn, 1986; Soo, 1991), where the goal is to provide database support for certain aspects of time. The essence of both approaches is to keep snapshots of the current state for subsequent analysis. For example, in a temporal database, facilities are provided to access the state of the system at some earlier time, t_i, so that it may be compared with the current state or a state at time t_k. But, we would argue, the maintenance of multiple snapshots of state does not constitute a representation of change itself — it only provides a means for the subsequent calculation of change with respect to a single independent variable, time.

Because time is widely used in many information management applications, it is accorded a special status in temporal databases. But, time is not the only independent variable of interest when data are analysed for change. The fact that time is treated specially in temporal databases implies that similar special treatment is necessary for all possible variables. However, it should be noted that, by definition, temporal databases do not include support for user-defined time, that is, time recorded as observations (Jensen *et al.*, 1992). So the extension of temporal database technology will not necessarily even provide a general solution to the representation of change.

If a database represents change itself, for example change in area, it should be possible directly to query the database, as in:

retrieve all elements x such that $\Delta area(x) > 125$

without sequentially accessing the area of all elements of interest and calculating $\Delta area$. That is, the query should not have to detect change. With this example, we are implicitly saying:

Assertion 1: A database *explicitly represents* a system property, or attribute, if query predicates can be formed in terms of the attribute and the query can be answered without sequentially scanning all the entities having that attribute.

Very loosely, this says that if every relevant database element must be accessed and processed to determine whether the predicate is satisfied, we would say that the property, or attribute, has not been explicitly represented, even though it may be implicitly derived from data that have been represented.

We have formulated this assertion to exclude claims that a database explicitly represents, for example, change in forest canopy just because it contains the path names of image files on which some process might extract regions of forest canopy to be compared with all other images. Most would agree that this does not represent a changing forest canopy attribute. However, our concept also asserts that no database explicitly represents a system attribute, such as *date*, *name*, or *count*, unless it is indexed. Few database systems index fully on every

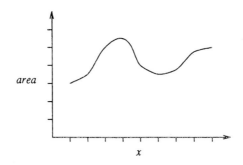

Figure 9.1 Area as a function of some variable x.

attribute. Equating explicit representation with indexing is not commonly accepted. Consequently, this assertion may be controversial. Nevertheless, we believe that it is correct. Except for very small databases, say with less than 1000 elements, retrieval by sequential search is simply impractical.

If change is to be explicitly represented, somehow the changing attribute must be indexed. We need to be a little careful here. We are speaking of changing attributes of a constant entity. We are not examining the more profound concept of change, in which entities change 'identity', for example, two distinct forest areas come together and are deemed to be one and are subsequently treated as such.

Before change can be represented, we must have a much clearer concept of what such change is. Note that the argument could be made that if a transaction-oriented database logs its transactions, as most do, then this log constitutes a representation of change. It does. But, even if the log itself is maintained as a portion of the database that can be queried, some additional processing to index the log will be required or it will have to be sequentially scanned.

What is Change?

Suppose we have a phenomenon that exhibits functional change, say the area of a region. The function can be graphed as in Figure 9.1. Here we have carefully avoided denoting the nature of the independent variable, x; it might be time, but it might also be average annual temperature, or rainfall or any of a number of possible variables. We can clearly see the change of area in this function, but we have not really represented it. Mathematically the change might be represented by taking its derivative as in Figure 9.2. There are two problems with using a derivative function as the representation of change. First, it is completely unintuitive. Few humans wishing to visualize the change of *area* as a function of x would ever choose Figure 9.2; we would always prefer to work directly with

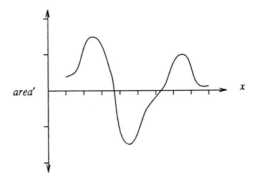

Figure 9.2 The derivative of the function of Figure 9.1.

Figure 9.1. Second, a derivative is technically not even a representation of change; it is a representation of the instantaneous rate of change. Change itself is better captured by the differential concept, that is $\Delta area(x)$, where this can be approximated by $\Delta area(x) \approx area'(x) \times \Delta x$. Deciding precisely what is meant by the representation of change is not easy.

Spatial Change

Here we limit our concern to the representation of spatial change as seen in the change of regions. These regions have presumably been designated by means specific to the particular research discipline. They may have been extracted from digital imagery; they may have been designated by hand drawing a boundary based on field observations or based on intuition. As database managers, we are unconcerned with the method of obtaining the region definition; we are only concerned with the representation of the regions and, especially, with the representation of their change. In this section we will examine in some detail two different possible representations of region change, and briefly discuss another.

Figure 9.3 shows very small portions of three spatial images taken from a study of shrub thicket growth on a barrier island off the Virginia coast (Shao *et al.*, 1992) conducted by the Department of Environmental Sciences, the University of Virginia. To emphasize the spatial aspect, each has been rendered as a binary (black or white) pixel bit map, although more data per pixel are present in the original. The spatial change in these figures is apparent. But, its perception exists only in the viewer's eye. Even though representations of these three spatial images are in our database, we still have no representation of the change itself. We cannot search the database with a query of the form

retrieve all *x* in hog_island_study
where *x* is a 'shrub thicket' and $\Delta area(x) > 125$.

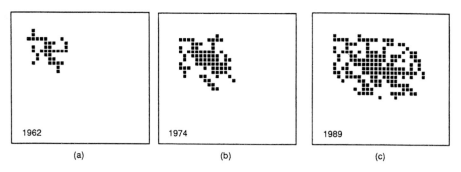

Figure 9.3 Pixels from three images of shrub thickets.

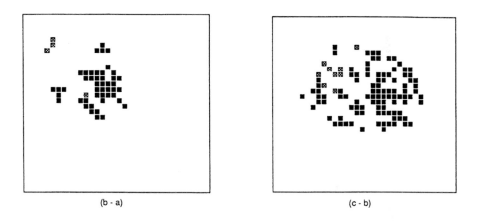

Figure 9.4 Pixel differences of images shown in Figure 9.3.

We could begin to capture the change by differencing the images. This will yield the two image differences *b–a* and *c–b* shown in Figure 9.4 in which the shaded pixels denote subtractions from the region and solid black pixels denote additions.

Differencing, the analogue of finite differences in numerical analysis, has many uses. For example, operating systems customarily employ differential files, that is, highly compacted files recording only the changes, if any, to user files, as a mechanism for recording daily file back-up. Given a checkpoint snapshot of any file, it may be restored to its state on any particular day by applying, in sequence, the daily differential files. We could do the same with spatial images. Given only the image of Figure 9.3a and the two differential images of Figure 9.4, we could reconstruct the remaining two spatial images Figure 9.3b and Figure 9.3c. These differential images clearly represent change. But our ability to issue the query:

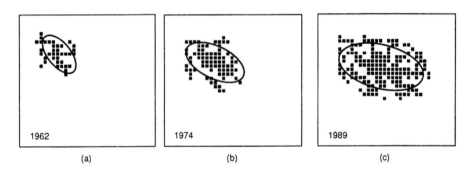

Figure 9.5 Approximating ellipses for areas shown in Figure 9.3.

retrieve all x in hog–island–study
where x is a 'shrub thicket' and $\Delta area(x) > 125$

remains as elusive as before. We can only retrieve *all* the differential images to calculate the change in area. According to Assertion 1, we have not yet explicitly represented change.

Differencing two spatial images can detect that change has taken place and, to some extent, the nature of the change. Differencing is a common means of change detection, which should not be confused with change representation, even though the two are frequently symbiotic partners. However, differencing at the pixel level is not without difficulty. The two images must be perfectly registered and this can be quite difficult, particularly when small rotations are involved (Mather, 1991). We should also note, in passing, that unlike the finite differences employed in numerical analysis, the image differences, which contain signed pixel values, are more complex than the original image operands.

An alternative approach to the representation of change of spatial regions can be based on the extraction of significant features from the regions and then the representation of the change of these features. For example, we could determine the centroid of each region, together with its area. Now we can simply record the centroid co-ordinates and area as real-valued functions of the independent variable x and represent their change appropriately.

In the process of extracting salient region features, information is lost. If regions are represented by only their centroid and area, all shape information has been lost. We have no indication of the extent of the region. Only translational change and area change can be captured. Although representation of regions by centroid and area features only may not be appropriate for many scientific studies, the concept of representing region change by representing the change of extracted features is worth further consideration.

A variant of this approach that we have been exploring is that of replacing the region itself by a best approximating ellipse, and then recording the change of these ellipses. In Figure 9.5, we have superimposed on each region an approximating ellipse whose area is the same as the original region. These

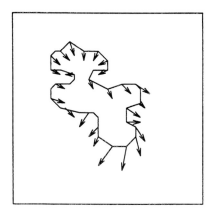

Figure 9.6 Vector field denoting change of a spatial boundary.

ellipses capture the general extent, position, orientation and area of the region, but lose specific shape detail. They seem to be appropriate approximations for certain kinds of relatively compact regions but may be inappropriate for other kinds of sparsely distributed regions. Assuming that they are appropriate for the spatial phenomena being studied, the advantages of representing regions by this, or any other feature extraction technique, are twofold. First, change can be represented as the change of real functions. In the case of approximating ellipses, there are five: the co-ordinates of the centre of the ellipse, the angle of the major axis, and the lengths of the major and minor axes. Second, these approximating features can be extracted from unregistered images, then adjusted independently by a mathematical change of co-ordinate system once the correct registration is determined. This method of representing change in terms of the change of approximating ellipses will be discussed more fully later.

So far, we have assumed a pixel definition of spatial regions. Defining regions by encoding their enclosing boundaries is also possible in many GIS systems (e.g., Environmental Systems Research Institute, 1989). With such a representation, the feature extraction approach is still feasible but direct differencing is not. It is possible to represent the change of curvilinear structures by a vector field, indicating the movement of elements composing the curve, as in Figure 9.6. But the technical difficulties involved in doing this (e.g., when a simply connected region either bifurcates or becomes multiply connected) seem significant.

Two Assertions

In this section we will make two more assertions. Based on the discussion of the preceding section we claim:

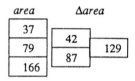

Figure 9.7 Observed area values and the finite difference structure.

Assertion 2: Spatial change can only be explicitly represented in terms of derived features. The primary rationale for this assertion is that it is difficult to formulate queries in any terms other than derived attributes, such as *area*. There are no query languages that support queries couched directly in region definitions themselves.

Consider again the three regions of Figure 9.5. Suppose we have represented each region by the five defining attributes of its approximating ellipse, together with an area attribute. To index the attribute $\Delta area$ in order to formulate queries involving change, we must represent $\Delta area$ quantities within the database. A way of doing this is to create a finite difference table, or triangle, as shown in Figure 9.7.

The finite difference representation proposed here differs from the usual notion of a difference scheme (Dahlquist and Bjorck, 1974) and finite differences as they are applied to numerical integration, differentiation and interpolation. In the usual formulation the difference operator, Δ, as applied to the sequence $<y_i>$ is given by

$$\Delta y_i = y_{i+1} - y_i$$

and the kth difference of the sequence y is given by

$$\Delta^k y_i = \Delta(\Delta^{k-1} y_i)$$

where $\Delta y_i = \Delta^1 y_i$. The values Δy_i and $\Delta^k y_i$ are often used in approximations to the derivative dy/dx and k^{th} derivatives $d^k y/dx^k$, respectively.

In our representation, the kth difference is given by

$$\Delta^k y_i = y_{i+k} - y_i.$$

So that for $k>1$ we have

$$\Delta^k y_i = \sum_{j=i}^{i+k} \Delta y_j.$$

The interpretation is that successive differences represent change over increasingly larger intervals. In our interpretation, the values Δy and $\Delta^k y_i$ denote the total change over the intervals, not an approximation to the rate of change; $\Delta^k y_i$ is a sum, not another difference.

One can now index each of the entries in the finite difference triangle $\Delta area$ to retrieve directly on the query predicate $\Delta area>125$. Clearly the entry, 129 in

Figure 9.7, satisfies the query predicate. Based on this observation, we are led to conclude the following:

Assertion 3: An explicit representation of change in a database must involve finite difference structures.

These triangular structures are rather different from the flat tables we associate with relational databases. They are most easily represented with object-based systems or with arrays.

Might it be possible to create a finite difference structure involving the regions themselves? Clearly, we could replace the three derived quantities to the left of Figure 9.7 with the images (a), (b), and (c) of Figure 9.3 and the two first finite differences to the right with (a) and (b) of Figure 9.4 (together with another second finite difference not shown). Now we have a direct representation of *region* and Δ*region*. It only remains to index these region representations for retrieval access. Currently we know of no way to do this completely effectively although there exist preliminary solutions, as in Nievergelt *et al.*, (1984) and Orlandic and Pfaltz (1991). It is these kinds of observations that lead us to question Assertion 2, even though they do not refute it.

Most importantly, our discussion has raised questions regarding the way we formulated queries earlier. It is possible that we only want to know *which* regions satisfied the change predicate Δ*area*>125, that is, to retrieve a set of region identifiers. But, it is much more reasonable to want also the interval over which the change occurred. Change is not a property of a single system state but rather a property of the difference between configurations. Consequently, a general query of this form should not return just the identity of the regions that changed. Rather, it should retrieve a set of more complex items. These retrieval items might, for example, be the region identifier together with the values of the independent variable x that bound the interval of change; or the two configurations that bound the interval of change; or a sequence of all the configurations constituting the interval of change.

This begins to raise query language issues that are beyond the scope of this paper. It suffices simply to observe that neither traditional query languages such as SQL nor the flat files of the relational model will be satisfactory for the general kinds of retrieval expected in an environmental database representing change. As noted above, we may have to turn to an object-based system such as ADAMS (Pfaltz, 1993; Pfaltz and French, 1993) which we have been developing at the University of Virginia. This object-based system includes a number of important features: support of subscripted identifiers that are more general than either vectors or arrays and can be used to implement sequences of elements (Pfaltz and French, 1990); provision for certain kinds of metadata in a very natural way; accommodation of heterogeneous, mixed media data; and support of dynamic schema modification (Pfaltz *et al.*, 1992) that can be exploited for the representation of change. The last feature appears to be of special importance, because large environmental databases cannot be locked into static representational structures. As environmental scientists discover new aspects to explore, the database supporting such inquiry must be able to adapt to it.

Feature Derivation

The derived features suggested by the preceding text have been relatively simple, obvious ones — *area* was explicitly used in our running example, *latitude, longitude, perimeter, connectivity* and *nbr_connected_components* could be others, if important for the study at hand. Single-valued quantities such as these are easily indexed for database retrieval. Construction of finite difference triangles is also straightforward if retrieval on the change of these derived properties is desired, although the space requirements may be considerable. It is clear that to create a complete finite difference triangle for a sequence of n elements would require $O(n^2)$ space to store all the differences.

A best approximating ellipse is a somewhat more complex derived property, but retrieval would still be on single-valued quantities or on the change of these values. Finite difference triangles on the centre co-ordinates could be used to query for drift; θ, the angle of the major axis, would be used to query for rotation; major and minor axis length would reveal change in elongation and general shape. It remains to be seen whether extraction of a best approximating ellipse really is feasible.

We first define a best approximating ellipse, E, for a region, R, in the following manner. For any E and R, let $diff^+$ denote $\int_{x \in E - R} dist(x, R) dx$ and let $diff^-$ denote $\int_{x \in R - E} dist(x, E) dx$. The quantities $diff^+$ and $diff^-$ can be regarded as weighted sums of elements in E but not R, and R but not E, where the weight is the shortest distance to R or E, respectively. We say E is a best approximation to R if

(1) $area(R) = area(E)$
(2) $diff^+ = diff^-$, and
(3) $diff^+$ is minimal.

If R is an asymmetric, connected region in the real plane, bounded by one or more simple closed curves, then there is a unique best approximating ellipse, E. The need for asymmetry can easily be seen by considering the case where R consists of two identical ellipses superimposed in the form of a cross. The condition of connectivity may turn out to be extraneous. Unfortunately, asserting existence gives no hint of how to determine E. Moreover, we are typically considering discretely quantitized regions. In this domain, uniqueness is no longer guaranteed.

Another approach that has been used in digital image processing is the Hotelling transform (Gonzalez and Wintz, 1977). This approach is based on statistical properties of the data. The idea is to consider each pixel as a random variable and find co-ordinate axes that point in the direction of maximum variance. Of course, these axes are also constrained to be orthogonal. This approach also yields ellipsoidal approximations. One drawback to this approach is the high computation cost. Because developing sound mathematical approximations appears to be computationally expensive, we are experimenting with several heuristic procedures for deriving an approximating ellipse. At present, the stability of these procedures is suspect; nevertheless we believe it is an approach that is worth developing further.

Derived features need not be as simple as these. The derived feature may be a fairly complex structure in its own right. For example, a representation of terrain surfaces using a collection of approximating polynomial expressions, $P_i(x,y)$, in two variables has been proposed by Pfaltz (1974). It was shown that these polynomial approximations could be rather easily generated, and that they could be piecewise continuous along the boundaries of their regions of validity. It is easy to represent polynomials of any degree by means of their coefficients in a database. Moreover, because these feature approximations are analytic, they have many nice properties. The slope at any point (x,y) is immediately available by symbolically differentiating $P_i(x,y)$ and evaluating. Local minima and maxima within an approximating patch can be obtained by solving two partial differential equations to find where $\nabla P(x,y) = 0$.

But, as powerful as this kind of derived feature is, it does not satisfy the concept of explicitly representing change as developed here. One cannot query the database to obtain points of equal elevation (i.e., $P(x,y) = c_0$) or maximal slope, or maximal change (elevation difference), without accessing each approximating polynomial patch in turn.

Conclusion

New directions in environmental data management have emphasized the shift in focus away from sheer data volume and toward information. The nature of change is one such kind of information. We have made three assertions that we believe will further this shift of focus. Assertion 1 observes that the vast amounts of environmental data, including those already on hand and those which are yet to be collected, must be easily and rapidly accessed if they are to be of real value.

There has also been a concurrent emphasis on metadata. This term denotes a multiplicity of concepts depending on the individual. However, it is becoming clear that useful data are more than just an archive of raw values collected in the field or by remote sensing apparatus. In assertion 2, we have taken this one step further and made it a requirement of change representation.

Finally, with the introduction of finite difference structures, we have suggested a new approach to the analysis of change. It is not clear that this is the only or the best such representation — indeed, directly representing change in a database is not easy nor well understood. But the ability to represent dynamically changing phenomena, in contrast to static systems, is essential if database technology is really to provide a computer representation of environmental behaviour.

Acknowledgement

This research was supported in part by Department of Energy grant DE-FG05-88ER25063 and by Jet Propulsion Laboratory contract 957721.

References

Dahlquist, G. and Bjorck, A., 1974, *Numerical Methods*, Englewood Cliffs, New Jersey: Prentice-Hall.

Environmental Systems Research Institute, 1989, *ARC/INFO Users Guide, The Geographic Information System Software*, Vols 1 and 2, Redlands, California: Environmental Systems Research Institute.

Gonzalez, R. C. and Wintz, P., 1977, *Digital Image Processing*, Reading, Massachusetts: Addison-Wesley.

Gray, J. and Reuter, A., 1993, *Transaction Processing: Concepts and Techniques*, Morgan Kaufmann. Calman, Los Altos.

Jensen, C. S., Clifford, J., Gadia, S. K., Segev, A. and Snodgrass, R. T., 1992, A glossary of temporal database concepts, *ACM SIGMOD Record*, **21**(3), 35–43.

Klahold, P., Schlageter, G. and Wilkes, W., 1986, A general model for version management in databases, *Proceedings, 12th International Conference on Very Large Data Bases*, pp. 319–27, Kyoto, Japan.

Mather, P. M., 1991, *Computer Applications in Geography*, New York: Wiley.

Nievergelt, J., Hinterberger, H. and Sevcik, K. C., 1984, The Grid File: An adaptable, symmetric multikey file structure, *ACM Transactions Database Systems*, **9**(1), 38–71.

Orlandic, R. and Pfaltz, J. L., 1991, *Q_0-trees: A Dynamic Structure for Accessing Spatial Objects With Arbitrary Shapes*, Technical Report IPC-91-10, Charlottesville, Virginia: Institute for Parallel Computation, University of Virginia.

Pfaltz, J. L., 1974, Computer representation of geographic surfaces, in Davis, C. J. and McCullagh, M. C. (Eds) *Geographical Analysis and Spatial Display*, pp. 210–30, New York: Wiley.

Pfaltz, J. L., 1993, *The ADAMS Language: A Tutorial and Reference Manual*, Technical Report IPC-93-03, Charlottesville, Virginia: Institute for Parallel Computation, University of Virginia.

Pfaltz, J. L. and French, J. C., 1990, Implementing subscripted identifiers in scientific databases, in Michalewicz, Z. (Ed.) *Statistical and Scientific Database Management*, pp. 80–91, Berlin: Springer.

Pfaltz, J. L. and French, J. C., 1993, Scientific database management with ADAMS, *Data Engineering*, **16**(1), 14–18.

Pfaltz, J. L., French, J. C., Grimshaw, A. S. and McElrath, R. D., 1992, Functional data representation in scientific information systems, in *International Space Year Conference on Earth and Space Science Information Systems (ESSIS)*, pp. 788–99, Pasadena, California.

Shao, G., Porter, J. H. and Shugart, Jr, H. H., 1992, Shrub thicket dynamics on Hog Island, Virginia, in *ARC/INFO Maps*, Redlands, California: Environmental Systems Research Institute.

Snodgrass, R. and Ahn, I., 1986, Temporal databases, *Computer*, **19**(9), 35–42.

Soo, M. D., 1991, Bibliography on temporal databses, *ACM SIGMOD Record*, **20**(1), 14–23.

SECTION III

Quality assurance/quality control

Expanding the focus of environmental research to encompass patterns and processes at relevant spatial (ecosystem to global) and temporal (years to centuries) scales typically requires the use of data collected by many investigators representing diverse scientific disciplines. Thus, it is not generally feasible for a single scientist to be involved in a comprehensive broad-scale or long-term project from start (experimental design, data collection) to finish (synthesis, modelling and publication). Consequently, as pointed out by Jelinski *et al.* in Chapter 4, there is a strong need to develop mechanisms for assessing and increasing data quality. The three chapters in Section III further elucidate the importance of this activity.

A database for long-term multivariate monitoring of the environment draws upon multiple sources for its data, which are collected and entered both automatically and manually. Each data set probably has its own idiosyncrasies that must be understood when assessing data quality. As a long-term data set accrues, the need for statistically based techniques for controlling its quality increases. Chapal and Edwards present examples from a research programme in which parametric and semiparametric regression was used in visualizing patterns in complex long-term data sets and in estimating or removing the effect of suspected disruptions or outliers.

The increased demand for broad-scale digital maps requires the development of adequate measures of map accuracy. However, at regional and even broader scales, the costs associated with determining map accuracy may be prohibitive. Moisen, Edwards and Cutler review the effect of spatial autocorrelation and cost on the relative efficiency of three unbiased sampling designs (simple random, systematic, cluster) for assessing the classification accuracy of remotely sensed data, and offer suggestions on how to improve the efficiency of such designs. In the final paper of this section, Chrisman maintains that a comprehensive GIS should exploit variations in data quality. He reviews emerging standards pertinent to the metadata required to evaluate the fitness of spatial data for use in environmental analysis, offers several suggestions for improvement and identifies research challenges related to dealing with spatial data of variable quality.

10

Automated smoothing techniques for visualization and quality control of long-term environmental data

Scott E. Chapal and Don Edwards

The philosophy of scientific data/information management, which posits that data are a valuable resource worthy of preservation, is vindicated only to the extent to which those data are used with confidence. Data of understood quality, free from introduced errors or biases, are critical for analyses that aim to isolate subtle trends or patterns related to disturbance, succession or ecosystem evolution. As data accrue, the demand for data-driven, statistically based quality control techniques increases. A data quality control method for long-term environmental data that emphasizes statistical data visualization is described. Traditional parametric modelling is compared with semiparametric smoothing techniques using the Akaike Information Criterion (AIC) for model selection. An interface in development fits a series of parametric and semiparametric models, selects the minimum AIC fit, returns graphical output and estimates the potential intervention points. This application is reviewed on the macrobenthos and nutrient chemistry data from North Inlet, South Carolina.

Introduction

Data users are typically unaware of the flaws and imperfections that can reside in even the simplest data sets. The potential for data quality problems increases dramatically with data set size and complexity. Many of the data that are collected to represent environmental parameters and ecological processes suffer from the problems posed by 'dirty data' (Coombs, 1986); they are often incomplete in terms of missing observations and inherently 'noisy' like the phenomena they represent. Errors and inaccuracies may be especially difficult to recognize in data that exhibit high variability. In order for environmental data management to be successful, error detection through quality assurance is mandatory. The problems created by poor data quality for analysis can range from loss of sensitivity to incorrect interpretations and conclusions.

The primary function of a data management system is to provide data of reliable quality for analysis, and should include routine protocols for the recognition and reduction of error at all stages of data collection, entry, verification and processing. Constructing a long-term multivariate monitoring database requires data from various sources, including data that are manually collected and entered, as well as data from computer-interfaced sensors and laboratory equipment. Data acquired from disparate sources require verification and quality assurance procedures tailored to the requirements of the data set. Data should retain the precision and accuracy of the measurement device, and input and processing errors should be minimized.

The integrity and utility of a database is dependent on the quality of its documentation. 'The test of adequate documentation is that it should contain sufficient information for a future investigator who did not participate in collecting the data to be able to use it for some scientific purpose' (Brunt and Brigham, 1992). The appropriate level of detail required is dependent on the complexity of the data set and the experimental design. Changes inevitably occur in research methodology, personnel, equipment and protocol during the course of a long-term monitoring study which can adversely impact the consistency of a data set. Although it is fundamentally important to document fully all aspects of the data generation process, in practice, the task may prove difficult to achieve, especially in a research environment that does not allocate resources for data set documentation. Data quality in long-term studies is critically dependent on thorough, complete and accurate documentation. If data are collected with a future purpose intended, the documentation must be prioritized throughout the study.

Quality Assurance in Environmental Data Management

The collection of long-term monitoring data presents a challenge for the implementation of an effective management and retrieval system. Quality assurance methods must be institutionalized if data integrity is to be expected as a primary product of the management system. For quality assurance methods to be effective, protocols should be incorporated into the infrastructure from entry to archival. Data verification strategies, including graphical and statistical summarization as well as complete documentation, are critical to successful quality control. A list of the categories used in the North Inlet database management system documentation (Table 10.1) illustrates the detail and diversity of information required to interpret a data set. These categories are designed to describe the data and research methods and provide an audit trail throughout data processing. In addition to providing the necessary information for interpretation of a data set, documentation is also extremely important for data quality assessment. The interconnected components of data management are illustrated in Figure 10.1, which highlights the interdependencies of documentation, routine graphical and statistical summaries, and data verific-

Table 10.1 Documentation categories taken from the North Inlet LTER long-term database illustrate the detail required for continued utility of data.

I. Data set descriptors
1. Title
2. Code
3. Index terms
4. Entry verification
5. Data set status
6. Latest update
7. Latest archive date
8. Documentation status

II. Research origin descriptors
9. Principal investigator(s)
10. Study purpose
11. Experimental design
12. Research methods
13. Site location and character
14. Data collection period
15. Associated researchers and projects

III. Access descriptors
16. Security restrictions
17. Storage location and medium
18. Contact persons

IV. Data set structure descriptors
19. File names and contents comprising the data set
20. Data type
21. Variable number, code, name, columns, format
22. Relational key indicator
23. Sorting hierarchy
24. Units of measurement
25. Rounding standards
26. Coded variable indicator and variable code definitions
27. Data anomalies
28. Missing data

V. Supplemental descriptors
29. Related materials
30. Computer programs

Note: These descriptors are designed to provide sufficient information for data retrieval by researchers who are not involved in the study. Mandatory documentation standards will be necessary for long-term, broad-scale research to succeed.

ation procedures for quality control. Given that long-term data sets are in a continuous state of dynamic change and growth, documentation must remain current to make data interpretable.

One of the principal sources of error in a data set occurs through data entry. Manual data entry errors can be reduced effectively with the use of a full-screen data entry program such as Easy Entry™ or a well-designed spreadsheet that

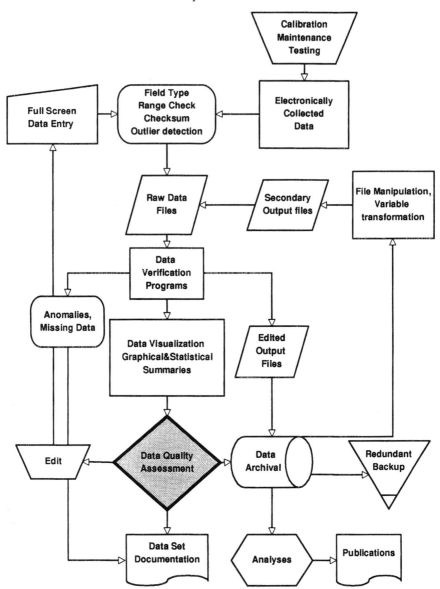

Figure 10.1 Flow diagram of the components of research data management taken from the North Inlet LTER, illustrating the integral importance of quality assurance.

Note: These procedures are generic enough to adapt to disparate data types. The products of the system are an archived database, documentation and analyses and publications.

mimics the actual data entry forms. Data entry programs can be written to prevent the input of inappropriate data types in a given field, to perform range checks and checksums, and to auto-duplicate header information to a series of records. With these automated features, entry can be made more efficient and less laborious and thus result in less error-prone data. Data verification can be facilitated by cross-checking hard copy output against the original data sheets, which is usually best accomplished by the technician or investigator who is most familiar with the data.

The issue of data quality problems was explored in *The Management of Electronically Collected Data Within the Long-Term Ecological Research Program* (Ingersoll and Chapal, 1992), which summarized the results of a survey of the LTER network in 1991. The most common elements of quality control in LTER climate installations include the implementation of redundant data collection systems and routine inspection of instrumentation and data (Ingersoll and Chapal, 1992; Ingersoll, 1993). Often, mechanical chart recorders are used to provide back-up data collection for electronically interfaced sensing equipment. Nevertheless, the correlation between the primary and back-up data is not always high, or easy to compute (Ingersoll and Chapal, 1992).

Routine procedures for data verification usually include range checks or outlier detection, graphical data inspection and various methods for detecting unusual data (outliers) in a series. In spite of vigilance regarding instrument calibration, sensor maintenance and routine data inspection, data quality problems may still arise and need to be addressed *post hoc*. Although every effort is made to ensure consistency in the record, unforeseen circumstances can cause interventions or changes in a series of data. Beyond the simplistic approaches of plotting data or performing outlier searches by flagging observations that exceed some threshold value, routine methods for data visualization and error detection in long-term environmental data remain relatively undeveloped.

To automate the process of visualizing data, we have written a collection of functions in the Splus^{TM} statistical programming language (Becker *et al.,* 1988; Chambers and Hastie, 1992; Hastie, 1992) that allows a user to fit families of models to data which have potential changes at known points in time (i.e., intervention points). The goal of this effort is to allow the user to specify the intervention dates and the parameters of interest, and to return a series of fitted models, a graphical representation of the data, and the fits and estimates of change (i.e., 'disruptions') at the intervention points. We chose two sample data sets (described below) from the North Inlet database for program development because they have identifiable intervention points and were relatively complete.

Case Studies

North Inlet estuary is located near Georgetown, South Carolina. The primary research area consists of a high salinity salt marsh separated from the Atlantic Ocean by barrier islands. Data have been collected since 1981 in an attempt to characterize the nutrient cycling, trophic levels, patterns of organic and inorganic processes, and disturbance events.

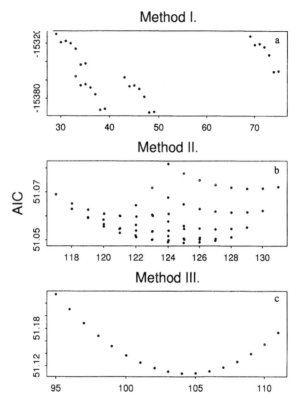

Figure 10.2 AIC (vertical axes) versus the number of model parameters (horizontal axes) for Method I (parametric) (a), Method II (seasonal semiparametric) (b), and Method III (adaptive semiparametric) (c) when given a range for their required input parameters.

Note: In this example, log 10 of total nitrogen was the parameter analysed. Method I was given a series of day groups and a range for the order of polynomial. For Method II, a range of effective degrees of freedom was specified for smoothing the date and another for smoothing the seasonal pattern. Method III required only a range of effective *df* for smoothing the date term. Each point represents a fitted model; the model with the smallest AIC was selected in each case.

Nutrient Chemistry — Daily Water Sample

Daily water samples have been collected at North Inlet since 1978 in order to monitor nutrient cycling within the estuary. Water quality parameters measured include total and dissolved nitrogen and phosphorous (Figures 10.3 and 10.4), dissolved ammonia and nitrate/nitrite, soluble reactive phosphorous, dissolved organic carbon, secchi disk depth, water temperature, salinity and suspended sediments. The samples were collected at 0.5 m depth at three locations in the estuary: Oyster Landing pier, Clam Bank dock and Town Creek station. Samples were filtered, then fixed, and frozen until analysed (Solorzano, 1969). In 1989, Hurricane Hugo caused extreme damage to the physical facilities. As a result of the destruction of the laboratory, the integrity of the daily nutrient chemistry data may have been compromised. When the hurricane struck on 23 September, backlogged samples were washed out of the laboratory freezers and into the adjacent forested wetland. The samples were recovered in the weeks thereafter and subsequently analysed, but they had been thawed out and some may have been subjected to slight evaporation. The parameters which may have been affected are total nitrogen and total phosphorous, both measured in μg at 1^{-1}. It is likely that abrupt shifts in the overall values of these parameters occurred at intervention dates 24 May and 23 September 1989; samples collected between these dates were subject to thaw effects.

Biweekly Macrobenthos Sampling

Macrobenthic (>0.5 mm) subtidal fauna populations have been sampled in the North Inlet estuary by collecting tidal creek sediment cores on a biweekly basis since 1981 (Figure 10.5). The suspected intervention points in these data resulted from a change in sampling protocol and two changes in the personnel who were responsible for sample counting and processing. During the first four years of the study, two cores, each 83.32 cm^2, were taken biweekly. The core size and depth changed on 18 January 1985; the new sampling regime consisted of eight cores, each 18.1 cm^2, taken biweekly. This change in sampling design was intended to reduce the variability attributed to patchiness and thereby improve the accuracy of the estimates of mean abundance by increasing the number of 'replicated' samples. The technicians responsible for counting and processing samples changed on 15 July 1985 and again on 20 December 1988. The three suspected intervention points were included in our analysis of these data. We chose to examine total macrobenthic organisms, which is compiled as a total of all of the categories of sorted macrobenthic organisms, as our initial test case. The analyses could be applied to individual taxonomic groups in the data set as well.

Statistical Methods

Basic Model

We have observations y_i, made at ordered times t_i, $i = 1,\ldots,n$. Time is assumed to be in day units numbered consecutively from the start date ($t = 0$). If there were no intervention points, a general model for an observation y at time t is

$$y = s(t) + \varepsilon$$

where $s(t)$ is an unknown 'signal' function and ε is a mean 0, constant variance error term. In our time series, error variance is dominated by sampling and

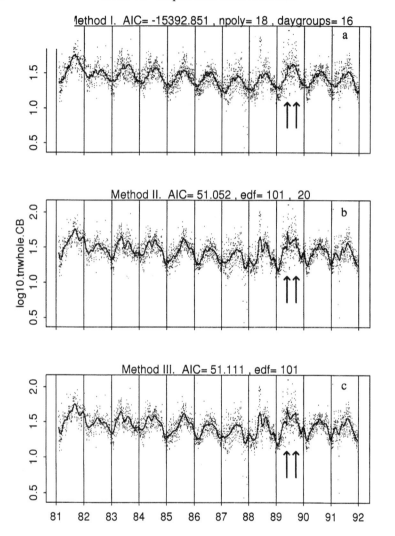

Figure 10.3 The results of the minimum AIC fits for total nitrogen at Clam Bank: (a) parametric method; (b) seasonal semiparametric method; (c) adaptive semiparametric method.

Note: The arrows indicate the potential interventions in the series (24 May 1989 and 23 September 1989). Methods II and III provide similar estimates of disruptions at the interventions in terms of direction and magnitude. Method I does not discern disruptions at the intervention points.

measurement errors; the mean 0, constant variance assumption is reasonable for all our examples, using y which is the logarithm of the original observation.[1] The 'signal' $s(t)$ at a specified time t thus represents the average obtained had one made a large number of measurements at time t and averaged them.

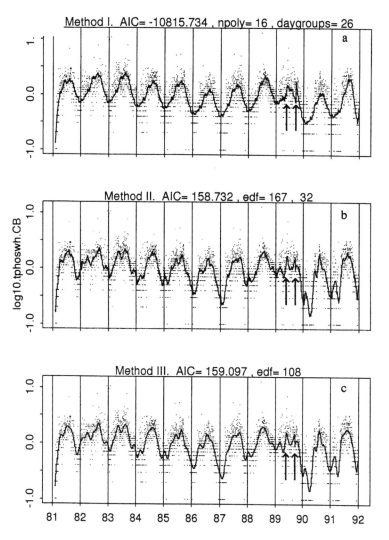

Figure 10.4 The results of the minimum AIC fits for total phosphorous at Clam Bank: (a) parametric method; (b) seasonal semiparametric method; (c) adaptive semiparametric method.

Note: The arrows indicate the potential interventions in the series (24 May 1989 and 23 September 1989). Methods II and III provide similar estimates of disruptions at the interventions in terms of direction and magnitude. Method I estimates a positive disruption for the second intervention, in contrast with the other two methods; this bias may result from an inflexible, underfitted model.

Suppose now there is a single (possible) intervention at a known time d during the sampling period. The simplest intervention model says

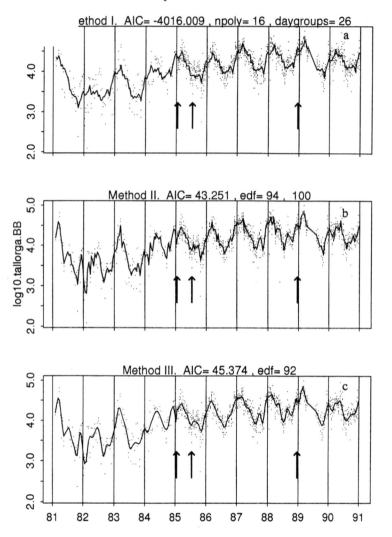

Figure 10.5 The results of the minimum AIC fits for total macrobenthic organisms at Bread and Butter Creek: (a) parametric method; (b) seasonal semiparametric method; (c) adaptive semiparametric method.

Note: The arrows indicate the potential interventions in the series (18 January 1985, 15 July 1985 and 20 December 1988). The three methods provide conflicting estimates (see Table 10.2) and may indicate near-zero disruptions.

$$y = \begin{cases} s(t) + \varepsilon & \text{if } t < d, \\ s(t) + \delta + \varepsilon & \text{if } t \geq d \end{cases} \tag{10.1}$$

The 'disruption' in the signal δ, which may be positive or negative (or 0 under the hypothesis of 'no intervention'), represents an abrupt, additive shift in the signal function at the time of the intervention. Model (10.1) can be written in a single line via the device of an indicator function:

$$I(t \geqslant d) = \begin{cases} 0 \text{ if } t < d \\ 1 \text{ if } t \geqslant d \end{cases}$$

which gives

$$y = s(t) + \delta I(t \geqslant d) + \varepsilon. \tag{10.2}$$

It can be verified that this is equivalent to model (10.1) by replacing I with 0 or 1 depending on whether t is $<$ or $\geqslant d$.

In general, we have k known time points $d_1, d_2, ..., d_k$ when interventions may have occurred. When time point d_j is passed, the signal shifts by an amount δ_j. This can be written

$$y = s(t) + \sum_{j=1}^{k} \delta_j I(t \geqslant d_j) + \varepsilon \tag{10.3}$$

The goal of the analysis is to estimate the disruptions $\delta_1, \delta_2, ..., \delta_k$, possibly to test the significance of these, and to create a 'clean' time series y^c to offer to users (we hasten to add that raw data y will *not* be lost). If the signal function $s(t)$ were known, these goals could be accomplished by multiple regression[2] of $y_i^* = y_i - s(t_i)$ on regressors $I_{i1} = I(t_i \geqslant d_1)$, $I_{i2} = I(t_i \geqslant d_2), ..., I(t_i \geqslant d_k)$. The difficulty lies in the fact that we must estimate simultaneously the unknown signal function $s(t)$ and the disruptions $\delta_1, \delta_2, ..., \delta_k$. The methods we consider represent three approaches to estimation of $s(t)$ in this setting.

Method I: A Purely Parametric Approach

A standard, widely used approximation to any unknown function $s(t)$ is a pth-order polynomial:

$$s(t) \cong \beta_0 + \beta_1 t + \beta_2 t^2 + ... \beta_1 t^p \tag{10.4}$$

Taylor's theorem from mathematics says that the approximation to $s(t)$ on the interval $[t_1, t_n]$ can be made arbitrarily sharp by choosing an order of polynomial p sufficiently large. In practice, very large values of p can cause fitting difficulties,[3] and the polynomial approximation will 'wiggle' in unnatural ways.

The dominant signal in our series is a loosely regular seasonal periodicity. Thus, the signal function might be decomposed into a sum of terms due to long-term trend and seasonal components

$$s(t) = l(t) + g(\text{julian}(t)) \tag{10.5}$$

where julian(t) is the day of year corresponding to date t. The trend function $l(t)$ might be well approximated by a moderate order polynomial ($p \leqslant 16$). The seasonal component function $g(\text{julian}(t))$ might well be approximated by a polynomial or by a step function of grouped julian dates (akin to removing a monthly or fortnightly mean). Since Method I represents a traditional approach, to be compared with methods based on more recent statistical developments (Methods II and III), we opt to group julian dates into m groups, fitting a separate intercept $\beta_{0g(i)}$ when a date t_i lies in group g, g = 1,..., m.

Our purely parametric model is thus,

$$y_i = \beta_{0g(i)} + \sum_{j=1}^{p} \beta_j\, t_i^j + \sum_{j=1}^{k} \delta_j\, I(t_i \geq d_j) + \varepsilon_i, \; i = 1,\ldots,n. \qquad (10.6)$$

Foreboding as this model may appear, it can in principle be fitted with any multiple regression program, given a sufficiently powerful computer. For example, in the Statistical Analysis System (SAS), one could create the julian-day grouping variable JDAYGRP, the regressors $T1, T2, \ldots TP$ corresponding to $t, t^2, \ldots t^P$, and the indicator functions $I_1, I_2, \ldots I_k$, and fit the model by

PROC GLM;
CLASS JDAYGRP;
MODEL Y = JDAYGRP $T1 \ldots TP\ I1 \ldots IK$.

The regression coefficients for $I1, \ldots IK$ are the estimated disruptions $\delta_1, \delta_2, \ldots, \delta_k$. Formal inferences on the true disruptions $\delta_1, \delta_2, \ldots, \delta_k$, follow immediately from standard multiple regression theory, if the model fits well (e.g., Neter *et al.,* 1990). The issue of choosing the order of the polynomial, p, and the number of julian-day groups, m, is a standard multiple regression model selection problem. The model degrees of freedom are $(m + p + k)$. Most model selection criteria weigh increases in model degrees of freedom versus decreases in the model sum of squares for error (sse). For example, the mean square error criterion recommends choosing m and p to minimize the mean square error, mse = sse$/[n - (m + p + k)]$. In the past decade, several new criteria have emerged. One of the most broadly applicable and widely used of these is Akaike's Information Criterion (Sakamoto *et al.,* 1986), which recommends (in our setting, i.e., independent, normal errors) choosing m and p to minimize:

$$\text{AIC} = n\,[\ln\,(\text{sse}/n)] + 2\,(m + p + k).$$

We have found the minimum-AIC choice of model (10.6) yields satisfactory results in most cases (see discussion). The purely parametric approach is meant to represent an 'old-fashioned' attack for our problem since it does not incorporate non-parametric regression or robust tools. These latter were actively researched in the 1980s, are still under some development, and are not incorporated as standard features in older statistics packages. The purely parametric approach has the advantage of being widely transportable in that the model can be fitted by virtually any statistical package, and also in that it yields straightforward inference on the disruptions $\delta_1, \delta_2, \ldots, \delta_k$, when the model fits well.

Semiparametric Methods

The approximation (Equation (10.4)) of $s(t)$ by a pth-order polynomial is referred to as a parametric approach because it assumes that $s(t)$ is determined by a finite number of unknown constants (which statisticians call parameters) $\beta_1, \beta_2,...,\beta_p$. The limitations of this approach led in the 1980s to an explosion of alternative methods for estimating $s(t)$ known as non-parametric regression methods, or simply 'smoothers'. The earliest smoothers were very crude estimators such as running means and running medians. For example, running medians with window size 3 estimates $s(t_p)$ by the median of y_{i-1}, y_i, and y_{i+1}, $i = 1,...,n$. In contrast, current non-parametric regression methods have become extremely sophisticated, and can be very valuable as 'pattern finders' in noisy data (e.g., ecological data). These methods include Cleveland's loess algorithms (Cleveland and Devlin, 1988) and penalized likelihood spline smoothers (Wahba, 1990). These methods are considered non-parametric because the final form of the estimate $s(t)$ is not representable by a closed form expression involving a finite number of parameters. The potential for application of locally weighted regression methods to ecological data has been explored recently by Trexler and Travis (1993).

Other methodological tools that saw considerable development in the 1980s are robustification techniques (Huber, 1981). These methods are designed (a) to identify observations ('outliers') whose y-values do not fit the signal+noise stochastic model (e.g., Equation (10.3)) and simultaneously (b) to downweight these observations appropriately in estimating the signal function $s(t)$. Standard (ordinary least squares) fitting of regression models can be adversely affected by these outliers if they are very severe and/or occur with sufficient frequency. Robust regression fitting algorithms are resistant to the effects of a limited number of errant data points. They are currently available in some, but not all, statistical software packages; they are an integral part of modern spline smoothers, including those we use in our Methods II and III.

Method II, the seasonal semiparametric approach, uses the model

$$y = l(t) + g(\text{julian}(t)) + \sum_{j=1}^{k} \delta_j\, I(t \geq d_j) + \varepsilon \qquad (10.7)$$

taking a non-parametric regression approach to estimation of the trend $l()$ and the seasonal component $g()$. This is a 'semiparametric' model because its deterministic kernel consists of both non-parametric terms, and $g()$, and the parametric regression indicator terms $\delta_1 I_{i1}, \delta_2 I_{i2}$, etc. Such models are a subcase of the generalized additive models described by Hastie and Tibshirani (1990), and they can be fitted and analysed in Splus^{TM} via the function gam() and its accessories (Hastie, 1992). We have used penalized likelihood spline smoothers for estimation of $l()$ and $g()$, each of which require specification of an 'effective degrees of freedom' value. These were chosen to minimize the vlaue of AIC (though the form of AIC differs for the spline smoother due to its use of weighted regression; one cannot compare AIC across methods).

Our last approach, Method III, is simply called an adaptive semiparametric approach. It returns to Equation (10.3) and models $s(t)$ as a single highly adaptive smoothing with very high effective degrees of freedom. We decided to consider this method when it became apparent that the seasonal variation was sufficiently inconsistent from year to year that there may be no advantage to modelling it separately as an additive deterministic term as in model (10.7). The effective degrees of freedom are again chosen to minimize AIC (Figure 10.2).

Results and Discussion

The common approaches for visualizing and summarizing long-term observational data tend to reduce the information to digestible components. Observational field data are often grouped into seasonal (monthly or biweekly) classes for interannual comparisons using traditional statistical tests. This approach may be effective for testing some hypotheses but does not take full advantage of the resolution of the data for discovering patterns or events. Subtle data quality problems are probably not discernible with simple regressions or means comparisons. Further, the assumption of regularity in seasonal pattern may or may not be true for a given measured parameter. It has become clear to us that the tools required to visualize long-term environmental data sets, which are often large and unwieldy, are not readily available. The methods we tested were designed to be data-driven approaches to modelling the signal and the potential intervention points in the data. The resolution of the data is preserved in the analyses, and the results provide a visual representation of a complex pattern.

Estimated disruptions at suspected intervention points for each of the three case studies were obtained by each of the three methods. The range of polynomials was restricted to 18th-order and smaller for the purely parametric approach because the larger models resulted in collinearity problems. For the total nitrogen series, using Method III (for example), the estimated effect of thawing on log(total nitrogen) is an abrupt upward shift in the signal function $s(t)$ in the amount of 0.2544 on 24 May 1989, with a corresponding downward return shift of 0.1490 on 23 September 1989 (Figure 10.3c) upon resumption of the usual frozen-sample analysis. It is not clear if the difference between these estimated shifts is statistically significant. The first disruption in log nitrogen, back-transformed by taking power 10, represents an 80 per cent increase in the mean value of total nitrogen in the thawed samples compared with what would have been obtained under normal processing procedures. Estimated disruptions using Method II agree loosely with those of Method III, but the purely parametric model (Method I) estimates are quite different. Since Method I is in actuality a conventional multiple regression model, 'p-values' for the significance of estimated disruptions are readily available (Table 10.2). We feel, however, that the user should regard the Method I estimates and p-values with scepticism, as their validity depends on the assumption that the final model chosen was

Table 10.2 Results from the three methods tested, showing estimated disruptions at intervention points tested.

	Method I	Method II	Method III
Nitrogen	**18 16**	**100 20**	**101**
K1	0.0732 (.0104)	0.2508	0.2544
K2	− 0.0272 (.3790)	− 0.1474	− 0.1490
Phosphorous	**16 26**	**167 32**	**108**
K1	0.0923 (.0554)	0.1158	0.1262
K2	0.3844 (.0001)	− 0.2037	− 0.2058
Total macrobenthos	**16 26**	**94 100**	**92**
K1	− 0.0519 (.5552)	0.1851	0.3664
K2	0.2187 (.0007)	0.0197	− 0.1146
K3	− 0.1341 (.0513)	− 0.1215	− 0.0772

*Note: K*1 = first intervention date, etc.
For Method I, estimates are shown with their associated *p*-values (in parentheses). For the semiparametric methods (II and III) only the estimates are given. For each variable tested, the minimum AIC models' chosen parameter values are given in bold: Method I — order of the polynomial term (1st), number of day groups (2nd); Method II — the effective degrees of freedom for smoothing the date term (1st) and the seasonal component (2nd); Method III — the effective *df* for smoothing the date term only.

sufficiently flexible to track the severe non-linearities in $s(t)$, and there is visual reason to doubt this. Tests of significance of disruptions using the non-parametric smoothers are not readily available.

The analysis of whole phosphorous in the daily water sample using the parametric model family revealed that the best fit was obtained using a 16th-order polynomial with 26 day groups (biweekly groups). The parametric fit suggests significant jumps at the intervention points (Figure 10.4a). The estimates (with standard errors) of the jumps are 0.0923(0.0482) and 0.3844(0.0497). The model fits appear to be similar at first glance; however, the parametric approach is less flexible at the extremes of the data range in certain years. Interestingly, the second disruption is portrayed as positive by the parametric model fit, but negative by both semiparametric methods. A symptom of inflexibility, which is universal to underfit regression models, is that it can produce serious bias in the estimated disruptions. We have observed false estimated disruptions at randomly placed intervention points using the parametric approach. Methods II and III also portray a mid-summer dip in the values of total nitrogen between the two intervention points, a pattern that occurs in some but not all years.

The results for total macrobenthos are less clear, and the three models provide conflicting estimates. This may be due to near-zero disruptions, in which case, estimated disruptions are random with mean zero. Method I estimates the first disruption as negative (though insignificant), while Methods II and III estimate it as positive (Table 10.2, Figure 10.5). Interestingly, the second disruption is estimated as positive by Method II and negative by Method III. The estimates agree, at least in direction, for the third intervention point. The inconsistency of the estimates suggests that the disruptions are not distinguishable from noise. Though there may be no detectable disruption in overall level of this series, it may be useful to test the variance of the residuals for disruptions especially at the date of the core size change. Method II may actually be suffering from a form of collinearity in so far as the seasonal term and the smoothed date trend may be, in a sense, redundant. This could be the cause of large effective degrees of freedom choices for the Method II models.

Conclusions

The problem of choosing a regression model (parametric or semiparametric) from a collection of candidate models has been discussed actively in the statistics literature in the past decade. Several reasonable criteria have been proposed. We elected to use Akaike's Information Criterion mainly because of acceptance in the literature on semiparametrics (Hastie and Tibshirani, 1990). We believe, however, by inspection of the obtained model fits, that the AIC semiparametric choices tend to overfit these series; the smoothers seem more flexible than necessary, particularly under Method II. The consequences of overfitting are not nearly as serious as those of underfitting; estimates of disruptions will be unbiased, but less precise, when models are overfitted. Nevertheless, we hope to investigate alternative criteria for choosing the models' effective degrees of freedom, day groups and polynomial order. The first alternative criterion we would consider is Schwarz's Bayesian Criterion, which in a conventional regression setting with normally distributed errors is given by

$$\text{SBC} = n \left[\ln \left(\text{sse}/n \right) \right] + \left(\text{model } df \right) \left[\ln(n) \right].$$

This differs from AIC in the second term, which increases the penalty of choosing larger model sizes, especially for large data sets. SBC chooses smaller models than AIC.

We believe that the semiparametric methods are, at least, not underfitting their series; this remains to be demonstrated, and all three approaches compared, by some objective test. A promising idea for such a test would be to create false disruptions at points in a real series to see if these methods can recover the false jumps accurately. Several disruptions of prespecified sizes would be inserted at randomly dispersed intervention dates, the methods applied, and this entire process iterated to estimate by Monte Carlo the bias and variance of the estimated disruptions under each method.

The semiparametric methods have, as an integral part of their fitting procedure, built-in robustification loops. These iteratively fit a weighted non-parametric regression, compute residuals and then re-compute weights for each observation to be used in the next fit. Outliers are down-weighted in the fitting process along the way. Besides yielding better estimates of $s(t)$, the final observation weights can be used to identify extreme values in the series, which may otherwise be hidden, simply by finding the observations with weights nearly zero. Some of these extreme values may, in fact, be unusual events of note, not simply data-in-error. We have no doubt that these outlier detection methods will out-perform range checks.

In conclusion, we believe these methods are very useful for visualizing pattern in complex data and for estimating/removing suspected disruptions. Semiparametric techniques are powerful by virtue of their inherent data-driven properties and may be sensitive to disruptions and other data quality problems. The common objection to their implementation is the subjectivity involved in selecting the degree of smoothing applied. The choice and application of an objective and appropriate criterion for model selection will be fertile ground for research.

The details of the implementation of these techniques can be made relatively user-transparent by using a graphical interface and object-orientated code. On modern workstations, the execution time required for iteration of these algorithms is quickly becoming non-limiting. Distributed processing and multi-processor operating systems will provide the basis for a truly real-time interactive data analysis environment. Visualization techniques embody unique potential for analysis and quality control, especially in light of the burgeoning volume of data in environmental science (Gosz, this volume; Stafford *et al.,* this volume). Automated, powerful tools, including statistical smoothing methods, will help bridge the chasm from data to information.

Notes

1. All logarithms here are base 10, except as denoted ln(), the logarithm base e.
2. With no intercept term.
3. The regressors $t, t^2, t^3, ..., t^p$ are severely collinear for large p. Effects of this kind can be reduced by centring t to $t^* = (t_i - \bar{t})/\text{sd}(t)$ prior to computing powers, or using orthogonal polynomials.

Acknowledgement

Thanks to Robert J. Feller and Elizabeth R. Blood for supplying data sets for analysis. The data for this study were acquired through the North Inlet Long-Term Ecological Research Program supported by the Ecosystem Study Program of the National Science Foundation, Grant DEB 8 012 165 (F.J. Vernberg, Principal Investigator). The thoughtful suggestions of several anonymous reviewers were much appreciated.

References

Becker, R. A., Chambers, J. M. and Wilks, A. R., 1988, *The New S Language,* Pacific Grove: Wadsworth & Brooks.

Brunt, James W. and Brigham, W., 1992, Data standards for collaborative research, in Gorentz, J. B. (Ed.) *Data Management at Biological Field Stations and Coastal Marine Laboratories,* pp. 12–18, W. K. Kellogg Biological Station, Michigan State University.

Chambers, J. M. and Hastie, T. J., 1992, *Statistical Models in S,* Pacific Grove: Wadsworth & Brooks.

Cleveland, W. S. and Devlin, S. J., 1988, Locally-weighted regression: An approach to regression analysis by local fitting, *Journal of the American Statistical Association,* **83,** 596–610.

Coombs, R. F., 1986, A management system for dirty data, *Software-Practice and Experience,* **16**(6), 549–58.

Hastie, T. J., 1992, Generalized additive models, in Chambers, J. M. and Hastie, T. J. (Eds) *Statistical Models in S,* pp. 249–307, Pacific Grove: Wadsworth & Brooks.

Hastie, T. and Tibshirani, R., 1990, *Generalized Additive Models,* London: Chapman and Hall.

Huber, P. J., 1981, *Robust Statistics,* New York: Wiley.

Ingersoll, R., 1993, *The Management of Electronically Collected Data Within the Long-Term Ecological Research Program — (1992): An update to the 1991 survey,* Seattle, Washington: LTER Network.

Ingersoll, R. and Chapal, S., 1992, *The Management of Electronically Collected Data Within the Long-Term Ecological Research Program,* Seattle, Washington: LTER Network.

Neter, J., Wasserman, W. and Kutner, M., 1990, *Applied Linear Statistical Models,* Homewood, Illinois: Irwin.

Sakamoto, Y., Ishiguro, M. and Kitagawa, G., 1986, *Akaike Information Criterion Statistics,* Tokyo: KTK Scientific.

Solorzano, L., 1969, Determination of ammonia in natural waters by the phenol hypochlorite method, *Limnology and Oceanography,* **14,** 799–801.

Trexler, J. J. and Travis, J., 1993, Nontraditional regression analyses, *Ecology,* **74**(6), 1629–37.

Wahba, G., 1990, *Spline Functions for Observational Data,* CBMS-NSF Regional Conference series, Philadelphia: SIAM.

11

Spatial sampling to assess classification accuracy of remotely sensed data

Gretchen G. Moisen, Thomas C. Edwards, Jr and D. Richard Cutler

Selecting a proper sample design is important in assessing the classification accuracy of remotely sensed data. However, design guidelines often do not account for spatial autocorrelation of map error or are based on simulations from a few select landscapes and cannot easily be generalized beyond those data sets. This study reviews the effect of spatial autocorrelation and cost on the relative efficiency of three unbiased sample designs within a stratified framework. These designs are simple random sampling, systematic sampling and cluster sampling. To illustrate the results, sampling simulations are performed on possible error patterns in a vegetation cover map of the Great Basin ecoregion. This map was generated from Landsat Thematic Mapper data by the US Fish and Wildlife Service's Gap Analysis Program.

Introduction

The accuracy of maps derived from remotely sensed data is important for the management and conservation of national and world-wide resources. Conservation efforts are increasingly being focused at the landscape or ecoregion level rather than at specific sites. While the goals of these efforts vary, most use a spatial component to organize data for analysis and interpretation. The US Fish and Wildlife Service's Gap Analysis Program (Scott *et al.*, 1993) uses vegetation classes and vertebrate species as indicators of biological diversity, and is designed to identify 'gaps' in the federal system of reserves. Once identified, gaps in the protection of biodiversity can be filled through land acquisition or changes in existing land-use practices.

Vegetation maps depicting the spatial distribution of vegetation cover classes are created using satellite imagery (e.g., Landsat Thematic Mapper) and other sources. These classes are derived by clustering individual pixels from the satellite imagery (Richards, 1986) and assigning them to vegetation cover classes based on field observations, aerial photographs and existing maps. Uses of these cover maps range from bioregional conservation planning at the ecoregion level to more finely focused management efforts by federal, state and private land managers. Because gap analysis data are organized around ecoregions, such as the Great Basin, maps can easily cover thousands of square kilometres. The sheer size of areas like the Great Basin poses immense logistical difficulties that complicate accuracy assessment. None the less, estimates of the spatial and classification accuracy of the vegetation cover classes are needed to assist land managers confronted with conflicting demands from user groups.

Assessing the accuracy of spatial databases has received considerable attention in the literature, as witnessed by Veregin's (1989a) extensive annotated bibliography. Bolstad and Smith (1992) discussed both positional and attribute accuracy in a resource management setting. Lunetta *et al.* (1991) discussed how error propagates through the map-making process, beginning with data acquisition and flowing through analysis, conversion and presentation of the final product. The taxonomy of error discussed by Veregin (1989b) also reflects the complexity of the issue. Various research agendas have been proposed for more creative error modelling strategies for geographic information systems (GIS). Chrisman (1989), Goodchild (1989), Openshaw (1989), Lanter (this volume) and others have recognized the unique problems posed by combining data of diverse nature, collected at different scales, and at various levels of error and uncertainty. Improved models of map error must be joined by better techniques to visualize spatial database accuracy. Beard *et al.* (1991) discussed the research needed to develop these techniques.

Although more comprehensive models of map error are needed, simpler measures of map accuracy are found in the remote sensing literature. Frequently, sample data are used to construct a contingency table or 'error matrix' from which many measures of thematic accuracy may be derived, including total percentage of pixels (or other sample units) classified correctly, percentage commission error by class, percentage omission error by class, and the Kappa statistic (Rosenfield, 1986; Story and Congalton, 1986; Congalton, 1991; Monserud and Leemans, 1992; Green *et al.*, 1993). Numerous studies have addressed the problem of choosing appropriate sample sizes and designs to assess classification accuracy (Berry and Baker, 1968; Hord and Brooner, 1976; Ginevan, 1979; Hay, 1979; Rosenfield, 1982; Congalton, 1991). Stehman (1992) clarified some common misconceptions about systematic designs and appropriate criteria for evaluating various sampling schemes. Congalton (1988a,b) discussed simple random, systematic and cluster sampling, and addressed the effect of spatial autocorrelation (dependency between neighbouring pixels) on sample design efficiency. Yet, questions remain about the choice of appropriate design, even for estimating something as simple as the percentage of misclassified pixels in a landscape. Many texts on finite population sampling theory discuss how cost may be considered when evaluating the relative efficiency of different designs (Kish, 1967; Cochran, 1977; Thompson, 1992). The cost of collecting information at locations randomly distributed across an entire state is substantially higher than collecting information at clustered locations. Cost functions can be used to compare costs of different sampling designs. Unfortunately, design guidelines in the remote sensing literature do not address the combined effect of spatial autocorrelation and cost on sample design efficiency.

The objective of this study is to illustrate the effect of spatial autocorrelation and a simple cost function on the relative efficiency of three sampling designs. The relative efficiencies of systematic and cluster sampling to simple random sampling are compared for estimating the proportion of misclassified pixels within a particular sampling stratum. These relationships are illustrated by performing sampling simulations on error patterns created from a vegetation cover map of the Great Basin ecoregion.

Three Sampling Designs Within a Stratified Framework

Cochran (1977) has discussed the advantages of stratified sampling. When assessing classification accuracy of remotely sensed data, an estimate of the percentage of misclassified units within each class is desired as well as an overall estimate of misclassification. Stratification allows subpopulation parameters to be estimated and can lead to increased precision by dividing a heterogeneous population into more homogeneous strata. Often, samples are drawn independently from each strata, which allows different sampling approaches to be taken for different strata. The spatial distribution of error within one stratum, such as riparian areas, may be quite different from that within another, such as grassland. A sample design that works well for one stratum may not be efficient for another. In practice, stratification may be used simply to allow the percentage of misclassified pixels to be estimated in each cover class. It may be more efficient to distinguish the strata after data have been collected, and account for covariance between strata. This problem, as well as the problem of sample units that straddle more than one stratum, were not addressed in this study. Rather, each of the designs discussed below is considered to be nested within independent strata.

Suppose there are H strata. Let N_h denote the total number of pixels in stratum h and let P_h be the proportion of misclassified pixels in stratum h. Then, $N = \sum_{h=1}^{H} N_h$ is the total number of pixels in the population, and if $W_h = N_h / N$, then $P = \sum_{h=1}^{H} W_h P_h$ is the proportion of pixels in the entire population that are misclassified. If p_h is an unbiased estimator of P_h, then

$$p = \sum_{h=1}^{H} W_h p_h \qquad (11.1)$$

is an unbiased estimator of P. Moreover, if $V(p_h)$ is the variance of p_h and p_l is an unbiased estimator of the proportion of misclassified pixels in stratum l, and $\text{Cov}(p_h, p_l)$ is the covariance between p_h and p_l, the variance of p is given by

$$V(p) = \sum_{h=1}^{H} W_h^2 \, V(p_h) + 2\sum_{h=1}^{H} \sum_{l>h} W_h W_l \text{Cov}(p_h, p_l) \qquad (11.2)$$

(Cochran, 1977). Note that the second sum in the right-hand side of the above expression is zero whenever the samples are drawn independently within the strata and generally is small compared with the first sum even if the samples are not drawn independently.

Within strata, sample data may be collected in a variety of ways. Simple random sampling, systematic sampling and cluster sampling are discussed at length in many texts, including Cochran (1977), Thompson (1991), and Kish (1967). To simplify the following discussion and simulations, only one stratum was considered. The subscript h was left on all equations to illustrate that results from individual strata can be combined using equations (11.1) and (11.2).

Relative Efficiency and Intracluster Correlation

Variables in natural landscapes often exhibit positive spatial autocorrelation. That is, data close together may be quite similar and cannot be considered independent. This dependence can extend in all directions and typically decreases with increasing distance. Spatial data pose unique problems because the usual statistical models that assume data independence no longer apply. Several texts address these problems, including Cressie (1991), Cliff and Ord (1981), Griffith (1988), and Ripley (1981, 1988).

Classification error in a remote sensing application can be thought of as a binary response. A pixel may be coded as a one if misclassified, zero if classified correctly. One measure of spatial autocorrelation for nominal variables is the join-count statistic developed by Moran (1948) and illustrated by Cliff and Ord (1981). This statistic was used by Congalton (1988a) to test for spatial autocorrelation in three difference images depicting classification error at the pixel level. In all three landscapes, spatial autocorrelation was found to be significant at all selected distances of one to 30 pixel lengths. This is not surprising given that classification error may be driven by landscape variables that exhibit strong spatial autocorrelation themselves. For example, certain vegetation types may be more difficult to classify than others resulting in error patterns similar to vegetation distributions. Error patterns similar to topographic features may arise when elevation, aspect and slope affect classification accuracy. Both gradual and abrupt transitions between vegetation types might also make classification difficult, resulting in boundary patterns of map error.

Spatial autocorrelation will affect the efficiency of systematic and cluster sample designs. The ratio of the variance of the proportion of misclassified pixels obtained under a simple random sample (srs) to the variance obtained from a sample of equivalent size under another design is known as 'relative efficiency'. Relative efficiency may be used to estimate how many more (or fewer) samples are required to ensure variance as small as that from simple random sampling. In systematic (sys) and cluster (clus) sample designs, relative efficiency is driven by intracluster correlation, ρ_h. (Systematic sampling can be thought of as a cluster sample with only one cluster.) If N_h is the total number of clusters, or primary units, in a stratum, and M_h is the number of pixels, or secondary units, within each cluster in that stratum, then relative efficiency can be expressed as

$$\frac{V_{srs}(p_h)}{V_{sys/clus}(p_h)} = \frac{1}{[1 + (M_h-1)\rho_h]} \tag{11.3}$$

where

$$\rho_h = \frac{\sum_{i=1}^{N_h} \sum_{j=1}^{M_k} \sum_{k>j} (p_{hij} - P_h)(p_{hik} - P_h)}{(M_h-1)(N_h M_h - 1)\sigma_h^2} \tag{11.4}$$

$$\sigma_h^2 = \frac{\sum_{i=1}^{N_h} \sum_{j=1}^{M_h} (p_{hij} - P_h)^2}{(N_h M_h - 1)}$$

In a remote sensing application, spatial autocorrelation is a function of distance between pixels. Intracluster correlation, however, is a function of cluster shape and size, as well as distance between pixels, and offers an easy way to relate spatial dependence to the relative efficiency of a systematic or cluster sample. When misclassified pixels form a clumped pattern in a landscape, spatial autocorrelation will be high between pixels that are very close. Spatial autocorrelation typically decreases with increasing distance between pixels, at a rate dictated by the nature of the spatial pattern. Similarly, intracluster correlation will be quite high for small clusters of connected pixels and will gradually decrease as cluster size increases, allowing greater distance between some of the pixels in the clusters. If classification errors exhibit periodicity that corresponds to the sampling interval in a systematic design, spatial auto-correlation at that distance will be high, as will be the intracluster correlation of the one cluster in the design. When elements within a cluster are dissimilar, ρ_h is small, the variance of the estimated proportion is low, and the relative efficiency shown in equation (11.3) is increased. Conversely, when ρ_h is large, variance under cluster or systematic sampling increases and the relative efficiency decreases.

It is often less expensive to collect data from clustered units than it is to collect data from the same number of units randomly distributed in a landscape. A cost function can be used to determine the number of clusters, n'_h, of M_h units that can be collected for the same cost as n_h units in a simple random sample. Multiplying equation (11.3) by the ratio $n'_h M_h / n_h$ gives a relative efficiency that weighs redundancy of information (where intracluster correlation is high) against larger quantities of data that become affordable through clustering. When $n'_h M_h$ is much larger than n_h for a given cost, the variance of a cluster sample may become smaller than that of a simple random sample, making the cluster design more efficient for a fixed cost.

Data

The US Fish and Wildlife Service's Gap Analysis Program is building a vegetation cover map of the Great Basin ecoregion using Landsat Thematic Mapper data. It would be ideal if we had populations of actual map error on which to run simulations to study the efficiency of various sample designs in estimating the proportion of misclassified pixels in this map. In their absence, however, we generated potential error patterns in several portions of the vegetation cover map of the Great Basin ecoregion. Three 7.5-min quadrangle maps (each measuring approximately 370 pixels × 470 pixels (11.1 km × 14.1 km)) were selected from the State of Utah. These included a mountain (MTN), canyon (CAN) and salt desert shrub (SDS) landscape. The MTN landscape, located in the Manti-La Sal National Forest of central Utah, is covered mostly by aspen, mountain shrub, spruce/fir, meadow and mixed conifer. The black brush, Pinyon pine/juniper, salt desert shrub and grass vegetation types cover the majority of the CAN landscape, located in Canyonlands National Park. A portion of the Colorado River, bounded by large cliffs and mesas, was also within this landscape. The SDS landscape is mostly covered with Pinyon pine/juniper and salt desert shrub vegetation types. It includes a large flat plain as well as rugged desert mountains. A different error model was applied to each of the three maps, ensuring that approximately 30 per cent of the pixels on each map were labelled as misclassified.

The first error population was constructed using data from the initial unsupervised classification of the Landsat Thematic Mapper data. Spectral distances between each pixel and the mean of the pixel's assigned class were available. Pixels with distances greater than the 70th quantile for that class represented greater probability of misclassification and were labelled as 1s (as opposed to 0s for correctly classified). The resulting map of 0s and 1s provided a potential error pattern on the canyon landscape and will be referred to as the CAN-1 population (Figure 11.1). '

The second error population was constructed by randomly labelling 50 per cent of the pixels that bordered two or more vegetative classes in the salt desert shrub landscape as misclassified, simulating difficulty in classifying areas near polygon boundaries (Figure 11.2, hereafter SDS-2).

The third error population was created to mimic the pattern of selected vegetation types. The Pinyon pine/juniper and salt desert shrub vegetation types composed approximately 70 per cent of the mountain landscape. All other vegetation types were labelled as misclassified, producing the clumped pattern in Figure 11.3 (hereafter MTN-3).

There are, of course, an unlimited number of spatial patterns of map error. The three potential error patterns generated here only serve to illustrate the relationships between intracluster correlation and design efficiency described in the preceding section.

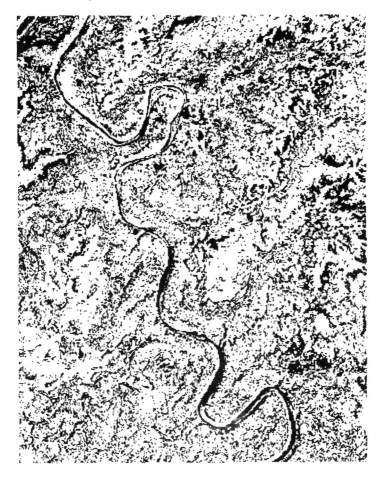

Figure 11.1 CAN-1 population.

Note: Pixels on a canyon landscape that had a spectral distance greater than 70 per cent of the cluster mean were labelled as misclassified and appear in black.

Simulations for Comparing Sampling Designs

Relative Efficiency of Systematic Sampling

For each of the three populations, CAN-1, SDS-2 and MTN-3, the efficiency of two systematic designs was evaluated relative to simple random sampling. The first systematic design, sys_1, was simply a square grid of points, with a random starting position. The second, sys_2, was an offset systematic design constructed by laying that same square grid over the landscape, but placing one sample point randomly within each square. The two designs are illustrated in Figures 11.4a and b. The size of the grid was allowed to increase from a spacing of five pixels

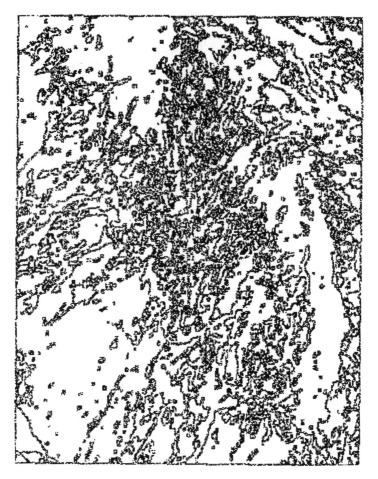

Figure 11.2 SDS-2 population.

Note: Fifty per cent of the pixels that bordered two or more vegetation classes on a salt desert shrub landscape were labelled as misclassified and appear in black.

to a spacing of 31 pixels in increments of two pixels. These grid sizes correspond to sample sizes ranging from about 4 per cent down to 0.09 per cent of the total population. Variance under simple random sampling, $V_{srs}(p_h)$, was calculated directly for each of the three populations using the formula:

$$V_{srs}(p_h) = \frac{P_h(1-P_h)}{n_h} \frac{(N_h-n_h)}{(N_h-1)}$$

where P_h is the true proportion of misclassified pixels in the population, N_h is the population size, and n_h the sample size. Variance under the first systematic design, $V_{sys1}(p_h)$, was calculated using the formula:

Figure 11.3 MTN-3 population.

Note: Vegetation classes comprising 30 per cent of the population on a mountain landscape were labelled as misclassified and appear in black.

$$V_{sys1}(p_h) = \frac{(1-f_h)}{n_h} \frac{\sum_{i=1}^{N_h} (p_{hi} - P_h)^2}{(N_h - 1)}$$

Here, N_h is the total number of systematic samples that can be taken from the population, p_{hi} is the proportion of misclassified pixels in any one systematic sample, and f_h equals n_h/N_h. Variance under the second systematic design, $V_{sys2}(p_h)$, was estimated for each sample size by averaging the variance over 1000 sample runs. That is,

$$V_{sys2}(p_h) = \frac{\sum_{i=1}^{1000} (p_{hi} - P_h)^2}{1000}$$

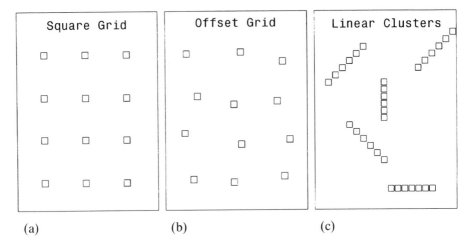

Figure 11.4 Examples of (a) square grid, (b) offset grid and (c) linear clusters.

Relative efficiency, expressed as the ratios of the variance under simple random sampling to the variances under both systematic designs, was plotted against grid spacing for each population and sample size.

Although an unbiased estimate of the variance of a systematic sample does not exist, several alternative methods are discussed by Wolter (1985). Often, in practice, the variance of a systematic sample is frequently estimated by assuming that the variable of interest is randomly distributed in space and applying the formula:

$$v_{srs}(p_h) = p_h(1 - p_h) \frac{(N_h - n_h)}{(n_h - 1)N_h}$$

Cochran (1977), Stehman (1992) and many others point out that when intracluster correlation for a systematic design is less than intracluster correlation for a simple random design, systematic sampling is more precise than simple random sampling. Using the estimate of variance under simple random sampling, $v_{srs}(p_h)$ will overestimate the variance in such a case. Conversely, when intracluster correlation for a systematic design is more than intracluster correlation for a simple random sample, systematic sampling is less precise than simple random sampling for a given landscape. Here, $v_{sys}(p_h)$ underestimates the variance, a more serious mistake. The consequences of estimating $V_{sys}(p_h)$ using $v_{srs}(p_h)$ were examined by averaging over 1000 sample runs for the CAN-1, SDS-2 and MTN-3 populations. That is,

$$V_{sys}(p_h) = \frac{\sum_{i=1}^{1000} V_{srs}(p_{hi})}{1000}$$

Relative Efficiency of Cluster Sampling

Cluster sample designs were constructed with linear clusters, 1 pixel wide, having north–south, north-east–south-west, east–west, north-west–south-east and random combinations of these four orientations (Figure 11.4c). Cluster lengths ranged from 3 pixels to 31 pixels. While any number of cluster shapes could be examined, linear clusters were chosen because they minimize the effects of spatial autocorrelation by spreading observations over further distances. For each population and cluster orientation, the intracluster correlation coefficient, ρ_h, was calculated using equation (11.4) and plotted against cluster size.

For a simple random sample, total cost was set equal to the product of the number of sample units (n_h) and cost of travelling to, and collecting data on, each of those units. That is,

$$\text{Cost} = n_h c_1 + n_h c_3 \tag{11.5}$$

where c_1 is the cost of travelling between sample units and c_3 is the cost of collecting data on the sample unit. For a cluster sample of n'_h clusters, or primary units, each containing M_h pixels within a cluster, or secondary units, total cost can be expressed as

$$\text{Cost} = n'_h c_1 + n'_h (M_h - 1) c_2 + n'_h (M_h - 1) b c_2 + n'_h M_h c_3 \tag{11.6}$$

where c_2 is the cost of locating and travelling between pixels within a cluster, and b is the proportional cost of c_2 to travel back across the pixel.

As an example, suppose we are interested in estimating the proportion of misclassified pixels in an 11×14-km quadrangle map. It might be reasonable to assume that it takes an average of 60 min to travel between clusters (c_1), an average of 5 min to locate and travel between pixels within a cluster (c_2), an average of 2.5 min to travel back across the pixel $(b = 0.5)$, and an average of 5 min to collect data at the pixel (c_3). Here, the ratio of time needed to travel within clusters to the time needed to travel between clusters is 1:12. However, a map the size of Utah may involve a much higher c_1 because of the increased distance between clusters.

Equating equations (11.5) and (11.6) and solving for the ratio of the number of units that can be sampled under a clustered design to the number under simple random sampling yields:

$$\frac{n'_h M_h}{n_h} = \frac{M_h (1 + c_3 / c_1)}{(1 + (1 + b)(M_h - 1) c_2 / c_1 + M_h c_3 / c_1)} \tag{11.7}$$

Multiplying equation (11.7) by equation (11.3) gives the relative efficiency with a cost consideration. For all simulations, c_3 was set equal to c_2 since it was believed that the time needed to determine the true class of a pixel would be about the same as the time needed to walk to it from a neighbouring pixel. For cost ratios of c_2 to c_1 ranging from 1:1 to 1:25, the relative efficiency of simple random sampling to linear cluster sampling for random cluster orientations was plotted against cluster size on each of the three landscapes. To see how well these relative efficiency patterns were mirrored at a small sample size, an estimate of the variance under cluster sampling, $v_{clus}(p_h)$ was computed using the formula:

$$v_{clus}(p_h) = \frac{(1-f_h)}{n_h} \frac{\sum_{i=1}^{n_h} (p_{hi} - P_h)^2}{(n_h - 1)}$$

This variance estimate was averaged over 1000 runs for each population and linear cluster size.

To investigate the sensitivity of the relative efficiency of the cluster design to imprecise estimates of ρ_h, a subset of the relative efficiency curves described in the preceding paragraph was bounded by curves produced by underestimating and overestimating ρ_h by 10–50 per cent in increments of 10 per cent for each of the linear cluster lengths with random orientation.

Results and Discussion

In Figures 11.5a–c, the relative efficiency of the systematic and offset systematic designs is plotted against grid spacing for each of the three populations. Where relative efficiency exceeds one, systematic sampling has a lower variance and is more efficient than simple random sampling. Conversely, when relative efficiency is less than one, systematic sampling is less efficient. For all three error populations, systematic sampling could be either more or less efficient than simple random sampling, depending on grid spacing. Because of this erratic pattern, it is difficult to make a general statement about the performance of the two designs on the three populations. Stehman (1992) found similar results after running simulations on eight potential error populations. Based on equation (11.3), relative efficiency of the systematic design mirrors the pattern of intracluster correlation for each particular grid spacing. The offset systematic design was less erratic than the systematic design and, in most cases, appeared more efficient than simple random sampling. However, these results cannot necessarily be generalized. As expected, the common practice of estimating variance under a systematic design using the simple random sampling formula overestimated the variance where systematic sampling was more efficient than simple random sampling, and underestimated it where systematic sampling was less efficient.

In Figures 11.6a–c, the intracluster correlation for various linear cluster lengths exhibited slight differences between the four cluster directions. The vegetation error model applied to the mountain landscape, MTN-3, had the highest intracluster correlation caused by larger blocks of misclassified pixels.

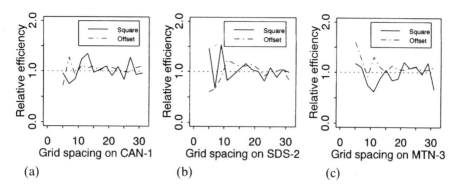

Figure 11.5 Relative efficiency of two systematic designs applied to the (a) CAN-1, (b) SDS-2 and (c) MTN-3 landscapes.

Note: The dotted line represents a relative efficiency of 1, where the variance under systematic sampling is equal to the variance under simple random sampling.

The error model based on the spectral distance between each pixel and the mean of the pixel's assigned class applied to the canyon landscape had the lowest intracluster correlation over all cluster directions. The relative efficiency of cluster sampling (at random directions) over increasing cost ratios (c_2:c_1), as shown in Figures 11.7a–c, raises some interesting points. For all three populations, when c_1 exceeded 15, cluster sampling was more efficient than simple random sampling for all cluster lengths of 30 or less. Maximums on some curves for very high c_1s were reached beyond the 30-pixel length. The shapes of the cost curves were driven by intracluster correlation and parameters in the cost function. Since the CAN-1 error pattern had the lowest intracluster correlation curves of the three populations, cluster sampling was more efficient than simple random sampling for relatively low c_1. The MTN-3 error pattern had the highest intracluster correlation curves and consequently required higher c_1 values in order for the relative efficiency of cluster sampling to exceed one. Sampling with a cluster length of approximately 23 and a cost ratio of 1:10 on the SDS-2 landscape is an example of where cluster and simple random sampling had approximately equal variances. The same patterns of relative efficiency were obtained by estimating relative efficiency in simulations using small sample sizes on each of the three populations.

Because intracluster correlation can only be approximated before sample information is collected, we considered how sensitive relative efficiency curves were to misspecification in ρ_h. In Figures 11.8a–c, the relative efficiency of cluster sampling over simple random sampling for cost ratios 1:2, 1:10 and 1:20 are shown bounded by a ± 20 per cent range in ρ_h. Cluster sampling given higher c_1 values remained more efficient, but lower c_1 values paired with intracluster correlations that were either quite high or only vaguely known allowed no guarantee that cluster sampling would be more efficient than simple random sampling. Prior information about the expected intracluster correlation can be

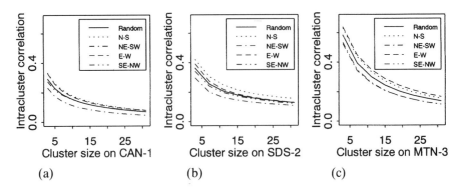

Figure 11.6 Intracluster correlation under different orientations of linear cluster designs applied to the (a) CAN-1, (b) SDS-2 and (c) MTN-3 landscapes.

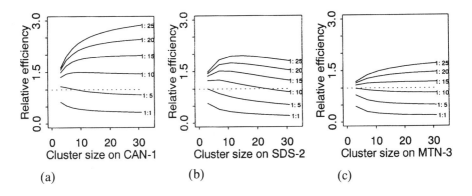

Figure 11.7 For cost ratios ranging from 1:1 to 1:25 the relative efficiency for a fixed cost of linear cluster designs applied to the (a) CAN-1, (b) SDS-2 and (c) MTN-3 landscapes.

Note: The dotted line represents a relative efficiency of 1, where the variance under a linear cluster design is equal to the variance under simple random sampling.

obtained by a limited preliminary sample, collected either on the ground or from aerial photographs.

Assuming that sampling units are clearly defined and easily identifiable, cluster sampling offers the opportunity for smaller variance at lower cost. This was true for the populations examined in this study under moderately high c_1 values and moderate uncertainty about intracluster correlation. Further benefit may be realized by subsampling clusters (Maxim and Harrington, 1984), or spacing pixels within clusters based on some knowledge of spatial autocorrelation at various distances. Certainly, the relative efficiency curves were sensitive to changes in intracluster correlation and cost function parameters. The point of

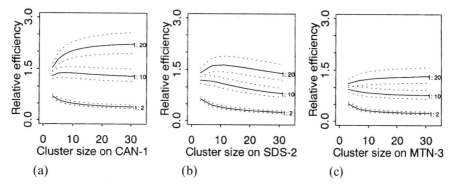

Figure 11.8 For cost ratios 1:2, 1:10 and 1:20, the relative efficiency for a fixed cost (solid lines) bounded by plus or minus 20 per cent uncertainty in intracluster correlation (dotted lines). Linear cluster designs were applied to the (a) CAN-1, (b) SDS-2 and (c) MTN-3 landscapes.

these illustrations is that, given some uncertainty in those parameters, sensible choices in sample designs may still be made.

The objectives of an accuracy assessment, however, may go beyond simply estimating the proportion of misclassified pixels within a vegetation class. As mentioned previously, numerous measures may be derived from the error matrix, including the total percentage of units classified correctly, percentage commission error by class, percentage omission error by class, and the Kappa statistic. The error matrix may be used to calibrate remotely sensed aerial estimates for misclassification bias (Czaplewski, 1992) or in modelling map error through generalized linear models (Chrisman, 1989). Analyses beyond the simple proportion of misclassified pixels, however, are often based on the sample design used to collect the data. Inference is based on specific sampling models. Clearly, data collected under clustered or systematic designs may be spatially autocorrelated, violating the assumption of independence in many standard remote sensing tests and analyses. However, recent work in generalized linear models for dependent data (McCullagh and Nelder, 1989) may provide the tools for analyses of error data collected under more cost-effective sampling schemes.

A more difficult problem arises when the sampling unit itself is in question. Individual pixels, or groups of pixels, may not be of practical interest or may not be easily identified in the field. Questions about the accuracy of polygons and their boundaries may become more pressing. As stated in the introduction, more complex models of map error are being developed. It seems unlikely that traditional finite population sampling theory will provide tools for gathering the necessary information in a cost-effective way. Sampling in a continuous domain (Stevens, 1993) with geometric point sampling (Grosenbaugh, 1958) seems a more flexible way to collect information on items as diverse as points, lines and polygons.

Conclusion

With the increasing demand for broad-scale maps comes the need for adequate measures of map accuracy. Unfortunately, the cost of assessing the accuracy of maps at the ecoregion scale may be prohibitive. To maximize their efficiency, sample designs should be selected based on a cost function and prior information about the spatial pattern of map error. Regardless of the scale of the sampling unit, efficiency may be gained through systematic sampling but is dependent on the intracluster correlation at a particular grid spacing. In addition, any efficiency gained using systematic sampling can be quite small compared with that gained by using cluster sampling with a fixed cost. The relative efficiency of a cluster design is driven by the intracluster correlation and a cost function for collecting data. Even limited knowledge about the spatial pattern of map error may make cluster sampling much more efficient than simple random sampling or systematic sampling for cost functions with high ratios of primary unit cost to secondary unit cost. While cluster sampling may be more cost effective for estimating the proportion of misclassified units in a map, techniques should be explored for estimating other measures of map accuracy using dependent data.

Acknowledgement

We thank all those who reviewed this manuscript for their useful comments. Ron Tymcio and Jim Arnott aided with illustrations. Collin Homer provided much appreciated help in guiding us through the intricacies of GIS. This work was supported by grant no. MON-OAS-92-504 from the US Fish and Wildlife Service's Gap Analysis Program.

References

Beard, K. M., Clapham, S. and Buttenfield, B. P., 1991, *Visualization of Spatial Data Quality*, Technical Paper 91-26, Santa Barbara, California: National Center for Geographic Information and Analysis.

Berry, B. J. L. and Baker, A. M., 1968, Geographic sampling, in Berry, B. J. L. and Marble, D. F. (Eds) *Spatial Analysis: A Reader in Statistical Geography*, pp. 91-100, Englewood Cliffs, New Jersey: Prentice-Hall.

Bolstad, P. V. and Smith, J. L., 1992, Errors in GIS: assessing spatial data accuracy, *Journal of Forestry*, **90**(11), 21-9.

Chrisman, N. R., 1989, Modeling error in overlaid categorical maps, in Goodchild, M. and Gopal, S. (Eds) *Accuracy of Spatial Databases*, pp. 21-34, Philadelphia: Taylor & Francis.

Cliff, A. D. and Ord, J. K., 1981, *Spatial Processes: Models and Applications*, London: Pion.

Cochran, W. G., 1977, *Sampling Techniques*, New York: Wiley.

Congalton, R. G., 1988a, Using spatial autocorrelation analysis to explore the errors in maps generated from remotely sensed data, *Photogrammetric Engineering and Remote Sensing*, **54**(5), 587–92.

Congalton, R. G., 1988b, A comparison of sampling schemes used in generating error matrices for assessing the accuracy of maps generated from remotely sensed data, *Photogrammetric Engineering and Remote Sensing*, **54**(5), 593–600.

Congalton, R. G., 1991, A review of assessing the accuracy of classifications of remotely sensed data, *Remote Sensing and the Environment*, **37**(1), 35–46.

Cressie, N. A. C., 1991, *Statistics for Spatial Data*, New York: Wiley.

Czaplewski, R. L., 1992, Misclassification bias in areal estimates, *Photogrammetric Engineering and Remote Sensing*, **58**(2), 189–92.

Ginevan, M. E., 1979, Testing land-use maps accuracy: Another look, *Photogrammetric Engineering and Remote Sensing*, **45**(10), 1371–7.

Goodchild, M. F., 1989, Modeling error in objects and fields, in Goodchild, M. and Gopal, S. (Eds) *Accuracy of Spatial Databases*, pp. 107–13, Philadelphia: Taylor & Francis.

Green, E. J., Strawderman, W. E. and Airola, T. M., 1993, Assessing classification probabilities for thematic maps, *Photogrammetric Engineering and Remote Sensing*, **59**(5), 635–9.

Griffith, D. A., 1988, *Advanced Spatial Statistics*, Boston: Kluwer.

Grosenbaugh, L. R., 1958, Point-sampling and line-sampling: probability theory, geometric implications, synthesis, Occasional Paper 160, US Department of Agriculture, Forest Service, Southern Forest Experiment Station.

Hay, A. M., 1979, Sampling designs to test land-use map accuracy, *Photogrammetric Engineering and Remote Sensing*, **45**(4), 529–33.

Hord, R. M. and Brooner, W., 1976, Land-use map accuracy criteria, *Photogrammetric Engineering and Remote Sensing*, **42**(5), 671–7.

Kish, L., 1967, *Survey Sampling*, New York: Wiley.

Lunetta, R. S., Congalton, R. G., Fenstermaker, L. K., Jensen, J. R., McGwire, K. C. and Tinney, L. R., 1991, Remote sensing and geographic information system data integration: error sources and research ideas, *Photogrammetric Engineering and Remote Sensing*, **57**(6), 677–87.

Maxim, L. D. and Harrington, L., 1984, On optimal two-stage cluster sampling for aerial surveys with detection errors, *Photogrammetric Engineering and Remote Sensing*, **50**(11), 1613–27.

McCullagh, P. and Nelder, J. A., 1989, *Generalized Linear Models*, 2nd Edn, New York: Chapman and Hall.

Monserud, R. A. and Leemans, R., 1992, Comparing global vegetation maps with the Kappa statistic, *Ecological Modelling*, **62**, 275–93.

Moran, P. A. P., 1948, The interpretation of statistical maps, *Journal of the Royal Statistical Society, Series B*, **10**, 243–51.

Openshaw, S., 1989, Learning to live with errors in spatial databases, in Goodchild, M. and Gopal, S. (Eds) *Accuracy of Spatial Databases*, pp. 21–34, Philadelphia: Taylor & Francis.

Richards, J. A., 1986, *Remote Sensing Digital Image Analysis: An Introduction*, New York: Springer.

Ripley, B. D., 1981, *Spatial Statistics*, New York: Wiley.

Ripley, B. D., 1988, *Statistical Inference for Spatial Processes*, New York: Cambridge University Press.

Rosenfield, G. H., 1982, Sample design for estimating change in land use and land cover, *Photogrammetric Engineering and Remote Sensing*, **48**(5), 793–801.

Rosenfield, G. H., 1986, Analysis of thematic map classification error matrices, *Photogrammetric Engineering and Remote Sensing*, **52**(5), 681–6.

Scott, J. M., Davis, F., Csuti, B., Noss, R., Butterfield, B., Caicco, S., Groves, C., Edwards, T. C., Jr, Ulliman, J., Anderson, H., D'Erchia, F. and Wright, R. G., 1993, Gap analysis: a geographic approach to protection of biological diversity, *Wildlife Monographs*, No. 123.

Stehman, S. V., 1992, Comparison of systematic and random sampling for estimating the accuracy of maps generated from remotely sensed data, *Photogrammetric Engineering and Remote Sensing*, **58**(9), 1343–50.

Stevens, D. L., 1993. Plot Configurations for sampling in continuous domains. 1993 Proceedings of the Statistics and the Environment Section, American Statistical Association.

Story, M. and Congalton, R. G., 1986, Accuracy assessment: a user's perspective, *Photogrammetric Engineering and Remote Sensing*, **52**(3), 397–9.

Thompson, S. K., 1992, *Sampling*, New York: Wiley.

Veregin, H., 1989a, *Accuracy of Spatial Databases: Annotated Bibliography*, Technical Paper 89–91, Santa Barbara, California: National Center for Geographic Information and Analysis.

Veregin, H., 1989b, *A Taxonomy of Error in Spatial Databases*, Technical Paper 89–12, Santa Barbara, California: National Center for Geographic Information and Analysis.

Wolter, K. M., 1985, *Introduction to Variance Estimation*, New York: Springer.

12

Metadata required to determine the fitness of spatial data for use in environmental analysis

Nicholas R. Chrisman

Much of the information managed for environmental analysis has a spatial component, which is critical in integrating one source with another. Recent standards efforts around the world (particularly the US Spatial Data Transfer Standard (SDTS)) specify the kind of information (data about the data or metadata) required to evaluate the data quality of some spatial data product. The SDTS requires that a data producer provide a Data Quality Report, which includes the lineage, positional accuracy, attribute accuracy, logical consistency and completeness of the data. Many of these elements of the report can be transmitted in the form of spatial distributions. A potential user should be able to peruse the Data Quality Report and render an informed judgement about the fitness of the data for the projected use. While SDTS sets important directions, it is not the only standard available. Environmental scientists must choose from a range of conflicting standards, each with their particular strengths. The Vector Product Format (designed for the public domain Digital Chart of the World), when compared with SDTS, shows particular improvements in implementing the original goals of including information to evaluate fitness for use. One key element is the treatment of metadata as spatial data in their own right.

Introduction

An initial perception of environmental *data* management is that it should focus its attention primarily on *the data*. However, once the masses of primary data are collected, it becomes quickly apparent that much management should operate at a more aggregate level, dealing with information that describes collections of data — often termed metadata (or data about data). Metadata, particularly those elements describing data quality, form the basis for making informed decisions regarding the fitness of a particular data source for a specified use.

The need for metadata might be best understood through a story recounted by Philip Gersmehl (1985). Gersmehl had compiled soils information for the continental USA in 1977. One outcome of this process was a map labelled

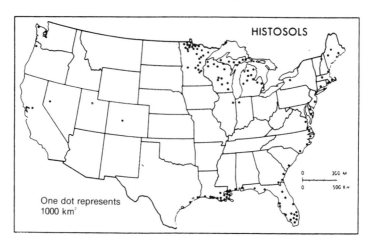

Figure 12.1 Histosols, soils with a surface layer of accumulated organic matter that is usually acid and wet.

Note: Reproduced from Gersmehl, 1985, p. 332, *Professional Geographer*, with permission of the Association of American Geographers.

'Histosols of the United States' (Figure 12.1). This map used the technique of dotmaps to show clusters of the distribution without specific locations. This technique is more often used for economic activity with legends such as '1 dot = 2000 cows'; clearly not all the 2000 cows are right under the dot. For this map, Gersmehl chose a dot size of '1 dot = 1000 km^2'. Most of the dots for histosols represented peat bogs in the glaciated regions of the north-eastern USA. In the Rocky Mountain region, three point symbols appeared (see Figure 12.1). One of these, representing true peats in the alpine portions of the region (from Montana to New Mexico), was located in Colorado. The other two dots represented saline mucks (saprists) in the dried lake beds of the Basin and Range Province. Through a set of steps involving other groups (and without Gersmehl's participation), these dots became the basis for a map titled 'Peat Energy Resources of the United States'. In this map (produced in a 1979 Department of Energy study of energy resources), the three point symbols were reinterpreted as potential peat energy zones (see Figure 12.2), though Gersmehl had not intended the point symbols to be taken literally as areas nor the histosol category to be taken as 'peat' in all circumstances. Gersmehl (1985, p. 331) reports a rumour that a federal 'Solar Energy Research Institute' was located in Colorado, in part, because it was thought, on the authority of this map, that peat deposits could be studied.

It is important to note that Professor Gersmehl had good information sources for most histosols; there were peat bogs in glaciated regions of many states incuding his home state of Minnesota. But the three dots in the Rockies were not based on the same detailed information; there was a good deal of guesswork in deciding the extent of peat in the alpine areas of the Rockies and the extent of

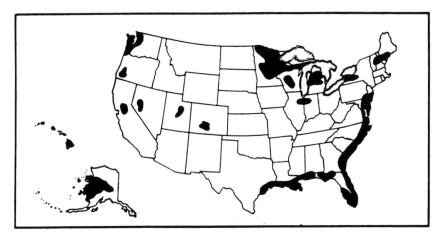

Figure 12.2 Peatlands of the United States, from a Department of Energy publication.

Note: Reproduced from Gersmehl, 1985, p. 333, *Professional Geographer*, with permission of the Association of American Geographers.

saprists. Professor Gersmehl followed the tradition of cartography in blending sources of radically different nature into a map. This tradition will be hard to break, even when the technology is now available to do so.

Gersmehl concluded his paper with three suggestions for responsibilities of cartographers:

> A person who puts information on a map has a duty to be fair to the data, to be clear to the map reader, and to try to anticipate the ways in which a third person may be affected by a foreseeable misinterpretation of the map. At the very least that third duty should include a resolute refusal to display or even imply any more accuracy and precision than we can justify. (Gersmehl, 1985, p. 334)

The Gersmehl 'parable' sets the scene for this dicussion. Modern information management has multiplied the opportunity for information to move from one user to another, eventually escaping the bounds of intended use. Professor Gersmehl is certainly not the first environmental scientist whose studies have been misinterpreted. With the increasing use of computer databases, the risks have been increased, and the nature of the problem has shifted. An understanding of error in geographic information has been recognized as a weak point in geographic information systems (GIS) for some time (Chrisman, 1984a; Burrough, 1986; Goodchild and Gopal, 1989).

At one time, environmental sciences retained a strong connection to fieldwork. Studies were performed by relatively small groups working in close connection to the landscape. The current world of environmental science is more complex as collections of individual studies are often mobilized to assemble a view of an ecosystem or even the whole world. In this process, the results of many separate studies become integrated, often in spatial context using some form of geographic analysis encapsulated by GIS software. The project that generates a particular piece of information is quite unlikely to conceive of all possible applications.

The conversion from the era of individual efforts to a period of collaboration increases the need for communication. Unfortunately, the emphasis in the development of 'global databases' has been placed on amassing the data, often to the exclusion of any other component. This chapter will develop a more complex role of information management in the process of integrating specific field studies into an ecosystem or global view. The key element is metadata, the kind of information that Gersmehl left off his original dotmap.

Gersmehl's call for responsibility sounds crucial, but it is based on assumptions built into cartographic media. A map maker does bear much of the responsibility, if the user is considered a passive 'bystander'. However, the nature of database use, in its modern form, alters the simple assignment of responsibility. Modern database management offers a more flexible range of tools unrestricted by the old pen-and-ink conventions of cartographic display (see Goodchild, 1988, for a treatment of these changes). The illusion of comprehensive, consistent coverage, while perhaps necessary for map publication, should be replaced by more openness about limitations in the data. The database user must bear as much responsibility as the originator because the user invokes many of the processing decisions and imparts much of the meaning. In an era of digital data, the original producer cannot foresee all potential users. Thus, it becomes crucial for a data producer to record important aspects about the data so that users may make informed judgements regarding fitness for use.

This chapter will present an overview of the requirements for metadata in environmental databases. First, a vision of the future will be presented, followed by a path to implement that vision. Some of the first steps have been taken by creating a variety of standards. Most of this discussion will focus on the component of environmental databases with a strong spatial character, and much of the literature will concern cartographic and GIS standards.

A Vision of Possible Environmental Databases

Some of the grand challenges to science involve understanding spatial distributions as a part of an earth system (Brown, this volume). Traditional disciplinary boundaries and comfortable scales of analysis must be replaced with new forms of scientific investigation. One specific example is the US Global Change Research Program, a congressionally established collaboration of federal agencies and research institutions. The recent *Draft Implementation Plan for the US Global Change Data and Information System* (GCDIS) (CEES, 1993) provides a full-fledged proposal for this effort. According to this proposal, over the next five to seven years, environmental scientists should be able to access a service (over Internet) that can serve as a clearinghouse/'card catalogue' for environmental information related to global change. A system such as GCDIS will organize data sources from many different disciplines and organizations. It encompasses a hierarchy of access from the simplest 'directory', through 'guide' and 'inventory' to 'browse' and eventually delivery of the full product. The user of GCDIS should be able to determine whether (or not) some particular data source could contribute to the investigation contemplated. Some key questions relate to the specific geographic region covered, the relevance of the measurements and concerns about the data quality. Prior to these recent developments, it was all too common to compile various databases from a collection of sources, losing the information about diverse origins and variable data quality.

GCDIS is just one proposal for spatial data directory services. The Federal Geographic Data Committee, with assistance from the National Research Council Committee on Mapping, is developing a plan for a National Spatial Data Infrastructure; many states and local jurisdictions in the USA have less technologically advanced inventories of data compiled. In Europe, there are a number of similar national and community-wide co-ordination efforts, but none with as ambitious an implementation plan as GCDIS.

The key question for the user of any data directory will be whether to acquire a particular database. In some cases it may be enough for data to be declared 'official' or 'certified' by the producer. In some fields of endeavour, the procedures are so clearly established that a grading scheme like 'Class II geodetic surveys' is sufficient because it is understood by all users. The recent wave of Total Quality Management and the ISO Standard 9000 efforts in the European Community focus on standardizing procedures (Slagle, this volume). Again, the producer assumes the responsibility.

But all the standardized procedures in the world cannot ensure that the product actually satisfies the user's needs. The question of whether to acquire a database revolves around a judgement that can best be termed 'fitness for use'. Does this information source serve the intended purposes sufficiently to be worth the effort to obtain and prepare for analysis? Some of the basic questions may turn out to be the same questions asked by ISO 9000 — questions of processes and procedures. Some of the information may fit into simple diagnostic numerical indices like the positional accuracy threshold of the National Map Accuracy Standard (NMAS). However, the key medium of a metadatabase for global environmental data is not a tabular database or a report in text. The individual data elements can be best portrayed on maps and analysed using a GIS.

Some institutions have begun to design and implement metadata systems (including NASA, Thieman, 1992; US Geological Survey, Holm and Scholz, 1992; the State of Florida, Stage, 1992), but much of this effort has been limited by various practical concerns. The GCDIS implementation plan calls for development of NASA's prototype as the basis for a decentralized system residing in a network of co-operating organizations. Before any such effort can encompass the diverse range of current information systems, there must be some agreement on common terms and standards.

Standards Efforts

There is no shortage of standards for the treatment of spatial data; the problem at the moment is choosing among competing incompatible standards. Standards for maps had been rather stable in the predigital era. The US National Map Accuracy Standard (Bureau of the Budget, 1947) set out a positional tolerance that 90 per cent of 'well-defined' points should be within $\frac{1}{50}''$ [0.5 mm] of their correct position. Curiously, maps that comply can proclaim 'This map complies with NMAS', while maps that do not comply say nothing. This standard seemed to be adequate for 40 years; a replacement has recently been developed (ASPRS, 1987). The old NMAS establishes a fixed threshold, a number that defines fitness for all possible uses. This form of standard, termed 'conformance to expectation', may work in periods of stable technology and centralized production, but the fitness for use doctrine is much more relevant in this era where users' needs are not known to the producers (Hayes and Romig, 1977).

In the realm of standards, most recent effort has focused on transfer of data. The major standards (such as SDTS (NIST, 1992), Digital Geographic Exchange Standard (DIGEST) (DGIWG, 1991), Vector Product Format (VPF) (DMA, 1992)) promulgated recently in the USA have been organized around a transfer format (or more properly a family of formats in each case). There are similar spatial data standards efforts in other countries (e.g., NTF in UK), but often, as in France and Australia, they rely upon a US standard as the starting point.

Each one of these distinct efforts includes some references to metadata of two kinds. As transfer formats, each provides a data dictionary and a database schema; information of a structural nature is often lumped with metadata. Some definitions of objects and relationships in the data dictionary relate directly to fitness for use, but much of the information involves technical details. The other portion concerns data quality, the component much more clearly linked to the judgement on fitness for use. The SDTS effort — in part because it endured for over a decade — set a basic framework for data quality that has been incorporated into the other US and international efforts. This chapter will focus on the data quality aspects, since these are a more central issue in judgements of fitness for use. Although critical to technical implementation, the data dictionary aspect of metadata is less of an impediment to data transfer because of the flexibility of modern software.

Review of SDTS Data Quality Specification

Although the basic thrust of SDTS focused on data exchange, a standard for data quality emerged from one working group of the National Committee for Digital Cartographic Data Standards (Chrisman, 1984b; NIST, 1992). The contents of that standard will serve as the organization of the next portion of this paper. The (nearly complete) contents of the data quality standard incorporated into SDTS will be inserted into this text with commentary interspersed.

> Digital cartographic data shall include a quality report. Where the spatial variation in quality is known, a quality report must record that variation.

This portion of the standard requires a producer to create a quality report with five mandatory sections. The statement about variations in data quality is particularly important. Databases compiled from multiple sources must record the spatial extent of the variability. SDTS does not actually provide tools to implement this requirement, but the principle is established.

> For those components of quality displaying spatial variation, a quality overlay system may be used. The producer of the quality report may choose to produce a comprehensive quality overlay describing all components of quality, or various components may be portrayed on separate overlays. When the quality report is issued on paper, the quality overlays appear as diagrams with text labels or thematic map depictions. In digital form, the overlays are encoded using the standards of the [SDTS format specification].

These quality overlays are meant to be encoded using the transfer standard, but the mechanism to link a specific data quality overlay to some other coverage is not particularly evident in the SDTS structure. One recent standard, the Vector Product Format (DMA, 1992) provides an explicit mechanism to associate a data quality coverage with any element in its storage hierarchy (database, library, coverage, tile).

> A quality report consists of five sections covering lineage, positional accuracy, attribute accuracy, logical consistency and completeness. Each section of the report will contain reference to temporal information.

During the development of the standard, the five components proved adequate in covering the important aspects. However, a specific section dealing with time was not included. In particular, there seemed to be no method to test for 'temporal accuracy' in maps without testing one of the other elements. For databases incorporating continuous traces from seismometers, hydrographs or similar instruments, there may need to be a distinct category for temporal accuracy.

Lineage
The lineage section of a quality report shall include a description of the source material from which the data were derived, and the methods of derivation, including all transformations involved in producing the final digital files. The description shall include the dates of the source material and the dates of ancillary information used for update.

Lineage is a key element. Most GIS have been deficient in recording the actual substance of the transformations performed (Lanter, this volume). Some raster packages (such as MAP II) record the origin of each map as a text comment field, but the information cannot be seen in its entirety, and the chain can be broken if just one map is deleted. Lanter (1993) has developed a lineage maintenance system for one of the full-service GIS packages, but it operates largely at the procedural level and does not record all details. To be complete, the spatial extent of alterations and the parameters of transformations must be recorded.

> The date assigned to a source shall reflect the date that the information corresponds to the ground, however, if this date is not known, then a date of publication can be used, if declared as such.

Lineage records 'database time' — the time when the operations occur. Yet the user also needs a sense of 'world time' — when the information relates to the original collection, not the publication date. Cartographic sources have been particularly remiss in displaying the publication date more prominently than the dates of the source material.

> Any database created by merging information obtained from distinct sources must be described at sufficient detail to identify the actual source for each element in the file. In these cases, either a lineage code on each element or a reliability overlay (source data index, etc.) will be required.

This paragraph provides an alternative to the quality overlay (source data index). Each object in the database can have its own lineage code. This option is the basis for the recent Automated Nautical Chart II product (Langran, 1993) and other efforts. The AVHRR data also include some form of lineage code in the ancillary data that report the angle of view, since AVHRR images are composites of pixels acquired over a number of days.

> The lineage section shall describe the mathematical transformations of coordinates used in each step from the source material to the final product....

The integrity of co-ordinate transformations is critical since GIS operates by overlaying features that occupy the same position. There are many pitfalls in various ellipsoids, datums and projection equations. The standard goes into further detail about projections and other transformations, a topic of considerable interest in the creation of global databases, since large distortions can occur through the use of map projections. The details of this specification are not critical here.

> The preferred test for positional accuracy is a comparison to an independent source of higher accuracy. The test must be conducted using the rules prescribed in the Spatial Accuracy Specifications for Large-Scale Topographic Maps (American Society of Photogrammetry, 1987). When the dates of testing and source material differ, the report shall describe the procedures used to ensure that the results relate to positional error, not to temporal

effects. The numerical results in ground units, as well as the number and location of the test points must be reported. A statement of compliance to a particular threshold is not adequate in itself. This test may only be applicable to well-defined points.

Positional accuracy is a critical part of data quality. The standard considers some less rigorous methods of determining accuracy (text not included in this chapter). Fundamentally, positional accuracy is measured as the distance between the location recorded and the 'true' position. A source of higher accuracy is operationally defined as one with a standard error one-third as large as the data tested. Unfortunately, this method only works, as the standard tersely admits, for 'well-defined' points. These standards are a relic of the old National Map Accuracy Standard (Bureau of the Budget, 1947) which restricted map representation to only those objects that could be 'plotted to .01 inch' [0.25 mm]. Thus, benchmarks, radio towers and road intersections may meet this criterion, but most features on vegetation maps and contour coverages would be excluded.

SDTS does not provide a test procedure for 'ill-defined' features (or fuzzy features). Fairly few environmental data consist of well-defined points, so testing procedures must be developed (Vonderohe and Chrisman, 1985). Some concerns can be remedied using the attribute accuracy tests below.

...Accuracy tests for categorical attributes can be performed by one of [3] methods. All methods shall make reference to map scale in interpreting classifications.

Deductive Estimate
Any estimate, even a guess based on experience, is permitted. The basis for the deduction must be explained. Statements such as 'good' or 'poor' should be explained in as quantitative a manner as possible.

A deductive estimate may be all that is available. It is important to distinguish the different methods used to develop a statement concerning accuracy. Still, any statement from the producer may help to inform the user.

Tests Based on Independent Point Samples
A misclassification matrix must be reported as counts of sample points cross-tabulated by the categories of the sample and of the tested material. The sampling procedure and the location of sample points must be described.

The *de facto* 'standard' in remote sensing is to test a classification essentially by throwing darts at it (Fitzpatrick-Lins, 1978; Congalton and Mead, 1983). The sampled points are cross-tabulated by the category reported and the ground-truth data. The sample can be organized on a random or stratified basis, though many accuracy studies reject points that fall near boundaries (often exactly the places with the classification error). In general, these tests assist in understanding total amounts of error but provide little assistance in determining the sources of errors.

Tests Based on Polygon Overlay
A misclassification matrix must be reported as areas. The relationship
between the two maps must be explained; as far as possible, the two sources
should be independent and one should have higher accuracy.

SDTS provides for an alternative form of testing based on overlaying two maps.
The source of higher accuracy could be a 'maplet' for some limited region of the
map tested (as the point samples only select limited areas to test). A maplet can
be seen as a very tight 'cluster' of point samples, with the clear understanding
that they are correlated. The advantage is that the polygon overlay method can
distinguish different sources of error through geometric characteristics of the
results (Chrisman and Lester, 1991).

Logical Consistency
A report on logical consistency shall describe the fidelity of relationships
encoded in the data structure of the digital cartographic data. The report shall
detail the tests performed and the results of the tests.

Logical consistency covers the axiomatic basis of the database since the fidelity
of relationships means compliance with some expected or axiomatic properties.
A database is not a simple bag of bits representing some measurements. It
organizes objects, relationships and a set of integrity constraints on the objects
and their relationships. Software often relies on the axioms, so logical consistency becomes a major issue in fitness for use. Logical consistency can be tested
on the basis of internal evidence (inside the database) without recourse to
'ground truth'.

For exhaustive areal coverage data transmitted as chains, or derived from
chains, it is permissible to report logical consistency as 'Topologically Clean',
under the condition that an automated procedure has verified the following
conditions [specified in SDTS].

A particular concern for vector spatial data is the topological integrity of the
embedding in the plane. The conditions specified ensure that the graph is planar
and properly labelled.

Completeness
The quality report must include information about selection criteria, definitions used and other relevant mapping rules. For example, geometric
thresholds such as minimum area or minimum width must be reported....

Completeness is the final section. Completeness may seem a bit redundant in this
scheme. In some cases, such as minimum mapping unit (the smallest object
mapped), completeness deals with the resolution or spatial grain, almost an issue
of positional accuracy. A test of the sizes of objects in the database would verify
whether any small features were included that were not intended, but there is no

simple test to verify those things not in the database. In some circumstances, completeness involves issues of attribute accuracy (essentially errors of omission), but the rules behind the procedures require another section. Completeness can be tested when there is some comprehensive external list of all objects that should appear. Such a test is not a test of the accuracy of each object but rather that each expected code appears somewhere, thus completeness cannot be subsumed in the attribute accuracy section.

Implementing SDTS Data Quality Reports

As with any standard, creating the SDTS document is not the end of the problem. It will take substantial effort to move the paper document into practice. There are a number of efforts in this direction. In several states, the coordination of geographic information has become a central issue (see Warneke, 1991 for compendium of state activities). Some states decided to foster data sharing through a state-wide index of GIS data. One example is Florida (Stage, 1992), where a comprehensive checklist has been under development. At Portland State University, a Northwest Spatial Data Index has been developed using a somewhat different list. Both these efforts used SDTS spatial data quality specifications as a starting point. There is also a federal attempt to create a content standard for spatial metadata. The US Geological Survey has assembled a draft *Content Standards for Spatial Metadata* (FGDC, 1992).

Vector Product Format

The Digital Chart of the World is a four CD-ROM database produced by the Defense Mapping Agency. To create this product, the contractor (ESRI) had to develop a new format. Existing formats, including SDTS and DIGEST, were oriented towards transfer of data, implying that a user would reformat it into the requirements of their system. A CD-ROM database must be used directly without reformatting. Direct use presents a more difficult challenge from the transfer problem. GCDIS and the environmental databases of the future that operate over networks should be more similar to the direct use approach, though they may begin simply as clearinghouses for transfer.

Data quality, while motivated under the framework of SDTS and DIGEST, was a critical element in VPF. Spatial coverages describing data quality can be attached at any level in the structure and examined by the browsing software VPFVIEW. The important result is seeing that data quality information requires its own spatial database; it cannot be confined to a text field. Eventually the metadata for large collections of spatial data will be accessed through spatial query systems.

Composite Indices

The SDTS specification may seem quite complicated after examining all the details. It certainly includes a diverse set of factors. A judgement of fitness will never be easy when confronted with all the peculiarities common in a large complex database. It is no wonder that a number of scientists have proposed simpler methods to handle decisions about data quality. Some professions boil down data quality to a single number. If your concern can focus exclusively on positioning, and the identity of points (and their date) present no trouble, then the root mean square error of the co-ordinates of test points should be the sole criterion. All the complexities of data quality form a classic application of multicriteria decision making. There have been some attempts to combine various factors into a composite scale. For example, Mead (1982) suggests an index based on the sum of 10 factors, each given similar weight.

In a recent article in *Environmental Management*, Costanza *et al.* (1992) offered another composite index. Pedigree (also termed grade) is calculated by a normalized sum of three integer scores (from 0 to 4) representing distinctly labelled ordinal evaluations of the quality of models, quality of data and degree of acceptance. There are many possible criticisms of this scoring system, but the intent is to combine the various factors into a single scale. Basically, they do not believe that users will look at all the 'wrinkles' in making a judgement of fitness for use. By using the term 'policy-relevant research' in their title, the implicit assumption seems to be that policy makers (and users) must be sheltered from the debates and difficulties of scientific data. In the long run, a single scale will not serve all potential users equally. In an attempt to simplify the situation, this index of pedigree may mix together factors that will have distinct importance to different users. Mead's (1982) proposed index would mix together 10 somewhat related factors, and certainly obtain a different score. While the SDTS specification involves many factors, it may be the most reliable method available to communicate the quality of information so that the user can make the decision on fitness for use.

Conclusion

From this overview, it should be apparent that data quality involves many factors. Any simplistic approach will not provide information to support a decision about fitness for use. The five categories in the SDTS Data Quality Report provide a reasonable starting point for map-based sources. For other forms of environmental data that record quantities on a continuous time trace or as samples at regular intervals, the SDTS scheme should be modified to treat temporal error and calibration separately.

Data quality information for a single site may be a simple paper report, but a complex global database will involve information from diverse sources and with a spatial distribution of quality parameters. Many critical scientific challenges require the integration of masses of data from highly diverse sources. The first step in data integration, maintaining the variation in data quality, can be served by the SDTS approach. The next step is to develop procedures that can handle information sources with large differences in resolution, accuracy and other key properties. This will require more attention to issues of spatial statistics and understanding the processes that lead to errors in spatial data. A full-function GIS must exploit the variations in data quality, not maintain the illusion of homogeneity and perfect knowledge.

References

American Society for Photogrammetry and Remote Sensing, 1987, Spatial accuracy standards for large scale topographic maps, *Photogrammetric Engineering and Remote Sensing*, **55**, 958–61.

Bureau of the Budget, 1947, *National Map Accuracy Standards*, Washington, DC: US Government Printing Office.

Burrough, P. A., 1986, Five reasons why geographical information systems are not being used efficiently for land resources assessment, *Proceedings AUTO-CARTO London*, **2**, 139–48.

Chrisman, N. R., 1984a, The role of quality information in the long-term functioning of a geographic information system, *Cartographica*, **21** (3), 79–87.

Chrisman, N. R., 1984b, Alternatives for specifying quality standards for digital cartographic data quality, in Moellering, H. (Ed.) *Report 4*, pp. 43–71, Columbus, Ohio: National Committee for Digital Cartographic Data Standards.

Chrisman, N. R. and Lester, M. K., 1991, A diagnostic test for error in categorical maps, *Proceedings AUTO-CARTO, London*, **1**, 330–48.

Committee on Earth and Environmental Sciences (CEES), 1993, *Draft Implementation Plan for the US Global Change Data and Information System*, Washington, DC: Office of Science and Technology Policy.

Congalton, R. G. and Mead, R. A. 1983, A quantitative method to test for consistency and correctness in photointerpretation, *Photogrammetric Engineering and Remote Sensing*, **49**, 69–74.

Costanza, R., Funtowicz, S. O. and Ravetz, J. R., 1992, Assessing and communicating data quality in policy-relevant research, *Environmental Management*, **16**, 121–31.

Defense Mapping Agency, 1992, *Vector Product Format, Military Standard 600006*, Washington, DC: Department of Defense.

Digital Geographic Information Working Group (DGIWG), 1991, *DIGEST: A Digital Geographic Exchange Standard*, Washington, DC: Defense Mapping Agency.

Federal Geographic Data Committee, 1992, *Draft Content Standards for Spatial Metadata*, Reston, Virginia: US Geological Survey.

Fitzpatrick-Lins, K., 1978, Comparison of sampling procedures and data analysis for a land use and land cover map, *Photogrammetric Engineering and Remote Sensing*, **47**, 343–51.

Gersmehl, P. J., 1985, The data, the reader, and the innocent bystander — a parable for map users, *Professional Geographer*, **37**, 329–34.

Goodchild, M. F., 1988, Stepping over the line: Technological constraints and the new cartography, *The American Cartographer*, **15**, 311–19.

Goodchild, M. F. and Gopal, S. (Eds), 1989, *The Accuracy of Spatial Databases*, London: Taylor & Francis.

Hayes, G. E. and Romig, H. G., 1977, *Modern Quality Control,* Encino, California: Bruce.

Holm, T. M. and Scholz, D. K., 1992, 'The global land information network', presentation at Information Exchange Forum on Spatial Metadata, Reston, Virginia, US Geological Survey.

Langran, G., 1993, Updating map and chart series in an automated production environment, *Cartography and GIS,* **20**, 107–12.

Lanter, D. P., 1993, A lineage meta-database approach towards spatial analytic database optimization, *Cartography and GIS,* **20**, 112–21.

Mead, D. A., 1982, Assessing data quality in geographic information systems, in Johannsen, C. and Sanders, J. (Eds) *Remote Sensing for Resource Management,* pp. 51–9, Ankeny, Iowa: Soil Conservation Society of America.

National Institute of Standards and Technology, 1992, Spatial data transfer standard, *Federal Information Processing Standard 173,* Gaithersburg, Maryland: National Institute of Standards and Technology.

Stage, D., 1992, A multi-agency management structure to facilitate the sharing of geographic data, *Proceedings of Information Exchange Forum on Spatial Metadata,* pp. 45–68, Reston, Virginia: US Geological Survey.

Thieman, J., 1992, 'Master data directory', presentation at Information Exchange Forum on Spatial Metadata, Reston, Virginia: US Geological Survey.

Vonderohe, A. P. and Chrisman, N. R., 1985, Tests to establish the quality of digital cartographic data: Some examples from the Dane County Land Records Project, *Proceedings AUTO-CARTO 7,* 552–9.

Warneke, L., 1991, *Compendium of State Government Activities in GIS,* Denver: Council of State Governments.

SECTION IV

Data sharing issues

In Section I, Gosz highlighted the importance of data sharing for addressing the complex, multidisciplinary issues related to global change, biodiversity and sustainability. He suggested that scientific progress in these areas will depend on the accessibility of numerous long-term databases from various agencies, educational and research institutions and individual scientists, and pointed out that significant changes will be required in the ways that environmental research is funded and performed. The three chapters in Section IV focus on many of the institutional and technical impediments to data sharing.

Porter and Callahan examine several of these institutional impediments and present the mechanisms whereby large, multi-investigator research projects have supported the creation, management and use of large, multisource databases. They claim that an inequity is created when users of shared databases derive greater benefits than the original contributors, and suggest several solutions. Evans addresses both the organizational and technical impediments to data sharing, evaluates existing software that can facilitate data sharing, offers several recommendations on the sharing of spatial environmental information, and identifies remaining research challenges. In the final chapter of this section, Slagle evaluates the impacts that information standards have had and will have on environmental information. He further suggests several ways to correct the current deficiencies in comprehensive standards.

13

Circumventing a dilemma: Historical approaches to data sharing in ecological research

John H. Porter and James T. Callahan

There is a fundamental dilemma embedded in database creation and management. At least on a perceptual level, the benefits derived from a database are greater for the user of the data than for the contributor of the data. Ultimately, the utility of a database depends on the quality of the data provided and the accessibility of the data to users. We develop a conceptual model for data sharing and examine the means by which large, multi-investigator research projects have provided for the creation, management and utilization of large, multisource databases. Also, based on a review of recent literature, we examine the 'speed of consumption' (the time between data generation and publication of results) of ecological data.

Introduction

The success of ecology as a science depends on development of environmental databases. No single individual is able to collect all the data needed to provide an integrated view of complex ecological systems. Success of shared ecological databases depends on the willingness of investigators and institutions to contribute and use data. However, there is a fundamental dilemma embodied in database creation if benefits derived from shared databases are larger for data users than for data providers. In this chapter we will examine the conditions under which an individual scientist should share data, and the ways in which institutions and research projects have promoted data sharing. Additionally, we will identify future avenues of opportunity for promoting database participation.

An Additive Model for Data Sharing

A simple additive model based on the time or energy budgets of individual investigators is used to develop a conceptual basis for understanding when an investigator should share data. Each investigator has a finite amount of time or energy available for research activities. Research activities include preparing publications (analysis and writing), collecting data, preparing data for analysis (data entry, quality assurance and quality control), and, if data are to be shared, preparing documentation. Some resources may be accessible (equipment, students and technicians) which supplement the time available for research. This yields a model of the form:

$$K = P(d) + C(d) + M(d) + U(d) - R - T \tag{13.1}$$

or, alternatively

$$P(d) = K + R - C(d) - M(d) - U(d) - T \tag{13.2}$$

Where

$P(d)$ is the time spent preparing publications using data d
K is a constant, the time available for research
R is the time made available through resources resulting from past work or
    commitments to current work = f (*papers, citations, funding*)
$C(d)$ is the time spent collecting data d
$M(d)$ is the time spent preparing documentation for data d
$U(d)$ is the time spent preparing data, d, for analysis
T is the time between when data are collected and first become available.

Thus, the amount of time available for analysis and writing (which yield tangible products of direct benefit to the investigator) is a function of the total time and resources available, less time spent collecting data, preparing data and preparing documentation.

The 'costs' in equation (13.2) vary depending on participation in shared databases. For an investigator who collects a data set and contributes it to a shared database, T is near zero, so

$$P(d) = K + R - C(d) - M(d) - U(d) \tag{13.3}$$

For a scientist who does not share data, $M(d)$ and T are typically near zero, so

$$P(d) = K + R - C(d) - U(d) \tag{13.4}$$

Finally, for a scientist who uses data from a database, but does not collect or contribute data $C(d)$, $U(d)$, and $M(d)$ are zero, so

$$P(d) = K + R - T \tag{13.5}$$

In an unconstrained data sharing system, data immediately become available to all participants ($T = 0$) and rewards (R) are independent of data contribution. In an unconstrained system, scientists who both collect data and contribute it to a database are at a distinct disadvantage. Scientists who collect but do not share data have a time advantage equal to $M(d)$, and scientists who use data but do not collect or contribute data have a time advantage of $C(d) + M(d) + U(d)$.

This inequity, favouring data users over data contributors, is the fundamental dilemma facing investigator-based environmental databases. It is not reasonable to expect scientists to act against their own best career interests, even if ultimately their actions benefit the scientific community and society in general.

The key to circumventing this dilemma rests with constraining data sharing systems to provide systematic rewards for data contribution. A stable system for data sharing can be maintained only if constraints are placed on the system so that use of the data either increases the R value for the data providers or if T is higher for data users than data providers. A necessary condition for a stable system is

$$R_p - C(d) - M(d) - U(d) \geqslant R_c - T \qquad (13.6)$$

where R_p and R_c are the rewards for data providers and data consumers, respectively.

The role of constraining data sharing to maintain a stable system falls to the information management systems. The policies that govern data access and the operation of scientific data management systems, either formal or implicit, are a necessary part of successful information management efforts. The remainder of this paper will focus on real-world data management policies to observe how they resolve the challenge posed by equation (13.6).

Information Management Policies

Ecology and evolutionary biology stand virtually alone among the environmental and environment-related sciences in the lack of some agency- or community-mandated data archiving and data sharing policy. This unanticipated occurrence was revealed by a conversational survey of colleagues in various programmes at the National Science Foundation (NSF) recently. Essentially all the physical environmental sciences programmes have such data policies, some specifically stated in NSF award instruments. Social and economic sciences programmes, and behavioural science programmes to a lesser extent, also manifest award-based requirements for data archiving and sharing. Other federal agencies supporting extramural research also have data archiving and sharing policies, some of which include centralized quality assurance/ quality control (QA/QC) protocols. For example, *The US Global Change Data and Information Management Program Plan* was published by the Committee on Earth Sciences of the Federal Coordinating Council for Science, Engineering and Technology (CES, 1992). This report details the overarching policy regarding data and data-related matters for the research sponsored under the US Global Change Research Program. The one significant time in the past when American ecologists and evolutionary biologists attempted to elaborate unified policies and protocols on data management was during the US International Biological Program (US/IBP). That attempt met with near complete failure from the outset, to the extent that data policies and protocols were never elaborated nor even agreed to in principle.

Gorentz and Hamilton (1992) surveyed 103 biological field stations and marine laboratories. Of 94 respondents to a question concerning proprietary rights to data access, only 11 reported that some sort of policy was in place at their institution. Despite the small number of policies in place, proprietary rights of data collectors were considered important by almost all respondents. Michener and Haddad (1992) identified development of a data access policy as one of six steps needed to implement a data management programme at a biological field station or marine laboratory.

Long-Term Ecological Research (LTER) sites are a conspicuous exception to the general lack of information management policies in ecology and evolutionary biology. An early recognition that long-term information management was a critical part of long-term research led to a requirement by the National Science Foundation that each LTER site conduct an active programme of data management. NSF left the specific forms of information management programmes and policies to the discretion of researchers at individual sites.

Guidelines for LTER data access policies were developed in 1990 (Table 13.1). Data access policies were discussed at length during the 1990 LTER Data Managers' Workshop (Michener *et al.,* 1990). Following that discussion, the LTER Coordinating Committee formed an *ad hoc* committee on data access. The committee was composed of principal investigators and data managers and was charged with producing guidelines for LTER site data management policies (Meyer *et al.,* 1990). The committee formulated guidelines based on principles listed by data managers and investigators during a joint workshop at the 1990 LTER All-Scientists' Meeting. The resulting guidelines were then electronically distributed to individual sites. The guidelines included 10 provisions that should be addressed by an individual site's data management policy. However, it left the investigators at each site to decide what specific policies should be implemented at their site.

The guidelines specify a framework of responsibilities for all participants in the system. Data providers are responsible for making data available, including contribution of data and adequate documentation. Responsibilities of data management staff include providing long-term archival storage, assuring continued availability and providing adequate security to ensure that investigators who collect data have first opportunity to exploit it. Data users are responsible for providing adequate acknowledgement, paying expenses related to data retrieval and refraining from reselling data. Additionally, there is a responsibility for adequate quality assurance and quality control, which is not

Table 13.1 LTER Guidelines for Site Data Management Policies issued in 1990 by the LTER ad hoc *Committee on Data Access.*

Each site should develop its own policy in consultation with its investigators and higher administrative units. The following provides general guidelines and rationale, but each site should be prepared to defend its own policy for data management through the site and peer review process. Bear in mind that the general policy of the Division of Biotic Systems and Resources (NSF) is that the data are public property one year after termination of the grant. The management policy should include provisions that assure:

A. the timely availability of data to the scientific community,
B. that researchers and LTER sites contributing data to LTER databases receive adequate acknowledgment for the use of their data by other researchers and sites receive copies of any publication using that data,
C. that documentation and transformation of data is adequate to permit data to be used by researchers not involved in its original collection,
D. that data must continue to be available even though an investigator were to leave the project through transfer or death,
E. that standards of quality assurance and quality control are adhered to,
F. that long-term archival storage of data is maintained,
G. that researchers have an obligation both to contribute data collected with LTER funding to the LTER site database and to publish the data in the open literature in a timely fashion,
H. that costs of making data available should be recovered directly or by reciprocal sharing and collaborative research,
I. that LTER data sets not be resold or distributed by the recipient, and
J. that investigators have a reasonable opportunity to have first use of data they collected.

specifically assigned to either data contributors or the data management staff.

LTER Site Information Management Policies

As of autumn 1993, 12 of 18 Long-Term Ecological Research (LTER) sites have adopted formal data access policies. An additional four sites have informal policies, and two sites have policies under development. Although data access was emphasized in the Guidelines for LTER Site Data Management Policies (Meyer *et al.*, 1990), most formal site policies extended beyond access issues. Many of the policies included provisions specifying the responsibilities of the site data management system (Table 13.2). When policies specified the form of administrative control of data management operations, control was always vested either in a committee consisting of investigators or jointly between a site's principal investigator(s) and data manager. A centralized model for database administration was the most common, with data and metadata residing in a central database. One site used a model wherein each investigator is responsible for archival storage of their own data and another site used a hybrid system where some data were maintained by individual investigators and some in a

Table 13.2 Responsibilities of the data management system and data providers at LTER sites, as stated in site data management policies.

Policies include provisions governing	Number of sites
Data management system responsibilities	
Who administers data management activities at site	4
Storage of data	
Centralized	7
Hybrid (some centralized, some investigator)	1
Investigator	1
Management of physical samples	1
Responsibilities of data providers	
Data entry	4
Quality assurance/quality control	6
Use of standard coding schemes (dates, locations)	2
Submission of data to site database	
Following publication	1
Within 1 year of collection	3

centralized system. Only one site had a specific policy regarding access to stored physical samples.

Despite vagueness in the LTER guidelines regarding specific responsibility for data quality, all the site policies which addressed the issue assigned primary responsibility to the data provider (Table 13.2). There was a general perception that quality control must begin with collection of the data and within an individual LTER site the data collector was most likely to have the specific expertise needed to assess data quality. Only four of the site policies specifically provided guidelines specifying when data should be submitted to the centralized site database. Of these, three specified that data should be submitted within one year of collection. Additionally two sites provided specific data coding standards for selected variables (e.g., dates and places) (Table 13.2).

Data access was an important part of all the policies (Table 13.3). At 12 sites, release of data required the permission of the investigator who collected the data, for at least some period of time. At two additional sites, the site principal investigator (PI) could also provide permission for data release. Three sites made data available following publication. Seven sites had special policies regarding 'baseline' monitoring measurements. These were typically available without requiring special permission or after a brief (one year) delay.

A majority of sites with formal policies incorporated time limits on how long data could be reserved (Table 13.3). These limits ranged from one year to five years, with a modal time limit of two years. For comparison, an informal survey of colleagues indicated that in the other environmental and environmentally related sciences, the allowable delay times between data generation and database entry (i.e., public accessibility) ranged from zero days (in the cases of closely related interdependent efforts such as oceanographic cruises) to two years (as a

Table 13.3 *Limitations on data access and responsibilities of data users, as stated in LTER site data management policies.*

Policies include provisions governing	Number of sites
Data access	
With permission of investigator (at least for a period of time)	12
Following publication	3
Only with permission of site PI	1
Time limits for restricted data access	
1 year	1
2 years	4
3 years	1
5 years	1
1 year after termination of grant	1
Special access for baseline or common site data (e.g., meteorological measurements)	7
Data user's responsibilities	
Proper citation	9
Cost recovery	4
Copies of publications	8
Data available for restricted purposes	
Scientific use only	3
No resale of data	4
Purpose specified by data requester	3

general maximum).

Responsibilities of data users were also widely delineated, with a majority of sites requiring both citation of data and copies of papers using LTER-generated data. Limits on data use varied from restrictions that data were for 'scientific use only', to restriction of data use to stated purposes (Table 13.3). Many sites also prohibited resale of data.

Comparison Between LTER and NASA Data Access Policies

Guidelines for data access in many NASA-sponsored projects are similar to those employed at LTER sites in that they impose some constraints on data access that benefits data providers. An example of such guidelines, for the SCAR-A project (McDougal, 1993), is given in Table 13.4. There are some significant differences between LTER and NASA policies. The first is that NASA policies do not place time limits on proprietary access, except as they affect the whole data archive, whereas LTER policies focus on specific data sets. This reflects the difference between the short-term nature of many NASA campaigns relative to the long-term perspective of LTER projects. Also, the NASA

Table 13.4 NASA SCAR-A Project Data Access Guidelines.

A.   Within project, all researchers have free access to all data.
B.   Data is proprietary until published or the entire archive is released.
C.   Unpublished data must be released, but the primary scientist for that data may
     demand coauthorship, or the right to publish a disclaimer.
D.   An individual may release only the data for which they are directly responsible.
E.   Scientists outside the project can access data only with the sponsorship of a
     project scientist and must be willing to reciprocate data sharing.

guidelines emphasize co-authorship rather than citation. Finally, unlike LTER policies, the NASA SCAR-A guidelines require reciprocal data sharing.

Facilitating Data Sharing

There are many ways in which data management systems can be used to facilitate data sharing. A primary way is by providing a framework within which the needed constraints can operate. Conditions for promoting data sharing are given by equation (13.6). Some components of equation (13.6) are not susceptible to manipulation. The time required to collect data ($C(d)$) and prepare metadata ($M(d)$), and rewards for data users (R_c) are typically outside the control of information management systems. However, there are three components (R_p, $U(d)$, and T), which can be manipulated. There are three primary ways in which constraint systems can operate. The first is to supplement the resources or rewards available to data providers (R_p). These can be financial, either as a requirement for research funding or as royalties at the time of data distribution. Alternatively, rewards can be in the form of credit such as authorship or citation. Current LTER and NASA policies exploit both strategies. Both NSF and NASA specify that data management activities take place as a condition of funding.

A potential pitfall of a top-down approach to supplementing R is that rewards are only indirectly associated with data use. A system emphasizing contribution of data, without providing ways of assessing the relative quality of contributed data, supplies little incentive for data providers to exceed minimum database standards. Inclusion of mechanisms to monitor data use, or better yet, an active peer-review process, which links data and metadata quality to supplementation of R, will promote a fuller participation by data providers. The royalty approach provides positive feedback for data contribution without a need for internal and external review. However, use of royalties is problematic when public funding is used. An alternative formulation of royalties is to use a 'barter' approach, where access to data archives is facilitated for, or restricted to, data contributors. A limited version of the 'barter' approach is implicit in LTER policies where within-project access guidelines can differ from those for external investigators

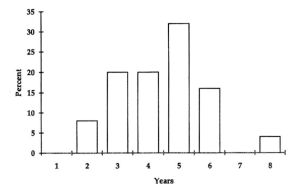

Figure 13.1 Time between termination of data collection and publication in Ecology *for 25 randomly selected articles which provided information on dates of data collection.*

and explicit in NASA policies where reciprocal data sharing is a requirement.

A second way in which data contribution can be facilitated is by providing resources specifically to reduce the time data contributors spend preparing data for analysis ($U(d)$). This can be done by placing technical resources, such as data-entry services, programs to perform QA/QC checking, training, computer access, disk storage and technical expertise at the disposal of data contributors. This is apt to be particularly effective for large data sets requiring extensive management, but is addressed in only a few data management policies because most policies focus primarily on data access, rather than data submission. None the less, this approach is widely applied within both LTER and NASA projects.

The final way information management systems can facilitate data sharing is, paradoxically, by increasing T, delaying the release of data. This, in effect, increases K (equation (13.1)), the total time for research activities, for data contributors relative to data users. As such, it reduces or eliminates the time advantage realized by data users in an unconstrained system. However, successful use of time limits is contingent upon selection of an appropriate delay factor (T). Lengthy delays have negative impact on the science, whereas brief delays yield too much of an advantage to data users.

As an aid in selecting an appropriate delay for ecological data, we examine the time period between data collection and publication for ecological data. Specifically, 34 articles from the journal *Ecology* were selected at random. Nine of the articles (primarily laboratory experiments) provided no details on the time of data collection, or used data that had been previously used for publication. For the remaining articles, the difference between when data collection terminated and the publication date was ascertained. The median publication time delay was 52 months, varying from a minimum of 18 months to a maximum of 89 months. Peaks in publication occurred in the five-year interval (Figure 13.1). The time delays in LTER policies span a similar range, with a modal value of two years.

Conclusion

Successful data sharing systems do not evolve spontaneously. The inequity between the benefits for the data provider versus the data user guarantee the failure of unconstrained information management systems. Only by constraining data exchanges can incentives for data providers be introduced into a system. In an idealized system, the distinction between data user and data provider will become blurred as individual investigators both contribute and use data from environmental databases. Such an ideal system can only come to fruition through the use of information management policies incorporating elements that increase the rewards for data contributors.

Acknowledgement

This chapter was supported by NSF grant DEB-9211772. William Michener and three anonymous reviewers provided suggestions on the form and content of the manuscript.

References

Committee on Earth Sciences (CES) of the Federal Coordinating Council for Science, Engineering and Technology, 1992, *The US Global Change Data and Information Management Program Plan,* 94 pp.

Gorentz, J. B. and Hamilton M. P., 1992, Summary of the workshop survey and pre-workshop demonstrations, in Gorentz, J. B. (Ed.) *Data Management at Biological Field Stations and Marine Laboratories,* pp. 29–41, Michigan: W. K. Kellogg Biological Station.

McDougal, J., 1993, 'Data Access', presentation at the SCAR-A Mission Planning Session, 27 April 1993.

Meyer, J. L., Hobbie, J. E., Magnuson, J. J., Michener, W. K., Stafford, S. and Porter, J. H., 1990, Guidelines for LTER site data management policies, electronic mail, LTER Network Office, 12 October 1990.

Michener, W. K., Brunt, J. W. and Nottrott, R. W., 1990, *Proceedings of the 1990 LTER data management workshop,* Report to the National Science Foundation, Seattle: LTER Network Office.

Michener, W. K. and Haddad, K., 1992, Data administration, in Gorentz, J. B. (Ed.) *Data Management at Biological Field Stations and Marine Laboratories,* pp. 4–14, Michigan: W. K. Kellogg Biological Station.

14

Sharing spatial environmental information across agencies, regions and scales: Issues and solutions

John Evans

Sharing spatial information across agencies, regions and scales is governed by the interaction of behavioural and technical 'catalysts' and 'inhibitors': wide-area networks and communications standards, incomplete data directories and semantic incompatibilities, as well as bureaucratic, institutional and legal forces. Information sharing is an active research topic not only in environmental science and geographic information systems, but also in organizational analysis, library science, computer hardware and software, and database management systems. Many findings from these fields are applicable to the sharing of spatial and environmental information among government agencies. The author evaluates existing software tools that address technical and organizational issues in concert (network browsers, data repositories and composite information systems), and recommends developing advanced network tools for sharing spatial information, creating a dynamic human and technical infrastructure, and collaborating with specialists in database management and computer-aided software engineering.

Introduction

Few would deny that sharing information is, in principle, a good thing to do. Peters (1987) has said that unless information is freely shared, 'it will be an impossible millstone around the neck of tomorrow's organization'. The motivation to share spatial and environmental information is especially strong wherever several organizations have related interests within a particular spatial extent (Nyerges, 1989a). Sharing spatial information across organizations presents several challenges, both technical and human. The following sections elucidate the key issues involved, review the current state-of-the-art of spatial information-sharing tools, and offer some directions for research in the near future.

Issues in Sharing Spatial and Environmental Information

To understand the issues involved in sharing spatial information across agencies, regions and scales, it is useful to divide them conceptually into predominantly technical issues and predominantly organizational issues. Key research areas lie in the interaction between these two types of issues (Evans and Ferreira, 1993).

Technical Catalysts and Inhibitors

Clearly, one's ability or willingness to share spatial information will always be closely tied to technology. The literature is full of examples of technological innovations to facilitate information sharing, and of unresolved difficulties despite these innovations. Many of these examples appear in the following paragraphs.

Wide-Area Networks

In recent years, wide-area computer networks have become nearly ubiquitous, affording substantial opportunities for information sharing. The first users of these networks were defence contractors and universities, followed by scientists and federal agencies. More recently, state and local agencies, private firms and the general public have begun to integrate wide-area networks into their work. The chief use of these networks, as measured by traffic volumes, is data sharing: during the month of April 1993, FTP (File Transfer Protocol) traffic along the Internet 'backbone' accounted for nearly three terabytes, or 44 per cent of all traffic (Merit, 1993). File transfer, remote log-in and electronic mail have shortened the effective distance to world-wide information resources. Remote data are brought even nearer by distributed file systems such as the Network File System (Sandberg *et al.,* 1985), Andrew File System (Morris *et al.,* 1986), and Prospero (Neuman, 1992), which make remote data files nearly indistinguishable from local ones. Kapor (1991) and Dertouzos (1991) use a 'marketplace' metaphor as they envision the impacts of wide-area networking in a few years. Clearly, maturing computer networks will be an important catalyst for sharing spatial and environmental information.

A Growing Acceptance of Standards

Throughout the information industries, there has been a growing realization that standards are sound business, as more and more users seek to mix and match hardware, software and information resources (Slagle, this volume). The trend toward open systems has facilitated information sharing by removing many incompatibilities in hardware interfaces (e.g., SCSI, the Small Computer

Systems Interface), communication protocols (TCP/IP, ASCII), operating systems (OSF-1), query languages (SQL), and graphical user environments (X Windows System) (Croswell and Ahner, 1990). Standard formats exist for images (GIF, the Graphics Interchange Format), database tables (DBF, the dBASE format), and even spatial data (SDTS, the Spatial Data Transfer Standard). Other, *de facto* standards have arisen from one firm's market share (e.g., Adobe PostScript) rather than from an industry consensus (e.g., Arc/Info export). Along with standards, software users have demanded extensive facilities for converting data among the different formats. Manageably few standard formats, query languages and interfaces, along with reliable conversion utilities, are other catalysts for sharing spatial information.

Information 'Deserts' and 'Oceans'

Focusing on data standards assumes that we already have the information on hand or know where to get it. Despite the advent of wide-area networks, many data suppliers either are disconnected from wide-area networks, lack a standard mechanism for informing potential users (Walker *et al.,* 1992), or prefer to advertise and distribute their products through more traditional means. Thus, the bottleneck in information sharing may be not knowing where to find useful information: the user 'thirsts' in an 'information desert'. In other cases, though information may exist on the network, no systematic directories exist to guide a user through the thousands of data sources: the user 'drowns' in an 'information ocean'. Both of these difficulties can be alleviated by electronic bulletin boards, newsgroups and mailing lists, whereby one can quickly obtain tips from respondents around the globe. As more suppliers come on line, the difficulty will not be mere access to information, but identifying and choosing relevant information of acceptable quality amidst the 'ocean' of data. The twin challenges of information 'deserts' and 'oceans' are an important inhibitor to information sharing.

Persistent Incompatibilities

Despite the growing role of data standards, one of the first issues to arise in information sharing is incompatibility. In his amusing article, Craig (1992) touches on many data incompatibilities and traces them to either the data's physical storage (e.g., tape formats, large data volumes), their syntactic organization (data formats necessitating conversion or repackaging), their quality and accuracy (e.g., geographic positioning and topology, missing or misleading labels), or their semantic interpretation (e.g., scale, datum, nominal vs. ordinal measurements, idiosyncratic or unclear definitions). The first two kinds of incompatibility, storage and syntax, are generally being solved by standards, even though the solutions (bigger disks, data conversion software) may not be easy to obtain. The last two difficulties, accuracy and semantics, are less easily resolved. First, the translation from one source to another is not easily automated. Second, discrepancies arise not from a simple lack of co-ordination but from legitimate differences in information requirements. Goodchild (1993) discusses the difficulty of sharing spatial information of different quality and at different map scales. As Guthrie (1992) has shown, land-use information is especially difficult to share among different agencies, due to incompatible definitions. Flowerdew (1991) reviews the problem of incompatible areal units (e.g., watersheds vs. voting districts), which Worboys and Deen (1991) have termed contextual semantic heterogeneity. Differences in accuracy and semantics are severe inhibitors to sharing spatial information across regions and scales and between agencies at different hierarchical levels (Frank, 1992).

Non-technical Catalysts and Inhibitors

According to the people actually involved in trying to share information, the most significant obstacles usually appear to be institutional, organizational or behavioural rather than technical (Croswell, 1989). Yet these issues have received less attention in research on information sharing. The human and organizational contexts can greatly inhibit information sharing — they can also be a powerful catalyst.

Bureaucratic and Cultural Forces

Sharing information with another party usually implies change: yet organizations often resist change, either because they do not know how to adapt to it, or have narrow or short-term goals that preclude organizational change. 'Inertia' of this sort leads to a reliance on tried-and-true methods (Calkins *et al.,* 1991), or to quick choices of hardware, software and data definitions to support a limited set of internal purposes (Craig, 1993). Tosta (1992) highlights the importance of 'adhocracies' designed for change (cf. Peters, 1987), and Pinto and Onsrud (1993) emphasize the role of shared, 'superordinate' goals among information-sharing parties. Negotiation (Obermeyer, 1993) and 'political savvy' (Tosta, 1992) are also crucial to establishing and maintaining information-sharing relationships. Cultural and political factors, such as antipathy for a particular group, a rejection of what is 'not invented here', or a fear of 'giving up ground' to another, can prevent sharing strategies that would otherwise be sound (Onsrud and Rushton, 1993).

A 'Motley Crew' and a Few 'Champions'

'Those people never know what they want!' Sharing spatial information can be difficult among a 'motley crew' of individuals with mismatched levels of expertise. Information requests from a novice can be hard to fill because they are vaguely worded, impossibly demanding or ignorant of the data's complexity. Having received the data, the novice may overestimate their accuracy and otherwise misuse them (August, 1991). Conversely, a novice supplier will tend to provide poorly documented data of unknown quality. In response, many organizations rely on a few technical experts ('power users') for GIS implementation and information sharing. The power user with a purely technical focus may have an impact for a time, but the organization as a whole remains 'novice' and vulnerable to personnel changes. However, a 'champion' — a power user who can navigate complex institutional relations, communicate a vision and involve the participation of others — can be a powerful catalyst for information sharing (Tosta, 1992).

Legal and Economic Issues

Legal and economic issues can be either inhibitors or catalysts for information sharing. Carter (1992) calls for greater legal protection of information as property so that some do not take credit for the work of others. Rhind (1992) argues that government-subsidized geographic information discourages sharing, and argues for appropriate pricing. Taupier (1993) reports that interorganizational sharing occurs when parties stand to gain substantial value from it. In contrast, Epstein (1993) has argued that free, open access to government information is fundamental to widespread information sharing. Onsrud (1993) and Porter and Callahan (this volume) review these and other legal aspects of sharing spatial information.

Spatial Information: Catalyst and Inhibitor

Research on technological and organizational aspects of information sharing in traditional information fields is helpful. However, such research can underemphasize the specific spatial nature of the information being shared among GIS systems, and its role as both a catalyst and inhibitor to information sharing. First, spatial relationships afford incentives to sharing that would not occur for simpler, alphanumeric information. For instance, compared with relational table look-ups on Social Security codes, overlaying spatial features based on location, geometry and topology draws on a much richer set of meaningful data relationships. Through GIS overlays, the utility of one piece of spatial data can increase dramatically when juxtaposed spatially with other pieces. Thus, by exploiting spatial relationships among data from multiple sources, GIS can result in a degree of synergism seldom found with non-spatial information. This effect is especially pronounced in the case of environmental information, where useful data come from many sources (for instance, studying non-point-source pollution requires much more than natural features), and many key relationships are based on proximity and topology.

The spatial nature of GIS information may afford incentives to sharing, but its complexity can also lead to several impediments. First, the GIS operations required (e.g., for a spatial overlay) are more complex and (at present) less easily automated than table look-ups. Second, spatial information is more prone to misunderstanding or misuse by outside users. For instance, it has a given map scale, or resolution, beyond which it becomes meaningless. Third, maintaining consistent, up-to-date geometry, topology and attributes on spatial information from multiple sources is difficult because each layer can change independently of others. Thus, relying on shared information means keeping track of spatial features, segmentations and spatial relationships and rebuilding them whenever changes occur in the underlying data — gradually, the beauty of spatial relationships can lead to an intractable mess.

Interaction Between Behavioural and Technical Issues

Research on organizational aspects of information sharing has tended to be isolated from technical issues. However, what users describe as 'turf battles' or 'bureaucratic inertia' may be the outcome of data producers and users balancing (implicitly or explicitly) the uncertain, intangible benefits of shared spatial information against the very tangible time, effort and expertise needed to make information sharing work. The reluctance of some organizations to use outside spatial information, and the refusal of others to give out spatial information, can be described not as power struggles to be solved apart from technical innovations but as behavioural symptoms of a problem intimately tied to the information's multidimensional nature, and thus to the ever-changing capabilities of current technology (Evans and Ferreira, 1993). Aside from a few recent innovations (described below), technical approaches to information sharing have tended to assume that information is freely shared with everyone, simple in structure, and the players are all equally experienced. Research efforts on information-sharing technology must address the organizational contexts of information-sharing efforts and the spatial nature of the data being shared. Because of the substantial interplay between technical and behavioural issues, the most fruitful research efforts will be those looking at behavioural and technical issues in concert, rather than in isolation one from another.

Review of Selected Information-sharing Implementations

In recent years, several innovative software tools have appeared that hold considerable promise for information sharing. In particular, they do not assume that all the information is easy to find or simple in structure, that all parties to an exchange are equally experienced, or that an organization is willing or able to change its procedures in order to share information. These tools fall into three categories: networked information discovery and retrieval, data repositories and composite information systems.

Networked Information Discovery and Retrieval

Several research groups have been developing tools for finding relevant information amidst a complex web of networked computers. In particular, the US National Science Foundation has established a Center for Networked Information Discovery and Retrieval (Dern, 1994). Schwartz *et al.* (1992) review developments in this area.

Archie, the 'archive server', tracks files available for 'anonymous FTP' at some 1100 archive sites around the Internet (Emtage and Deutsch, 1992; Schwartz *et al.,* 1992). Using Archie's graphical interface, a user first queries Archie to see which site has a particular file, and then retrieves the file (by anonymous FTP, at the press of a mouse button) from a site that holds it. Archie has proved to be a simple, effective means of finding publicly accessible files anywhere on the Internet.

WAIS (Wide Area Information Servers) (Kahle, 1991) provide information search and retrieval based on a public protocol (Lynch, 1991), full-text indexing and several simple client interfaces. By supplying a few keywords and a list of data sources to search, a user receives 'headlines' of files containing those keywords, ranked by their frequency of occurrence, and can retrieve the full text (or image, or sound) of a file. Unlike Archie, WAIS do not use a global index but tie together indexes at many sites through a 'directory of servers'. The US government has begun to investigate WAIS for information management (McClure *et al.,* 1992); Christian and Gauslin (1992) report on efforts to adapt WAIS to spatial information.

Other tools exist for finding and retrieving useful information on a large network: WorldWideWeb, Gopher, Knowbots, etc. (cf. Schwartz *et al.,* 1992). Due in part to the simplicity of these tools, many of them are becoming interoperable: for instance, a WAIS source named dynamic-Archie presents WAIS queries to an Archie server.

Spatial Data Browsers

In the last few years, many researchers have responded to the particular challenge of locating relevant spatial data. Most of their approaches have been based on simple spatial metadata, that is, structured descriptions of the theme, extent, scale and accuracy of spatial data sets.

For instance, MRRL (Walker *et al.,* 1992) is a browser to spatial data sets that allows both the data owners and secondary users to describe the information in their own terms. The system identifies 'close' matches to a user's query using a keyword thesaurus and a measure of the overlap between the location and time frame of the data set vs. that of the query. Vrana (1992) has developed the

Spatial Data indeX (SDX), which cross-references data sets based on their overlap with known boundaries. FINDAR, at Australia's National Resource Information Center, provides both a map-based user interface and a relational cross-index among spatial partitions (Shelley and Johnson, 1991). Evans *et al.* (1992) describe MITmapper, a map-based interface to a list of spatial data sets identified by their spatial extent and format. Other efforts include the Global Land Information System (GLIS) at the US Geological Survey's EROS Data Center (Holm and Scholz, 1992); the Global Change Master Directory at the US National Aeronautic and Space Administration (NASA) (Thieman, 1992); and the X GeoBrowser (XGB) at Columbia University (Menke *et al.,* 1991). They are all freely accessible via the Internet (see reference notes). Medyckyj-Scott *et al.* (1991) present several other spatial data browsers.

Data Repositories

Researchers in database management systems (DBMS) and computer-aided software engineering (CASE) have been trying for years to integrate and manage software engineering resources. The outcome has been 'information resource dictionary systems', or data repositories. These are systems for recording and analysing the definitions of data entities and the semantic relationships among them. These entities and relationships constitute the information's schema, or enterprise model (cf. Sen and Kerschberg, 1987). Data repositories also assist in locating and retrieving information, tracking the usage of information resources and administering interdependent data (Jones, 1992). Building a repository is usually associated not with information sharing but with information resource management as a whole (Yates, 1990). A repository is not a central storage space for data; instead, it contains descriptions of the semantic content, inter-dependencies and access paths of an organization's information resources, which are usually distributed over networks.

The spatial data browsers described earlier are useful for finding and retrieving information across a network. As systems for storing and retrieving metadata, they are considered data dictionaries for spatial information. However, these browsers differ from data repositories in two important ways. First, the entities they describe are entire data sets rather than individual features. Thus, they focus on data formats and overall spatial extent, and omit many important data semantics and relationships at the feature level. Second, they are passive in that they simply present a screenful of metadata for the user to decide whether to use the data set, and how to retrieve, convert and interpret it. A repository is an active system that can provide metadata to a software application in response to the application's query, and thus function as a part of a larger software and information environment.

Narayan (1988) discusses general data dictionary concepts, and Marble (1991) extends these concepts to spatial information. Many repository products conform to the Information Resource Dictionary Systems (IRDS) standard, described by Law (1988). Robinson and Sani (1993) apply IRDS to the management of spatial information. Repository technology, and IRDS in particular, may hold considerable promise for sharing spatial and environmental information, since a standard vocabulary for describing spatial information may be a more attainable goal than standardizing the information itself (Slagle, this volume). Interestingly, IRDS has seen relatively little implementation thus far. Newman (1992) suggests that the IRDS suffers from its own breadth and complexity — that few can grasp its many layers of abstraction and generalization. In contrast, Sayani (personal communication) argues that implementation has been sluggish because the IRDS is too simple — that its underlying *binary* Entity-Relationship model fails to express important n-*ary* data relationships (that is, relationships involving more than two entities).

Composite Information Systems

Finally, no discussion of the state-of-the-art in information sharing would be complete without mentioning composite, or federated, information systems. A composite information system is the result of building a repository spanning autonomous databases in multiple organizations. It presents the user with a single schema describing the data entities, their definitions and their relationships; it interprets subqueries of the user's query for each component database; and it assembles the responses returned by each component database manager. Sheth and Larson (1990) and Litwin *et al.* (1990) review the state-of-the-art in federated and interoperable database designs and Templeton *et al.* (1987) describe in detail one example of a composite information system. Baker and Broadhead (1992) present a design for composite information systems of spatial data. Composite information systems usually follow a non-intrusive approach that allows organizations to share information while still maintaining whatever data model, interface or definitions best suit their internal needs. Composite information systems can also restrict outside access to a partial view of an organization's data. A persistent challenge with composite information systems, even after all the communications filters are in place, is recognizing and reconciling data semantics, or context, across autonomous databases (Wang and Madnick, 1989; Siegel and Madnick, 1991). Nyerges (1989a, b) addresses this issue for spatial data: he emphasizes the integration of schemas between different sources of spatial data.

In summary, composite information systems offer considerable promise for sharing information with its correct interpretation, without imposing significant procedural changes or allowing unrestricted access to an organization's private information. However, composite information systems do not really exist — that is, not outside sheltered experimental environments. None the less, despite their newness and complexity, they may still be thought of as a template for interoperable database systems. Information-sharing systems can be measured as to their non-intrusiveness, the autonomy of their component databases, the expression and communication of data semantics, etc. The next section develops these measures further.

A Framework for Classifying Information-sharing Developments

Schwartz *et al.* (1992) suggest four measurement axes for resource discovery tools: granularity (the basic search unit, e.g., filenames vs. words), distribution (how many of the data and metadata are distributed vs. centrally stored), interconnection topology (what links exist between related pieces of data), and data integration scheme (how each system is populated with useful information). These dimensions emphasize system issues, rather than data-centred differences. This may be appropriate for informal 'resource discovery' tools. However, for ongoing interorganizational sharing of spatial information, an alternative set of dimensions is required, one more focused on the data's content and structure. Mackay and Robinson (1992) offer two scales for designing spatial information-sharing systems: functionality (from simple file transfer to fully integrated multidatabases) and autonomy (from tightly coupled databases with a global schema to independent systems). The following paragraphs suggest a categorization according to the level of standardization, the coupling of information-sharing parties, and the level of connectivity.

Levels of Standardization vs. Degrees of Structuring

All of the information-sharing tools reviewed earlier depend on a shared vocabulary. For instance, FTP transfers often use the ASCII convention to interpret alphanumeric characters as they are being transferred; a WAIS uses a full-text index to find words requested by the user in a set of documents. Other applications use standards for image files (e.g., GIF, tiff, etc.), rows-and-columns tables (e.g., dBASE), or database queries (SQL). Increasingly structured, semantically rich information demands correspondingly intelligent tools for exploiting and sharing the information: from FTP for file transfer character-by-character, to WAIS for full-text searches, to repository tools for descriptions of the information's meaning. This qualitative scale is similar to Schwartz's granularity measure (Schwartz *et al.*, 1992), except that the distinction here is not simply the size of the basic unit (character, word, file) but the amount of information encapsulated.

Loosley vs. Tightly coupled Information Sharing

Another way to measure different solutions to information sharing is the amount of participation required of information-sharing parties. To what degree must they conform in their procedures, data formats, data definitions or metadata in order to accommodate information sharing? How much must they change these aspects of their information? At one extreme, homogeneous databases are centrally maintained and must all conform exactly to the global schema. At the other extreme, the ideal composite information system is non-intrusive. This defines a scale similar to the autonomy measure proposed by Mackay and Robinson (1992), except that it measures required changes rather than flexibility. This is because the focus here is on tying together existing databases rather than designing new ones from scratch.

Levels of Connectivity: Physical, Logical, Semantic

Finally, tools for information sharing differ in the degree of connectivity among data sources. In general, an information-sharing system requires three kinds of connectivity: physical (communication protocols, access and security), logical (reconciling data models and formats), and semantic (reconciling definitions, identifying equivalent concepts, resolving ambiguities) (Wang and Madnick, 1989). The ideal composite information system scores high on all three of these scales; however, a very useful tool could be built that consisted only of physical and logical connectivity, while another might focus entirely on semantic reconciliation, with no communications protocols, etc. These scales are similar to the 'functionality continuum' in Mackay and Robinson (1992), which places physical connectivity at the bottom of the scale, and full multidatabase functionality at the top.

Recommendations for Research

Reviewing, evaluating and comparing systems for sharing information leads to charting future research directions. The following paragraphs present four recommendations for research on sharing spatial and environmental information.

There's a Network Out There: Let's (Really) Use It

First, perhaps an obvious point: wide-area networks hold great promise for effective information sharing. These networks now support not only traditional activities such as file transfer, electronic mail and remote log-in, but an increasing variety of high-level client-server applications as well. The infrastructure to support basic data sharing has been in place for some time; now a critical mass of tools and expertise is in place for more sophisticated uses of the network. Christian and Gauslin (1992) provide an example of what may be fruitful research: learning about advanced information-sharing tools (e.g., WAIS) and adapting them to spatial information.

Create a Dynamic Infrastructure of Technology and People

Many years will pass before the tools are in place to provide anything like a composite information system to the average state environmental agency. Even in the long term, the key to effective information sharing will not be a single tool but an infrastructure of networks, interoperating tools and people able to use them effectively. Most of the information-sharing tools mentioned here have emerged in the last two years; in two more years, the landscape will look very different. The network and its services are dynamic and will grow 'organically' (i.e., over time): 'the issue is not how to build such a network, but rather how it is allowed to grow' (Spackman, 1990). Because the human context of information sharing is crucial to its success, research in this area ought to be anchored in the experience of real organizations. A seminal example is that of the World Weather Watch, which consists of scientists and data managers around the globe co-operating for the common goal of weather forecasting (Rasmussen, 1992).

Is Spatial Information Special?

Earlier we made the assertion that spatial information is different from traditional information, owing to the richness of spatial relationships and the complexity of using and maintaining spatial information with a limited scale and many interdependencies. This topic has been discussed elsewhere (Anselin, 1989), but the answer will change over time based on the capabilities of current technology, both in the GIS field and in other information fields. The question is more than just theoretical; it affects how well GIS researchers communicate with database management specialists and understand their contributions, and it shows GIS researchers how to involve database theorists, knowledge engineers and others in co-operative research. Re-examining this question periodically, in the light of current technology, will benefit these interdisciplinary efforts.

GIS People Should Work More Closely With DBMS and CASE People

GIS specialists have a lot to learn from researchers in database management and computer-aided software engineering (CASE). First, GIS research on information sharing would benefit from looking at the problem not as file exchange and conversion but as querying information entities across a network, following a heterogeneous information systems approach (Mackay and Robinson, 1992). Specialists in database management systems (DBMS) are making important progress in this area. Second, specialists in computer-aided software engineering are making progress in the design and implementation of data repositories and composite information systems. These developments may help in representing the content of spatial data features and the relationships among them, and in converting or interpreting them to suit a variety of contexts. Thinking about spatial information sharing in new ways (e.g., as information entities and queries rather than transfers of bytes) may provide novel solutions.

Summary and Conclusion

The problem of information sharing is complex, and has no easy solutions. It is governed by many technical and organizational factors, and by the richness and complexity of spatial relationships. Several promising information-sharing tools have emerged in recent years. These tools differ in their level of standardization, the degree of coupling between data-sharing parties and the level of connectivity attainable. These tools provide a basis for research on information sharing, which will involve sophisticated, yet nearly transparent, use of wide-area networks. An important research focus will be to encourage appropriate use of information by encapsulating and expressing the content, meaning and accuracy of information, and not just data formats and sizes. The work will need to extend today's research on spatial metadata management by encompassing information resource management tools.

In addition, the success of new technologies will depend on how well they are tied to their human and organizational context and to the special incentives and difficulties of working with spatial information. Thus, behavioural research related to the new technologies will be needed in order to gain a better understanding of the organizational, legal and economic forces that govern information sharing, and the interactions between human and technical factors. Many pieces of the information-sharing problem are now in place. In the near future, mastering this complex web of interactions will lead to effective sharing of spatial and environmental information. The challenging research required will be more than offset by the resulting benefit to environmental managers, scientists and policy makers in the coming decades.

References

Anselin, L., 1989, *What is Special about Spatial Data? Alternative Perspectives on Spatial Data Analysis,* NCGIA Technical paper 89-4, Santa Barbara, California: National Center for Geographic Information and Analysis, University of California. *Note:* Email: ncgiapub@ncgia.ucsb.edu (128.111.105.65).

August, P. V., 1991, Use vs abuse of GIS data, *Journal of the Urban and Regional Information Systems Association (URISA),* 3(2), 99–101.

Baker, T. and Broadhead, J., 1992, Integrated access to land-related data from heterogeneous sources, in Salling, M. J. and Brown, P. (Eds) *Proceedings of the 1992 Annual Conference of the Urban and Regional Information Systems Association (URISA),* Vol. III, pp. 110–21, Washington, DC: URISA.

Calkins, H., Epstein, E., Estes, J., Onsrud, H., Pinto, J., Rushton, G. and Wiggins, L., 1991, *Sharing of Geographic Information: Research Issues and a Call for Participation, Initiative 9 specialists' meeting, Feb. 1992,* Santa Barbara, California: National

Center for Geographic Information and Analysis, University of California. *Note:* Email: ncgiapub@ncgia.ucsb.edu (128.111.105.65).

Carter, J. R., 1992, Perspectives on sharing data in geographic information systems, *Photogrammetric Engineering and Remote Sensing,* **58**(11), 1557–60.

Christian, E. J. and Gauslin, T. L., 1992, Wide Area Information Servers (WAIS), *Information Systems Developments,* April 1992, 92–6.

Craig, W. J., 1992, Why we couldn't get the data we wanted, *Journal of the Urban and Regional Information Systems Association,* **4**(2), 71–8.

Craig, W. J., 1993, Why we can't share data: institutional inertia, in Onsrud, H. J. and Rushton, G. (Eds) *Sharing Geographic Information,* pp. 97–106, New Brunswick, New Jersey: Center for Urban Policy Research, Rutgers University.

Croswell, P. L., 1989, Facing reality on GIS implementation: Lessons learned and obstacles to be overcome, in Salling, M. J. and Gayk, W. F. (Eds) *Proceedings of the 1989 Annual Conference of the Urban and Regional Information Systems Association (URISA),* Washington, DC: URISA.

Croswell, P. L. and Ahner, A., 1990, Computing standards and GIS: A tutorial, in Salling, M. J., Bamberger, W. J. and Trobia, G. (Eds) *Proceedings of the 1990 Annual Conference of the Urban and Regional Information Systems Association (URISA),* Vol. II, pp. 88–105, Washington, DC: URISA.

Dern, D. P., 1994, *The Internet Guide for New Users,* Ch. 12, pp. 416–18, New York: McGraw-Hill.

Dertouzos, M. L., 1991, Building the information marketplace, *Technology Review,* **94**(1), 28–40.

Emtage, A. and Deutsch, P., 1992, *Archie* — an electronic directory service for the Internet, in *Proceedings of the Winter 1992 USENIX Conference,* pp. 93–110, Berkeley, California: USENIX Association.

Epstein, E. F., 1993, Control of public information, in Onsrud, H. J. and Rushton, G. (Eds) *Sharing Geographic Information,* pp. 271–80, New Brunswick, New Jersey: Center for Urban Policy Research, Rutgers University.

Evans, J. D. and Ferreira, F., Jr, 1993, Sharing spatial information in an imperfect world: interactions between technical and organizational issues, in Onsrud, H. J. and Rushton, G. (Eds) *Sharing Geographic Information,* pp. 395–406, New Brunswick, New Jersey: Center for Urban Policy Research, Rutgers University.

Evans, J. D., Ferreira, F., Jr and Thompson, P. S., 1992, A visual interface to heterogeneous spatial databases based on spatial metadata, in Bresnahan, P., Corwin, E. and Cowen, D. (Eds) *Proceedings of the 5th International Symposium on Spatial Data Handling,* Columbia, South Carolina: International Geographical Union (IGU), Humanities and Social Sciences Computing Laboratory, University of South Carolina.

Flowerdew, R., 1991, Spatial data integration, in Rhind, D. W., Maguire, D. J. and Goodchild, M. F. (Eds) *Geographical Information Systems: Principles and Applications,* Vol. 1, pp. 375–87, London: Longman.

Frank, A. U., 1992, Acquiring a digital base map: A theoretical investigation into a form of sharing data, *Journal of the Urban and Regional Information Systems Association (URISA),* **4**(1), 10–23.

Goodchild, M. F., 1993, Sharing imperfect data, in Onsrud, H. J. and Rushton, G. (Eds) *Sharing Geographic Information,* pp. 363–74, New Brunswick, New Jersey: Center for Urban Policy Research, Rutgers University.

Guthrie, J., 1992, What I Z (zed) is not what you Z (zee): Shared data depends on shared meaning, in Salling, M. J. and Brown, P. (Eds) *Proceedings of the 1992 Annual Conference of the Urban and Regional Information Systems Association (URISA),* Vol. III, pp. 122–43, Washington, DC: URISA.

Holm, T. M. and Scholz, D. K., 1992, The Global Land Information System, in *Report of the Information Exchange Forum on Spatial Metadata,* pp. 207–19, Washington,

DC: Federal Geographic Data Committee, US Geological Survey. *Note:* GLIS is accessible by *telnet* to glis.cr.usgs.gov (152.61.192.54). For an X Windows graphical interface to GLIS, *telnet* to xglis.cr.usgs.gov (152.61.192.37).

Jones, M. R., 1992, Unveiling repository technology, *Database Programming and Design,* **5**(4), 28–35.

Kahle, B., 1991, An information system for corporate users: Wide Area Information Servers, *On-line Magazine,* **15**(5), 56–62.

Kapor, M., 1991, *Building the Open Road: The NREN As Test-Bed For The National Public Network,* Internet Request for Comments (RFC) No. 1259. *Note:* Available by anonymous *ftp* from ftp.internic.net (198.49.45.10).

Law, M. H., 1988, *Guide to Information Resource Dictionary System Applications: General Concepts and Strategic Systems Planning,* NBS Special Publication SP 500–152, Gaithersburg, Maryland: National Institute of Standards and Technology (NIST).

Litwin, W., Mark, L. and Roussopoulos, N., 1990, Interoperability of multiple autonomous databases, *Association for Computing Machinery (ACM) Computing Surveys,* **22**(3), 267–93.

Lynch, C. A., 1991, The Z39.50 information retrieval protocol: An overview and status report, *Computer Communication Review* (ACM SIGCOMM newsletter), **27**(1), 58–70.

Mackay, D. S. and Robinson, V. B., 1992, *Towards a Heterogeneous Information Systems Approach to Geographic Data Interchange,* Discussion Paper 92/1, Ontario, Canada: Institute for Land Information Management, University of Toronto.

Marble, D. F., 1991, The extended data dictionary: A critical element in building spatial databases, in *Proceedings of the Eleventh Annual ESRI User Conference,* pp. 169–77, Redlands, California: Environmental Systems Research Institute.

McClure, C. R., Ryan, J. and Moen, W. E., 1992, *Identifying and Describing Federal Information Inventory/Locator Systems: Design for Network-based Locators,* Final Report to the Office of Management and Budget, The National Archives and Records Center, and the General Services Administration, Syracuse, New York: School of Information Studies, Syracuse University.

Medyckyj-Scott, D. J., Newman, I. A., Ruggles, C. L. N. and Walker, D. R. F. (Eds), 1991, *Metadata in the Geosciences,* Loughborough, UK: Group D Publications.

Menke, W., Friberg, P., Lerner-Lam, A., Simpson, D., Bookbinder, R. and Kerner, G., 1991, A voluntary, public method for sharing earth science data over Internet using the Lamont view-server system, *Eos, American Geophysical Union Transactions,* July 1991, pp. 409–14. *Note:* The XGB client is available by anonymous *ftp* from lamont.1dgo.columbia.edu (129.236.10.30).

Merit/NSFNET Network Information Center, 1993, NSFNET statistics by service, April 1993. *Note:* ASCII files available by anonymous *ftp* from nis.nsf.net (35.1.1.48).

Morris, J. H., Satyanarayanan, M., Conner, M. H., Howard, J. H., Rosenthal, D. S. H. and Smith, F. D., 1986, Andrew: a distributed personal computing environment, *Communications of the Association for Computing Machinery,* **29**(3), 184–201.

Narayan, R., 1988, *Data Dictionary: Implementation, Use, and Maintenance,* Englewood Cliffs, New Jersey: Prentice-Hall.

Neuman, B. C., 1991, The Prospero file system, a global file system based on the Virtual System Model, *Computing Systems: The Journal of the USENIX Association,* **5**(4), 407–32.

Newman, I. A., 1991, Data dictionaries, Information Resource Dictionary Systems, and metadatabases, in Medyckyj-Scott, D. J., Newman, I. A., Ruggles, C. L. N. and Walker, D. R. F. (Eds) *Metadata in the Geosciences,* pp. 69–83, Loughborough, UK: Group D Publications.

Nyerges, T. L., 1989a, Information integration for multipurpose land information systems, *Journal of the Urban and Regional Information Systems Association (URISA)*, **1**(1), 27–38.

Nyerges, T. L., 1989b, Schema integration analysis for the development of GIS databases, *International Journal of Geographical Information Systems*, **3**(2), 153–83.

Obermeyer, N. J., 1993, Reducing inter-organizational conflict in order to share geographic information, in Onsrud, H. J. and Rushton, G. (Eds) *Sharing Geographic Information*, pp. 125–34, New Brunswick, New Jersey: Center for Urban Policy Research, Rutgers University.

Onsrud, H. J., 1993, Role of law in impeding and facilitating the sharing of geographic information, in Onsrud, H. J. and Rushton, G. (Eds) *Sharing Geographic Information*, pp. 259–70, New Brunswick, New Jersey: Center for Urban Policy Research, Rutgers University.

Onsrud, H. J. and Rushton, G. (Eds), 1993, *Sharing Geographic Information*, New Brunswick, New Jersey: Center for Urban Policy Research, Rutgers University.

Peters, T. J., 1987, *Thriving on Chaos: Handbook for a Management Revolution*, New York: Knopf.

Pinto, J. K. and Onsrud, H. J., 1993, Sharing geographic information across organizational boundaries: A research framework, in Onsrud, H. J. and Rushton, G. (Eds) *Sharing Geographic Information*, pp. 45–62, New Brunswick, New Jersey: Center for Urban Policy Research, Rutgers University.

Rasmussen, J. L., 1992, The World Weather Watch: A view of the future, *Bulletin American Meteorological Society*, **73**(4), 477–81.

Rhind, D. J., 1992, Data access, charging, and copyright and their implications for geographical information systems, *International Journal of Geographical Information Systems*, **6**(1), 13–30.

Robinson, V. B. and Sani, A. P., 1993, Modeling geographic information resources for airport technical data management using the Information Resources Dictionary System (IRDS) standard, *Computer, Environment, and Urban Systems*, **17**, 111–27.

Sandberg, R., Goldberg, D., Kleiman, S., Walsh, D. and Lyon, B., 1985, Design and implementation of the Sun Network File System, in *Proceedings of the Summer 1985 USENIX Conference*, pp. 119–30, Berkeley, California: USENIX Association.

Schwartz, M. F., Emtage, A., Kahle, B. and Neuman, B. C., 1992, A comparison of Internet resource discovery approaches, *Computing Systems, the Journal of the USENIX Association*, **5**(4), 461–93.

Sen, A. and Kerschberg, L., 1987, Enterprise modeling for database specification and design, *Data and Knowledge Engineering*, **2**, 31–58.

Shelley, E. P. and Johnson, B. D., 1991, Towards a national directory of natural-resources data, in Hudson, G. R. T. and Tellis, D. A. (Eds) *Proceedings of National Conference on the Management of Geoscience Information and Data, Adelaide, 22–25 July 1991*, Session 1, pp. 19–29, Adelaide: Australian Mineral Foundation.

Sheth, A. P. and Larson, J. A., 1990, Federated database systems for managing distributed, heterogeneous, and autonomous databases, *Association for Computing Machinery (ACM) Computing Surveys*, **22**(4), 183–236.

Siegel, M. and Madnick, S. E., 1991, *A Metadata Approach to Resolving Semantic Conflicts*, Sloan School working paper 3252-91-MSA. Cambridge, Massachusetts: Sloan School of Management, Massachusetts Institute of Technology.

Spackman, J. W. C., 1990, The networked organisation, *British Telecommunications Engineering*, **9**, 7–15.

Taupier, R., 1993, Comments on the economics of geographic information and data access in the Commonwealth of Massachusetts, in Onsrud, H. J. and Rushton, G. (Eds) *Sharing Geographic Information*, pp. 245–58, New Brunswick, New Jersey: Center for Urban Policy Research, Rutgers University.

Templeton, M., Brill, D., Dao, S. K., Lund, E., Ward, P., Chen, A. L. P. and MacGregor, R., 1987, Mermaid: A front end to distributed heterogeneous databases, in *Proceedings of the Institute of Electrical and Electronics Engineers (IEEE)*, **75**(5), 695–708.

Thieman, J., 1992, The Global Change Master Directory, in *Report of the Information Exchange Forum on Spatial Metadata*, pp. 221–36, Washington, DC: Federal Geographic Data Committee, US Geological Survey. *Note:* Accessible via *telnet* to nssdca.gsfc.nasa.gov (128.183.36.23) (username nssdc).

Tosta, N. J., 1992, Data sharing, *Geo Info Systems,* **7**, 32–4.

Vrana, R., 1992, Design and operation of a meta-GIS, in *GIS '92 — Working Smarter, Proceedings, International Symposium on Geographic Information Sytems (6th) February 10–13, 1992,* Vancouver, British Columbia: Polaris Learning Associates, Forestry Canada.

Walker, D. R. F., Newman, I. A., Medyckyj-Scott, D. J. and L. N. Ruggles, C. L. N., 1992, A system for identifying relevant data sets for GIS users, *International Journal of Geographical Information Systems,* **6**(6), 511–27.

Wang, R. and Madnick, S. E., 1989, Facilitating connectivity in composite information systems, *Data Base,* **20**(3), 38–46.

Worboys, M. F. and Deen, S. M., 1991, Semantic heterogeneity in distributed geographic databases, *ACM SIGMOD Record,* **20**(4), 30–4.

Yates, H. J., 1990, Information resource management, *British Telecommunications Engineering,* **9**, 22–7.

15

Standards for integration of multisource and cross-media environmental data

Rodney L. Slagle

Scientific and measurement standards and conventions are routinely developed for and incorporated into environmental programmes. Information systems designed to support these programmes also require standards. These information standards, which affect immediate and long-term use of the data, may rely extensively on the science involved and may place requirements on data content (e.g., metadata, taxonomic codes, location data and measurement methods). In addition, standards are necessary for addressing 'non-scientific' issues that can dramatically affect the long-term utility and availability of the data (e.g., database management systems, data access, computer architecture and data repositories). Several types and categories of current and developing information standards are discussed, and their impact on environmental information is evaluated. Several national organizations and federal agencies are actively involved in developing standards. Environmental information issues, however, are complex, and national approaches to standards development can be slow in producing results. National and international standards that incorporate one scientific methodology will be difficult to establish, if not undesirable in their restriction. National and international standards addressing information systems and other related issues will require rapid development, approval and dissemination to remain current and relevant. Interim solutions, such as establishing linkages and cross-relationships between complementary standards, may be most feasible. Current status of environmental standards development within national and international organizations is discussed, and an evaluation of future requirements is presented.

Introduction

Traditionally, scientific communication has relied on information exchange through written documents. These documents present scientific concepts, theories and conclusions and may include summaries of the empirical data that support the conclusions. During the past 10 years to 15 years, information exchange has begun to include the compendium of data, where the data themselves are viewed as a product (EPA, 1993). Data and information exchange have now become integral to conducting environmental research and management.

In an attempt to understand the complex linkages within the earth's ecosystems, environmental researchers and managers are expanding their applications to a holistic, cross-media (e.g., air, water, soil and biological) approach. Global change initiatives and the associated complex problems have increased the demand for cross-media data from multiple sources (e.g., federal agencies, universities, private sector, etc.). These requirements for cross-media data also exist at a more fundamental level. For example, a watershed manager may need access to data on stream flow and chemistry, snow pack level and chemistry, forest harvest and silviculture, and agricultural activity.

An additional driving factor for the exchange of environmental data is the high cost of data collection. Budget constraints and policies require organizations to collaborate in data collection and use. Current information management approaches produce incompatible and inconsistent heterogeneous databases that require extensive effort to identify, acquire, comprehend and integrate.

In general, all standards are developed in response to a common, recurring problem (Wood, 1989). The use of information standards is a fundamental step in addressing the problem of efficient data integration for environmental applications. Establishing national and international standards for appropriate and critical components of environmental data and information systems will provide the infrastructure for efficient use of this costly resource.

Status of Environmental Information Standards

Although some standards exist, no co-ordinated effort is focused on establishing comprehensive environmental information standards. These standards must be implemented by the developers and managers of data. This task is complex, because ownership of environmental data is spread over hundreds of local, county, state, federal, institutional and private organizations. In addition, the process of developing an industry-wide standard takes three to five years and can be very expensive. Even under these circumstances, however, environmental managers and scientists appear willing to compromise to achieve data exchange (e.g., NASA Master Directory).

When data ownership is distributed and development time significant, the financial responsibility for the development process becomes an issue. It is also very likely that those who implement standards may see only small benefits from their direct efforts. The high costs and considerable effort required make it important to focus standardization by careful selection of priority standards and through cost–benefit analyses.

In summary, the problems encountered in the exchange of data are becoming increasingly common, recurring issues. Standards impart a structured process to the design, development and management of environmental data. It is this structure, if well designed, that creates a cost-efficient approach to the multiple use of cross-media, multisource data.

Types of Standards/What to Standardize

Standards for environmental information fall into two general categories: (1) standards that incorporate the scientific approach, and (2) technology/architecture standards. Both types of standards are subject to continuous research, which complicates the process of developing and formalizing them. Therefore, standardization must be viewed as part of the ongoing research process. In assessing where to apply standards, it is important to consider the status of activities. Effective standards usually formalize consensus and current operational designs; however, standards that promulgate new designs and concepts can facilitate rapid scientific progress (Stonebraker, 1989).

Standards Incorporating a Scientific Approach

Environmental science encompasses a broad spectrum of disciplines such as chemistry, biology, meteorology, oceanography and geology. Each discipline must encapsulate its scientific approach in the data that it generates. The following subcategories of standards are common to all environmental disciplines:

- measurement methods;
- data coding schemes;
- minimum data elements;
- location data;
- scientific documentation;
- quality assurance;
- programme objectives.

The following sections address information standards issues related to each of these subcategories.

Measurement Methods

Historically, certain scientific disciplines have evolved standard measurement methods more than others. For example, considerable effort has gone into the development of standard analytical chemistry measurement methods. These well-documented methods have been formalized by several organizations (e.g., American Society for Testing Methods (ASTM), US Environmental Protection Agency (EPA)). Their standardization benefited from associated work by other industries (e.g., medical, manufacturing, material testing) that contributed large capital expenditures. Other environmental measurement methods, particularly those addressing biological parameters, are not so well defined; however, generally accepted practices are usually employed (e.g., flora and fauna taxonomy). Even when different methods are used to obtain the same 'measurement', it is common for the science to develop methods for comparing and converting the data. The critical need is for environmental databases to incorporate and link methods documentation to the data elements. In addition, measurement method standards must be responsive to improvements and modifications in the extensive research programmes generating the data.

Data Coding Schemes

Even with the use of standardized measurement methods, there are significant impediments limiting efficient interorganizational data exchange. One significant problem is the use of different approaches to 'coding' scientific results. Data coding schemes are an abstraction or classification of the data. Standardizing these coding schemes is very useful in the exchange of data. For example, the chemical industry has developed a Chemical Abstract Service (CAS) registry that assigns a unique serial number to chemical compounds. The biological profession, however, has not developed this level of standardization. Although biological taxonomy is a mature science, the coding of taxonomic data varies greatly within and across organizations. For example, different US Department of Agriculture Forest Service regions use different coding schemes for the same tree species (Ritters *et al.*, 1991). Thus, when compiling a national forest inventory every 10 years, the effort required to unify the data is considerable. The expense and effort are compounded when forest tree species data collected by the Bureau of Land Management (BLM), National Park Service, state forestry agencies, universities and private forest products companies must be incorporated.

To address this basic need to standardize biological taxonomic coding, EPA, the United States Geologic Survey (USGS), the National Oceanographic and Atmospheric Administration (NOAA), the Smithsonian Institution and others are developing a standard taxonomy database system. This system is based on a serial number coding scheme and will address problems of reclassification (taxonomic changes) and incorporate the necessary reference information.

Recognizing the importance of standard coding schemes, the National Institute of Standards and Technology (NIST) developed the Federal Information Processing Standard publication (FIPS) 19-1, Catalog of Widely Used Code Sets, and several other FIPS addressing standard coding schemes. Unfortunately, none of these coding standards adequately addresses scientific concepts. It is critical to note that standard coding schemes can act as interoperability links between various schemes. In other words, implementation of a standard coding scheme does not require uniformity, but rather establishes a common 'mapping' set.

Minimum Data Elements

The secondary use of environmental data is often dependent on what data elements (i.e., attributes, parameters or variables) the data set contains. Where appropriate, references to the location (i.e., point, site or region) of the observations presented and the date and time of data collection are essential to the long-term usefulness of the data. Occasionally, locations and date/time may not be applicable or desirable (i.e., confidentiality). A standard could be established requiring that these parameters be included in all databases with allowance for documented exceptions.

Establishing standard minimum data elements that address specific scientific disciplines is a much more complex and difficult problem. Few organizations are amenable to adding measurements that are not collected specifically for the primary use of the data. This added cost of non-essential data collection and subsequent management is often prohibitive. Yet, if a goal is to maximize the secondary use of the data then this additional effort may be required. A feasible approach to solve this problem is to develop, maintain and distribute a recommended minimum set of data elements that will aid new programmes in evaluating the potential cost–benefit of additional data collection. This minimum set could identify organizations that have interest in the specific elements so that these organizations may have the opportunity to contribute resources to its collection. Because of the relationship between the data elements, these minimum sets should be developed and organized along scientific disciplines, such as the EPA's Minimum Set of Data Elements for Groundwater (EPA, 1989).

Location Data

Location information is often critical to data integration. The accuracy of these spatial data is as important as their inclusion in the data set. A critical issue for secondary users of the data is that measurement methods and estimates of accuracy are included in the data set or documentation. Certain environmental programmes may require more accurate measurements than others, but maximum error goals can be established (EPA, 1992). In addition, the identification and documentation of the location reference grid is of specific importance as it establishes the location framework for all measurements. A major change was made in one such grid established for the North American Datum (NAD) in the change from NAD 1927 to NAD 1983.

Location data often represent sampling points; however, more complex features also must be described (e.g., data from transects and information about a polygon or region). These data must be included and represented in a format that makes them usable by a geographic information system (GIS) and, if possible, in a database management system (DBMS). Complex polygons or regions are not easily managed in a DBMS unless they are predefined (e.g., state, county). Several FIPS have been established to define and code commonly used regions, and a proposed FIPS defines a national standard, the Spatial Data Transfer Standard.

Scientific Documentation

Scientific documentation (i.e., metadata) is a vital component of any environmental programme and provides background information required for credible use of the data. Integration of data from multiple sources and across environmental media (e.g., air, soil, surface water, groundwater) is much simpler if certain components of this documentation have been standardized. Several efforts are ongoing to co-ordinate and standardize this information. The NASA Master Directory effort attempts to structure database directories and inventories that provide general information on what data are available. This effort has resulted in a commonly accepted Directory Interchange Format (NASA, 1991). The NASA Catalog Interoperability Workgroup is attempting to co-ordinate and standardize the content of and systems that manage scientific metadata and catalogues (e.g., database abstracts, data collection methods, data processing methods).

An objective of standardizing and managing scientific metadata is to link this documentation tightly with the actual database itself. Problems arise when trying to implement this approach in current commercial DBMS. The documentation is primarily text-based information, and extensive text linkages are not handled well by DBMS. In addition, this information usually takes the form of scientific documents; summarizing and standardizing this complex scientific information is difficult.

Several federal agencies, as well as the NASA co-ordinated efforts, are addressing this common problem. Compromises by the scientific and environmental communities will be required to standardize this information in a timely manner.

Quality Assurance

Quality assurance (QA) information is of much interest to primary and secondary data users. Considerable effort is expended in its collection and analysis. Although QA processes are themselves being standardized (e.g., International Standards Organization (ISO) 9000), the relevant issue here is designing and documenting this information and establishing a standard approach to linking/integrating it into databases.

The QA information available depends on the QA programme that was established for the data collection and processing. Efforts by ISO and the EPA's Quality Assurance Management Services Division to standardize the approaches used in these QA programmes will provide the foundation for common design, management, reporting and display of quality information. Establishment of minimum QA attributes — precision and accuracy of measurements and aggregated results — is only the beginning.

Programme Objectives

The objectives of an environmental programme establish the information goals that are often fulfilled through data collection based on sampling designs. Inherent in all sampling designs are accompanying constraints and limitations on the use of the data (e.g., experimental designs versus monitoring designs), and integration of data is hindered by incompatible design constraints. Even in the design of monitoring programmes, the site selection can be based on an unbiased process or it can be specifically focused on a known environmental problem (i.e., biased selection).

Because it is not possible to establish a single sampling design that meets the information objectives of all programmes, standardization should establish a suite of designs with specified integration characteristics. These integration characteristics would define, at the outset, which designs have compatible constraints. With this information, an environmental scientist or programme manager will better understand the limitations and opportunities of secondary uses of the data. In addition to predefined classes of sampling designs, there is a need for new statistical techniques and tools that facilitate integration of data with differing constraints and limitations (Brown, this volume; Stafford *et al.,* this volume).

The most promising area in standardizing programme objectives has little to do with formalized standards. Expanded communication and better collaborative efforts between managers and designers of large data-intensive environmental programmes will help to establish complementary designs. Again, it should be noted that the standardization objective is not uniformity but the designed integration of different approaches.

Technology/Architecture Standards

Managing and processing environmental data requires extensive use of electronics and computational tools, which necessitates consideration of technology and architecture topics applicable to environmental information:

- networks;
- database management systems;
- database documentation;
- data repositories.

To a large extent, the environmental industry must rely on technology developed by other industries. The environmental industry will not be able to fund the necessary development of new capabilities but will need to capitalize on the investments of other industries. This collaboration and leveraging of technology will realize significant cost reductions in the critical tools needed to process environmental data.

Networks

The defence and business industries have been the primary funding sources for the technology development necessary to support routine electronic information exchange. Today, the functionality of national and international electronic networks and systems is expanding (e.g., Internet, WAIS, Gopher, Archie), and their interoperability (e.g., TCP/IP, DECNET) has made them more available and useful. The environmental industry is just beginning to realize these capabilities and to capitalize on these resources.

Although the environmental community has been aware of the need for sharing data, the design and implementation of organized solutions have been slow. Traditional processes of locating relevant data through a 'network' of contacts and associates, followed by a request for a copy of the data and documentation, is still a common, if not predominant, procedure. Telecommunication networks provide the pipeline for routine, unattended data access. Planning for the expanded use of these networks is crucial to realizing data integration opportunities.

Efforts are under way to establish and expand a national information infrastructure highway (e.g., HR1757, the Higher Performance Computing and High Speed Networking Application Act of 1993) that provides an accelerated access and distribution mechanism. These efforts are critical to the success of routine data exchange.

Environmental information networks will capitalize on industry developments; however, physical access to data is not enough. Once given access to data, the effort required to understand the data, from the date management and scientific perspectives, can be a most formidable problem. This is where the scientific standards addressed earlier become so important.

Database Management Systems

In addition to electronic networks, significant progress has been made on the software tools used to manage and store data. Commercial DBMS have advanced significantly in their functionality and capability, as has the support for distributed access and interoperability between systems. Several organizations (e.g., ANSI, NIST, ISO) have been involved in development of standards that are implemented in commercial DBMS. The ANSI X3H4 committee is primarily responsible for the formalization of the Standard Query Language (SQL), both the 1989 and 1992 versions, and NIST has developed FIPS-127, which standardizes the SQL conformance test. This is a dynamic area of standards development, and current efforts are under way to establish standards for the object-oriented extensions to SQL; however, there currently is no consensus (Celko, 1993). Additional requirements are being developed to extend SQL to include spatial query operators. The interoperability of SQL and its implementation in high-powered statistical and spatial data analysis tools that link directly to sophisticated DBMS also improve the capability to process large, diverse environmental data sets.

Although the environmental industry will make extensive use of DBMS, environmental information systems are likely to rely primarily on commercial systems. The EPA has contributed to standardization of DBMS use in terms of defining data element naming conventions (Newton, 1992a). With NIST, the EPA is developing a draft standard for an interoperable name set (i.e., one that every group can map to their own names) to facilitate understanding the contents of a data set.

Database Documentation

Database documentation is often embedded in the DBMS; however, environmental applications often require documentation beyond that provided by commercial systems. Each data element in a database represents a scientific measurement or a coding of some scientific abstraction. The detailed description or attribution of those data elements is critical to their extended and secondary use. Again, EPA and NIST have developed a draft standard for data element attributes (Newton, 1992b) that is based on an ISO Committee draft on this topic.

Besides the data element attribute documentation, the complete database system needs to be documented to capture the database design, data flow diagrams and other elements. One approach to designing and developing this information is Enterprise Architecture Planning (Spewak, 1993) and a method of structuring the information gathered is the Zackman Framework (Zackman, 1987). These procedures attempt to define standard approaches to and structures for the complex set of requirements and design documents that are used in mapping an intricate information system.

Data Repositories

Environmental databases are becoming very complex information systems, with linkages across multiple years of data from multiple sources, and crossing multiple environmental media. The planning and design required for a large monitoring programme, university research department, or federal agency branch takes several years to formulate and document. One structured approach is termed the data repository, or Information Resources Dictionary System (IRDS). These systems contain and manage the compendium of 'computer science' metadata (e.g., process flow diagrams, entity relationship diagrams). The EPA Office of Information Resources Management, Information Management and Data Administration Division (OIRM IM/DA) has worked with NIST, the American National Standards Institute (ANSI), and others to standardize this repository design, which resulted in FIPS-156. This continuing work is expected to expand the concepts to include the compendium of enterprise planning information and knowledge resources (e.g., scientific metadata). Current efforts in this area have been termed a Global Information Resources Dictionary (GRID) model for unified metadata representation and management (Hsu *et al.,* 1991).

The implementation of standard repositories of environmental databases can be an important step in the maintenance, exchange and integration of data. EPA has suggested that each environmental organization (e.g., federal agencies, universities) produce a conceptual data model that is registered and maintained in on-line repositories.

Current National and International Efforts

Several national and international organizations and federal agencies are actively involved in developing required priority standards. Just a few of these organizations and their activities are presented.

International Standards Organization

The International Standards Organization (ISO) is an international coalition of national standards organizations. The USA has representation (NIST, ANSI, in the ISO; however, the European Community (EC) is a principal driver of the approach and products that are developed. Several ISO products are of interest to the environmental community. First, ISO has a significant focus on Quality Systems (i.e., ISO 9000 (Nadkarni, 1993)) that generally apply to all product and service producers. Much of the approach is consistent with Total Quality Management (TQM); however, ISO has documented a standard, certifiable process. ISO is also addressing the QA of software (i.e., ISO 9000-3, Software Standards/Guidance (Bloor, 1993)) and takes the approach that 'software development needs to be viewed as any other product, not as an art form'. ISO has also developed a standard that addresses definitions and terminology (i.e., ISO 8402). These standard terminology approaches can be very useful if extended to environmental sciences and information.

National Institute for Standards and Technology

The National Institute for Standards and Technology (NIST), formally the National Bureau of Standards (NBS), is involved in collaborative efforts to define, develop and formalize several standards that affect environmental information. The following FIPS data standards are relevant to environmental information:

- Representation for Calendar Date and Ordinal Date for Information Interchange (FIPS 4–1);
- Codes for the Identification of the States, the District of Columbia and the Outlying Areas of the United States, and Associated Areas (FIPS 5–2);
- Counties and Equivalent Entities of the United States and its Possessions, and Associated Areas (FIPS 6–4);
- Metropolitan Statistical Areas (FIPS 8–5);
- Congressional Districts of the United States (FIPS 9);
- Catalog of Widely Used Codes Sets (FIPS 19–1);
- Guidelines: Codes for Populated Places, Primarily County Divisions, and Other Location Entities of the United States and Outlying Areas (FIPS 55–2);
- Representations of Local Time of Day for Information Interchange (FIPS 58–1);
- Representation of Geographical Point Locations for Information Interchange (FIPS 70–1);
- Hydrologic Units (FIPS 103).

These standards are useful in development of information systems; however, they are far from comprehensive. NIST continues to expand this list and to work with other agencies to define required environmental information standards.

Environmental Protection Agency

The EPA Office of Information Resources Management (OIRM) has primary responsibility within the agency for addressing information standards. OIRM provides leadership in developing new agency standards as well as in collaborating with other federal agency efforts, and is responsible for the EPA Data Standards Program. The manager of the OIRM Information Management/Data Administration Division is a member of the ANSI X3H4 committee that is addressing database management standards (e.g., SQL).

In other areas, EPA has established internal standards and policies. The EPA publication *EPA Catalog of Data Policies and Standards Policy — July 1991* (EPA, 1991) serves as a reference and provides documentation of current EPA standards. This publication currently provides information on:

- federal data-related laws, regulations, policies and guidelines;
- EPA data-related policies;
- EPA agency-wide data standards;
- federal data standards;
- international data policies, guidelines and standards.

Other standards and policies will be added to this document as they are developed and approved by EPA.

Conclusion

Common resistance to the call for standardization includes the arguments that it will stifle creativity, stop the development of new techniques and create a monoculture of antiquated systems. Standards, however, only attempt to address an efficient handling of a common, recurring issue; the value of data beyond its immediate intended use is not commonly clear to researchers and information support staff. In addition, information managers working in isolation often design redundant data and documentation systems to meet their immediate needs. Only over time, with organized standardization efforts, will these inefficiencies be corrected.

Several opportunities and directions exist for correcting the current deficiency in comprehensive standards for environmental information. First, effort should be focused on establishing scientific and technology/architecture standards that are developed for or applied to environmental information. A comprehensive map of standards categories, more detailed than the one used in this paper, could be used to chart the required areas of need and establish priorities for their development. A cost–benefit approach to prioritizing these efforts should be taken. This effort must be collaborative, and all stakeholders must be represented (e.g., federal and state agencies, universities, institutions, private sector). These efforts could be led and co-ordinated by current national or international standards organizations (e.g., NIST and ISO). Such an organization could act as an accreditation board to certify that a specific information system complies with established standards. Funding for this effort could be distributed among participants, thereby reducing the individual burden. Using a phased priority approach to developing standards and capitalizing on standards developed through investments from other industries (e.g., networks, DBMS technology) will significantly reduce costs.

Standardization efforts often fail after a standard has been completed and approved. To avoid this, funding sources could include requirements to conform to and implement all information standards. A process of applying for exception and variances would allow for review and provide leverage for compliance. In addition, all larger programmes should initiate a standards programme that co-ordinates their standards efforts and provides mechanisms to interact with national and international standards organizations. An up-front acknowledgement of the importance of sharing environmental data and commitment to data publication and network access and distribution should be included in mission statements by all data-producing organizations. In the long term, standards are one part of a framework for an integrated, distributed national/international environmental information system.

References

Bloor, R., 1993, Mission critical view, *DBMS Client Server Computing,* **6**, 10–12.

Celko, J., 1993, SQL explorer, *DBMS Client Server Computing,* **6**, 16–17.

EPA, 1989, *Environmental Protection Agency Order 7500.1 — Minimum Set of Data Elements for Groundwater,* Washington, DC: Environmental Protection Agency.

EPA, 1991, *Agency Catalog of Data Policies and Standards,* 21M-1019, Washington, DC: Environmental Protection Agency.

EPA, 1992, *Locational Data Policy Implementation Guide,* EPA/220/B-92/008, March 1992, Washington, DC: Environmental Protection Agency.

EPA, 1993, *EMAP Information Management Strategic Plan: 1993-1997,* Washington, DC: Environmental Protection Agency.

Hsu, C., Bouziane, M., Ratter, L. and Yee, L., 1991, Information resources management in heterogeneous, distributed environments: A metadata-base approach, *IEEE Transactions on Software Engineering,* **17**(6), 604–25.

Nadkarni, R., 1993, ISO 9000 Quality management standards for chemical and process industries, *American Chemistry,* **65**(8), 387–95.

NASA, 1991, *Directory Interchange Format Manual,* Version 4.0, NSSDC, December 1991, No. 91-32, Greenbelt, Maryland: NASA Goddard Space Flight Center.

Newton, J., 1992a, 'Draft standard for data element naming conventions', unpublished draft, National Institute for Standards and Technology and Environmental Protection Agency.

Newton, J., 1992b, 'Draft standard for data element attribution', unpublished draft, National Institute for Standards and Technology and Environmental Protection Agency.

Riiters, K., Papp, M., Cassell, D. and Hazard J. (Eds), 1991, *Forest Health Monitoring Plot Design and Logistics Study,* Research Triangle Park, North Carolina: US EPA, Office of Research and Development.

Spewak, S., 1993, *Enterprize Architecture Planning,* Wellesley, Massachusetts: QED Publishing Group.

Stonebraker, J., 1989, Future trends in database systems, *IEEE Transactions on Knowledge and Data Engineering,* **1**(1), 33–43.

Wood, H., 1989, Information technology standards: What determines success?, *Computer Magazine,* **22**, 67–8.

Zackman, J., 1987, A framework for information systems architecture, *IBM Systems Journal,* **26**(3), 276–92.

SECTION V

Databases for broad-scale research

Developing atmospheric circulation models, biosphere models and other broad-scale models designed to represent environmental processes at regional to continental, or global scales has generally been difficult, and the resulting models have frequently been inadequate. Reasons for these problems include the improper scale (spatial, temporal and spectral) of many existing databases, inappropriate level of aggregation (either too fine, or, more frequently, too coarse), inadequate metadata, and the fact that databases developed for one purpose are generally being forced into new applications. Furthermore, many existing models were developed from the standpoint of the physical sciences (e.g., atmospheric sciences, oceanography) and do not adequately incorporate the influences that the earth's biota have on 'physical' processes.

Section V represents an attempt to examine many of the issues related to design, development and utilization of databases for broad-scale research and includes pertinent examples related to vegetation mapping, atmospheric circulation, soils and land cover. In the first chapter of the section, Dungan, Peterson and Curran provide examples of regression and two geostatistical methods for mapping and quantifying vegetation by a combination of point-based and whole-area approaches. They emphasize that the specific consequences of choosing a particular mapping method (accuracy, ease of parametrization, etc.) and the specific goals of the map should be considered. Hayden documents the need to abandon the old paradigm that climate is controlled by vegetation only at the microscale. He further provides a comprehensive analysis of the types and resolution of data that will be required to represent adequately the feedback loops between vegetation and climate for General Circulation Models. In a similar vein, Potter, Matson and Vitousek present a framework for spatial modelling and evaluation of potential aggregation errors associated with global gridded data sets. Specifically, they present sensitivity and scaling analyses that incorporate attributes of a soil database in an ecosystem model of global primary production and soil microbial respiration, and identify several needs and challenges related to research on global change.

Reed et al. indicate that scientists working at ecosystem to global scales typically develop independent data sets to satisfy their specific needs, thereby duplicating the efforts of others, but adapting different standards and formats. They describe a prototype database on land cover characteristics that supports translation to various classification schemes used in global models, thereby potentially solving many of these problems.

In the final chapter of the section, Kineman and Phillips discuss the role of characterization databases with respect to global modelling and synthesis. They emphasize the need to develop rigorous standards for data publication and other mechanisms that will improve the scientific and technical quality of data and information.

16

Alternative approaches for mapping vegetation quantities using ground and image data

Jennifer L. Dungan, David L. Peterson and Paul J. Curran

A major challenge facing ecologists studying the earth as a system is the mapping of vegetation quantities over large areas. In the absence of remotely sensed data, only point-based interpolation methods, such as kriging other distance-related methods, are possible. The availability of remotely sensed data allows the use of whole-area approaches, using classification, statistical or physical models. Geostatistical methods such as co-kriging and conditional simulation may allow the combination of point-based and whole-area approaches to exploit more fully the available information. These alternatives may have very different consequences depending on the purpose of the vegetation map, as shown by an example constructed from a 36 km^2 segment of imaging spectrometer data. The consequences of these methods must be considered before such maps are provided to geographic databases.

Introduction

A major challenge facing ecologists studying the earth as a system is the mapping of vegetation quantities over large areas. Maps of these quantities, such as leaf area index (LAI) or absorbed photosynthetically active radiation (APAR), are needed to parametrize biogeochemical cycle and climate models (Avissar and Verstraete, 1990; Henderson-Sellers, 1991). Such maps may also be useful in themselves for monitoring regional or global status and changes in vegetation quantities.

LAI, APAR and other vegetation quantities are continuous variables that have zero or positive finite values for every point or area on the terrestrial surface. To create maps of such variables, Burrough (1987) identified whole-area and point-based approaches. Whole-area approaches use inexpensive, plentiful data and apply transforms that relate those data and the variable of interest to the area. Point-based approaches are based on sampling the variable to be mapped and use an interpolation scheme to create the map. Both approaches have been used for the regional and global mapping of vegetation quantities (Lieth, 1975; Ludeke *et al.*, 1991; Chong *et al.*, 1993; Merrill *et al.*, 1993).

In the historical development of global primary productivity maps, for example, plot measurements were assumed to be representative of large, homogeneous areas containing a particular vegetation type. The plot measurements were weighted by area to produce global average estimates. Lieth (1975) suggested that the accuracy of these estimates increased as the number of assumed homogeneous areas increased. This method can be thought of as a type of nearest-neighbour interpolation, because each land area is assigned the value of the sample contained within it and is at this scale essentially point-based. Another point-based approach has been to use interpolated climate data as input to models that predict primary productivity (Rosenzweig, 1968; Friedling-stein *et al.*, 1992). In contrast, approaches utilizing remotely sensed images are essentially whole-area approaches. Traditionally, classification methods have been used to assign a qualitative category to each pixel in an image for such purposes as land-cover mapping. To map quantitative variables, regression methods or inversion of physically based models have been utilized (Pearson *et al.*, 1976; Johnson, 1978; Curran, 1987; Rosema *et al.*, 1992).

Point-based methods for mapping that have not been used extensively for mapping vegetation quantities are geostatistical estimation techniques. These methods exploit the autocorrelation that is usually present in the spatial distribution of natural phenomena in order to increase the accuracy of estimates between sample locations (Journel, 1986). Kriging is the geostatistical estimation method for a single variable; when more than one variable is being used for estimation it is referred to as co-kriging. Geostatistical simulation (known as conditional simulation or stochastic imaging) is an alternative method of producing maps (Journel and Alabert, 1989). Both co-kriging and conditional simulation may allow the combination of point-based and whole-area approaches to exploit more fully the available information.

The choice of a mapping method has critical implications for accuracy, representation of spatial pattern and scale, and for correlation with variables in other maps or layers. The requirements of producing a map should be clearly understood before a method is selected. One requirement may be the accurate estimation of the central tendency or the whole-map statistics of the variable of interest (such as mean, variance, minimum, maximum and/or quantiles). Or, the goal may be to measure the correspondence between two variables of interest, such that their correlation is estimated accurately. Correlation with other maps or layers may be particularly important when such maps are being used in a geographic information system or for modelling purposes. Or, it may be critical that error at any location does not exceed some specific threshold. Subsets of the map may be of particular interest; in which case, local accuracy is paramount. If the map is to be input to some non-linear transfer function, such as a deterministic process model, the most important features may be the extreme values. Representation of connectivity would be important if the transfer function includes spatial dependence, such as a hydrologic model. An auxiliary goal may be to provide estimates of the confidence with which the quantity is mapped. No one method is optimum for all purposes — trade-offs in the choice of a method must be considered before such maps are provided to geographic databases.

This chapter presents an example of three alternatives for creating maps that combine point and whole-area information. The example uses an artificial data set constructed from Airborne Visible Infrared Imaging Spectrometer (AVIRIS) data, a multispectral scene from a 36-km^2 forested area on the coast of Oregon. The data are used to construct an analogy to a situation where both direct vegetation measurements and remotely sensed data are available. Regression, co-kriging and conditional simulation are applied to the data and evaluated on a number of criteria. The spatial implications of these different mapping methods are also presented. While these are by no means the only such methods, the example is used to compare two geostatistical methods that have not found wide use with remotely sensed images and a regression method that has been applied frequently in the remote sensing field.

Three Alternatives to Using Remote Sensing for Mapping Vegetation Quantity

The Regression Method

In the remote sensing literature, reports of studies abound in which radiance measures of selected pixels are compared with ground measurements of some attribute of vegetation. These measurements include woody biomass, LAI and mean diameter at breast height of forests; and LAI, percentage cover and green biomass of agricultural crops (Asrar *et al.*, 1985; Jensen and Hodgson, 1985; Badwhar *et al.*, 1986; Vujakovik, 1987; Peterson *et al.*, 1987; Ardo, 1992; Cohen and Spies, 1992; Curran *et al.*, 1992). Correlation coefficients and sometimes regression equations describing relationships are reported in these studies. These relationships have been used only occasionally for direct prediction or mapping (Tucker *et al.*, 1985; Curran and Williamson, 1987; Wessman *et al.*, 1988; Running *et al.*, 1989). In these studies the ground measurements are in a sense used to 'calibrate' the sensor for mapping the variable of interest. The calibration equation is applied to all pixels containing the appropriate land cover type to produce a map of the variable.

Regression, or in fact any statistical method, may be seen as a naive method for using remote sensing to map vegetation quantities on the earth's surface. Rosema *et al.* (1992) state that 'Regression analysis generally does not involve all the relevant parameters and will never provide a relationship of more general validity. In fact, regression analysis cannot be considered a very suitable tool [for the purpose of mapping].' The inversion of physically based models such as Rosema *et al.* (1992) describe is perhaps more realistic but does require significantly more knowledge about the scene.

The regression method depends on a close relationship between remotely sensed data and the variable of interest and on the consistency (or stationarity) of that relationship across the region being mapped. It is not generally an 'exact' method, in that the values estimated using the regression equation will not be exactly the same as the sample values. All pixels or sample units are taken to be independent of their neighbours, so spatial autocorrelation is preserved only as a side-effect of applying a transformation to what may well be autocorrelated values. Curran and Williamson (1986) state that, if this assumption of independence is made, the number of ground samples required for calibration usually becomes undesirably large. Uncertainty bounds could be given to each estimated value of a vegetation quantity by the standard error of prediction from the regression analysis, but these bounds would not vary spatially.

The Co-kriging Method

Geostatistical techniques such as co-kriging have potential for exploiting spatial correlation, which most other methods, including the regression method discussed here, do not. In co-kriging, values at unsampled locations are estimated using a linear combination of values from sampled locations and values of another related variable. This method is especially useful in cases where data for the variable of interest (herein referred to as the primary variable) are scarce and data for a related variable (herein referred to as the secondary variable) are more plentiful (Journel and Huijbregts, 1978). It is a point-based interpolator, except in those cases where data for the related variable are available on an exhaustive basis, such as a complete grid. Then it could be considered a hybrid of point-based and whole-area methods.

Co-kriging has been used in several studies in which one variable has been interpolated using sparse measurements of that variable and a second, more thoroughly sampled variable (Yates and Warrick, 1987; Leenaers *et al.*, 1990; Nash *et al.*, 1992). The method has only recently been investigated for merging ground measurements and remotely sensed data. Atkinson *et al.* (1992) discuss an application of co-kriging using primary measurements of green leaf area index, dry matter and percentage cover of clover in agricultural fields, with vegetation indices from a field radiometer as the secondary variable. They found that co-kriging was about nine times more efficient than kriging for creating maps of percentage cover of clover. The comparison of efficiency was based on the sizes of the maximum estimation variances, one relative measure of the overall error in the resulting map. They concluded that co-kriging with remotely sensed data was useful because it is easier to obtain radiometric measurements than to destructively sample vegetation. Bhatti *et al.* (1991), also working in an agricultural area, were interested in soil organic matter, soil phosphorus concentration and grain yield as primary variables, and a linear transform of Landsat Thematic Mapper bands 4 and 5 as the secondary variable. They too found a reduction of the estimation variance when comparing co-kriging with kriging and recommended the use of co-kriging with remotely sensed data for resource mapping. Gohin and Langlois (1993) used co-kriging to map sea surface temperature (SST) from direct measurements of SST and information from bands 4 and 5 of the NOAA-9 Advanced Very High Resolution Radiometer (AVHRR). In this study, many of the remotely sensed data were not used for interpolation — only those pixels collocated with *in situ* temperature measurements were utilized because of extensive areas of cloud. The authors suggest that the combination of both sources of temperature information would improve the results over those using *in situ* or satellite sensor measurements alone, though the accuracy relative to other methods was not evaluated.

The co-kriging estimator at an unsampled location is a linear combination of nearby samples as follows:

$$p(\mathbf{x}_0)^* = \sum_{k=1}^{N_1} \lambda_k p(\mathbf{x}_k) + \sum_{j=1}^{N_2} w_j s(\mathbf{x}_j)$$

where $p(\mathbf{x}_0)^*$ is the estimate at location $\mathbf{x}_0$, $p(\mathbf{x}_k)$ are the N_1 nearby primary values weighted with factors λ_k, and $s(\mathbf{x}_j)$ are the N_2 nearby secondary values weighted with factors ω_j. The weights are derived from a set of normal equations describing minimum expected squared error and non-bias, as in kriging (Journel and Huijbregts, 1978). These conditions make co-kriging a 'best linear unbiased estimator' (Goldberger, 1962). Alternatives are available for setting constraints on the weights to ensure unbiased estimates. The traditional constraints are to set $\sum_{k=1}^{N_1} \lambda_k = 1$ and simultaneously $\sum_{j=1}^{N_2} \omega_j = 0$. N_1 and N_2 are defined by the 'search neighbourhood', or the distance beyond which sample values are not thought to be relevant to the unsampled value.

The application of co-kriging, as with any geostatistical method, begins with the development of models of spatial covariance, which are used in the calculation of the weights, λ_k and ω_j. For co-kriging with two variables, three covariance models are needed: one for the primary variable, one for the secondary variable and one describing their cross-correlation. Further, the models must be described such that any covariance matrix that is created from them is positive semi-definite (Christakos, 1984). The resulting model, the so-called linear model of coregionalization (Journel and Huijbregts, 1978), is sometimes difficult to develop in practice because there is no guarantee that experimentally estimated covariances will correspond with this requirement, an outcome of mathematical theory rather than natural law. The requirements for this model are described more fully in Myers (1982) and Goulard and Voltz (1992).

The appeal of co-kriging in comparison with kriging lies in its potential for using plentiful secondary information, such as remotely sensed data, to increase the accuracy of estimation at locations where the primary variable has not been measured (Leenaers *et al.*, 1990; Stein *et al.*, 1991). This potential increase in accuracy comes at the cost of more covariance modelling, which is usually the most time-consuming step for the investigator carrying out such an analysis. Whether the additional effort is worthwhile depends on the strength of the relationship and the spatial cross-correlation between the two variables.

Like kriging and most other interpolators, co-kriging results in a smoothed map (Lam, 1983) and therefore will not reproduce fine-scale features or the spatial covariance models that were used in its construction. It is an exact method in that values at sample locations are preserved in the estimated map. The kriging models of error have been used in the past to create maps of uncertainty. Webster and Oliver (1990) cite this prediction of error to be one of the primary advantages of kriging methods. Journel (1986) cautions against the use of kriging variance as a spatial predictor of error, as it depends on data locations rather than data values. Journel and Rossi (1989) and Rossi *et al.* (in press) show examples of the lack of relationship between kriging variance and actual local error.

The Conditional Simulation Method

Rather than providing one map composed of linear minimum error variance estimates, as the kriging methods do, conditional simulation is a geostatistical method that involves a stochastic component resulting in many 'equally probable' maps. Each unsampled location is therefore not described by its expected value, but by a distribution created from information about the neighbouring samples and the spatial covariance. Simulations are 'conditional' to the data, in the sense that they come from a probability model of the variable at an unsampled location 'given' the surrounding data. Monte Carlo selections from each distribution at every point produce an arbitrary number of maps that each reproduce the sample data at their measured locations and spatial covariance deemed representative of the region. One of the main features of conditional simulation, the attempt to reproduce spatial patterns, makes it more suitable for some mapping purposes than kriging. In addition, it characterizes the uncertainty about the spatial distribution of the variable of interest (Journel and Alabert, 1989).

The first conditional simulation method was the turning bands algorithm (Matheron, 1973; Journel, 1974). Other algorithms have since been developed, including sequential methods that are more flexible for a variety of applications. Journel and Isaaks (1984), Dowd (1992) and Rossi *et al.* (1993) present case studies using sequential simulation. The sequential indicator method (Journel and Alabert, 1989; Gomez-Hernandez and Srivastava, 1990) is a non-parametric method that is particularly suited to the addition of information about a second related variable, such as one measured by remote imaging instruments, but is not yet widely represented in the published literature.

Sequential indicator simulation is founded on the properties of 'indicators' (Journel, 1983), binary random variables defined as

$$I\{Z(x);k\} = \begin{cases} 1 \text{ if } Z(x) \leqslant k \\ 0 \text{ otherwise} \end{cases}$$

where $Z(\mathbf{x})$ is a continuous random variable, a function of the spatial location $\mathbf{x}$; and $I\{\cdot\}$ is the indicator for k, some threshold value. Marginal and joint distributions are conveniently characterized from moments of indicator variables, since

$$E\ [I\{z(\mathbf{x});k\}\rangle = Prob\{Z(\mathbf{x}) \leqslant k\}$$
$$E\ [I\{z(\mathbf{x});k\}I\{z(\mathbf{x}+h);k\}\rangle = Prob\{Z(\mathbf{x}) \leqslant k, Z(\mathbf{x}+h) \leqslant k\}$$

where $E\ [\cdot]$ is the expectation and $Prob\{\cdot\}$ is the probability. Kriging indicator variables to obtain their expected values effectively estimates the probabilities of the continuous variable $Z(\mathbf{x})$, which can be used to construct conditional cumulative distribution functions (cdfs). The stochastic component of simulation draws values at random from these cdfs.

Journel and Zhu (1990) have developed a sequential indicator method, which they call the Markov–Bayes algorithm, that uses two variables. The Bayesian component of this algorithm is common to all sequential simulation methods and arises from Bayes's theorem about conditional probability. The theorem states that the probability that a value is less than a threshold value *given* the sample data and all other available *a priori* information is equal to the joint probability of not exceeding a threshold value *and* the occurrence of sample data divided by the marginal probability of the occurrence of the sample data. Devroye (1986) explains how the sequential application of this theorem can be used to describe joint probability. The Markov component of the method is that, at any given location, primary data take precedence over secondary information. By choosing appropriate thresholds, effectively discretizing the continuous variables, and transforming the primary and secondary data into indicators for these thresholds, the requisite conditional probability distributions can be developed.

The incorporation of a secondary variable in a conditional simulation method has the potential to provide more information, as in co-kriging. However, this information must be accompanied by the modelling of secondary and cross-covariances for all indicator thresholds. The Markov–Bayes algorithm reduces the burden of modelling by using the correlations between the primary and secondary indicator variable as calibration parameters to derive cross-covariance models from the primary covariances.

Each map produced by conditional simulation is said to be as probable as any other given the sample. Therefore the set of all simulations, taken jointly, describes the uncertainty arising from the interpolation or extrapolation of the sample. The set may be summarized to provide mean, median or other estimates from the cdfs. These summaries may be used in risk assessment, sample design or other applications where spatial estimates of uncertainty are needed (Journel and Alabert, 1989; Rossi *et al.*, 1993).

An Example

As an example of the differences between the regression, co-kriging and conditional simulation methods for using ground measurements and remotely sensed data for mapping vegetation quantities, a data set was artificially constructed so that results could be compared objectively using a single 'true' map. The data came from the Airborne Visible Infrared Imaging Spectrometer (AVIRIS), which collects data in 224 10-nm bands between 400 nm and 2400 nm (Vane, 1988). Data collected on 14 August 1990 during the Oregon Transect Ecosystem Research Project (Peterson and Waring, in press) of a 36-km^2 area on the Oregon coast were used. A scene of a 300×300 pixel area was chosen that was composed almost entirely of vegetated land cover, mostly conifer forest with small areas of road and clearcut containing regrowth. No visible clouds existed in the scene, and the data were generally of high quality.

Values representing radiance in $\mu Wcm^{-2}nm^{-1}sr^{-1}$ from an average of seven near-infrared bands of the AVIRIS scene were selected to represent the vegetation variable of interest, the primary variable. The image of the primary variable can then be considered the 'true' map; that is, the actual spatial distribution of a vegetation quantity on the ground as represented with a 20-m pixel size (the spatial sample support of AVIRIS). This image was the average of seven bands representing approximately 776 nm to 844 nm, on the near-infrared plateau of vegetation spectra. Radiance from a second spectral region, comprising an average of eight bands from 2128 nm to 2207 nm, was chosen to represent the secondary variable, as an analogue to remotely sensed data of the same area. These secondary data are related in some way to the primary variable, but not through a simple linear transformation.

These two bands were chosen from different regions of the spectrum to avoid the high correlation that most adjacent bands have. Figure 16.1 shows grey-scaled displays of the true map and the secondary variable map. A scatterplot showing the relationship between the pixels in the two images is shown in Figure 16.2. The linear correlation coefficient (r) of the 90 000 primary data with the secondary data is 0.42. A random sample of 300 pixels was taken from the true map to represent actual measurements as might be taken in a ground sampling campaign. The sample therefore represents 0.3 per cent of the true map. The locations of the samples were chosen to be spatially random by selecting values from a uniform distribution using a pseudorandom number generator (Park and Miller, 1988). These locations are plotted in Figure 16.3. The 300 sample values of primary data and the 90 000 secondary data were used to create maps of the primary variable using regression, co-kriging and conditional simulation methods.

(a)

2300

Radiance
$\mu W\ cm^{-2}nm^{-1}sr^{-1}$

500

2000 m

(b)

115

Radiance
$\mu W\ cm^{-2}nm^{-1}sr^{-1}$

0

2000 m

Figure 16.1 Grey-scaled images from bands of AVIRIS data used to construct data set:
(a) near-infrared band, used as 'true' map of primary variable; (b) middle infrared band,
used as map of secondary variable.

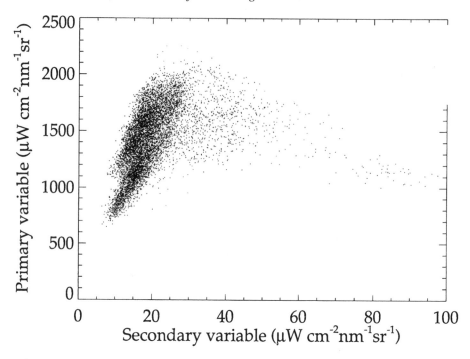

Figure 16.2 Scatterplot of primary and secondary data.

Application of Regression

A straightforward application of the regression method includes the development of a linear regression model from the sample primary data and collocated
secondary data, and use of this model to estimate values of the primary variable
for all pixel locations in the image. Figure 16.4 shows a regression model
developed from the 300 sample locations. The model, with two outliers removed
(for $s(x_k) > 80$) is

$$p(x_k)^* = 23.466 s(x_k) + 944.09$$

where $p(x_k)^*$ is the estimated value of the primary variable at location x_k, $s(x_k)$ is
the secondary data value at the same location, and r is 0.55. Applying this
function to all the secondary data produces the map shown in Figure 16.5.

Application of Co-kriging

Both geostatistical methods described here make use of a stationary random
function model — a conceptual model that cannot be tested. The essence of the
requirements of the random function model is that the spatial covariance model

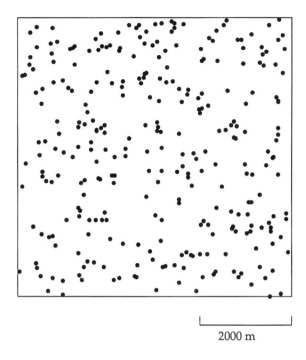

2000 m

Figure 16.3 Locations of 300 sample pixels from the 90 000 pixels of the true map.

Note: Symbols representing locations do not represent pixel size, but are enlarged for visibility.

is assumed to be applicable to the whole region. Another, stricter requirement is that the means do not fluctuate very much throughout the region. One way of checking that the data are consistent with a stationary random function model is to examine the mean of each pair of data values as a function of their distance apart. These means were plotted as a function of distance for the sample in Figure 16.6. The lack of fluctuation in the means indicates that a stationary random function may be a reasonable model to apply to the data.

One of the consequences of the stationary model is that spatial covariance can be expressed as semivariance (γ), which is the traditional measure of autocorrelation used in geostatistics (Curran, 1988). Semivariance as a function of lag ($\mathbf{h}$), the distance vector between pairs of data used to calculate semivariance, is called the semivariogram. In this example, semivariogram models for the primary variable ($\gamma_p(\mathbf{h})$) and the secondary variable ($\gamma_s(\mathbf{h})$) and their cross-semivariogram $\gamma_{ps}(\mathbf{h})$) were needed for co-kriging.

To develop ($\gamma_p(\mathbf{h})$, semivariogram values were calculated from the 300 sample data. As with any estimator, the values from the sample will differ from the values for the population. These differences are evident in comparing the omnidirectional (computed in all directions) semivariogram calculated from the

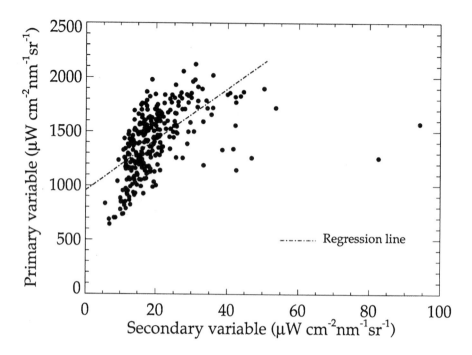

Figure 16.4 Scatterplot of primary samples with collocated secondary data.
Note: Linear regression line represents fit with two outliers removed.

samples with the four directional semivariograms calculated from the true map
(Figure 16.7a). Despite these differences and the noisiness of the sample
semivariogram, the model developed from the sample data alone follows closely
the directional semivariograms. Four directional semivariograms computed
from the exhaustive sample of the secondary variable are shown in Figure 16.7b.
The cross-semivariograms calculated on just the collocated primary and secondary data are shown in Figure 16.7c. Models were chosen to fit these experimental
semivariograms such that they could be combined in a linear model of
coregionalization as follows:

$$\gamma_p(\boldsymbol{h}) = 2 + 25000\,Exp_{10}(\boldsymbol{h}) + 70000\,Sphr_{60}(\boldsymbol{h})$$
$$\gamma_s(\boldsymbol{h}) = 2 + 20\,Exp_{10}(\boldsymbol{h}) + 90\,Sphr_{60}(\boldsymbol{h})$$
$$\gamma_{ps}(\boldsymbol{h}) = 1 + 400\,Exp_{10}(\boldsymbol{h}) + 1400\,Sphr_{60}(\boldsymbol{h})$$

where $Exp_{10}(\boldsymbol{h})$ represents an exponential model with a range of 10 pixels
(200 m), and $Sphr_{60}(\boldsymbol{h})$ represents a spherical model with a range of 60 pixels
(1200 m). The range is the distance beyond which the semivariogram levels off
and reaches its sill. The spherical and exponential models are two basic models
that are described in detail in Isaaks and Srivastava (1989).

2300

Radiance
$\mu W\ cm^{-2}nm^{-1}sr^{-1}$

500

2000 m

Figure 16.5 Map predicted with regression.

The algorithm used for co-kriging is from Deutsch and Journel (1992). The unbiasedness constraint used was the 'traditional' constraint, that is $\sum_{k=1}^{N}\lambda_{k}= 1$ and $\sum_{j'=1}^{N_{i}}\omega_{j}= 0$. A search neighbourhood of 60 pixels was used to define which values would be weighted for the estimate. The resulting image is shown in Figure 16.8.

Application of Conditional Simulation

The Markov–Bayes algorithm was used for conditional simulation of the primary variable. The parametrization of this algorithm is different from that of the co-kriging method, as it uses sequential indicator simulation (Deutsch and Journel, 1992). For the purposes of this example, the nine deciles were chosen to discretize the primary variable and generate the indicators. Without the simplified method of generating cross-semivariograms allowed by the Markov assumption, 81 (the number of thresholds squared) semivariogram models would be needed. Instead, only nine models were needed for the first to the ninth decile of the primary variable distribution. The experimental indicator semivariograms and the fitted models are shown in Figure 16.9. The parameters of the exponential models used are presented in Table 16.1. Journel and Posa (1990) show that, in theory, the sill of the pth quantile indicator must be equal to

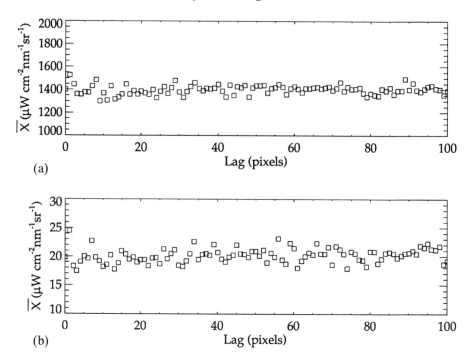

(a)

(b)

*Figure 16.6 Plots of means of pairs at **h** pixels apart: (a) of primary variable sample; (b) of secondary variable sample.*

$p(1 - p)$. The experimental semivariograms appeared to be in agreement with this equality and the fitted model sills approached the theoretical values.

All the secondary and cross-indicator semivariogram models were inferred using the calibration procedure described in Journel and Zhu (1990). This procedure is described in terms of covariances instead of semivariograms as follows:

$$C_{pks_j}(\mathbf{h}) = B(k)C_{p_k}(\mathbf{h}) \forall \; \mathbf{h}$$
$$C_{s_j}(\mathbf{h}) = B^2(k)C_{p_k}(\mathbf{h}) \forall \; \mathbf{h} > 0$$
$$= |B(k)| \, C_{p_k}(\mathbf{h}), \; \mathbf{h} = 0$$

where $\mathbf{h}$ is the lag, $C_{s_j}(\mathbf{h})$ is the secondary covariance for the jth indicator, $C_{pks_j}(\mathbf{h})$ is the cross-covariance of the kth primary indicator and the jth secondary indicator, and $B(k)$ is the difference between the mean secondary indicator where the primary indicator is 1 (above threshold) and the mean secondary indicator where the primary indicator is 0 (below threshold). Zhu (1991) found that $B(k)$ values should exceed 0.2 if the secondary variable will add appreciable information; values in this example ranged from 0.22 to 0.39. Results from the four simulations generated using the Markov–Bayes algorithm are shown in Figure 16.10.

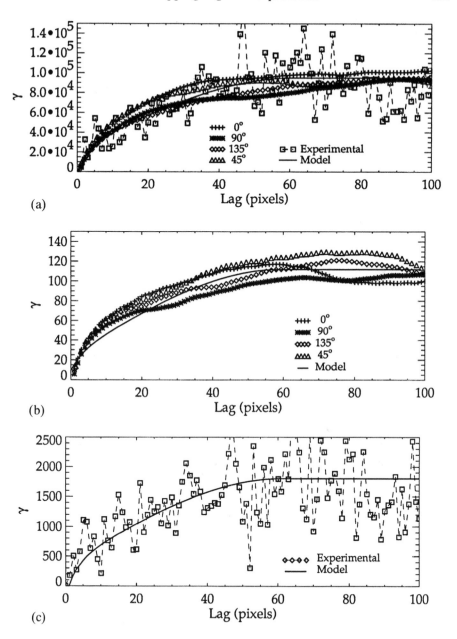

Figure 16.7 Semivariograms used for co-kriging. (a) *experimental semivariogram is calculated from 300 samples of primary data. Directional semivariograms are calculated from true map. Model is fitted to the experimental variogram;* (b) *directional semivariograms from map of secondary variable. Model is fitted to represent omnidirectional semivariogram;* (c) *experimental cross-semivariogram from collocated primary sample data and secondary data. Model fitted to represent experimental cross-variogram.*

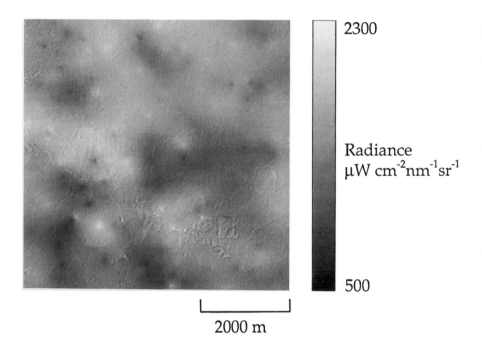

2300

Radiance
$\mu W\ cm^{-2}nm^{-1}sr^{-1}$

500

2000 m

Figure 16.8 Map predicted with co-kriging.

Comparison of Results

Table 16.2 shows the whole-map statistics from each method's results compared with the reference values of the true map. The regression method estimated the mean most accurately, but all methods came within 2 per cent of the true mean. The regression method estimated the quartiles and the minimum and maximum least accurately. The whole-map standard deviations estimated by both the co-kriging and regression methods are underestimated by more than 23 per cent of the true standard deviation; the simulations resulted in equally serious over-estimation. The co-kriged map had the strongest correlation with the true map ($r = 0.69$) and the smallest root mean square error (RMSE). The simulations were slightly less correlated with the true map but had by far the highest RMSE.

Summary statistics are not the only means of comparing results from the mapping methods. The spatial patterns can be examined through use of images. Contour maps, the traditional way of presenting kriged results, are not suited to this comparison since the results of regression and conditional simulation are not smoothed and show finer spatial structure. Grey-scaled images, shown in previous figures, can be compared directly. Figure 16.11, showing a 50 × 50 subset from the three methods, shows a more detailed view of this comparison.

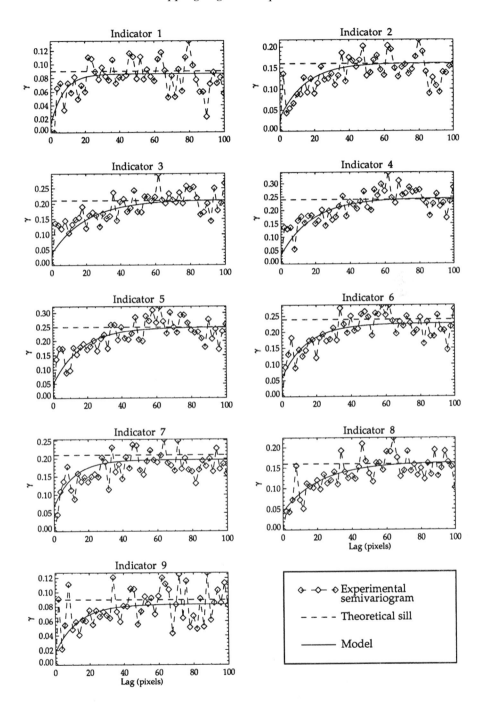

Figure 16.9 Semivariograms used for conditional simulation.

Table 16.1 *Set of exponential models of the form* $\gamma(\boldsymbol{h})=C(0)+C(1)Exp_r(\boldsymbol{h})$ *used for indicator vario-grams.*

Indicator	$C(0)$	$C(1)$	r
1	0.012	0.07	7.28
2	0.032	0.13	14.88
3	0.039	0.17	20.32
4	0.033	0.21	18.25
5	0.049	0.20	16.50
6	0.058	0.17	14.78
7	0.067	0.13	11.31
8	0.044	0.12	21.98
9	0.017	0.07	13.32

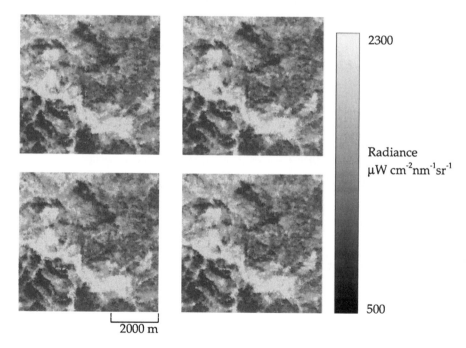

2300

Radiance
$\mu W\ cm^{-2}nm^{-1}sr^{-1}$

500

2000 m

Figure 16.10 Four conditional simulations.

The subset shows that regression was particularly insensitive to the spatial variance in this local area. The smoothing effect of co-kriging is particularly evident, whereas the simulations show a closer correspondence with the spatial pattern of the true map, though they are noisier.

Table 16.2 Comparison of whole-map summary statistics for the results of the regression, co-kriging and two of the many possible simulations.

Map	Mean	s	Min	p = 0.25	p = 0.5	p = 0.75	Max	ρ	RMSE
True	1409.4	312.5	504.7	1182.0	1421.0	1644.7	2463.7	–	–
Regression	1410.7	237.7	974.3	1279.2	1353.0	1477.0	3646.0	0.42	303.3
Co-kriging	1418.8	226.7	692.8	1264.2	1429.4	1589.7	2289.3	0.69	225.2
Simulation 1	1428.6	388.7	643.3	1108.3	1395.3	1693.2	2120.3	0.61	317.9
Simulation 2	1425.9	387.6	643.3	1106.5	1392.9	1696.2	2120.3	0.62	313.8

Note: Statistics are mean, standard deviation, minimum, 25th, 50th and 75th quartiles, maximum, correlation with true map, and root mean square error. Units in $\mu W cm^{-2} nm^{-1} sr^{-1}$ save for ρ (dimensionless).

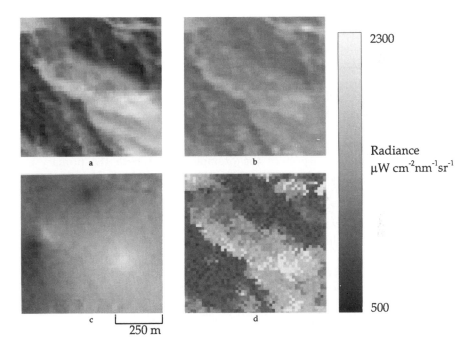

2300

Radiance
$\mu W\ cm^{-2} nm^{-1} sr^{-1}$

500

250 m

Figure 16.11 50 pixel × 50 pixel subsets from: (a) the true map; (b) the map predicted with regression; (c) the map predicted with co-kriging; (d) a single simulated map.

Discussion

The example described in this chapter is realistic in some respects. Often, no strong, one-to-one relationship exists between remotely sensed data and vegetation quantities measured on the ground. Correlation coefficients reported in the literature are often no greater than in this example. The example indicates that estimation has the potential to be improved by utilizing spatial information in addition to the correlation information, though the quality of results is sensitive to the adequacy of the semivariogram models and stationarity assumptions, and the strength of the relationship between the primary and secondary variables.

An unrealistic aspect of this example is that 300 ground samples is an unusually large number of sample data, especially for forest canopies, where sampling is often difficult. While geostatistical techniques have sometimes been cited as reducing the number of samples required to achieve a required level of accuracy (Webster and Burgess, 1984), they do require robust estimation of spatial covariance. This example shows how noisy such experimental spatial covariance measures can be, even with a large number of samples. Three hundred was an arbitrary number chosen for this example; an analysis could be performed to determine the optimal number required to achieve a given co-kriging variance (McBratney and Webster, 1983). Such an analysis would not reveal how well the sample could be used to model the semivariogram.

In real campaigns, samples are often not located to represent complete spatial randomness, but are chosen for logistical or historical reasons. These less objective ways of choosing samples can result in statistics that are not representative of the whole region. Samples clustered in areas of extreme values will naturally influence all statistics, including semivariograms, that are used for the mapping method (Deutsch, 1989).

Another unrealistic aspect of the example is that both the primary and secondary data are available on the same support, that is, 20 m. Identical sample supports are rarely the case in mapping experiments. One of the difficulties in using AVHRR data for regional and global mapping of vegetation amount, for example, is that direct vegetation measurements are only available at supports much smaller than the 1.1 km minimum of an AVHRR pixel.

The purpose of this example was to illustrate some of the potential implications for different mapping methods using remote sensing. The specific results of the comparison cannot be taken as absolute because there are several variants on the algorithms that are possible. For instance, the application of regression here uses one simple linear model for the whole map. There are alternative applications of co-kriging, for example, Isaaks and Srivastava (1989) suggest that the traditional constraint on co-kriging weights unnecessarily reduces the influence of the secondary variable and instead recommend the use of $\sum_{k=1}^{N_1} \lambda_k + \sum_{j=1}^{N_2} \omega_j = 1$. The Markov–Bayes algorithm is one of several conditional simulation methods that can incorporate more than one variable. The differences found among resulting maps will of course not only be influenced by the choice of an approach but the knowledgeable application of a particular algorithm (Englund, 1990).

A critical assumption that is inherent in the application of all three methods is stationarity. While the application of regression does not call for a stationary random function model, decisions must be made concerning the populations for which a particular regression model is applicable. The stationary random function models that underlie the development of geostatistical estimators require that similar decisions must be made for the spatial covariance models developed from the data. In the example used here, the primary variable was assumed to represent one population across the 36 km^2 field. The assumption will usually not be useful for regional and especially global mapping, where major heterogeneities must first be identified and modelled separately.

Despite the fact that the example is from a small region with a small sample support, methods could be applied to large regions with large support depending on the availability of data. For the geostatistical methods, a critical step is the development of covariance models that capture the spatial variance of the variable of interest. These models are dependent not only on the variable itself, but also on the support with which it is represented. If information were available about the relationship between spatial covariance of data with large support and that of data with small support, either through theoretical (Zhang *et al.*, 1990) or experimental (Woodcock *et al.*, 1988; Rubin and Gomez-Hernandez, 1990) means, it could be used to scale up knowledge from local to regional scales.

Turner *et al.* (1989) describe several measures for evaluating the performance of deterministic or physically based spatial simulation models. They state that fractal dimension, nearest neighbour probabilities, edge lengths, spatial predictability and goodness-of-fit at several spatial resolutions are some of the characteristics of a spatially distributed variable that may be important in ecological studies. While these measures were not compared in this example, it is evident from the maps that only some of them would be consistent across methods. Others, such as fractal dimension and edge lengths, would probably not be appropriate to use on any of these estimated maps.

As digital cartography and geographic information systems mature, more need is seen for the development of measures of accuracy, error or uncertainty for spatial data (Goodchild, 1988). For data expressed on a consistent sample support, a method that appears to have much promise for the description of uncertainty is conditional simulation. However, much research is yet needed on the implications of various simulation methods and on efficient ways to implement and utilize them.

Conclusion

An example of regression and two geostatistical methods was discussed as an analogy to the mapping of vegetation quantities using a combination of point-based and whole-area approaches. Regression, co-kriging, and conditional simulation produced widely different results in estimating the characteristics and spatial pattern of a known map. The best choice of mapping approach will depend on the purpose for which the map is intended. The regression method, while the most inaccurate in the example, is the easiest to parametrize. Co-kriging was the most accurate estimator but was inferior to conditional simulation for reproducing the spatial pattern of the true map. Both geostatistical methods have potential for application to real data. Evaluation should be accomplished by looking at alternative descriptions, a sort of exploratory data analysis of the results. This can be done by using displays of maps, histograms, scatterplots and summary statistics. The specific consequences of a given mapping method and the goals of the map should always be considered in the development of maps of vegetation quantity.

Acknowledgement

Thanks to André Journel of Stanford University, Richard Rossi of Pacific Northwest Laboratories, and Susan Benjamin and Lee Johnson of NASA Ames Research Center who generously reviewed this contribution. Comments by four anonymous reviewers were thorough and thought-provoking, and led to an improved manuscript. The work was supported by NASA RTOP No. 462-21-62-10.

References

Ardo, J., 1992, Volume quantification of coniferous forest compartments using spectral radiance recorded by Landsat Thematic Mapper, *International Journal of Remote Sensing*, **13**, 1779–86.

Asrar, G., Kanemasu, E., Jackson, R. and Pinter, P., 1985, Estimation of total above-ground phytomass production using remotely sensed data, *Remote Sensing of Environment*, **17**, 211–20.

Atkinson, P., Webster, R. and Curran, P. J., 1992, Co-kriging with ground-based radiometry, *Remote Sensing of Environment*, **41**, 45–60.

Avissar, R. and Verstraete, M., 1990, The representation of continental surface processes in atmospheric models, *Reviews of Geophysics*, **28**, 35–52.

Badwhar, G., MacDonald, R., Hall, F. and Carnes, J., 1986, Spectral characterization of biophysical characteristics in a boreal forest: Relationship between Thematic Mapper band reflectance and leaf area index for aspen, *IEEE Transactions on Geoscience and Remote Sensing*, **24**, 322–6.

Bhatti, A. U., Mulla, D. and Frazier, B., 1991, Estimation of soil properties and wheat yields on complex eroded hills using geostatistics and Thematic Mapper images, *Remote Sensing of Environment*, **37**, 181–91.

Burrough, P., 1987, Multiple sources of spatial variation and how to deal with them, in Chrisman, N. (Ed.) *Proceedings, AUTO-CARTO*, **8**, 145–54.

Chong, D., Mougin, E. and Gastellu-Etchegorry, J., 1993, Relating the Global Vegetation Index to net primary productivity and actual evapotranspiration over Africa, *International Journal of Remote Sensing*, **14**, 1517–46.

Christakos, G., 1984, On the problem of permissible covariance and variogram models, *Water Resources Research*, **20**, 251–65.

Cohen, W. and Spies, T., 1992, Estimating structural attributes of Douglas-fir/western hemlock forest stands from Landsat and SPOT imagery, *Remote Sensing of Environment*, **41**, 1–17.

Curran, P. J., 1987, Airborne multispectral scanner data for estimation of dye dispersion from sea outfalls, *Proceedings of the Institution of Civil Engineers*, **83**, 213–41.

Curran, P. J., 1988, The semivariogram in remote sensing: An introduction, *Remote Sensing of Environment*, **24**, 493–507.

Curran, P. J., Dungan, J. and Gholz, H., 1992, Seasonal LAI in slash pine estimated with Landsat TM, *Remote Sensing of Environment*, **39**, 3–13.

Curran, P. J. and Williamson, H., 1986, Sample size for ground and remotely sensed data, *Remote Sensing of Environment*, **20**, 31–41.

Curran, P. J. and Williamson, H., 1987, Airborne MSS data to estimate GLAI, *International Journal of Remote Sensing*, **8**, 57–74.

Deutsch, C., 1989, DECLUS: A FORTRAN program for determining optimum spatial declustering weights, *Computers & Geosciences*, **145**, 325–32.

Deutsch, C. and Journel, A. G., 1992, *GSLIB: Geostatistical Software Library*, New York: Oxford University Press.

Devroye, L., 1986, *Non-Uniform Random Variate Generation*, New York: Springer Verlag.

Dowd, P., 1992, A review of recent developments in geostatistics, *Computers & Geosciences*, **17**, 1481–1500.

Englund, E., 1990, A variance of geostatisticians, *Mathematical Geology*, **22**, 417–55.

Friedlingstein, P., Delire, C., Muller, J. F. and Gerard, J. C., 1992, The climate induced variation of the continental biosphere: A model simulation of the last glacial maximum, *Geophysical Research Letters*, **19**, 897–900.

Gohin, F. and Langlois, G., 1993, Using geostatistics to merge *in situ* measurements and remotely-sensed observations of sea surface temperature, *International Journal of Remote Sensing*, **14**, 9–19.

Goldberger, A., 1962, Best linear unbiased prediction in the generalized linear regression model, *Journal of the American Statistical Association*, **57**, 369–75.

Gomez-Hernandez, J. and Srivastava, R., 1990, ISIM3D: An ANSI-C three-dimensional multiple indicator conditional simulation program, *Computers & Geosciences*, **16**, 395–440.

Goodchild, M., 1988, The issue of accuracy in global databases, in Mounsey, H. and Tomlinson, R. (Eds) *Building Databases for Global Science*, pp. 31–48, London: Taylor & Francis.

Goulard, M. and Voltz, M., 1992, Linear coregionalization model: Tools for estimation and choice of cross-variogram matrix, *Mathematical Geology*, **24**, 269–86.

Henderson-Sellers, A., 1991, Developing an interactive biosphere for global climate models, *Vegetatio*, **91**, 149–66.

Isaaks, E. and Srivastava, R. M., 1989, *An Introduction to Applied Geostatistics*, Oxford: Oxford University Press.

Jensen, J. and Hodgson, M., 1985, Remote sensing forest biomass: An evaluation using high resolution remote sensor data and loblolly pine plots, *Professional Geographer*, **37**, 46–56.

Johnson, R., 1978, Mapping of chlorophyll a distributions in coastal zones, *Photogrammetric Engineering and Remote Sensing*, **44**, 617–24.

Journel, A., 1974, Geostatistics for conditional simulation of ore bodies, *Economic Geology*, **69**, 527–45.

Journel, A., 1983, Nonparametric estimation of spatial distributions, *Mathematical Geology*, **15**, 445–68.

Journel, A., 1986, Geostatistics: Models and tools for the earth sciences, *Mathematical Geology*, **18**, 119–40.

Journel, A. and Alabert, F., 1989, Non-Gaussian data expansion in the earth sciences, *Terra Nova*, **1**, 123–34.

Journel, A. and Huijbregts, C., 1978, *Mining Geostatistics*, London: Academic Press.

Journel, A. and Isaaks, E., 1984, Conditional indicator simulation: Application to a Saskatchewan uranium deposit, *Mathematical Geology*, **16**, 685–718.

Journel, A. and Posa, D., 1990, Characteristic behaviour and order relations for indicator variograms, *Mathematical Geology*, **22**, 1011–26.

Journel, A. and Rossi, M., 1989, When do we need a trend model in kriging? *Mathematical Geology*, **21**, 715–38.

Journel, A. and Zhu, H., 1990, *Integrating Soft Seismic Data: Markov–Bayes Updating, an Alternative to Co-kriging and Traditional Regression*, Technical Report 3, Stanford, California: Stanford Center for Reservoir Forecasting.

Lam, N.-N., 1983, Spatial interpolation methods: A review, *The American Cartographer*, **10**, 129–49.

Leenaers, H., Okx, J. and Burrough, P., 1990, Comparison of spatial prediction methods for mapping floodplain soil pollution, *Catena*, **17**, 535–50.

Lieth, H., 1975, Historical survey of primary productivity research, in Lieth, H. and Whittaker, R. (Eds) *Primary Productivity of the Biosphere*, pp. 147–66, New York: Springer Verlag.

Ludeke, M., Janecek, A. and Kohlmaier, G., 1991, Modelling the seasonal CO_2 uptake by land vegetation using the global vegetation index, *Tellus Series B-Chemical and Physical Meteorology*, **43**, 188–96.

Matheron, G., 1973, The intrinsic random functions and their applications, *Advances in Applications of Probability*, **5**, 439–68.

McBratney, A. and Webster, R., 1983, How many observations are needed to estimate the regional mean of a soil property? *Soil Science*, **135**, 177–83.

Merrill, E., Bramblebrodahl, M., Marrs, R. and Boyce, M., 1993, Estimation of green herbaceous phytomass from Landsat MSS data in Yellowstone National Park, *Journal of Range Management*, **46**, 151–7.

Myers, D., 1982, Matrix formulation of co-kriging, *Mathematical Geology*, **14**, 249–57.

Nash, M., Toorman, A. and Wierenga, P., 1992, Estimation of vegetation curves in an arid rangeland based on soil moisture using co-kriging, *Soil Science*, **154**, 25–36.

Park, S. and Miller, K., 1988, Random number generators: Good ones are hard to find, *Communications of the ACM*, **31**, 1192–201.

Pearson, R., Tucker, C. and Miller, L., 1976, Spectral mapping of shortgrass prairie biomass, *Photogrammetric Engineering and Remote Sensing*, **42**, 317–23.

Peterson, D., Spanner, M., Running, S. and Teuber, K., 1987, Relationship of Thematic Mapper Simulator data to leaf area index of temperate coniferous forests, *Remote Sensing of Environment*, **22**, 323–41.

Peterson, D. and Waring, R., in press, Overview of the Oregon Transect Ecosystem Research Project, *Ecological Applications*.

Rosema, A., Verhoef, W., Noorbergen, H. and Borgesius, J., 1992, A new forest light interaction model in support of forest monitoring, *Remote Sensing of Environment*, **41**, 23–41.

Rosenzweig, M., 1968, Net primary productivity of terrestrial communities: Prediction from climatological data, *American Naturalist*, **102**, 67–75.

Rossi, R. E., Borth, P. W. and Tollefson, J. J., 1993, Stochastic simulation for characterizing ecological spatial patterns and appraising risk, *Ecological Applications*, **3**, 719–735.

Rossi, R., Dungan, J. and Beck, L., in press, Kriging in the shadows: Geostatistical interpolation for remote sensing, *Remote Sensing of Environment*.

Rubin, Y. and Gomez-Hernandez, J., 1990, A stochastic approach to the problem of upscaling of conductivity in disordered media — theory and unconditional numerical simulations, *Water Resources Research*, **26**, 691–701.

Running, S., Nemani, R., Peterson, D., Band, L. and panner, M., 1989, Mapping regional forest evapotranspiration and photosynthesis by coupling satellite data with ecosystem simulation, *Ecology*, **70**, 1090–101.

Stein, A., Startitsky, I. and Bouma, J., 1991, Simulation of moisture deficits and areal interpolation by universal co-kriging, *Water Resources Research*, **27**, 1963–73.

Tucker, C., Vanpraet, C., Sharman, M. and Ittersum, G. V., 1985, Satellite remote sensing of total herbaceous biomass production in the Senegalese sahel: 1980–1984, *Remote Sensing of Environment*, **17**, 233–49.

Turner, M., Costanza, R. and Sklar, F., 1989, Methods to evaluate the performance of spatial simulation models, *Ecological Modeling*, **48**, 11–18.

Vane, G., 1988, First results from the Airborne Visible/Infrared Imaging Spectrometer (AVIRIS), *Proceedings of the SPIE: Imaging Spectroscopy II*, **834**, 166–74.

Vujakovic, P., 1987, Monitoring extensive 'buffer zones' in Africa: An application for satellite imagery, *Biological Conservation*, **39**, 195–208.

Webster, R. and Burgess, T., 1984, Sampling and bulking strategies for estimating soil properties in small regions, *Journal of Soil Science*, **35**. 127–40.

Webster, R. and Oliver, M., 1990, *Statistical Methods in Soil and Land Resource Survey*, Oxford: Oxford University Press.

Wessman, C., Aber, J., Peterson, D. and Melillo, J., 1988, Remote sensing of canopy chemistry and nitrogen cycling in temperate forest ecosystems, *Nature*, **335**, 154–6.

Woodcock, C., Strahler, A. and Jupp, D., 1988, The use of variograms in remote sensing, II: Real digital images, *Remote Sensing of Environment*, **25**, 349–79.

Yates, S. and Warrick, A., 1987, Estimating soil water content using co-kriging, *Soil Science Society of America Journal*, **51**, 23–30.

Zhang, R., Warrick, A. and Myers, D., 1990, Variance as a function of sample support size, *Mathematical Geology*, **22**, 107–21.

Zhu, H., 1991, 'Modeling mixture of spatial distributions with integration of soft data', unpublished PhD thesis, Stanford University, Stanford, California.

17

Global biosphere requirements for general circulation models

Bruce P. Hayden

During the 1980s, atmospheric scientists discovered that General Circulation Models (GCMs) of the atmosphere failed to predict contemporary climates adequately, unless the biosphere was properly specified. GCM-climate sensitivity to the global pattern of surface roughness, which is largely a function of vegetation, is manifested as a twofold variation in the surface wind velocity. Changes in the surface roughness specification also made a major difference in divergence of mass and energy in the low latitudes and, therefore, rainfall patterns in these regions. GCM temperature-field sensitivity to evapotranspiration, in contrast with the 'bucket' evaporation in early models, was + 25°C over mid-latitude continents, and the intensity of global pressure-fields in the same study was 10mb, a very large effect. GCM sensitivity to land–cover albedo is clearly evident in the transition between dry and moist convection in the low latitudes and, thus, realized rainfall. Sensitivity experiments using a doubled atmospheric carbon dioxide showed new patterns of climate, which imply new patterns of vegetation and thus changed surface roughness, evapotranspiration and albedo. Results of the GCMs of the 1980s clearly indicate that the old paradigm that says vegetation controls climate only at the microscale must be abandoned. The biosphere in the current generation GCMs is a static boundary condition, but clearly it should not be. The feedback loops with vegetation controlling climate are not yet on the GCM drawing boards; they will require new Global Biosphere Databases at a spatial resolution greater than those of the next generation of GCMs.

Introduction

During the 1980s, atmospheric scientists were using General Circulation Models (GCMs) of the atmosphere first designed in the 1960s to project the consequences of doubling atmospheric carbon dioxide ($2\text{X}CO_2$). It is now generally recognized that the flux of greenhouse gases from the biosphere modifies climate and that this changed climate may alter the flux of greenhouse gases from the biosphere. Sensitivity analyses of GCMs revealed that the models are highly sensitive to attributes of the biosphere: surface roughness, evapotranspiration and albedo. If the biosphere is not properly specified, the models have significant errors in estimating contemporary climate. In addition to these now well-known

Table 17.1 Increased vegetation cover impacts on important climatic attributes.

Increase	Decrease
Near-infrared	Visible albedo
Solar energy absorption	Infrared emission
Aerodynamic roughness	Surface wind speed
Turbulence intensity	Runoff
Evapotranspiration	Erosion
Relative humidity	Bowen ratio
Minimum temperature	Maximum temperature
Organic aerosols	Inorganic aerosols
Specific humidity	
Equivalent potential temperature	
Upward motions(?)	
Clouds (?)	
Rainfall (?)	

Source: After Anthes, 1984.

feedbacks from vegetation to climate, other connections are today recognized as important: marine phytoplankton release of dimethyl sulphide, which becomes cloud condensation nuclei (Andreae, 1980); terrestrial vegetation as a source of cloud condensation nuclei (Went, 1961); electric charges generated by the biosphere (Went, 1962); and terrestrial and marine vegetation production of atmospheric ice nuclei (Schnell and Vali, 1972).

The notion that vegetation cover and its products released into the atmosphere play important roles in atmospheric dynamics gained ground in the 1980s. Anthes (1984) evaluated the role of vegetation in atmospheric dynamics; his findings are summarized in Table 17.1. He found that vegetation cover altered both short- and long-wave radiation budgets; aerodynamic energy and mass exchanges between the biosphere and the atmosphere; turbulence; hydrologic balances; humidity; temperature; and biogenic aerosols; and thus visibility. Anthes' idea was that a planned landscape with specified land cover would result in a more desirable climate.

Avissar (1992) lists the following vegetation characteristics as essential in atmospheric models: surface roughness, leaf area index, vegetation height, albedo, transmissivity, emissivity and temperature. In addition, he lists the following properties of the soil: density, roughness, water content at saturation, soil water potential, hydraulic conductivity at saturation, albedo and emissivity, as necessary to the models. Some of these needed parameters will have to come from observation (ground or satellite), and others can only be obtained from ecosystem models. Given this new reality of the importance of the biosphere in atmospheric modelling, the time has come to accelerate the development of General Biosphere Models (GBMs) that can interface with GCMs and to begin the construction of Global Biosphere Databases (GBDs) at appropriate scales of temporal and spatial resolution for these models.

The necessity for these new efforts is partly established on the basis of GCM sensitivity tests. Shortcomings of GCMs, when compared with observational climatology, have resulted in such studies and an elucidation of key biosphere attributes needed in current generation of GCMs.

Lessons from GCM Sensitivity Tests

Early analyses of the climate models revealed that there was no evidence of forecast skill at regional and subregional scales. For example, the UK Meteorology Office model made the Sahara wetter than Scotland. With a fundamental faith in the first-principal equations that are the heart of these models, researchers began to look seriously at the boundary conditions of the model as one likely source of model inadequacy. The boundary conditions with the greatest uncertainty were associated with the biosphere. Two problems were apparent. Estimates of the geographic variation in the thermodynamic and dynamic attributes of the surface vegetation of earth were poorly known, subject to societal (i.e., land use) changes, and difficult to scale up to the coarse scale (5° latitude x 5° longitude) used by the GCMs. To determine which attributes of the biosphere were important in model behaviour, a series of sensitivity tests were performed on biosphere boundary conditions (Charney et al., 1977; Sud and Smith, 1984; Cunnington and Rowntree, 1986; Lavel and Picon, 1986; Sud and Molod, 1988; Lean and Warrilow, 1989). These tests were designed to answer the following generic question: how large a climate change results from a specified change in a boundary condition of the biosphere? Three general areas of sensitivity testing were conducted: biosphere surface roughness (Sud and Smith, 1985), biosphere evapotranspiration (Shukla and Mintz, 1982), and biosphere albedo (Charney *et al.,* 1977).

Surface Roughness

About 3.1 W m^{-2} of the 345 W m^{-2} energy from the sun goes to the production of kinetic energy (i.e., generation of winds; Oort, 1964; Paltridge, 1979). The earth as a thermodynamic system is dissipative. By means of friction, the 3.1W m^{-2} are returned as heat and eventually released to space as terrestrial radiation. A large fraction of the 3.1W m^{-2} used to make the winds blow is dissipated by the friction offered by terrestrial vegetation and waves at sea. For example, wind speed over the oceans adjacent to the UK average 12 mph while the average wind speed over England and Scotland is only 6 mph. The vegetation slowed the wind to half. The roughness offered by vegetation slows the winds as surely as the sun sets them in motion. To put into perspective the importance of the 3.1 W m^{-2}, consider that a doubling of atmospheric carbon dioxide is a perturbation of only 2.2 W m^{-2}. Improper specification of the surface roughness offered by terrestrial vegetation results in large GCM model errors (Sud and Smith, 1985).

Table 17.2 Surface roughness used in GCMs.

CGM Model	Z_0 (m)
BMRC	0.17
CSIRO (4-level)	Constant drag coefficient used
CSIRO (9-level)	0.016
University of Melbourne	0.30
AES/CCC Version II	Variable drag coefficient used
ECMWF CY29 and CY39	Spatially variable Z_0
LMD	Spatially variable Z_0
Hamburg GCC	Spatially variable Z_0
MRI	0.45
JMA	Spatially variable Z_0
UKMO	Spatially variable Z_0
GFDL I and II	0.17
GISS	Spatially variable Z_0
GLAS	0.45
NCAR/CCM1	0.25
NMC University of Maryland	Spatially variable Z_0
UCLA/GLA	Constant Z_0
UCLA/CSU	Spatially variable Z_0

Source: After Garratt, 1993.

Sud and Smith (1985) investigated the role of surface roughness specification in desert areas on the numerical simulations of the atmosphere using the GLAS GCM. GCMs often generate rainfall in excess of that observed in desert regions. Sud and Smith used a sensitivity analysis approach to assess the role of surface roughness. The control-model run set surface roughness to 0.45 m everywhere. This value of surface roughness is typical of contemporary GCMs (Table 17.2). The experiment-model run used a surface roughness of 0.0002 m for the Sahara Desert and 0.45 m for all other land areas. Where the desert surface roughness was properly included, Saharan rainfall was reduced in relation to the control runs. In addition, the experimental model also positioned the intertropical convergence zone at about 14° which is closer to its normal position of 10° N in July, thus correcting earlier model versions on the position of this important climatic boundary. Sud and Smith note that in desertification the land becomes less rough, further promoting desertification. In a second paper, in 1985, Sud and Smith looked into the sensitivity of the Indian monsoon to variations in surface roughness. They found that reducing the surface roughness from 0.45 m to 0.0002 m resulted in a reduction of the strength of the monsoon and a reduction in rainfall. Most GCMs use either a constant surface roughness or drag coefficient in all terrestrial regions or have a variable roughness at the scale of 5° latitude by 5° longitude (Table 17.2).

Surface roughness is calculated from vertical profiles of wind speed over the terrain of interest. Direct measurements are few. In general, surface roughness is proportional (in a non-linear sense) to vegetation height. Surface roughness is also dependent on wind speed, as vegetation becomes streamlined at higher wind speeds. GCMs, as they are currently constituted, do not generate such extreme winds. Speeds in the range of the mean wind speed are typically used in the model, and singular estimates of surface roughness are used for given land covers. The general relationship between vegetation height (*h*) and surface roughness (Z_o) based on empirical studies is

$$\log Z_o = 0.997 \log h - 0.833$$

This relationship is well tested for agricultural crops but may give overestimates for forests. None the less, by using published vegetation heights by species, it is possible to estimate surface roughness for land-cover classes. We have used Olson's classification (Olson *et al.,* 1983) and the published literature on estimates of surface roughness for various vegetation types to generate global databases and maps of vegetation surface roughness. Olson's maps of vegetation are constructed at a map element size of 0.5° latitude by 0.5° longitude. While we are convinced that the refinement of the specification of surface roughness values to Olson's vegetation classes might be improved, the resulting maps appear reasonable.

Evapotranspiration

Evapotranspiration from land surfaces loads the atmosphere with water vapour, reduces the sensible heating of the surface layer of the atmosphere, and provides a means of energy transport to higher altitudes and over great horizontal distances. Shukla and Mintz (1982) ran the GLAS GCM with maximum soil moisture (and thus maximum evapotranspiration), and with a dry soil and no evapotranspiration. In the absence of evapotranspiration loading of the atmosphere, precipitation on all continents was reduced by 50 per cent or more. Temperatures over the continents were 25°C hotter without evapotranspiration. Pressure fields were fundamentally altered. Anticyclones over the subtropical oceans were 15 mb to 20 mb higher in central pressure and land areas were as much as 15 mb lower in pressure. Surface windfields were thus fundamentally different in the evapotranspiration case compared with the no-evaporation case. GCM models require adequate estimates of evapotranspiration, and this need may be best provided by means of ecosystem models as part of the GCM. Currently three types of models are used: bucket, BATS (Henderson-Sellers, 1990), and SiB (Sellers *et al.,* 1986). The bucket model is just simple evaporation from a free water surface and was used in early GCM models.

Future GCM models will need to incorporate changes in vegetation cover type and quality to make adjustments in evapotranspiration estimates as the model is running. As climate changes, the vegetation cover also changes; as a result, the parametrizations for BATS or SiB require adjustment as these changes occur. In preparation for these interactive biosphere–atmosphere models of the future, ecologists will have to develop databases on BATS and SiB parameters for each vegetation and landcover type at a spatial resolution of about 1° latitude by 1° longitude.

Table 17.3 Changes in evapotranspiration (E) and precipitation (R) for specified changes in albedo (A).

Reference	A	E (mm/day)	R (mm/day)
Charney *et al.* (1977)	0.21	– 0.8	– 2.0
Chervin (1979)	0.27		– 1.7
Carson and Sangster (1981)	0.20	– 0.95	– 1.2
Sud and Fennessey (1982)	0.16	– 0.4	– 0.6
Sud and Smith (1985)	0.06	– 1.6	
Cunnington and Rowntree (1986)	0.06		– 0.75
Lavel and Picon (1986)	Increase	Decrease	Decrease
Sud and Molod (1988)	Decrease		Increase
Lean and Warrilow (1989)	0.05	– 0.2	– 0.75
Mylne and Rowntree (1991)	0.10	– 0.1	– 0.2

Source: After Garratt, 1993.

Albedo

Charney (1975) used NASA's GISS General Circulation model to test the sensitivity of surface albedo on the atmospheric circulation over the Sahel of Africa. He found that increasing the albedo north of the Intertropical Convergence Zone (ITCZ) from 14 per cent to a more realistic 35 per cent resulted in a southward shift of the ITCZ and a 40 per cent decrease in rainfall in the Sahel during the rainy season. Studies by Chervin (1979), Sud and Fennessy (1984), Carson and Sangster (1981), and Mylne and Rowntree (1991) found the same results. On average, they found a 20 per cent decline in rainfall for each 0.1 increase in albedo. Rowntree (1983) also found that a 0.1 increase in albedo reduced evaporation by 0.65 mm/day. Garratt (1993) summarized the changes induced in evaporation and precipitation resulting from changes in model albedo. These changes are summarized in Table 17.3.

Albedos of most natural land cover classes are not known through direct observation. Pielke and Avissar (1990) summarized published albedos for natural surfaces (see Table 17.4). In landscapes with a mixture of vegetation and exposed soils, albedos depend on soil albedos that are larger than vegetation albedos and that may vary from 0.20 to 0.60. Clearly the greatest potential for albedo change is found in the arid and semi-arid regions on seasonal and decadal time scales, and in areas subject to cultural land use changes and variations in agricultural practices.

Biogenic Feedbacks on Climate

Surface roughness, albedo and evapotranspiration are 'real-time' parameters in the climate models. The dynamic equations require these terms to be updated at

Table 17.4 Albedos for natural surfaces.

Ground cover	Albedo
Tropical fields	0.20
Dry steppe	0.25
Tundra and heather	0.15
Tundra	0.18
Orchards	0.15–0.20
Deciduous forest (winter)	0.15
Deciduous forest (summer)	0.20
Mixed hardwoods (summer)	0.18
Rainforest	0.15
Pine, fir, and oak forest	0.18
Coniferous forest	0.10–0.15
Red pine forest	0.10–0.27

Source: After Pielke and Avissar, 1990.

each model time-step. Other aspects of biosphere forcing on the atmosphere involve the accumulation of products of biogenic processes. For example, the biosphere produces a wide variety of gases that are radiatively active and alter earth's long-wave radiation budget (i.e., greenhouse gases). In addition, the biosphere produces materials that contribute to cloud drop formation and ice crystal formation, and alter atmospheric reflection and scattering of solar radiation.

Dimethyl Sulphide

Charlson (1987) proposed a link between emission of dimethyl sulphide (DMS) by marine phytoplankton, cloudiness and climate. In the simplest form, their model indicates that warmer ocean temperatures result in higher DMS production by marine phytoplankton. From the gaseous DMS in the atmosphere non-sea salt sulphates are produced, which serve as cloud condensation nuclei and cause increased cloudiness. The cloudiness reduces solar load on the oceans and cools the waters. This positive feedback system is offered as a biospheric, negative-feedback control on marine climate. Marine atmospheres are, in comparison with continental air masses, depauperate in condensation nuclei. DMS, in marine phytoplankton, apparently serves an osmoregulatory role (Dickson *et al.,* 1982; Reed, 1983; Vairavamurthy *et al.,* 1985). Dinoflagellates and coccolithophores are the major marine producers of DMS (Andreae, 1980). Ayers *et al.,* (1991) found that cloud condensation nuclei in the marine atmosphere showed the same seasonal cycle as phytoplankton DMS production.

High DMS production in tropical waters indicates a temperature role, but production of DMS is not independent of available nutrients and the southern oceans are relatively rich in DMS (McTaggart and Burton, 1992). DMS derivatives are also found in the Vostock ice cores and exhibit a 20 000 year periodicity, indicating that DMS is modulated at well-known climate time scales. It is not possible from these data to state whether the DMS is directly involved as a climate moderator or simply as a respondent to climate changes of a different origin. Charleson (1987) proposes a planetary albedo modulation by marine phytoplankton by means of DMS production.

While the details of biogenic DMS's role in climate is currently the subject of extensive research, the significance to next generation climate models needs to be identified. As next generation GCMs approach the 1° latitude by 1° longitude grid-cell resolution, physically based cloud droplet microphysical processes will be incorporated into the models to achieve a significant improvement in cloud and rainfall function. At that time, temporal and spatial variations in cloud condensation nuclei like DMS will be required. It is likely that satellite-based sea surface temperature measurements and satellite-based measurements of ocean surface chlorophyll will permit a synthetic estimate of DMS loading of the atmosphere. It should then be possible to develop predictive functions to produce synthetic data suitable for GCMs and for testing hypotheses.

Biogenic Cloud Condensation Nuclei

In continental regions, cloud condensation nuclei (CCN) are readily available and are of both biogenic (Woodcock and Gifford, 1949; Went *et al.,* 1967) and anthropogenic origin (sulphur dioxide from combustion processes forms atmospheric sulphates, which are excellent condensation nuclei). Haze droplets, which arise from biogenic hydrocarbons as well, and anthropogenic hydrocarbons are excellent cloud condensation nuclei (Junge, 1951). These nuclei are counted as Aitken nuclei. Suitable condensation nuclei in the lowest one-half kilometre of the atmosphere number in the order of 23,000 per cc. In the next half-kilometre aloft, only half as many nuclei are available and the decline continues upward. There are 10 times as many condensation nuclei over the land as over the oceans. While it is not believed that cloud condensation nuclei are limiting to cloud formation and development over land areas, the role of the biosphere in cloud condensation nuclei should be investigated as the cloud microphysical processes will be included in the next generation of GCMs. There is evidence that both cloudiness and cloud brightness are dependent on the number of CCN present in the atmosphere (Williams *et al.,* 1980).

Biogenic Ice Nuclei

Most of the ice nuclei (IN) that exist in the atmosphere arise from decomposition of organic matter (Schnell and Vali, 1972; Vali *et al.,* 1976). The temperatures at which freezing occurs in the atmosphere depend on the specific nature of the organic decomposition products available. Without biogenic ice nuclei, cloud drop freezing occurs around $-25°C$ if sea salts or mineral kaolinite are available. Freezing at warmer temperatures requires biogenic ice nuclei. Such nuclei have been shown to cause freezing at temperatures as warm as $-2°C$ or $-3°C$. Schnell and Vali (1973) showed that the freezing temperatures of ice nuclei varied from biome to biome, with the coldest freezing temperatures in the tropical forests and the warmest in the high-latitude forests. Some marine phytoplankton have been shown to produce ice nuclei (Schnell, 1975; Schnell and Vali, 1976), but generally the marine air has few ice nuclei. Rising motions of the air within clouds is greatly aided by freezing of cloud drops. These high velocities are required to produce charge separation and electrical discharges; this explains why electric discharges at sea are uncommon compared with terrestrial regions. The magnitude of the production of ice nuclei from organic decomposition has also been shown to have a seasonal variation (Lander *et al.,* 1980; Schnell *et al.,* 1980).

When the next generation of GCMs are built they are likely to have higher spatial resolution (1° latitude by 1° longitude) and include cloud microphysics. With these developments there will be an increased need to include a biosphere-specific spectrum of ice nuclei by freezing temperature. New research and a global database on ice nuclei quantity and quality will be required.

Biogenic Hydrocarbons

Much of the haze produced arises from terpene and hemiterpene production by vegetation (Duce, 1983). Biogenic haze alters solar radiation budgets (Herman *et al.,* 1971), serves as condensation nuclei, transfers electric charge from the surface to the atmosphere and acts as greenhouse gases. Production of non-methane hydrocarbons depends on the composition of the vegetation, primary productivity and temperature. In arid regions, with average relative humidity less than 65 per cent, biogenic hydrocarbons are active greenhouse gases (Filippov and Mirumyants, 1970; Georgiyevskiy, 1973; Dianov-Klokov and Ivanov, 1981) and elevate nocturnal temperatures as much as $20°C$. Went (1960) suggested that blue haze was of biogenic origin and that terpenes and hemiterpene were the principal hydrocarbons involved. He estimated that solar radiation was reduced as much as 10 per cent by biogenic hazes. Herman *et al.* (1971) showed that biogenic hydrocarbons at the top of the troposphere reduced solar radiation over Hawaii by 1 per cent and indicated that the bulk of the aerosols was of continental origin. Went (1964, 1966) and Went *et al.* (1967) showed that these hydrocarbons are counted as Aitken nuclei. Furthermore, they acted as cloud condensation nuclei and gave rise to charged particles in the atmosphere (Went 1961, 1962). Hydrocarbon production rates need to be monitored and modelled for various land cover and vegetation classes. Such information will be needed when microphysical processes become included in GCMs of the future.

The Need for a New Biosphere Geography

The requirements of current generation GCMs, especially those that include a vegetation canopy, include specifications of albedo, surface roughness, soil moisture and evapotranspiration. Surface resolution in these models is about 5° latitude × 5° longitude. Next generation GCMs using greatly expanded computational power will have grid cells in the order of 1° latitude × 1° longitude. The need for specification of vegetation canopy type and albedo, surface roughness, soil moisture and evapotranspiration will no longer permit a simple biome-scale parametrization. Much greater detail on geographic variation in vegetation cover with a specification of the annual progression of phenological stages will be required. At the scale of 1° latitude × 1° longitude, the resolution of the GCMs would be of the same order as the existing network of Class A weather stations in the USA. Higher resolution models are unlikely for quite some time. Even at this scale (75 km) the vegetation cover is often heterogeneous, and a proper specification of needed parameters will require mapping and new classifications of vegetation.

Many of the parameters needed for future generation climate models cannot be observed directly and will require ecosystem models, General Biosphere Models (GBMs), to provide them. Generation by models of carbon dioxide, methane and hydrocarbon production will be needed from models of soil carbon stores like CENTURY (Parton *et al.*, 1987). In the same fashion, primary production and nutrient-cycling models at the global scale (e.g., GEM) will also be required. In some cases, GBMs utilize statistical transform functions to relate observables like the normalized difference vegetation index (NDVI) to CO_2 production (Tucker *et al.*, 1986) or NDVI to leaf area index (LAI) (Running *et al.*, 1989). To meet the needs of the ecological community to understand the global biosphere across scales from patch to continent an adequate global biosphere database is required. A means to generate future climate states will be essential to generating projections of the future state of the biosphere. To meet the needs of the climate modelling community to project future climate states from subregional to global scales, an adequate global biosphere database is required. The interfacing of these joint needs establishes these two communities as partners in filling this data matrix.

Scale

At the present time, climate models are coarse in scale. Climate model grid-cells are about 5° latitude by 5° longitude and climate-model time-steps are hourly. In contrast, ecosystem models and observational networks are spatially fine scale (less than 1 min latitude by 1 min longitude) and temporally coarse scale (monthly or annual). Class A weather station input data used in climate models are characterized by station spacing of about 1° latitude by 1° longitude, and it is unlikely that climate models will be designed with higher resolution. The fine-

Table 17.5 Data needs and data sources.

Variable needed	Likely data sources
Surface roughness	Classification model data
Canopy albedo	Classification model data and satellite data
Vegetation height	Classification model data
Evapotranspiration	Dynamic model data
Soil organic matter content	Dynamic model data
Soil moisture	Station data and satellite data
CCN production	Classification model data and satellite data
DMS production	Statistical model data
IN production	Statistical model data
Leaf area index	Classification model data and satellite data
NMHC production	Classification model data
Transmissivity	Classification model data
Phytoplankton productivity	Satellite data
NDVI	Satellite data
Emissivity	Classification model data
Soil density	Classification model data
Land surface roughness	Classification model data
Sand, silt and clay %	Classification model data
Soil depth	Classification model data
Soil water content at saturation	Classification model data
Soil water potential	Station data and satellite data
Hydraulic conductivity	Classification model data
Soil albedo	Classification model data
Soil emissivity	Classification model data

scale heterogeneity of the biosphere is substantial, and most biogeographers view the appropriate resolution for inventory and model development between 1 km by 1 km to 1° latitude by 1° longitude or finer.

Data Sources

The detailed data that will be needed for future modelling efforts will come from model output statistics of dynamic ecosystem models, parametrization within vegetation or land cover classification models, statistical regression models using surrogate variables and direct satellite observation. The variables that will be needed and the likely sources of data to meet these needs are listed in Table 17.5.

Conclusion

The thesis of this contribution is that the biosphere exerts significant controls on climate at all scales and that, without proper specification of these controls, the products of atmospheric GCMs will continue to be inadequate. While others have called for better biogeographies and vegetation and land cover classifications, the call here is for classifications that are designed to meet the needs of linkages to GCM requirements. Calibration data across vegetation types for albedo, surface roughness, evapotranspiration fluxes and other variables in Table 17.5 still require new research. It is also necessary to expand research on the longer term, slower biosphere controls of climate: biogenic trace gases, complex hydrocarbons, cloud condensation nuclei and ice nuclei. The production of these biogenic materials varies with biome and land cover types. When GCMs contain code for the dynamic processes associated with cloud microphysics, the need for knowledge about biogenic production of CCN and IN will become acute. At the present time, cloud and rain mechanisms are simple parametrizations. The ecologists who will be part of the global change research adventure in the decades ahead will need to understand fully the controlling mechanisms of the biosphere on the atmosphere.

Acknowledgement

Assistance in the development of some of the GIS materials in this contribution was provided by Patrick Halpin at the University of Virginia. Support for the time to develop this synthesis contribution was provided by DEB 92-11772.

References

Andreae, M. O., 1980, The production of methylated sulfur compounds by marine phytoplankton, in Trudinger, P. A., Walter, M. R. and Ralph, B. J. (Eds) *Biogeochemistry of Ancient and Modern Environments,* Berlin: Springer.

Anthes, R., 1984, Enhancement of convective precipitation by mesoscale variations in vegetation covering in semi-arid regions, *Journal of Climate and Meteorology,* **23,** 541–54.

Avissar, R., 1992, Conceptual aspects of a statistical-dynamical approach to represent landscape subgrid-scale heterogeneities in atmospheric models, *Journal of Geophysical Research,* **97**(D3), 2729–42.

Ayers, G. P., Ivey, G. P. and Gillett, R. W., 1991, Coherence between seasonal cycles of dimethyl sulphide, methanesulphonate and sulphate in marine air, *Nature,* **349,** 404–6.

Carson, D. J. and Sangster, A. B., 1981, The influence of land-surface albedo and soil moisture on GCM circulations, *Numerical Experimentation Programme Report No. 2,* 5.14–5.21, Bracknell: UK Meteorological Office.

Charlson, 1987, Oceanic phytoplankton, atmospheric sulfur, cloud albedo and climate, *Nature,* **326,** 655–61.

Charney, J. G., 1975, Dynamics of deserts and drought in the Sahel, *Quarterly Journal of the Royal Meteorological Society,* **101**(428), 193–202.

Charney, J. G., Quirk, W. J., Chow, S. H. and Kornfield, J., 1977, A comparative study of the effects of albedo change on drought in semi-arid regions, *Journal of Atmospheric Science,* **34,** 1366–85.

Chervin, R. M., 1979, Response of the NCAR GCM to changed land surface albedo, *Report of the Joint Ocean Climate Conference Study on Climate Models: Performance, Intercomparison and Sensitivity Studies,* **1,** 563–81.

Cunnington, W. M. and Rowntree, P. R., 1986, Simulations of the Saharan atmosphere — dependence on moisture and albedo, *Quarterly Journal of the Royal Meteorological Society,* **112,** 971–99.

Dianov-Klokov, V. I. and Ivanov, V. M., 1981, Relative importance of extinction mechanisms in the 8–13 micron window under various meteorological conditions, *Izvestiya, Atmospheric and Oceanic Physics,* **17**(6), 430–5.

Dickson, D. M., Wyn Jones, R. G. and Davenport, J., 1982, Osmotic adaptation in *Ulva lactuaca* under fluctuating salinity regimes, *Planta,* **155**, 409–15.

Duce, R. A., 1983, Organic material in the global troposphere, *Reviews of Geophysics and Space Physics,* **21**(4), 922–5.

Filippov, V. I. and Mirumyants, S. O., 1970, The variation of the spectral coefficients of radiation attenuation by hazes in the spectral region of 0.59–13 micrometers, *Izvestiya, Atmospheric and Oceanic Physics,* **6**(6), 372–4.

Garratt, J. R., 1993, Sensitivity of climate simulations to land-surface and atmospheric boundary-layer treatments — A review, *Journal of the Climate,* **6**, 419–49.

Georgiyevskiy, Y. S., 1973, Variations of the spectral extinction coefficients in the transparency windows, *Izvestiya, Atmospheric and Oceanic Physics,* **9**(6), 655–60.

Henderson-Sellers, A., 1990, The 'coming of age' of land surface climatology, *Global and Planetary Change,* **82**, 291–319.

Herman, B. M., Browning, S. R. and Curran, R. J., 1971, The effect of atmospheric aerosols on scattered sunlight, *Journal of Atmospheric Sciences,* **28**, 419–28.

Junge, C., 1951, Nuclei of atmospheric condensation, in T. F. Malone (Ed.) *Compendium of Meteorology,* Boston, Massachusetts: American Meteorological Society.

Lander, G., Morgan, G., Nagamoto, C. T., Solak, M. and Rosinski, J., 1980, Generation of ice nuclei in the surface outflow of thunderstorms in Northeast Colorado, *Journal of Atmospheric Sciences,* **36**, 2484–94.

Lavel, K. and Picon, L., 1986, Effect of a change of the surface albedo of the sahel on climate, *Journal of Atmospheric Sciences,* **43**, 2418–29.

Lean, J. and Warrilow, D. A., 1989, Simulation of the regional climatic impact of amazon deforestation, *Nature,* **342**, 411–13.

McTaggart, A. R. and Burton, H., 1992, Dimethyl sulfide concentrations in the surface waters of the Australasian Antarctic and Subantarctic Oceans during an austral summer, *Journal of Geophysical Research,* **97**(C9), 14407–12.

Mylne, M. F. and Rowntree, P. R., 1991, Modelling the effects of albedo change associated with tropical deforestation, *Climatic Change,* **21**, 317–43.

Olson, J. S., Watts, J. A. and Allison, L. J., 1983, Carbon in live vegetation of major world ecosystems, Publication No. 1997, Oak Ridge, Tennessee: Oak Ridge National Laboratory, Oak Ridge Tennessee Environmental Sciences Division.

Oort, A. H., 1964, On estimation of the atmospheric energy cycle, *Monthly Weather Review,* **92**, 483–93.

Paltridge, G. W., 1979, Climate and thermodynamic systems of maximum dissipation, *Nature,* **279**, 630–1.

Parton, W. J., Schimel, D. S., Cole, C. V. and Ojima, D. S., 1987, Analysis of factors controlling soil organic matter levels in Great Plains grasslands, *Soil Science Society of America Journal,* **51**, 1173–9.

Pielke, R. A. and Avissar, R., 1990, Influence of landscape structure on local and regional climate, *Landscape Ecology,* **4**, 133–55.

Reed, R. H., 1983, Measurement and osmotic significance of B-dimethyl sulphoniopropionate in marine microalgae, *Marine Biology Letters,* **4**, 173–81.

Rowntree, P. R., 1983, Simulation of atmospheric response to soil moisture anomalies over Europe, *Quarterly Journal of the Royal Meteorological Society,* **109**, 501–26.

Running, S. W., Nemani, R. R., Peterson, D. L., Band, L. E., Potts, D. F. and Pierce, L. L., 1989, Mapping regional forest evapotranspiration and photosynthesis by coupling satellite data with ecosystem simulation, *Ecology,* **70**(4), 1090–1101.

Schnell, R. C., 1975, Ice nuclei produced by laboratory cultured marine phytoplankton, *Geophysical Research Letters,* **2**(11), 500–3.

Schnell, R. C. and Vali, G., 1972, Atmospheric ice nuclei from decomposing vegetation, *Nature,* **236**, 163–5.

Schnell, R. C. and Vali, G., 1973, World-wide source of leaf-derived freezing nuclei, *Nature,* **246**, 212–13.

Schnell, R. C. and Vali, G., 1976, Biogenic ice nuclei, Part I: Terrestrial and marine sources, *Journal of Atmospheric Sciences,* **33**, 1554–64.

Schnell, R. C., Worbel, B. and Miller, S. W., 1980, Seasonal changes and terrestrial sources of atmospheric ice nuclei at Boulder, Colorado, *AIMPA Commission International de Physique des Nuages,* **1**, 42–6.

Sellers, P. J., Mintz, Y., Sud, Y. C. and Dalcher, A., 1986, A simple biosphere model (SiB) for use within general circulation models, *Journal of Atmospheric Sciences,* **43**, 505–31.

Shukla, J. and Mintz, Y., 1982, Influence of land-surface evapotranspiration on the Earth's climate, *Science,* **215**, 1498–1501.

Sud, Y. C. and Fennessy, M. J., 1982, A study of the influence of surface albedo on July circulation in semi-arid regions using the GLAS GCM, *Journal of Climatology,* **2**, 105–25.

Sud, Y. C. and Fennessy, M. J., 1984, Influence of evaporation in semi-arid regions on the July circulation: A numerical study, *Boundary Layer Meteorology,* **33**, 185–210.

Sud, Y. C. and Molod, A., 1988, A GCM simulation study of the influence of Saharan evapotranspiration and surface-albedo anomalies on July circulation and rainfall, *Monthly Weather Review,* **116**, 2388–2400.

Sud, Y. C. and Smith, W. E., 1984, Ensemble formulation of surface fluxes and improvement in evapotranspiration and cloud parameterization in a GCM, *Boundary Layer Meteorology,* **29**, 185–210.

Sud, Y. C. and Smith, W. E., 1985, The influence of surface roughness of deserts on July circulation, *Boundary Layer Meteorology,* **33**, 15–49.

Tucker, C. J., Fung, I. Y., Keeling, C. D. and Gammon, R. H., 1986, Relationship between atmospheric CO_2 variations and a satellite-derived vegetation index, *Nature,* **319**, 195–9.

Vairavamurthy, A., Andreae, M. O. and Iverson, R. L., 1985, Biosynthesis of dimethyl sulfide and dimethylpropiothetin by *Hymenomonas carterae* in relation to sulfur source, *Limnology and Oceanography,* **30**, 59–70.

Vali, G., Christensen, M., Fresh, R. W., Galyan, E. L., Make, L. R. and Schnell, R. C., 1976, Biogenic ice nuclei, Part II: Bacterial sources, *Journal of Atmospheric Sciences,* **33**, 1565–70.

Went, F. W., 1960, Blue hazes in the atmosphere, *Nature,* **187**, 641–3.

Went, F. W., 1961, Organic matter in the atmosphere as an energy supply for lightning, *Science,* **134**, 1437.

Went, F. W., 1962, Thunderstorms as related to organic matter in the atmosphere, *Proceedings of the National Academy of Sciences,* **48**(3), 309–16.

Went, F. W., 1964, The nature of Aitken condensation nuclei in the atmosphere, *Proceedings of the National Academy of Sciences,* **51**, 1259–67.

Went, F. W., 1966, On the nature of Aitken condensation nuclei in the atmosphere, *Tellus,* **18**(2), 549–56.

Went, F. W., Simmons, D. B. and Monzingo, H. N., 1967, The organic nature of atmospheric condensation nuclei, *Proceedings of the National Academy of Sciences,* **58**, 69–74.

Williams, M. D., Treiman, E. and Wecksung, M., 1980, Plume blight visibility modeling with a simulated photograph technique, *Journal of the Air Pollution Control Association,* **30**(2), 12130.

Woodcock, A. H. and Gifford, M. M., 1949, Sampling atmospheric sea-salt nuclei over the oceans, *Journal of Marine Research,* **8**, 177–97.

18

Evaluation of soil database attributes in a terrestrial carbon cycle model: Implications for global change research

Christopher S. Potter, Pamela A. Matson and Peter M. Vitousek

This paper presents sensitivity and scaling analyses using soil database attributes in an ecosystem model of global primary production and soil microbial respiration. The CASA (Carnegie–Ames–Stanford Approach) Biosphere model uses satellite imagery (Advanced Very High Resolution Radiometer and solar radiation), along with climate history (monthly temperature and precipitation) and soil attributes (texture, carbon and nitrogen contents and inundation) from global data sets as GIS input variables. A framework is summarized for spatial modelling and evaluation of potential aggregation errors associated with global gridded data sets. Soil carbon transformations predicted by the model are influenced by moisture effects on microbial activity and soil texture effects on the efficiency of heterotrophic respiration. We tested the assumption that the quality of global soil databases is critical to prediction of ecosystem controls on carbon cycling at large spatial scales. Model sensitivity analysis suggests that predicted soil carbon storage is highly sensitive to texture. A spatially uniform, fine-texture setting resulted in the highest soil carbon pool size; this trend was consistent over the entire global gradient of climate conditions. Using the FAO Soil Map of the World, the model estimates that on a world-wide basis more than 40 per cent of the surface SLOW carbon pool is stored in tropical forest and savanna biomes. Addition of soil inundation effects on microbial activity resulted in a 2 per cent increase in global soil carbon storage, with the most important changes in the needleleaf evergreen forest biome. Improvement of process understanding of organic soil at depth and texture characterization in tropical ecosystems are identified as priorities for global change research.

Introduction

The atmospheric composition of greenhouse gases is strongly affected by the cycling of carbon and nitrogen (C and N) in terrestrial ecosystems. The pool of carbon stored in soils and plant litter exceeds that in the atmosphere by about twofold (Houghton *et al.,* 1990). Turnover of carbon in these pools, accompanied by transfer of radiative trace gases from soils and vegetation to the atmosphere, is broadly controlled by temperature, moisture and soils characteristics. Global changes in climate and land use over the next 50 to 100 years may significantly alter controls on net ecosystem carbon balance (Jenkinson *et al.,* 1991).

Regional and global models of terrestrial biogeochemistry have incorporated a certain degree of climate and soil control over ecosystem C fluxes (examples include Raich *et al.,* 1991; McGuire *et al.,* 1991; Potter *et al.,* in press). These models rely, however, on a single soils data source, the *Soil Map of the World* (SMW; FAO/UNESCO, 1971), for characterization of spatial patterns in soil particle size distribution and moisture-holding capacity. Consequently, their model predictions may be strongly influenced by existing inaccuracies in the SMW representation of spatial patterns in texture classes and rooting depth. While previous studies have analysed the mapping accuracy of soil databases and procedures of data aggregation for soil-water state modelling purposes (Wösten *et al.,* 1985; Bouma, 1986), few if any tests have been conducted to examine the sensitivity of modelled ecosystem C and N fluxes to prescribed properties in the SMW. In this paper, we examine ecosystem process model sensitivity to aggregated global data drivers, with focus on soil texture and carbon contents as test cases.

Texture is one of the several important variables that control carbon transformations in soils. Numerous studies have shown that low microbial turnover rates are associated with fine-textured soils (Sorenson, 1981; Schimel, 1986; Gregorich *et al.,* 1991); higher clay fractions lead to increased soil carbon stabilization (Parton *et al.,* 1987). There is also a strong, texture-dependent functional relationship between organic matter turnover rate and the degree of moisture-filled soil pore space (Linn and Doran, 1984). Inundation and soil moisture saturation are additional factors that may be associated with accumulation of carbon in wetland soils (Clymo, 1984; Gorham, 1991).

Simulation modelling of ecosystem effects on global biogeochemical cycles often involves application of controls verified at small scales to predict large-scale patterns (Rastetter *et al.,* 1992). Conceptually, a generic model developed for one ecosystem can be applied to all grid cells in a regional or global data set (Schimel and Potter, in press). These types of gridded spatial models operate under the assumption that the 'lumped' estimate of a driving variable is representative of its entire grid cell coverage. Because limited information is available on the frequency distributions of SMW texture classes at the sub-1° latitude–longitude level, lumped soil properties are generally prescribed in global model experiments.

With regard to these scaling issues, a closely related question for global change research is whether efforts aimed at improvement of land data sets such as the SMW will potentially lead to substantial changes in biogeochemical model predictions. The objective of this simulation modelling study was to evaluate the sensitivity of soil C storage to global data set characterizations of soil texture and inundation. We conducted tests using the CASA (Carnegie–Ames–Stanford Approach) Biosphere model (Potter *et al.,* in press), a terrestrial ecosystem model based on global satellite and surface data, to address the need for improved characterization in global soil data sets. Our tests included a comparison of global model runs using spatially uniform soil texture values, calculation of a probabilistic range of soil C storage using scaling factor frequency distributions, and effects of wetland inundation on soil carbon storage.

Model Structure

The CASA-Biosphere model is described in detail by Potter *et al.* (in press). It is a gridded spatial model that represents major ecosystem carbon and nitrogen transformations driven by remote sensing inputs from the Advanced Very High Resolution Radiometer (AVHRR) and long-term average climate data sets. The model runs on a monthly time interval to simulate seasonal patterns in carbon fixation, biomass and nutrient allocation, litterfall, soil nitrogen mineralization and trace gas emissions. Input and pool variables correspond to global gridded data sets at 1° latitude–longitude resolution. A schematic representation of database and ecosystem model integration is shown in Figure 18.1.

The CASA Biosphere framework for spatial modelling studies combines a generalized ecosystem model of biogeochemical cycling with a geographic information system (GIS), probability function generation and public domain and commercial software for data visualization and analysis (Figure 18.2). Initialization and output data sets are stored as raster map arrays in the GIS — Geographic Resources Analysis Support System (GRASS; Shapiro *et al.,* 1992). Integration of dynamic model variables with the GIS brings with it the capability for map layer overlay, weighting or averaging output data by reference map sets (e.g., biome type, latitude zone or soil attribute), and selective analysis of regions or grid cells of interest. We followed the vegetation classification of Dorman and Sellers (1989) for biome-based analyses. Hierarchical data format (HDF; Fortner, 1992) is used as the file standard for data transfer. Our spatial modelling system links the GIS to simulation model code by calling raster data files from the GRASS directory to the memory of the computer workstation for dynamic floating point computations. As results are produced, output data files are written to disk in a GIS-compatible format. For the experiments described in this study, the frequency distribution component of the system was merely a series of hypothetical scaling factor inputs to the GIS, which are sufficient for description of discrete category variables like soil texture. For non-discrete GIS variables such as climate and radiation drivers, we have developed Monte Carlo simulation components (King *et al.,* 1989) for use in future probability function studies.

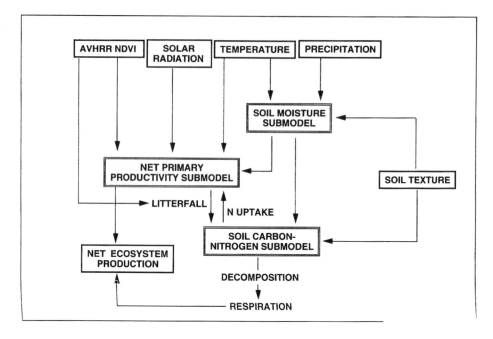

Figure 18.1 CASA Biosphere model database and submodel integration scheme.

Note: Global climate data sets are combined with soil texture settings to compute monthly water balance, which controls NPP and soil microbial activity. Soil texture (SC particle size fractions) also controls the turnover rate of soil microbes and the fraction of carbon lost as CO_2 from microbes in transfer to the SLOW pool.

In the CASA Biosphere ecosystem model, net primary productivity (NPP) and litterfall are estimated using remotely sensed data from the AVHRR sensor. Global data sets of solar insolation and the AVHRR-Normalized Difference Vegetation Index (NDVI) provide the basis for monthly estimates of intercepted radiation. The model computes a global light-use efficiency term that is attenuated at each grid cell by monthly moisture and temperature stress factors.

Major soil carbon–nitrogen pools and transformation processes connecting them at the ecosystem level are shown in Figure 18.3. For flux calculations, microbially mediated maximum carbon transformation rates are adjusted according to a series of non-dimensional (range of 0–1) scalars related to air temperature, soil moisture, substrate quality (N and lignin contents) and soil texture. The model is designed to simulate C and N fluxes only in surface soil layers (to 30 cm depth). Production of CO_2 at a grid cell location (x) during month t results from microbially mediated decomposition of plant and soil organic residues, as shown in equation (18.1).

$$CO_2(x,t)_i = C(x,t)_i \cdot k_i \cdot W_s(x,t) \cdot T_s(x,t) \cdot (1 - M_\varepsilon) \qquad (18.1)$$

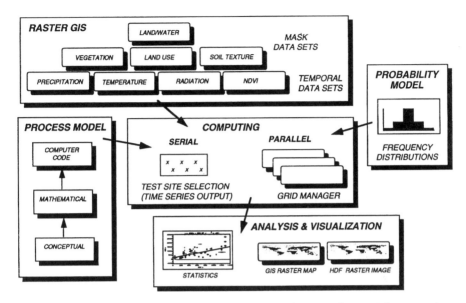

Figure 18.2 Framework for spatial modelling and evaluation of potential aggregation errors in environmental information systems.

where

$C(x,t)_i$ = carbon content of pool i;

k_i = maximum decay rate constant of pool i;

$W_s(x,t)$ = a scalar for the effect of soil moisture content on decomposition;

$T_s(x,t)$ = a scalar for the effect of temperature on decomposition;

$M\varepsilon$ = carbon assimilation efficiency of microbes; set at a spatially uniform value of 0.45.

Carbon is transferred from litter to microbial pools, and subsequently to soil carbon pools, according to equation (18.1) but using M_ε in place of (1-M_ε).

Following the CENTURY model structure proposed by Parton *et al.* (1987), soil C is stored in three pools: microbial (MIC), SLOW and OLD, which are separated by their respective turnover times. Microbial C turns over at a rate of one to several years, depending on the biome type. SLOW C turns over at a rate of 10 years to >75 years, whereas carbon in the OLD pool may remain in the soil for thousands of years.

The CASA Biosphere soil moisture submodel is a one-year 'bucket' formulation based on previous regional and global surface hydrology simulation studies (Mintz and Serifini, 1981; Vörösmarty *et al.*, 1989; Bouwman *et al.*, 1993). Soil texture controls several aspects of the water balance and C:N transformations. In soil moisture calculations, wilting point (WPT (x)) and field capacity (FC(x)) were derived from particle size relationships described by Saxton *et al.* (1986) for the five major texture classes represented in the SMW

Table 18.1 Texture attributes in the FAO SMW classes.

No.	Class	% Clay	% Silt	% Sand
1	Coarse	9	8	83
2	Coarse/medium	20	20	60
3	Medium	30	33	37
4	Medium/fine	48	25	27
5	Fine	67	7	17

(Zobler, 1986) (Table 18.1). These settings are combined with generalized soil rooting depths for forest (2.0 m) and other (1.0 m) vegetation classes (Vörösmarty *et al.*, 1989) to produce global maps of WPT and FC. Texture also controls the rate of soil drying due to evapotranspiration losses. Certain soil types in the SMW were treated as special cases. Lithosols were assigned to a shallow soil class of 27 per cent FC (total soil volume) with rooting depth of 10 cm (Vörösmarty *et al.*, 1989). Organic soils (histosols) were assigned to the coarse–medium texture class (Bouwman *et al.*, 1993).

Soil texture influences soil carbon transformations in two direct ways. First, microbial turnover rate decreases in fine texture soils according to the scalar E_T shown in equation (18.2) (Ladd *et al.*, 1981; Sorenson, 1981; Gregorich *et al.*, 1991).

$$E_T(x) = 1 - (0.75 \cdot SC(x)) \tag{18.2}$$

where $SC(x)$ is the silt plus clay fractions.

Second, the fraction of carbon lost as CO_2 from soil microbes during transfer to the SLOW C pool ($SLOW_f$) decreases as the SC content increases according to the relationship specified in equation (18.3) (Van Veen *et al.*, 1984; Parton *et al.*, 1987).

$$SLOW_f(x) = 0.85 - (0.68 \cdot SC(x)) \tag{18.3}$$

Soil nitrogen transformations are tied to carbon fluxes following the basic structure of several previous models (Jenkinson and Rayner, 1977; McGill *et al.*, 1981; Parton *et al.*, 1987). Plant residue inputs are divided initially into subpool *metabolic* (rapidly decomposing cytoplasmic constituents and nucleic acids) and *structural* (slowly decomposing cellulose plus lignin) fractions according to residue lignin-to-nitrogen content ratio (McGill *et al.*, 1981; Parton *et al.*, 1987). All residue lignin is assumed to reside in structural fractions.

For initialization of soil pools, we assumed that most of the C and N that turns over on decadal-to-century time scales is in the upper 0.3 m of the soil. Hence, we set initial soil C and N pool contents (g m^{-2}), using 50 per cent of the

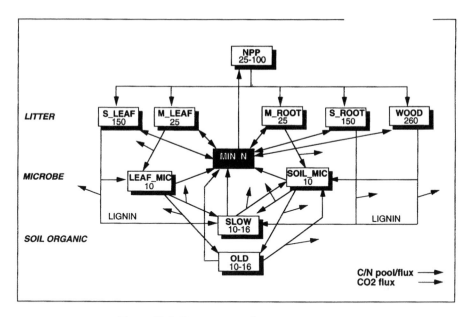

Figure 18.3 Ecosystem carbon–nitrogen model.

Note: Carbon pools in white with C:N ratios, nitrogen pools in black, C and N fluxes in solid arrows, CO_2 production in stippled arrows. Levels of litter, microbe (MIC), and soil organic (SLOW and OLD) pools are shown. Structural (S) and metabolic (M) pools are shown for leaf and root litter.

totals provided by Post *et al.* (1985), to 1 m depth for the Holdridge life zone classification of Leemans (1990). The CASA model does not consider cycling of the remaining C and N in deeper soil layers. For initialization only, 30 per cent of the total soil C and N contents in the top 0.3 m are allocated to the OLD pool, with the remainder allocated between soil microbial and SLOW pools.

Nitrogen transformations are stoichiometrically related to C flows (Figure 18.3). Fluxes from litter and soil to microbial pools and from microbial pools back to soil pools occur in proportion to C assimilation rates so that fixed C:N ratios for the various recipient organic matter pools are maintained (Parton *et al.*, 1987). Immobilization flows may occur at rates necessary to meet critical pool C:N ratios (levels to which litter accumulates N until release occurs). Mineralization inputs to the common mineral N pool (MINN) are equal to the difference between total decomposition-based N flux and the amounts needed to meet combined recipient pool C:N ratios.

Model Experiments and Results

Sensitivity Analysis of Soil Texture on Soil C Storage

To test the influence of texture on soil carbon pools, a series of model runs to steady state (less than 1 per cent change) were made, each with soil texture set at a spatially uniform value for the globe that corresponds to one of the five major classes in the SMW. Exceptions to the uniform settings were lithosols, which remained unchanged from the original distributions in Zobler (1986). Organic soils (histosols) were not treated explicitly in this analysis. The result of these runs were five global output data sets, each of which represents soil C storage for a single texture setting. While modifications in soil texture and predicted moisture content would affect NPP in the fully coupled version of the CASA Biosphere model, we confined effects of texture changes to the soil C:N submodel in order to simplify interpretations of test results and to focus exclusively on below-ground processes.

We selected the SLOW C pool for a comparative analysis because it consistently made up 65–85 per cent of total soil carbon storage in model estimates (Potter *et al.,* in press). GIS routines were used to compute average SLOW C pool size (g m^{2} to 0.3 m depth), annual temperature and total annual precipitation over 1° latitude zones. Climate calculations included terrestrial grid cells only. Results showed that SLOW C storage is highly sensitive to soil texture (Figure 18.4). Uniform fine-texture soil settings resulted in highest SLOW C pool storage; this is the case over the entire global gradient of climate conditions. Comparison of average SLOW C estimates showed that the difference between uniform fine-texture and coarse-texture settings is only slightly more pronounced at low latitudes; for example, at the equator, storage in fine-textured soils is 1.5 times greater than storage in coarse-textured soils, compared with a 1.4-fold difference near 60°N. Texture does influence the proportion of global soil C stored at various latitude zones; for uniform fine-texture settings, the high-latitude peak was 10 per cent greater than the low-latitude peak, whereas the differences between high- and low-latitude peaks at medium and coarse settings were 8 per cent and 13 per cent, respectively.

We constructed a composite global map of SLOW C storage (Figure 18.5) by mapping the CASA Biosphere model SLOW C estimates from each of the spatially uniform texture settings to their corresponding geographic locations in the five classes of the original FAO SMW (Zobler, 1986). Global SLOW pool size for surface soil layers in the composite data set is 300 Pg C. Over 40 per cent of this total is stored in tropical forest and savanna biomes. Another 35 per cent is located in northern coniferous forest and tundra ecosystems. For the remainder of this study, the data set shown in Figure 18.5 will be referred to as the composite SLOW C map.

As discussed by Potter *et al.* (in press), comparison of our steady-state estimates of SLOW C pools (Figure 18.5) to initial model conditions, the latter of which matches the geographic distribution for life zones as reported by Post

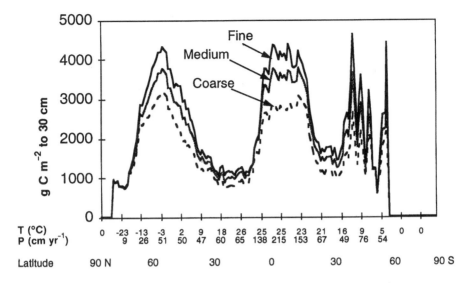

Figure 18.4 Sensitivity of carbon storage in the SLOW C pool to uniform texture settings over the global latitude range.

Note: Annual average temperature and total precipitation values are included on the abscissa.

et al. (1985), indicates that the model predicts somewhat lower soil carbon storage at high-to-mid latitudes (90–30° N) and higher pools in tropical (20° N–20° S) latitudes. Part of the difference may be explained by the fact that we modelled only the upper 0.3 m of the soil, while Post *et al.* (1985) considered C storage to 1 m soil depth. Distribution of C storage in upper soil layers does not necessarily reflect soil carbon pools below 0.3 m depth, where there is relatively poor understanding of organic matter dynamics. Nevertheless, previous comparisons suggest that the greatest difference between tropical and temperate soils is found in deep profile layers (Sanchez *et al.,* 1982).

Texture Scaling Factor Distributions

In scaling up model estimates, simple linear aggregation of an independent variable may lead to inaccurate representation of non-normally distributed spatial data (King *et al.,* 1989; Henderson-Sellers and Pitman, 1992). Serious error propagation problems can arise when aggregated estimates from either averaging subgrid cell attribute values, or assigning a 'most common' class value are used in non-linear functional responses without consideration of subgrid cell spatial variation.

The previous experiment demonstrated that soil C pool estimates in the CASA Biosphere model are strongly affected by texture settings. In that

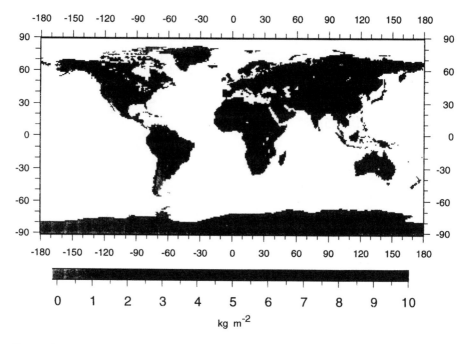

Figure 18.5 Composite distribution of SLOW C predicted by the CASA Biosphere model using global soil texture settings from Zobler (1986).

experiment, a single texture value was assumed to be representative of the entire grid cell coverage of interest. Such 'lumping' is generally followed for global models because of limited information on parameter variance at the subgrid cell level. In contrast, a probabilistic approach to grid-based spatial modelling uses frequency distributions of major input variables for statistical scaling analyses (King *et al.,* 1989). In extended range modelling (Luxmoore *et al.,* 1991), the land surface is treated as a collection of grid elements each with its own set of ecological and edaphic characteristics that can be represented by frequency distributions. For example, soil input values can be drawn from measured (or hypothetical) frequency distributions for the edaphic property of interest. Comparison of simulation results from various input distributions provides valuable information about possible aggregation errors in a lumped parameter model.

The objective of this second experiment was to establish a probabilistic range of SLOW C pool sizes for comparison with the composite SLOW C map from the first experiment in which texture was treated as a lumped parameter. Consequently, we proceeded with tests to examine how several hypothetical texture frequency distributions at the subgrid (1°) cell level would affect global soil C pool sizes.

In the SMW developed by Zobler (1986), each 1° cell is characterized by a dominant soil unit, which occupies the largest area map unit. *Associated* soil types cover 20–30 per cent of a cell map unit, whereas *included* soil types cover 5–10 per cent of a cell map unit. The qualitative reliability of the SMW is divided into three classes. Class I units were based on actual soil surveys; Class II units were based on field reconnaissance of topography, geology, vegetation and climatic data; Class III units were based on general information from the local literature. Zobler (1986) pointed out that the effective spatial resolution of the SMW will vary geographically as a function of actual field survey coverage. Following FAO guidelines, Zobler (1986) defined texture according to the percentage of sand:silt:clay for the dominant soil unit. If more than one dominant class was specified by FAO files, texture was assigned by averaging multiple values from the SMW texture triangle.

Frequency distributions for this experiment were developed using scaling theory and the similar media concept (Miller and Miller, 1956), which holds that if λ_i, is the attribute value for the *i*th soil component and λ_x is the attribute value of the reference soil (in this case, the SMW texture class for grid cell *x*), then the dimensionless scaling factor (α) can be defined as

$$\alpha = \lambda_i / \lambda_x \tag{18.4}$$

The probability density function for a set of normally distributed scaling factors with a mean (μ) of one and standard deviation σ is given by

$$f(\alpha) = 1/(\sigma(2\pi)^{0.5}) \exp[-(\alpha - \mu)^2 / 2\sigma^2] \tag{18.5}$$

Certain field studies have suggested that probability density functions of α used to scale soil moisture properties are better approximated by the log-normal distribution (Warrick *et al.,* 1977; Sharma and Luxmoore, 1979). Hence, the probability density function for a log-normal distribution is

$$f(\alpha) = 1/(\alpha\sigma_n(2\pi)^{0.5}) \exp[-(1/2\sigma_n^2)(ln(\alpha) - \mu_n)^2] \tag{18.6}$$

where
$$\sigma_n = (ln((\sigma/\mu)^2 + 1))^{0.5} \tag{18.7}$$

and
$$\mu_n = ln(\mu) - 1/2\alpha_n^2 \tag{18.8}$$

It is generally acknowledged that comparative studies of actual soils cannot fully satisfy certain assumptions of the similar media concept, largely because of variability in porosity and tendencies to shrink and swell (Warrick *et al.,* 1977). Nevertheless, the α term does provide a useful index for theoretical tests of presumed spatial variability in soil properties.

We derived normal (N) and log-normal (L) frequency distributions for the five major SMW texture classes for use as scaling-factor probability densities (Sharma and Luxmoore, 1979). To do so, it was necessary to make assumptions concerning the value of σ for soil texture distributions over the grid cell size of interest. A range of standardized σ values from 0.1 to 0.6 has been reported from field survey studies (Nielsen *et al.*, 1973; Peck *et al.*, 1977). Therefore, we selected two levels of σ, expressed as coefficients of variation (CV) at 0.2 and 0.5, for comparisons of model sensitivity to scaling N and L factor probability densities (Figures 18.6a and 18.6b). These σ levels were chosen as reasonable ends of the spectrum for spatial variation in soil attribute coverage. At the low end of the spectrum, variance might be related to the small plot or watershed scale; at the high end, variance might be related to landscape-to-regional scales, which potentially cover several ecosystems and soil groups. From a similar perspective, low variance may reflect a single *dominant* map unit class, whereas high variance suggests more than one dominant map unit class for a grid cell.

Probabilistic distribution maps were created using hypothetical frequency distributions applied to the five spatially uniform texture data sets from the first model experiment. Scaling factor intervals were set at mid-points between the five texture classes according to their respective SC content (Zobler, 1986). By weighting the five uniform data sets, each according to the fractional probability densities for texture scaling factor intervals, four probabilistic data sets (one each for N and L distributions at two levels of σ) were produced.

Each probabilistic data set represented a different sub-1° grid cell frequency distribution of soil texture. Among the four global SLOW pools (Pg C) that result from the hypothetical probability functions, the L-0.5 distribution shows the greatest deviation from the total SLOW pool computed from the composite SLOW C map in Experiment 1 (Table 18.2). Compared with the composite SLOW C map, the L-0.5 distribution results in global reduction of 2 per cent (6 Pg C) in storage. Under the L-0.5 distribution, the greatest changes in SLOW C pool are in broadleaf evergreen forest and savanna (classes 1 and 6; −3 to −4 per cent), broadleaf shrub (classes 8 and 9; −2 per cent), and cultivation (class 12; −3 per cent) vegetation types. The N-0.5 frequency distribution tends to increase SLOW C storage slightly in mid-to-high latitude forest (classes 4 and 5) and tundra (class 10) vegetation types.

Soil Texture in the Tropics

Results from the previous two experiments suggest a strong influence of fine-textured soils on soil C storage. Because 66 per cent of all fine-textured and 48 per cent of medium-to-fine textured soils are located between 30°N and 30°S latitude (SMW; Zobler, 1986), an experiment was devised to further investigate texture controls on SLOW pool C storage in tropical zones.

Conventional wisdom holds that ferrasols (oxisols in the US Soil Taxonomy system) are common over large areas of the tropics. Comparing vegetation maps with the SMW, ferrasols are seen to support a wide range of ecosystems, from tropical savannas, like the *cerrado* of Brazil, to rain forests. The SMW shows that these soils occupy 21 per cent of the total tropical land mass and are found mostly in South America (40 per cent) and Africa (18 per cent) (Zobler, 1986).

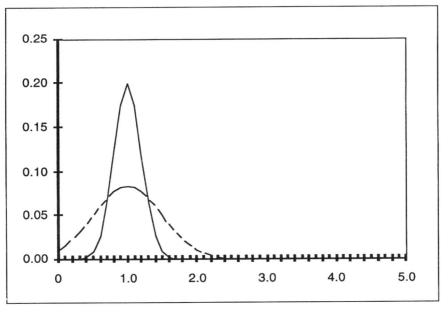

(a)

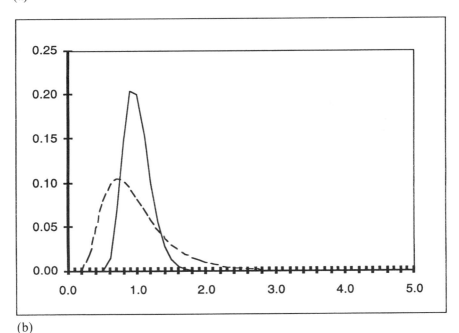

(b)

Figure 18.6 Probability density functions for texture scaling factors at two standard deviation (σ) levels: (a) normal distributions, (b) log-normal distributions.

Note: Solid lines represent CV = 0.2, dashed lines CV = 0.5.

Table 18.2 Soil carbon storage (SLOW pool) in biomes under the lumped FAO-SMW (Composite) texture settings and four hypothetical probability density functions (CV = 0.2 and 0.5) for texture scaling factors.

Class	Description	Composite	Normal		Log-normal	
			0.2	0.5	0.2	0.5
1	Broadleaf evergreen tress	74	74	74	74	72
2	Broadleaf deciduous trees	8	8	8	8	8
3	Broadleaf and needleleaf trees	16	16	17	16	16
4	Needleaf evergreen trees	52	53	54	53	53
5	Needleleaf deciduous trees	23	23	23	23	23
6	Broadleaf trees with ground cover	50	50	50	50	49
7	Perennial grasslands	11	11	11	11	10
8	Broadleaf shrubs with grasslands	6	6	6	6	6
9	Broadleaf shrubs with bare soil	8	8	8	8	8
10	Tundra	13	13	14	13	13
11	Bare soil and desert	5	5	5	5	5
12	Cultivation	33	33	33	33	32
	TOTAL	300	301	303	300	294

Note: Units are Pg C; biome types from Dorman and Sellers (1989).

More recent survey information suggests that these figures overestimate the importance of ferrasols; as little as 12 per cent of the total tropical land mass may actually be covered by ferrasols (Richter and Babbar, 1992). A revised soils map of Brazil, based on surveys conducted by EMBRAPA, indicates that ferrasol coverage should be reduced from 67 per cent to around 39 per cent over the Brazilian Amazon. While the extent of ferrasols is probably overestimated, acrisols (ultisols in the US Soil Taxonomy system), which cover about 10 per cent of the tropical regions in the SMW, are probably underestimated. Consequently, many soils of the tropics may be less weathered than indicated by the SMW.

According to the SMW texture map, over 90 per cent of global grid cells in the ferrasol category are classified as having SC fractions greater than 60 per cent (Zobler, 1986). While this sort of medium-to-fine texture classification would indicate relatively high moisture-holding capacity (30–50 per cent by volume; Saxton *et al.,* 1986), ferrasol clays have been shown to form very durable aggregates, such that their moisture-holding capacities are actually closer to that of coarse-textured soils. For example, Cochrane *et al.* (1985) reported that one half of the Amazonian soils they mapped had water-holding capacities <7.5 per cent by volume, a level typical of very sandy soils. Even in upland ferrasols, plant-available water storage capacity may be less than 10 per cent by volume.

To compensate for these potentially misleading classifications, Bouwman *et al.* (1993) made several texture adjustments to Zobler's (1986) SMW for use in a model of soil nitrous oxide fluxes. Vertisols and ferrasols were assigned to the medium-texture FC class, whereas andosols were assigned to the fine-texture FC class. These same adjustments were adopted for the first version of the CASA Biosphere model (Potter *et al.,* in press).

In this experiment, we examined the sensitivity of CASA Biosphere model predictions for soils of the tropics by setting moisture-holding capacity to a coarse-texture level (FC $\sim$ 20 per cent by volume; Saxton *et al.*, 1986) for all ferrasols. This setting is lower than the medium texture (FC $\sim$ 30 per cent by volume) prescribed by Bouwman *et al.* (1993). To simplify interpretations of results, the effect of the SC fraction on microbial turnover rate (as shown in equation (18.3)) was not altered in this model experiment. This would be an unrealistic assumption if microbial turnover rates in ferrasols were actually typical of fine-textured soils, despite low soil moisture-holding capacities.

Model simulation results suggest that reassignment of ferrasols to a coarse-texture class leads to a loss of 13 Pg C from the SLOW pool in tropical zones, compared with the composite SLOW C map from Experiment 1. On a global basis, this represents a decline of 4 per cent. Most of the loss (8 Pg C) occurs in areas covered by the broadleaf evergreen forest (class 1) vegetation type. Total SLOW C contents decrease by 11 per cent over this vegetation type; average SLOW C content decreases from 4.3 kg C m^{-2} to 3.8 kg C m^{-2}. Substantial losses are also seen in savanna vegetation (broadleaf trees with ground cover, class 6), which declines 8 per cent with this adjustment of ferrasols.

Soil Inundation Effects

In addition to texture, drainage capacity and wetland inundation may affect soil C storage. A global data set of fractional inundation of wetland soils is available for modelling studies (Matthews and Fung, 1987). It specifies inundation of 1° × 1° grid cells on a percentage area basis. In this experiment, we simulated the effect of saturated moisture conditions on soil carbon turnover rates by setting the W_s scalar in equation (18.1) to a fixed value of 0.3 for all monthly time steps. The 0.3 level is consistent with effects on organic matter decomposition for conditions of maximum moisture availability and minimum oxygen availability (Parton *et al.*, 1992; Potter *et al.*, in press). At grid cells in the Matthews and Fung (1987) data set where the fraction of area inundated (ω) was greater than zero, the model was run with this saturated soil moisture setting. For grid cells with 0 per cent area inundated, soils were assumed to be predominately well drained, such that if rainfall inputs exceed field capacity, runoff occurs with no time lag (Vörösmarty *et al.*, 1989) and the W_s term is computed according to a well-drained soil moisture status over the full range of 0–1 (Potter *et al.*, in press).

Carbon pools for grid cells where ω is greater than zero were computed as the weighted sum of two model run results: (a) total soil carbon (microbial biomass, SLOW and OLD pools) with impeded drainage effects (W_s = 0.3) and (b) total soil carbon without impeded drainage effects (W_s computed as in Potter *et al.*, in press), multiplied by ω and (1 − ω), respectively. Because the CASA model is not designed to simulate deeper soil carbon dynamics, this experiment tested changes in storage of soil carbon from surface and upper rooting zone only.

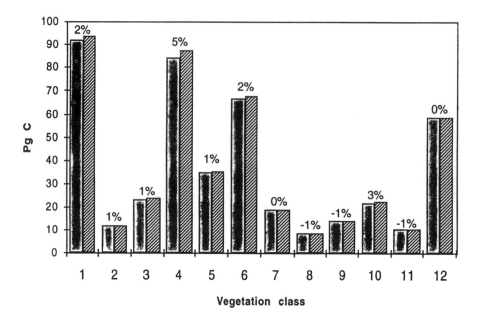

Figure 18.7 Soil inundation effects on total carbon storage (hatched bars).

Note: Shaded bars are biome totals based on data without effects of soil inundation on decomposition from Potter *et al.* (in press). Percentage change is shown for each biome class.

Simulated inundation has a negligible (<0.1 per cent change) effect on steady-state global microbial respiration fluxes. There is, however, an increase of 7.6 Pg C in the size of the total soil carbon pool as a result of the hypothesized wetland effects on microbial activity; this represents a global increase of about 2 per cent in soil C storage. The net increase is chiefly the result of changes in the SLOW pool. The largest proportional increase on a biome basis is found for needleleaf evergreen forests (class 4; 3.8 Pg C, 5 per cent), tundra (class 10; 0.6 Pg C, 3 per cent), and savanna (class 6; 1.3 Pg C, 2 per cent) (Figure 18.7). It should be noted that a portion of the savanna vegetation class is located where seasonal flooding is indicated in the inundation data set.

Discussion

The CASA biosphere framework for analysis of biogeochemical cycles integrates satellite imagery with historical climate and soil attributes (texture, inundation, C and N contents) from global, gridded data set inputs. The ecosystem process approach used in conjunction with large-scale land databases can help bridge the gap in scales that exists between biosphere models and the very local data used to calibrate them. The approach has some limitations (as discussed by Potter *et al.,* in press) but also has important advantages. For example, the use of a satellite-derived vegetation index gives the model extensive access to intra- and inter-annual variability, including certain aspects of agriculture and land use change. Because the model emphasizes scaling at the process rather than the biome level, the results are less sensitive to the quality and quantity of data characterizing any single ecosystem type.

The CASA Biosphere model estimate for soil carbon with a mean residence time of less than 100 years (SLOW pool of 300 Pg C) is slightly less than one-half the total carbon content in the atmosphere; SLOW pool size is about 100 times greater than the annual increase in atmospheric pools that results from fossil fuel emissions and deforestation (Houghton *et al.,* 1990). Hence, global processes with the potential to accelerate mobilization of soil carbon are of particular concern. Land use and climate change are possible triggers for large losses of soil carbon to the atmosphere.

The analysis presented in this study confirms the hypothesis that predicted ecosystem C storage on a global level is highly sensitive to soil texture settings. Model behaviour suggests that conversion of relatively fine- to coarse-textured soils could result in large transfers of CO_2 from the land to the atmosphere. Our experience is that the SMW must be used with caution in global carbon cycle modelling. Because the extent of fine-texture coverage is a major factor determining soil C storage, especially at low latitudes, reducing uncertainty in the global carbon budget will depend to some degree on (1) better large-scale mapping of soil texture and (2) improved understanding of the functional response of soil C fluxes to microbial activity and mineral reactions over a range of soil texture conditions.

The combination of high NPP (> 1 kg C m^{-2} y^{-1}) and fine soil texture consistently results in SLOW C storage of greater than 4 kg m^{-2} in the surface layers (to 0.3 m depth). Accumulation of SLOW C surface pools >4 kg m^{-2} at latitudes above 40°N also seems to be associated with high annual rainfall. There are two major latitude zones (from 70°N to 35°N and from 10°N to 25°S) where large pools of soil C accumulate (Figure 18.4). Both zones are bounded by steep dry–wet transitions in the polar-to-equatorial direction.

Single grid cell tests of internal (linear and non-linear) model controls on potential soil C storage revealed that texture-dependent moisture-holding capacity has a relatively small influence on SLOW C pool size. For simulations where potential soil C storage is high, the efficiency of microbial respiration, expressed as the fraction of carbon lost as CO_2 from soil microbes during transfer to the SLOW pool (equation (18.3)), is the primary functional response to texture settings. While the linear form of this function may reduce potential aggregation error in large-scale estimates (Rastetter *et al.,* 1992), further field studies are needed to test the broad applicability of the functional response represented in equation (18.3). Although Parton *et al.* (in press) conclude that the commonly observed correlation between soil clay content and carbon pool size is chiefly the result of increases in their CENTURY model 'passive' (residence time > 500 years) organic matter pool, our global scale modelling suggests that there may also be substantial effects on the 'slow' pool size (residence time >25 years to 100 years). While the CENTURY and CASA Biosphere modelling approaches differ in scope and objective, they lead to similar conclusions on the coupled effects of soil texture and cultivation on carbon cycling.

Based on the experiment using frequency distributions of scaling factors, it appears that there are proportionally large contributions of fine-textured soils to estimates of soil C pool size at regional and global scales. The greatest proportional difference, on both a total area and a SLOW pool size (g C m^{-2}) basis, between the composite SLOW C map and the L-0.5 data set can be attributed to weighting adjustments at grid locations where dominant texture is characterized as fine. For cells characterized as fine textured by Zobler (1986), the L-0.5 distribution includes a relatively high proportion of coarse-to-medium texture soils in the probabilistic 'fine' grid cell estimate, which results in as much as a 4 per cent mean reduction in SLOW C size for certain vegetation types. Under our scaling assumptions, this dilution effect is more important in determining the overall probabilistic estimate of C pool size than is the effect of including comparatively carbon-rich fine soils in final 'coarse' grid cell totals. On a global scale, however, it is difficult to determine whether soil texture and associated effects on soil C storage are best represented by normally or by log-normally distributed scaling factors. Whereas some detailed survey information is available for soil water states (Bouma, 1986), it is not clear that distributions in texture-mediated moisture characteristics reflect distributions of other soil physical properties that control microbial processes (Nielsen *et al.*, 1973).

While the prime focus of this study has been on soil texture effects on model predictions, this outlook addresses but one aspect of soil C responses to climate or land use change. Carbon stored and cycled in organic soils (histosols) may be of equal or greater importance to global biogeochemical fluxes. Wet tundra soils are estimated to accumulate from 0.03 Pg C yr^{-1} to 0.12 Pg C yr^{-1} on a world-wide basis (Oechel *et al.*, 1993); these same authors imply that future net carbon losses from high-latitude soils may result from indirect effects of enhanced drainage, aeration and altered water table, rather than from direct effects of climate warming. Hence, from a global, gridded modelling perspective, critical global change aspects of wetland carbon cycles may be difficult to represent, in part because of the coarse resolution of data input layers. The information needed to model drainage and water table dynamics exists chiefly at the basin and landscape scale and is not readily scalable to the 1° grid cell level. Furthermore, the application of ecosystem models that were designed for non-inundated sites to wetland studies must be highly qualified, since such formulations consider nutrient fluxes in surface soils only and mean residence times of ≤ 1000 years for stored carbon.

Conclusion

Based on the CASA Biosphere modelling experience, it appears that improved predictions of soil effects on ecosystem carbon balance will depend more on accurate characterization of the dominant 'lumped' texture class and improved knowledge about process-level controls on microbial respiration than on inclusion of detailed information on frequency distributions of associated soils. This is especially true in tropical zones, where carbon-rich, fine-textured soils are the most common class shown in the SMW. More fieldwork is needed to improve our understanding of the functional responses of carbon fluxes to soil clay contents in a variety of ecosystems. This study also highlights the importance of possible changes in soil C storage with climate-driven drying or anthropogenic conversion of wetlands at mid-to-high latitudes. Field and model experiments that deal with carbon transport and storage at depth in all biome types are needed.

Acknowledgement

Steven Klooster provided all programming assistance. Thanks to Chris Field and Jim Randerson for valuable discussions on model design. This work was supported by grants from a NASA EOS-IDS project (Sellers–Mooney), and from NASA's Earth System Science Modeling and Satellite Data Analysis Program in Ecosystems and Land-Atmosphere Interactions (ref. no. 2539-MD/BGE-0019). Graphics support was provided by the Numerical Aerodynamic Simulation facility at the NASA-Ames Research Center.

References

Bouma, J., 1986, Using soil survey information to characterize the soil-water state, *Journal of Soil Science,* **37**, 1–7.
Bouwman, A. F., Fung, I., Matthews, E. and John, J., 1993, Global analysis of the potential for N_2O production in natural soils, *Global Biogeochemical Cycles,* **7**(3), 557–97.
Clymo, R. S., 1984, The limits of peat bog growth, *Philosophical Transactions of the Royal Society of London,* Series B, **303**, 605–54.
Cochrane, T. T., Sanchez, L. G., de Azevedo, L. G., Porras, J. A. and Garver, C. L., 1985, *Land in Tropical America, Vol. 1: A Guide to Climate, Landscapes, and Soils for Agronomists in Amazonia, the Andean Piedmont, Central Brazil and Orinoco,* Cali. Columbia: Centro Internacional de Agricutura Tropical (CIAT).
Dorman, J. L. and Sellers, P. J., 1989, A global climatology of albedo, roughness length and stomatal resistance for Atmospheric General Circulation Models as represented by the Simple Biosphere Model (SiB), *Journal of Applied Meteorology,* **28**, 833–55.
FAO/UNESCO, 1971, *Soil Map of the World,* 1:5 000 000, Paris: UNESCO.

Fortner, B., 1992, *The Data Handbook,* Champaign, Illinios: Spyglass.

Gorham, E., 1991, Northern peatlands: Role in the carbon cycle and probable responses to climatic warming, *Ecological Applications,* **1**(2), 182–95.

Gregorich, E. G., Voroney, R. P. and Kachanoski, R. G., 1991, Turnover of carbon through the microbial biomass in soils with different textures, *Soil Biology and Biochemistry,* **23**(8), 799–805.

Henderson-Sellers, A. and Pitman, A. J., 1992, Land-surface schemes for future climate models: Specification, aggregation and heterogeneity, *Journal of Geophysical Research,* **97**(D3), 2687–96.

Houghton, J. T., Jenkins, G. T. and Ephraums, J. J. (Eds), 1990, *Climate Change: The IPCC Scientific Assessment,* Report of the Intergovernmental Panel on Climate Change, Cambridge: Cambridge University Press.

Jenkinson, D. S., Adams, D. E. and Wild, A., 1991, Model estimates of CO_2 emissions from soil in response to global warming, *Nature,* **351**, 304–6.

Jenkinson, D. S. and Rayner, J. H., 1977, The turnover of soil organic matter in some of the Rothamsted classical experiments, *Soil Science,* **123**, 298–305.

King, A. W., O'Neill, R. V. and DeAngelis, D. L., 1989, Using ecosystem models to predict regional CO_2 exchange between the atmosphere and the terrestrial biosphere, *Global Biogeochemical Cycles,* **3**(4), 337–61.

Ladd, J. H., Oades, J. M. and Amato, M., 1981, Microbial biomass formed from ^{14}C- and ^{15}N-labeled plant material decomposition in soils in the field, *Soil Biology and Biochemistry,* **13**, 119–26.

Leemans, R., 1990, *Possible Changes in Natural Vegetation Patterns Due to a Global Warming,* Laxenburg Working Paper WP-08, Laxenberg, Austria: International Institute for Applied Systems Analysis.

Linn, D. M. and Doran, J. W., 1984, Effect of water-filled pore space on carbon dioxide and nitrous oxide production in tilled and nontilled soils, *Soil Society of America Journal,* **48**, 1267–72.

Luxmoore, R. J., King, A. W. and Tharp, M. L., 1991, Approaches to scaling up physiologically based soil-plant models in space and time, *Tree Physiology,* **9**, 281–92.

Matthews, E. and Fung, I., 1987, Methane emission from natural wetlands: Global distribution, area, and environmental characteristics of sources, *Global Biogeochemical Cycles,* **1**(1), 61–86.

McGill, W. B., Hunt, H. W., Woodmansee, R. G. and Reuss, J. O., 1981, A model of the dynamics of carbon and nitrogen in grassland soils, *Ecological Bulletin (Stockholm),* **33**, 49–116.

McGuire, A. D., Melillo, J. M., Joyce, L. A., Kicklighter, D. W., Grace, A. L., Moore III, B. and Vörösmarty, C. J., 1992, Interactions between carbon and nitrogen dynamics in estimating net primary production for potential vegetation in North America, *Global Biogeochemical Cycles,* **6**(2), 101–24.

Miller, E. E. and Miller, R. D., 1956, Physical theory for capillary flow phenomena, *Journal of Applied Physics,* **27**, 324–32.

Mintz, Y. and Serafini, Y., 1981, *Global Fields of Soil Moisture and Land Surface Evapotranspiration,* NASA Technical Memorandum 83907, Research Review 1890/81, pp. 178–80, Greenbelt, Maryland: NASA Goddard Flight Center.

Nielsen, D. R., Biggar, J. W. and Erh, K. T., 1973, Spatial variability of field measured soil-water properties, *Hilgardia,* **42**, 215–59.

Oechel, W. C., Hastings, S. J., Vourlitis, G., Jenkins, M., Riechers, G. and Grulke, N., 1993, Recent change of Arctic tundra ecosystems from a net carbon dioxide sink to a source, *Nature,* **361**, 520–3.

Parton, W. J., Schimel, D. S., Cole, C. V. and Ojima, D. S., 1987, Analysis of factors controlling soil organic matter levels in Great Plains grasslands, *Soil Science Society of America Journal,* **51**(5), 1173–9.

Parton, W. J., McKeown, B., Kirchner, V. and Ojima, D., 1992, *CENTURY Users Manual,* Fort Collins, Colorado: Natural Resource Ecology Laboratory, Colorado State University.

Parton, W. J., Schimel, D. S., Ojima, D. S. and Cole, C. V., in press, A general model for soil organic matter dynamics: Sensitivity to litter chemistry, texture and management, *Soil Science Society of America Journal.*

Peck, A. J., Luxmoore, R. J. and Stolzy, J. L., 1977, Effects of spatial variability of soil hydraulic properties in water budget modeling, *Water Resources Research,* 13, 348–54.

Post, W. M., Pastor, J., Zinke, P. J. and Stangenberger, A. G., 1985, Global patterns of soil nitrogen storage, *Nature,* 317, 613–16.

Potter, C. S., Randerson, J. T., Field, C. B., Matson, P. A., Vitousek, P. M., Mooney, H. A. and Klooster, S. A., in press, Terrestrial ecosystem production: A process model based on global satellite and surface data, *Global Biogeochemical Cycles.*

Raich, J. W., Rastetter, E. B., Melillo, J. M., Kicklighter, D. W., Steudler, P. A., Peterson, B. J., Grace, A. L., Moore III, B. and Vörösmarty, C. J., 1991, Potential net primary production in South America: Application of a global model, *Ecological Applications,* 1(4), 399–429.

Rastetter, E. B., King, A. W., Cosby, B. J., Hornberger, G. M., O'Neill, R. V. and Hobbie, J. E., 1992, Aggregating fine-scale ecological knowledge to model coarser scale attributes of ecosystems, *Ecological Applications,* 2(1), 55–70.

Richter, D. D. and Babbar, L. I., 1992, Soil diversity in the tropics, *Advances in Ecological Research,* 21, 315–89.

Sanchez, P. A., Gichuru, M. P. and Katz, L. B., 1982, Organic matter in major soils of the tropical and temperate regions, *Transactions of the 12th International Congress of Soil Science* (New Delhi), 1, 99–114.

Saxton, K. E., Rawls, W. J., Romberger, J. S. and Papendick, R. I., 1986, Estimating generalized soil-water characteristics from texture, *Soil Science Society of America Journal,* 50, 1031–6.

Schimel, D. S. 1986, Carbon and nitrogen turnover in adjacent grassland and cropland ecosystems, *Biogeochemistry,* 2, 345–57.

Schimel, D. S. and Potter, C. S., in press, Process modeling and spatial extrapolation, in Matson, P. and Harriss, R. (Eds) *Methods in Ecology: Trace Gases,* Cambridge, Massachusetts: Blackwell.

Shapiro, M., Westervelt, J., Gerde, D., Larson, M. and Brownfield, K. R., 1992, *GRASS 4.0 Programmer's Manual,* Champaign, Illinois: US Army Construction Engineering Research Laboratory.

Sharma, M. L. and Luxmoore, R. J., 1979, Soil spatial variability and its consequences on simulated water balance, *Water Resources Research,* 15(6), 1567–73.

Sorenson, L. H., 1981, Carbon–nitrogen relationships during humification of cellulose in soils containing different amounts of clay, *Soil Biology and Biochemistry,* 13, 313–21.

Van Veen, J. A., Ladd, J. N. and Frissel, M. J., 1984, Modelling C and N turnover through the microbial biomass in soil, *Plant and Soil,* 76, 257–74.

Vörösmarty, C. J., Moore III, B., Grace, A. L., Gildea, M. P., Melillo, J. M., Peterson, B. J., Rastetter, E. B. and Steudler, P. A., 1989, Continental scale models of water balance and fluvial transport: An application to South America, *Global Biogeochemical Cycles,* 3(3), 241–65.

Warrick, A. W., Mullen, G. L. and Nielsen, D. R., 1977, Scaling field-measured soil hydraulic properties using similar media concept, *Water Resources Research,* 13, 355–62.

Wösten, J. H. M., Bouma, J. and Stoffelsen, G. H., 1985, The use of soil survey data for regional soil water simulation models, *Soil Science Society of America Journal,* 49, 1238–44.

Zobler, L., 1986, *A World Soil File for Global Climate Modeling,* NASA Technical Memorandum 87802, Greenbelt, Maryland: NASA.

19

Designing global land cover databases to maximize utility: The US prototype

Bradley C. Reed, Thomas R. Loveland, Louis T. Steyaert, Jesslyn F. Brown, James W. Merchant and Donald O. Ohlen

One of the most pressing problems in global climate and ecosystem studies is a lack of adequate land cover data. Staff from the United States Geological Survey's EROS Data Center and the University of Nebraska have developed a United States prototype for a proposed global land cover characteristics database derived from 1-km Advanced Very High Resolution Radiometer satellite data. A total of 159 seasonally distinct spectral/temporal land cover classes were labelled according to their constituent vegetation types rather than forcing the vegetation complexes into a predefined classification scheme. The database contains attributes that characterize each land cover class, including elevation, climate attributes and biophysical parameters derived from the normalized difference vegetation index (NDVI). The database design permits convenient translation to a variety of land cover classification schemes commonly used in global scale models. This approach allows scientists to continue using classification schemes with which they are comfortable, eliminates duplicating land cover database development, and provides a degree of uniformity in the development of parallel but distinct databases.

Introduction

Over the past decade, a number of efforts have been made to map land surface characteristics to support atmospheric general circulation models (GCMs), mesoscale meteorological models, hydrologic models and ecological models. While these applications require some basic land surface parameters (such as cover type), they also require specific biophysical parameters, methods for aggregating parameters and data at specific scales (Kemp, 1992; Steyaert, 1992) (Table 19.1). As a result, land surface data sets have been created at a variety of scales with unique classification and aggregation schemes. Most of these data sets are derived from existing maps and translated into the necessary classification schemes to drive particular models. Biophysical parameters (such as roughness, albedo and leaf area index) for the models are usually assigned as constants, according to vegetation type. This dependence on the vegetation classification scheme makes it difficult to create a land surface data set that can be adapted to the broad range of global modelling applications.

Table 19.1 Diverse land cover data requirements of land–atmosphere interaction models.

Type of Model	Classification scheme	Spatial scale	Biophysical attributes
Atmospheric GCMs:			
NASA/GLA	SiB[1]	4°× 5°	SiB set
COLA	Simple SiB	4.5°× 7.8° 1.8°× 2.8°	SSiB set
NCAR	BATS[2]	2°× 4°	BATS set
Mesoscale models:			
CSU-RAMS	LEAF[3]	Nested grids 1, 10, 40 km	LEAF set, NDVI[4]
PSU-NCAR MM4	BATS	Nested grids 4, 12, 36 km	BATS set
Hydrologic models:			
Watershed Precip/runoff	Basic classes	2.5, 5, 10 km	
Agricultural Chemical runoff	Anderson Level II	County, 1 km	
Ecosystem models:			
RHESSys/biome BGC	Basic biomes	1–50 km	NDVI
CENTURY ecotone	Anderson Level II	1–50 km, 1 km	Greenness seasonality
Equilibrium vegetation Biogenic emissions	Key species (oak, hickory, etc.)	20 km	NDVI

Notes:
[1] Simple biosphere model.
[2] Biosphere atmosphere transfer scheme.
[3] Land ecosystem atmosphere feedback.
[4] Normalized difference vegetation index.

Recent land surface classifications derived from satellite data have shown improvements in large-area characterization, providing relatively high spatial resolution, uniform grid size and current, rather than historic, land cover conditions (Tucker *et al.*, 1985; Townshend *et al.*, 1987). They are still, however, tied to specific classification schemes and have insufficient resolution for certain applications. This chapter presents a multiple-use concept in the design, development and utilization of a prototype global land cover characteristics database derived from 1-km Advanced Very High Resolution Radiometer (AVHRR) satellite imagery and ancillary data developed by the staff at the US Geological Survey's (USGS) EROS Data Center (EDC) and the University of Nebraska-Lincoln (UNL).

Land Surface Needs for Selected Models

Many scientific studies of global change require land cover data. The following discussion includes representative models and typical land cover requirements, but it is by no means an exhaustive list. For more extensive model reviews see Beach (1987), Baker (1989), Henderson-Sellers and McGuffie (1987), Sklar and Costanza (1990) and Goodchild *et al.* (1993). The application areas discussed here include atmospheric mesoscale models and GCMs, water resources assessments and ecological models.

Atmospheric Mesoscale and General Circulation Models

Mesoscale models such as the regional atmospheric modelling system (RAMS) and the mesoscale model version 4 (MM4) require land surface data. RAMS uses the land-ecosystem-atmosphere-feedback (LEAF) model (Lee, 1992) for its land surface parametrization scheme. The LEAF model requires several land surface parameters including cover type, leaf area index, suface albedo, canopy height and other attributes for each land cover class. The MM4 model uses the biosphere–atmosphere transfer scheme (BATS) for its land surface parametrization (Dickinson *et al.*, 1986). The BATS parameters are similar to those of LEAF, and both models use the BATS land cover classification scheme.

For the National Center for Atmospheric Research (NCAR) GCMs, BATS is usually applied over regions with grid cells of $2° \times 4°$ latitude/longitude with the previously defined land surface parameters. Other GCMs may use other land surface parametrizations, such as the simple biosphere (SiB) model (Sellers *et al.*, 1986) and simplified SiB (SSiB) (Xue *et al.*, 1991). For a coarse-resolution GCM, SiB uses land surface parameters unique to each SiB land cover class to calculate water and energy exchange fluxes for grid cells on the order of $4° \times 5°$ latitude/longitude (Sellers and Dorman, 1987). SSiB requires similar calculations, but for region sizes that range from $4.5° \times 7.8°$ grids to $1.8° \times 2.8°$ grids (Xue *et al.*, 1991).

Water Resources Assessments

Hydrologic models such as precipitation runoff models typically require land cover, soils and terrain information to define homogeneous hydrologic response units (HRUs) for model computations. The definition of these regions typically is based on relatively simple land cover classes such as bare soil, grasses, shrubs and trees. Agricultural chemical runoff models, such as the areal non-point-source watershed environmental response simulation (ANSWERS), require more detailed land use information (corn, soybeans, pasture, feedlots, etc.) for small land units such as hectares (Beasley and Huggins, 1982). More general non-point-source pollution runoff studies may require a general Anderson Level II land use and land cover classification calculated at 1-km or county levels or both (Anderson *et al.*, 1976).

Ecological Models

The CENTURY model is an ecosystem model that simulates the temporal dynamics of soil organic matter and plant production in grazed grasslands (Parton *et al.*, 1987, 1988; Burke *et al.*, 1991). Land cover, land use and monthly climate data are key elements of this model. Depending on the application, the CENTURY model may use the Anderson Level II land cover classification scheme (and other detailed land use data) for a regional 1-km resolution analysis, or BATS land cover classes aggregated to a 50-km grid for a national analysis.

The Regional Hydrological Ecosystem Simulation System (RHESSys) model operates on land cover data at the biome level (grasses, shrubs, coniferous and deciduous forests) at 1-km to 60-km grid cell sizes. Unique land cover attributes for each cover class are used with satellite-derived leaf area index estimates (derived from NDVI) and daily weather data as part of the biophysical scheme (Nemani and Running, 1989; Running, 1990).

Ecosystem data, including land cover and seasonal vegetative activity, are necessary for inventories of biogenic hydrocarbon emissions on a global scale. Current biogenic emissions inventories are typically modelled at around $4° \times 5°$ resolution, with efforts under way to estimate emissions on a $1° \times 1°$ global grid (Graedel, 1992). Most inventories are conducted on an annual basis but have large variability at seasonal and shorter time scales. Land cover data and estimates of seasonal vegetative activity are needed to assist these inventories.

Problems in Current Land Cover Databases

Because of the wide variety of scales, classification schemes and derived land cover parameters that are employed, current global land cover databases (e.g., UNESCO, 1973; Olson and Watts, 1982; Matthews, 1983) cannot fill the needs of global environmental research. Scientists typically must select a land cover framework based on availability rather than on suitability for the problem. A spatial land cover framework that is readily available and widely used is not necessarily a good choice if it is used for purposes other than those for which it was developed or intended (Peplies and Honea, 1992).

Past efforts at global database developed have involved combining existing source maps with dissimilar classification schemes into coarse-scale mosaics (UNESCO, 1973). The source maps are sometimes out of date and usually require a translation in class names between the map classification and the desired classification scheme. In the case of the UNESCO map, problems often

Table 19.2 *Biome classes developed for the Simple Biosphere Model.*

Class	Description
1	Tropical evergreen broadleaf
2	Temperate evergreen seasonal broadleaf
3	Cold-deciduous forest, with evergreens
4	Tropical/subtropical evergreen needleleaf forest
5	Cold-deciduous needleleaf
6	Tropical/subtropical drought-deciduous woodland
7	Tall/medium/short grassland with shrub cover
8	Evergreen broadleaf sclerophyllous forest, woodland, winter rain
9	Evergreen broadleaf shrubland/thicket, evergreen dwarf shrubland
10	Evergreen needleleaf or microphyllous shrubland/thicket
11	Desert
12	Cultivation
13	Ice

Source: S. Los (personal communication).

result from the use of the map as a source for new global mapping efforts (e.g., Olson and Watts, 1982; Matthews, 1983; Dickinson *et al.*, 1986).

Procedures to create appropriate land surface data for model input typically include reclassification or aggregation of existing vegetation information. Background knowledge is required to designate accurately any grid cell as one of the classes in a classification scheme. Unfortunately, the level of information available for performing this task is often insufficient. For example, in the SiB scheme, the world is divided into 13 simplified biomes (Table 19.2). These biomes were originally derived mainly from Küchler's potential natural vegetation (Küchler, 1983), with input from Matthews's cultivation data (Dorman and Sellers, 1989). However, an updated version of the model uses biomes derived almost entirely from Matthews's scheme (S. Los, personal communication), linking this system back to the vegetation framework designed by UNESCO.

One type of land cover may be classified very differently in various classification schemes, making translations or linkages difficult or impossible. For example, chaparral may be translated into very different classification bins within the BATS, SiB and Anderson schemes (Table 19.3). The main vegetative constituents of chaparral (in California) are annual grasses, manzanita, oak, pinyon pine and juniper. None of the three classification schemes provides an ideal definition for this land cover. None of the three schemes accounts for the tree component (pinyon and juniper), and the BATS scheme does not include the grass understorey.

Table 19.3 California chaparral translated into three classification schemes.

System	No.	Description
SiB[1]	7	Tall/medium/short grassland with shrub cover
BATS[2]	17	Deciduous shrub
LU/LC[3]	32	Shrub and brush rangeland

Notes:
[1] Simple biosphere model.
[2] Biosphere atmosphere transfer scheme.
[3] Land use and land cover.

Scale Issues

The issue of scale is a recurring theme in global and regional environmental monitoring and modelling. Attention is turning to the importance of scale as increasing focus is directed on the environmental impacts of global warming and other phenomena that occur over large regions (Hayden, this volume; Stafford *et al.*, this volume). Since much of the current knowledge of ecosystems and land processes has been generated from small-area studies, it is increasingly important that researchers focus on scaling up their field observations from sites to regions (Burke *et al.*, 1991). This will require geographic databases at intermediate scales appropriate for regional analysis, and comparable databases that can be used at coarse global scales.

Three key questions need to be addressed for selecting spatial scale and source materials (Malingreau and Belward, 1992): (1) can patterns detected at a given measurement scale be linked with specific environmental processes?; (2) when measurement scales are not attuned to the intrinsic scale of a specific environmental process, can inferences be made regarding these processes?; and (3) can the scale of a database be made compatible with the measurement scale of the image sources used in its preparation?

With land cover classification schemes, scale directly influences the categorical detail that can be mapped accurately. Large-area analyses frequently must compromise spatial and, thus, categorical detail to achieve large-area consistency. As the minimum map units increase in size, the spatial heterogeneity within each grid cell increases and, as a result, more general levels of categorization must be used to define landscape characteristics reliably and consistently. Many land cover classification schemes recognize the relationship between scale and class detail through the use of hierarchical categories (UNESCO, 1973; Anderson *et al.*, 1976).

There is increasing recognition that greater sophistication and improved predictability of GCMs can only occur when larger scale modelling of subgrid scale processes is common. For example, failure to consider interspersion of

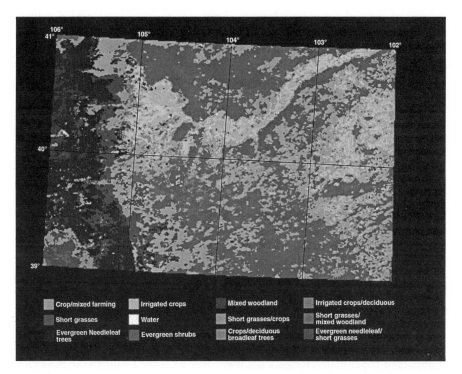

Figure 19.1 Spatial heterogeneity within a BATS GCM grid cell (2°×4° latitude, longitude) in the central USA.

lakes in GCMs leads to an underestimate of overall wetness in grid elements (Henderson-Sellers and Pitman, 1992). Figure 19.1 illustrates the spatial heterogeneity within a BATS GCM grid cell (2° × 4° latitude, longitude) in the central USA.

A land cover database should permit users to perform aggregations with a variety of methodologies and at any scale the model requires. Many models use the 'most common' element in a specified grid cell for aggregating global data (Dickinson *et al.*, 1986). Significantly different results are obtained using an averaging of the input parameters in a grid cell (Henderson-Sellers and Pitman, 1992). The averaging method may result in derived parameters that do not actually occur in nature, but this approach retains more of the information provided in the source data.

An important practical consideration is related to the temporal and spatial scale of environmental models. As models simulate phenomena at increased scales, there are many associated costs, including a need for the development of more sophisticated models that couple integrated processes occurring at detailed scales, greater computational demands and more expense in the development of databases. There is a direct relationship between temporal and spatial scales; as the spatial scale of a database increases (and spatial detail increases), it is more difficult and costly to increase the temporal interval of the spatial sample (Peer, 1990).

Demand for Database Flexibility

Ideally, a land cover database should permit users to perform temporal, spatial and categorical aggregations with a variety of methodologies and at any scale. In cases where models require site-specific inputs for large areas, measures or estimates of surface biophysical paramaters must be made at coarse scales. Fortunately, while coarse-resolution sensors such as AVHRR look nominally at 1-km areas, with heterogeneous cover, studies have shown that indices such as the NDVI from AVHRR are appropriate for estimating regional biophysical parameters (Aman *et al.*, 1992).

Seasonal Land Characteristics Database Concept

Staff members of the USGS's EROS Data Center and the University of Nebraska-Lincoln have developed a United States prototype for a proposed global land cover characteristics database derived from 1-km AVHRR satellite data. In all, 159 seasonally distinct spectral/temporal land cover classes were labelled according to their constituent vegetation types and productivity (Loveland *et al.*, 1991; Brown *et al.*, 1993).

AVHRR data have been used previously for developing large-area land cover classifications. These efforts have included using subsampled, multitemporal NDVI for classifying land cover for Africa (Tucker *et al.*, 1985) and South America (Townshend *et al.*, 1987) with ground resolutions of 15–20 km. While the use of satellite data has resulted in improvements in scale, data consistency and repeatability of methods, the resulting land cover classifications still have limitations for various modelling applications because they depend on a single classification scheme with relatively coarse spatial resolution.

The conterminous US land cover characteristics database represents an unconventional approach to meeting information needs of the environmental assessment and modelling community. The database uses the concept of seasonal land cover regions as a framework for presenting the temporal and spatial patterns of vegetation in the USA. The regions are composed of relatively homogeneous land cover associations (for example, similar floristic and physio-gnomic characteristics) that exhibit unique phenology (i.e., onset, peak and seasonal duration of greenness), and have common levels of primary production.

Rather than being based on precisely defined mapping units in a land cover classification scheme, the spectral/temporal land cover regions represent common sets of landscape conditions and serve as summary units to which both descriptive and quantitative variables may be attached as attributes. The attributes may be considered as spreadsheets of regional conditions, and they permit updating, calculating or transforming the entries into new parameters or classes. This provides the flexibility for using the land cover characteristics database in a variety of models without extensive modification of model inputs.

The concept of a land cover characteristics data set based on seasonal land cover regions is founded in classical geography. In geography, regionalization is a procedure that investigates many characteristics and elements that compose areas, and extracts those that are significant. Regions represent space on the surface of the earth that are distinctive in terms of features, events or processes (Peplies and Honea, 1992). Seasonal land cover regions may, therefore, be thought of as models that can be used to break down complex landscapes into spatial patterns that can control the measurement, interpolation and extrapolation of land parameters.

US Land Characterization Database Components

The prototype US land characterization database carries with it a number of quantitative and descriptive attributes for the land cover regions. All sources of information used in the database development are retained as individual database elements in order to maintain a flexible tool set for analysis. The temporal elements of the database, derived from AVHRR data, provide a quantitative description of the regions. These seasonal characteristics include 1990 biweekly NDVI statistics (mean, mode, median and standard deviation), AVHRR channel statistics, and phenological parameters (onset and duration of greenness and time of peak greenness) for each class. The NDVI statistics provide information on the temporal changes in vegetation condition, and the AVHRR channel statistics provide information on spectral and thermal characteristics of the land surface. Other elements in the database include elevation, land use and land cover, ecological descriptions, climate and soil.

The seasonal land cover regions represent three broad components of the landscape: (1) common mosaics of vegetation and land cover, (2) common patterns of seasonality (onset, peak and length of green season) and (3) relative levels or patterns of annual net primary production (ANPP). The latter is possible since research has shown that NDVI is related to biophysical processes, including net primary production, leaf area, evapotranspiration and carbon fluxes (Tucker and Sellers, 1986; Tucker *et al.*, 1986; Box *et al.*, 1989; Ludeke *et al.*, 1991; Spanner *et al.*, 1990). While absolute quantification of ANPP using NDVI is not yet possible, regions based on NDVI for a full growing season are tied to relative levels of annual net primary production.

Figure 19.2 illustrates the importance of seasonality. In this example, two different regions with similar land cover have slightly different seasonal

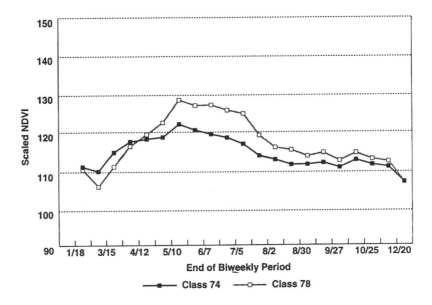

Figure 19.2 NDVI temporal profile for land cover characteristics database classes 74 and 78 (big sage, rabbitbrush, wheatgrass and fescue).

properties. Both classes are composed of mosaics of big sage, rabbitbrush, wheatgrass and fescue. While the land cover mosaic and the onset of greenness are nearly the same, the NDVI levels and the length of the greenness periods differ. Since the integral of NDVI over a growing season has been demonstrated to approximate ANPP, class 78 can be expected to have higher ANPP than class 74 (Goward *et al.*, 1985). The seasonal profile unique to each class serves as a temporal scale that can be used to represent biweekly or monthly variations in the biogeochemical processes of each region.

Translations from the land cover types described in the database are made to the land cover classification schemes commonly used in global environmental modelling, including BATS, SiB and Anderson Level II.

Applications

A series of experiments is under way to evaluate the utility of the land cover characteristics database. These experiments involve participants from universities and government research centres working on land surface parametrizations for atmospheric mesoscale and global climate models; water resources assessments; and ecological models, including ecosystem dynamics, biogeochemical cycles and biogenic emissions. Preliminary findings of these experiments illustrate the flexibility of the land cover characteristics database and suggest improvements for updating the database. Selected examples from these experiments include the following.

Atmospheric Mesoscale and General Circulation Models

One experiment incorporated the land cover characteristics database into research on the role of landscape variability in mesoscale meteorological systems in the US Great Plains (Pielke *et al.*, 1993). The land cover data, converted to predominant BATS class within 10-km and 40-km nested grid cells, and NDVI data (to estimate LAI) were used to parametrize and test the LEAF model for enhancing RAMS. Further model simulations showed that anomalously high NDVI values were associated with local severe storm outbreaks (Lee, 1992). These model simulations for a single date of study in eastern Colorado were significantly more realistic than simulation runs based on bare soil or natural grasslands. This fact suggests the significance of land cover as a determinant for weather and climate in the semi-arid Great Plains.

The land cover characteristics database was also evaluated for use in the SiB model (Xue, personal communication). Some major differences were noted between land cover characteristics SiB translation and the original SiB land cover data set. More forest land cover versus wheat was noted in the south-east USA than in the original version and more semi-desert versus desert in Nevada. Also, additional spatial complexity and richness of data were noted in the SiB translation. These differences have implications for improving leaf area index estimates and for modifying other derived parameters.

Water Resources Assessments

The land cover characteristics database is being used as part of a multiscale assessment of climate impact on water resources in the Susquehanna River Basin Experiment (Smith *et al.*, 1993; Lakhtakia, personal communication). This is a coupled modelling effort (hydrologic and land surface processes) for watershed-level precipitation runoff estimates. The hydrologic model is driven by conventional meteorological, soil and existing vegetation data. This is fed into the mesoscale model 4 (MM4), which uses the BATS land cover translation. Results from the MM4 are, in turn, fed into the watershed model. This modelling effort operates on grid cell domains of 4 km, 12 km and 36 km. The relatively fine resolution of land cover data permit the investigation of different methods of aggregation and are providing more spatial detail than had been previously available.

Ecological Models

The land cover characteristics database was evaluated for its utility in the CENTURY model to study the interrelationships between climate and grassland growth, land use and soil carbon formation in a portion of the Great Plains (Burke, personal communication). A translation to the Anderson Level II land cover classification scheme for a 1-km analysis was used; the database was further refined to create new land classes not found in the original data, such as continuously cropped wheat, fallow wheat, irrigated corn and rainfed corn. Accuracy assessments of the land cover classes, based on expert knowledge of the area, were positive. The spatial richness of the land cover data was also said to be a significant improvement over other available data for the region.

The land cover characteristics database was also successfully used as part of the RHESSys model for the State of Montana (Nemani and Running, personal communication). The 159 land cover classes were translated into four types (deciduous, coniferous, grasses and shrubs) and assigned to 60-km cells using the predominant class. Based on comparisons with test site data, the investigators concluded that the land cover data were suitable for this system to assess CO_2 warming scenarios, land cover and soil carbon budget relationships. A few categorical errors were found in the data set (confusion between irrigated crops and coniferous forest), but the data were found to be useful. Future work will use a modified land cover classification using physiognomic traits of the vegetation (canopy height, leaf type, leaf longevity) and will investigate scale and aggregation questions within GCM cells.

In another study, biogenic hydrocarbon emissions rate factors based on field data were assigned to various wooded landscapes (for example, oak-hickory forests) as defined in the land cover characteristics database (Guenther *et al.*, personal communication). The modelling also used seasonally integrated NDVI to estimate leaf area index and biomass for each forest class. The land cover characteristics data were coupled with more detailed US Department of Agriculture (USDA) Forest Service forest inventory and analysis (FIA) data for a multiscale analysis. The land cover data were said to simulate significant landscape-level variations not represented by more general schemes. The researchers caution, however, that biomass and species composition estimates taken from the land cover database are a first approximation and should be used only where detailed data are not available.

The data are also being used for similar applications by the US Environmental Protection Agency (EPA). The EPA found the data to be an improvement over other existing data, particularly in the western USA (Pierce *et al.*, personal communication). However, they found limitations in the eastern USA in heavily urban areas and in regions where the landscape is very heterogeneous within 1-km areas.

Plans

The development of a conterminous US land cover characteristics database was initiated as a prototype to test the conceptual format and methods for developing a global land cover characteristics database. The intention was to expand the project to produce a global 1-km land cover characteristics database if an evaluation of the technical and conceptual validity of the prototype data set was positive, and the required source data (AVHRR time-series, digital elevation model and ecoregions) were available. The second issue is being addressed as the necessary data are rapidly becoming available. A joint USGS, NASA, NOAA, International Geosphere-Biosphere Programme (IGBP) and European Space Agency project to assemble a daily global 1-km AVHRR data set for at least an 18-month period spanning 1992–3 is providing this critical data set (IGBP, 1992). The test and evaluation activities described in this chapter, along with an independent, ground-based accuracy assessment, are providing the information needed to determine the technical and conceptual validity of the land cover characterization concept. Assessments, to date, have been positive.

Planning is under way to initiate the development of a Western Hemisphere land cover characteristics database in late 1993, with completion of a preliminary database scheduled for late 1994. Following this, plans are to continue, continent by continent, to characterize the remaining land masses. The priorities for the eventual completion of the effort will depend on recommendations of the science community.

Conclusion

The design of special-purpose land surface classification schemes is driven primarily by the need to simplify the land cover input for model efficiency. The prototype USGS land cover characteristics database is driven by the vegetation cover as represented by the AVHRR sensor. The resulting land cover characteristics are detailed and flexible, and tests have demonstrated that the data can be aggregated into a variety of model classification schemes. This allows cross-tabulation and linkages between schemes. The database is a bridge between the extreme detail of the earth's surface and the extreme generalization of the land cover as portrayed in biophysical or climate models.

In general, the strength of a land cover characteristics data set based on seasonal land cover regions is that it provides a suite of landscape variables, rather than a single name, to represent land cover. This ultimately provides consistency in data quality, reduces redundancy in a database creation, serves several scientific purposes and represents temporal dynamics of landscapes. Specific models currently use similar, but typically different, categorizations unique to an organization, discipline or application. A land cover characteristics database, then, may provide the flexibility to adapt to an application, rather than force the application to be adapted to a database. Without standardized land cover data, users will continue separate initiatives to satisfy their needs, thus duplicating each other's efforts and creating multiple data sets containing similar information, but with different standards and data formats.

Acknowledgement

Work was performed under US Geological Survey contract 1434-92-C-40004. The University of Nebraska's involvement was partially supported by the US Environmental Protection Agency grant X007562-01.

References

Aman, A., Randriamanantena, H., Podaire, A. and Frouin, R., 1992, Upscale integration of normalized difference vegetation index: The problem of spatial heterogeneity, *IEEE Transactions on Geoscience and Remote Sensing*, **30** (2), 326–38.

Anderson, J. R., Hardy, E. E., Roach, J. T. and Witmer, R. E., 1976, *A Land Use and Land Cover Classification System for Use with Remote Sensor Data*, Professional Paper 964, Reston, Virginia: US Geological Survey.

Baker, W. L., 1989, A review of models of landscape change, *Landscape Ecology*, **2** (2), 111–33.

Beach, T., 1987, A review of soil erosion modeling, in Brown, D. A. and Gersmehl, P. J. (Eds) *File Structure Design and Data Specifications for Water Resources Geographic Information Systems*, Ch. 9, pp. 9/1–9/19, Special Report No. 10, St Paul, Minnesota: Water Resources Research Center, University of Minnesota.

Beasley, D. B. and Huggins, L. F., 1982, *ANSWERS (Areal Nonpoint Source Watershed Environmental Response Simulation): User's Manual*, EPA-905/9-82-001, Chicago, Illinois: US Environmental Protection Agency.

Box, E. O., Holben, B. N. and Kalb, V., 1989, Accuracy of the AVHRR Vegetation Index as a predictor of biomass, primary productivity and net CO_2 flux, *Vegetatio*, **80**, 71–89.

Brown, J. F., Loveland, T. R., Merchant, J. W., Reed, B. C. and Ohlen, D. O., 1993, Using multisource data in global land cover characterization: Concepts, requirements, and methods, *Photogrammetric Engineering and Remote Sensing*, **59** (6), 977–87.

Burke, I. C., Kittel, T. G. F., Lauenroth, W. K., Snook, P., Yonker, C. M. and Parton, W. J., 1991, Regional analysis of the Central Great Plains: Sensitivity to climate variability, *BioScience*, **41** (10). 685–92.

Dickinson, R. E., Henderson-Sellers, A., Kennedy, P. J. and Wilson, M. F., 1986, *Biosphere–Atmosphere Transfer Scheme (BATS) for the NCAR Community Climate Model*, NCAR Technical Note NCAR/TN-275+STR, Boulder, Colorado: National Center for Atmospheric Research.

Dorman, J. L. and Sellers, P. J., 1989, A global climatology of albedo, roughness length and stomatal resistance for atmospheric general circulation models as represented by the Simple Biosphere Model (SiB), *Journal of Applied Meteorology*, **28**, 833–55.

Goodchild, M. F., Parks, B. O. and Steyaert, L. T., (Eds), 1993, *Environmental Modeling Geographic Information Systems*, New York: Oxford University Press.

Goward, S. N., Tucker, C. J. and Dye, D. G., 1985, North America vegetation patterns observed with the NOAA-7 Advanced Very High Resolution Radiometer, *Vegetatio*, **64**, 3–14.

Graedel, T. E., 1992, The IGAC activity for the development of global emissions inventories, in *Emission Inventory Issues in the 1990s*, Special Conference Proceedings, pp. 140–7, Pittsburgh, Pennsylvania: Air and Waste Management Association.

Henderson-Sellers, A. and McGuffie, K., 1987, *A Climate Modelling Primer*, Chichester, UK: Wiley.

Henderson-Sellers, A. and Pitman, A. J., 1992, Land-surface schemes for future climate models: Specification, aggregation and heterogeneity, *Journal of Geophysical Research, D Atmospheres*, **97** (3), 2687–96.

International Geosphere–Biosphere Programme, 1992, *Improved Global Data for Land Applications*, IGBP Global Change Report No. 20, Ed. J. R. G. Townshend, Stockholm, Sweden: IGBP.

Kemp, K. K., 1992, Spatial models for environmental modeling with GIS, in Bresnahan, P., Corwin, E. and Cowen, D. (Eds) *Proceedings of the 5th International Symposium on Spatial Data Handling*, Vol. 2, pp. 524–33, Columbia, South Carolina: International Geographical Union (IGU), Humanities and Social Sciences Computing Laboratory, University of South Carolina.

Küchler, A. W., 1983, World map of natural vegetation, in *Goode's World Atlas*, 16th edn, pp. 16–17, Chicago: Rand McNally.

Lee, T. J., 1992, 'The impacts of vegetation on the atmospheric boundary layer and convective storms', unpublished PhD dissertation, Colorado State University, Fort Collins, Colorado.

Loveland, T. R., Merchant, J. W., Ohlen, D. O. and Brown, J. F., 1991, Development of a land-cover characteristics database for the conterminous US, *Photogrammetric Engineering and Remote Sensing*, **57**, 1453–63.

Ludeke, M., Janecek, A. and Kohlmaier, G. H., 1991, Modelling the seasonal CO_2 uptake by land vegetation using the Global Vegetation Index, *Tellus*, **43B**, 188–96.

Malingreau, J. P. and Belward, A. S., 1992, Scale considerations in vegetation monitoring using AVHRR data, *Journal of Remote Sensing*, **13** (12). 2289–2307.

Matthews, E., 1983, Global vegetation and land use: New high-resolution databases for climate studies, *Journal of Climate and Applied Meteorology*, **22**, 474–87.

Nemani, R. R. and Running, S. W., 1989, Estimation of regional surface resistance to evapotranspiration from NDVI and Thermal-IR AVHRR data, *Journal of Applied Meteorology*, **28**, 276–84.

Olson, J. and Watts, J. A., 1982, *Major World Ecosystem Complexes (map, scale = 1:30M)*, Oak Ridge, Tennessee: Oak Ridge National Laboratory.

Parton, W. J., Cole, C. V., Stewart, J. W. B., Ojima, D. S. and Schimel, D. S., 1988, Simulating regional patterns of soil C, N, and P dynamics in the US central grasslands region, *Biogeochemistry*, **5**, 109–31.

Parton, W. J., Schimel, D. S., Cole, C. V. and Ojima, D. S., 1987, Analysis of factors controlling soil organic matter levels on grasslands, *Soil Science Society of America Journal*, **51**, 1173–9.

Peer, R. L., 1990, *An Overview of Climate Information Needs for Ecological Effects Models*, Research Triangle Park, North Carolina: US Environmental Protection Agency, Atmospheric Research and Exposure Assessment Laboratory.

Peplies, R. W. and Honea, R. B., 1992, 'Some classic regional models in relation to global change studies', unpublished report, San Diego, California, 88th Annual Meeting, Association of American Geographers.

Pielke, R. A., Lee, T. J., Kittel, T. G. F., Baron, J. S., Chase, T. N. and Cram, J. M., 1993, The effect of mesoscale vegetation distribution on the hydrologic cycle and regional and global climate, *Conference on Hydroclimatology: Land-Surface/Atmosphere Interactions on Global and Regional Studies*, 73rd Annual AMS Meeting, Anaheim, California, pp. 82-7, Boston, Massachusetts: American Meteorological Society.

Running, S. W., 1990, Estimating terrestrial primary productivity by combining remote sensing and ecosystem simulation, in Hobbs, R. J. and Mooney, H. A. (Eds) *Remote Sensing of Biosphere Functioning*, pp. 65–86, New York: Springer.

Sellers, P. J. and Dorman, J. L., 1987, Testing the Simple Biosphere Model (SiB) using point micrometeorological and biophysical data, *Journal of Climate and Applied Meteorology*, **26**, 622–65.

Sellers, P. J., Mintz, Y., Sud, Y. C. and Dalcher, A., 1986, The design of a simple biosphere model (SiB) for use within General Circulation Models, *Journal of Atmospheric Science*, **43** (6), 505–31.

Sklar, F. H. and Costanza, R., 1990, The development of dynamic spatial models for landscape ecology: A review and prognosis, in Turner, M. G. and Gardner, R. H. (Eds) *Quantitative Methods in Landscape Ecology*, pp. 239–88, New York: Springer.

Smith, C. B., Lakhtakia, M. N., Capehart, W. J. and Carlson, T. N., 1993, Initialization of soil-water content for regional-scale atmospheric prediction models, *Conference on Hydroclimatology*, 73rd AMS Annual Meeting, Anaheim, California, Boston, Massachusetts: American Meteorological Society.

Spanner, M. A., Pierce, L. L., Running, S. W. and Peterson, D. L., 1990, The seasonality of AVHRR data of temperate coniferous forests: Relationship with leaf area index, *Remote Sensing of Environment*, **33**, 97–112.

Steyaert, L. T., 1992, Integrating geographic information systems and environmental simulation models, in *ASPRS/ACSM/RT '92 Technical Papers*, Vol. I: *Global Change and Education*, pp. 233–43, Washington, DC: ASPRS/ACSM/RT.

Townshend, J. R. G., Justice, C. O. and Kalb, V. T., 1987, Characterization and classification of South American land cover types using satellite data, *International Journal of Remote Sensing*, **8**, 1189–1207.

Tucker, C. J., Fung, I. Y., Keeling, C. D. and Gammon, R. H., 1986, Relationship between atmospheric CO_2 variations and a satellite-derived vegetation index, *Nature*, **319**, 195–9.

Tucker, C. J. and Sellers, P. J., 1986, Satellite remote sensing of primary production, *International Journal of Remote Sensing*, **7**, 1395–1416.

Tucker, C. J., Townshend, J. R. G. and Goff, T. E., 1985, African land cover classification using satellite data, *Science*, **227**, 233–50.

UNESCO, 1973, *International Classification and Mapping of Vegetation*, Paris: UNESCO.

Xue, Y., Sellers, P. J., Kinter, J. L. and Shukla, J., 1991, A simplified biosphere model for global climate studies, *Journal of Climate*, **4**, 345–64.

20

Global environmental characterization:
Lessons from the NOAA–EPA Global
Ecosystems Database Project

John J. Kineman and Donald L. Phillips

This chapter discusses the need for better application of descriptive scientific methods to build integrated databases required by global change research. The concept of environmental characterization is cited as an appropriate paradigm for defining this need. The role of characterization databases with respect to modelling and synthesis is discussed. An existing five-year inter-agency project to develop a public global change database is described. Preliminary lessons from this project imply a great need to bridge the gap between voluminous yet largely untested 'off the shelf' data, and the more focused, quantitative information needed for co-ordinated global research. Furthermore, it is apparent that more rigorous publication traditions for data (analogous to those for literature), including open scientific review and testing of data sets, are necessary to improve the scientific and technical quality of descriptive information.

Introduction

As a consequence of the inauguration of the International Geosphere–Biosphere Program (IGBP) and various national programmes focusing on global change, geoscience has experienced revolutionary changes. These changes appear as new priorities for research, in the philosophy of earth science itself, and, as discussed here, in how we think about and manage scientific information.

The goal of the global change effort, as written by the National Aeronautics and Space Administration (NASA) Committee on Earth System Science (1988), is:

To obtain a scientific understanding of the entire Earth system on a global scale by describing how its component parts and their interactions have evolved, how they function, and how they may be expected to continue to evolve on all time scales.

and,

> To develop the capability to predict those changes that will occur in the next decade to century, both naturally and in response to human activity. (NASA Earth Systems Science Committee, 1988)

Assembling such a holistic picture of natural and anthropogenic change from today's compartmentalized, disciplinary knowledge is a staggering task that will require collaboration and communication on a scale never before attempted. Information resources will need to be shared over a much broader base than ever before so that multiple studies can proceed in parallel, accessing and contributing to a common information pool (Evans, this volume). While various aspects of a common data system are currently being developed and implemented both nationally and internationally, integrating data resources to serve a growing multidisciplinary information need and interfacing the result with modelling remains a problem. Improved integration methods, associated technologies and a greater priority for information synthesis are needed.

Limitations of Shared Information

The needs of global multidisciplinary science create almost paradoxical requirements: a common information pool that is driven by the needs of research — an independent (almost theory neutral) database that can be used for testing and expanding our awareness of phenomena.

Ideally the development and testing of global change models shares a complementary relationship with the design of experiments and observations. The collection and distribution of global data must therefore be done with sound guidance from the scientific community. Nevertheless, because it takes time to collect and manage a network of global data, data collections designed for one purpose will encounter different demands in the future. As a result, a common pool of data cannot be optimally designed for the uses it is likely to encounter. It is thus especially important that the global database be designed in the most robust way possible, as an unbiased yet focused representation of the global environment that can be tailored to many uses.[1]

Figure 20.1 shows a concept of information flow from raw observations, where there is currently a strong emphasis on remote sensing, to the final predictive synthesis that is called for in global change research. Modelling, with its concomitant data needs, is generally recognized as the activity that will support such synthesis. However, many of the needed input variables are not directly observed or well defined. It follows that an information base must be developed from raw data sources and validated independently in its own right. Our experience with the currently available public database clearly indicates that more definitive and manageable representations of critical variables are needed

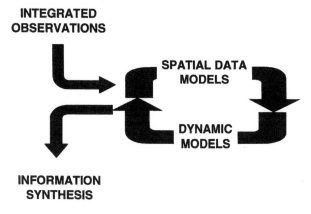

**INTEGRATED
OBSERVATIONS**

**SPATIAL DATA
MODELS**

**DYNAMIC
MODELS**

**INFORMATION
SYNTHESIS**

Figure 20.1 From data to information.

as an important step between data dissemination and modelling. Such analytical activities are common within independent modelling groups; however, it is our belief that these efforts must become better linked between groups, and more commonly shared. Due to the increasing complexity of global science, there is much to learn in defining a definitive set of key variables, and it seems unlikely that individual groups can accomplish this working alone.

For these reasons, it is helpful to define the effort of producing integrated scientific databases (derived from existing data sources) in disciplinary terms. This approach should couple the database with modelling priorities, yet allow its independent development and validation. Aside from differences in scale and focus, such an approach is represented in the concept of ecological characterization developed in the 1970s by the US Fish and Wildlife Service, and defined as:

> ... a study to obtain and synthesize available environmental data and to provide an analysis of the functional relationships between the different components of an ecosystem and the dynamics of that system ... it is simply a structured approach to combining information from physical, chemical, biological, and socioeconomic sciences into an understandable description of an ecosystem. (Watson, 1978)

According to this definition, characterization is primarily descriptive and generally directed toward management goals. While it is not designed for research purposes as such, it is organized by conceptual models of an ecosystem (i.e., the results of research). The models determine priorities for data integration and information synthesis. Characterization can be a major step toward synthesis, which draws together theory and observations relating to various environmental change and human action scenarios. This approach, which we may generalize to global scales as environmental characterization,[2] may have much in common with the prediction and policy goals for global change research in general.

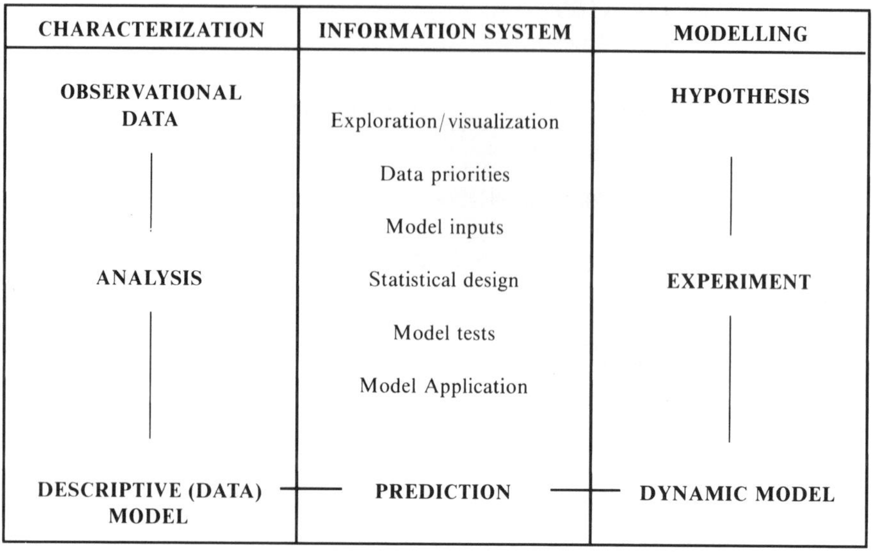

CHARACTERIZATION	INFORMATION SYSTEM	MODELLING
OBSERVATIONAL DATA	Exploration/visualization	**HYPOTHESIS**
	Data priorities	
	Model inputs	
ANALYSIS	Statistical design	**EXPERIMENT**
	Model tests	
	Model Application	
DESCRIPTIVE (DATA) MODEL	**PREDICTION**	**DYNAMIC MODEL**

Figure 20.2 Characterization and modelling as complementary activities.

Methods

Environmental characterization methods, as presented here, involve integrating a broad array of existing information. Modelling, on the other hand, emphasizes the development and testing of theory, which generally dictates a more selective consideration of phenomena and test data. Under the demands of global change research, these inherently complementary activities are becoming more complex, increasing the need for specialization.

Figure 20.2 illustrates a complementary relationship between descriptive studies and theory-based modelling or explanation. Just as modelling depends on characterization data for boundary conditions and empirical testing, characterization depends on the requirements and results of modelling to establish descriptive priorities, to provide sampling design criteria and to present current scientific information in usable form. Epistemologically, prediction requires their balanced combination.

Environmental characterization, as an integration of current information, follows traditional scientific methods of empirical validation. Characterization supports the modelling community by providing inputs and boundary conditions for models, and by providing variables derived from data for independent empirical tests. In a broader view, the characterization database becomes part of a goal-directed exercise to support resource management, assessment and policy. Yet as a synthesis of data, information and knowledge about a system, it can also provide a basis for exploration and more advanced hypotheses that will help drive critical research and the next generation of models.

Characterization Databases

Because of this increasingly important role that the global change information system must play, data management has received new interest and new definitions. For example, the preparation of needed databases is considered by the US Global Change Research Program (CES, 1992) to be a prerequisite for global change research; historically, it has often been treated as an afterthought.

It is our belief that global change research will depend heavily on the construction of integrated databases and information systems that are flexible enough to meet diverse needs. For example, the US Global Change Program emphasizes the lead role of modelling and the need for better theory, while also emphasizing the 'prerequisite' for observational data. The US Committee on Earth Sciences (CES) states that 'The empirical climate record and general circulation models have made obvious the lack of comparable, detailed ecological databases', and that 'Data sets must be organized and assimilated so that model simulations of present and past environments can be compared to nature as a test of model accuracy'. The climate–biosphere interaction is cited as one of four major research areas where 'primary data are a prerequisite' for understanding global change (see Hayden, this volume). The CES states that this understanding must be developed using existing data, and that it must be tested and refined experimentally (CES, 1989, 1991).

Useful data sets do exist and should be analysed more effectively to reveal the richness of phenomena that may be represented only partly by current theory (and less so by current models). Today, with the rapid advances in computers and geographic information systems (GIS), geographically compatible and prioritized databases with data and error analysis capabilities can be constructed according to the needs of environmental characterization and comparative studies. This involves combining and comparing existing multidisciplinary data to represent important spatial patterns and temporal trends, along with capabilities for data comparison, data synthesis and quantitative assessments of confidence based on statistical design information (i.e., metadata).

Discussion

The Global Ecosystems Database Project

The Global Ecosystems Database Project (GEDP) was designed with the foregoing considerations in mind, both to meet the specific present needs of its sponsors and to explore methods in building an integrated characterization database, using GIS concepts as an integrating philosophy. The GEDP was an interagency project between the National Oceanic and Atmospheric Administration (NOAA) National Geophysical Data Center (NGDC) in Boulder,

Colorado, and the US Environmental Protection Agency's (EPA) Environmental Research Laboratory in Corvallis, Oregon. The EPA provided funding for the project to improve the availability and integration of data that can support modelling, and to publish EPA data and model results. EPA scientists also provided guidance on data priorities with respect to their modelling interests: effects of global climate change on ecosystems; ecosystem feedbacks to climate through water, energy and carbon fluxes; and assessment of biosphere management options to conserve and sequester carbon.

The database released by the project in 1992 (NOAA–EPA GEDP, 1992) contained 14 global data sets integrated to a common set of nested grid sizes and a common grid origin. These included two 10-min gridded versions of the NOAA 16-km Global Vegetation Index (NGDC, 1992; EDC and NGDC, 1992) aggregated to a monthly time-step, two long-term average monthly climate data sets on 0.5° and 1° grids (Legates and Willmott, 1989; Leemans and Cramer, 1991), four land cover class data sets ranging from 10-min to 1° grids (Matthews, 1983; Wilson and Henderson-Sellers, 1985; Leemans, 1989; Olson, 1992), two 1° gridded methane data sets (Lerner *et al.* 1989; Matthews and Fung, 1989), four 1° gridded soils data sets (Matthews, 1983; Wilson and Henderson-Sellers, 1985; Staub and Rosenzweig, 1987; Webb *et al.,* 1991), one 10-min gridded surface elevation and terrain data set (FNOC, 1985), and a 1-min vector boundaries data set (Pospeschil, 1988). The data processing and quality control for these data sets are described in documentation that accompanies the CD-ROM (Kineman and Ohrenschall, 1992).

In 1993, a similar number of global data sets were added along with a US regional database. Data sets added include new versions of the Global Vegetation Index composited to a 10-min gridded 48-month time-series, a nominal 2.5° gridded Sea Surface Temperature data set, a 1° gridded snow cover and depth data set, two potential vegetation data sets at 10-min and 0.5° grids, an improved urban coverage from the US Navy 10-min gridded terrain data, solar insolation on a 1° grid, an improved 5-min gridded elevation data set, a socio-economic database by country, a forest practices data set, simulation runs from five General Circulation Models (GCM) and selected site data from a grasslands ecosystem model. In addition, a 1-km resolution US regional database was added, including a regional water-balance model (CO_2 doubling) database with temperature, precipitation, wind, vapour pressure, potential evapotranspiration, soil characteristics, wind energy and several land cover class data sets.

Scale Considerations

The database focuses on monthly time-series and averages with spatial resolutions from about 1 km to 100 km. As shown in Figure 20.3, the overall context of environmental characterization based on such data would be intermediate in scale between more site-specific process studies, that is, 'patch and landscape scales' (using definitions from the IGBP), and the coarser scale 'regional' to 'global' modelling (e.g. GCM).

PATCH to LANDSCAPE		LANDSCAPE/REGIONAL		REGIONAL to GLOBAL	
Resource management & site studies		Global change characterization		Global simulation modelling	
10 – 100 m	1 – 4 km	4 –16 km	50 – 100 km	100 km	500 km

Figure 20.3 Intermediate scale of global characterization data.

Data integration activities at this intermediate scale can be used to refine coarse global predictions as part of the scaling down process. For example, in a data set provided by EPA scientists for the 1993 GEDP database release, climate data at a 1-km resolution were used to refine GCM predictions to produce a US regional database and to drive a regional water-balance model. Conversely, intermediate-scale data can be used to extrapolate site data to global distributions. For example, vegetation redistribution models (e.g., CENTURY, MAPSS) provide predictions for specific sites, which are then extrapolated to global distributions using an available vegetation classification data set. These extrapolations are then aggregated to GCM scales using area calculations. The validity of such scaling is a topic of considerable interest that can now be tested.

GIS Linkage

The GEDP database has been produced in a relatively generic GIS format (Clark University's IDRISI software). Operability within at least one GIS that is research orientated, suitable as a common basis for data comparison and widely tolerant of multiple data types was viewed as a means of maintaining compatibility with the greater global change and GIS communities. This approach also affords the opportunity to standardize data structures and GIS designs. Other software has been developed to improve platform independence and accessibility by different systems; however, we believe that GIS linkage is most important for encouraging common methods.

The database has been published on CD-ROM (as will be future releases) at minimal cost, with thorough documentation (Kineman, 1992; Kineman and Ohrenschall, 1992). By publishing an integrated geographically structured database within a common analytical environment for comparing and combining data sets, we believe that we can facilitate better scientific support for global change studies in general than would be possible by distributing isolated data sets with often incompatible measurements and statistical representations. The traditional data dissemination concept of standardized formats addresses only the technical accessability of data.

The GIS approach and characterization philosophy allow this project to support the necessary interactions between integration activities that can be done at a data centre, modelling activities performed at diverse research centres, and a broad public user community. Specifically, the project supports and/or facilitates:

- multiple applications of global change information (e.g., research, education);
- adaptation of digital information to various experimental requirements, including modelling;
- development of methods that allow the comparison and validation of data sets;
- improved information priorities by providing an assessment of currently available data;
- wider sharing of data and methods among users;
- open scientific review within a widely distributed global science community. The feedbacks from this community, in turn, improve database design and content.

Results

Many examples of database user developments can be provided as a result of this project, including interpolation methods, improved documentation, quality studies, new data sets and new analytical techniques. Also, by maintaining a distinction between database development and software development, it has been possible to encourage and support improvements in GIS technology for global science that parallel and feed back to the database development.

Besides providing a useful integration of existing data sets, many direct improvements in data have also resulted from this work. For example, a generalized version of the NOAA Global Vegetation Index (GVI) (see NGDC, 1992), provided in the GVI prototype (1989), remains one of the most consistent and usable versions of the GVI generally available. This was used successfully for land cover classification in Asia (Gaston *et al.,* in press) and analysis of seasonal and long-term characteristics in the GVI data using principal components analysis (Eastman and Fulk, 1993a,b). Improvements in the Olson World Ecosystems database (Olson, 1992) reduced some of the errors in previous versions distributed separately, and improved many of the classes. To provide data on compatible, nested grids, some data sets were resampled, for example, the Legates and Willmott Average Monthly Surface Air Temperature and Precipitation data set (Legates and Willmott, 1989), which was on one of the standard grids adopted in the GED, but was registered by grid-cell centroid, rather than edge (the convention adopted for the project). In each case where the data set was altered to meet project conventions, methods were considered carefully, usually in consultation with the original investigators, and documented in the database manual. Also, each data set incorporated into the database has undergone a significant level of quality analysis, often resulting in expanded documentation.

Indirect data developments have also been stimulated by the project. For example, through the production and use of the Monthly Generalized GVI, we discovered registration errors in the NOAA weekly NDVI and other biweekly versions we were also able to document sensor drift characteristics and the high degree of variability and noise in GVI source data that make it unsuitable for many kinds of spatial analyses without some form of filtering and noise-reduction techniques. Data flagging or masking techniques that have been common in many data sets have also been shown to be inappropriate for this kind of database, which benefits more from having independent quality indicators that can be analysed or applied separately. These lessons have been provided as design inputs to several current projects being conducted elsewhere to develop improved GVI and AVHRR time series. One of these is at the University of Maryland where a recalibrated time series of GVI is being produced, and another is a NASA–NOAA Pathfinder project to reprocess all of the AVHRR channel data to improve interannual and intersatellite compar-ability. Some of these and other special versions of the GVI and AVHRR channel data are being distributed by NGDC on separate CD-ROMs as part of its overall programme in global change data.

In another example, registration and other errors discovered in the global 5-min elevation data set (Edwards, 1989), along with known artefacts in the US Navy 10-min elevation data (FNOC, 1985), have led to several efforts to develop improved global topography data at NGDC, and inputs to the higher resolution, international project called 'Globe', which aims to deliver a 1-km elevation data set in 1995. One of the projects at NGDC is Terrain Base, which is integrating and quality testing numerous data sets from the best available sources to produce a revised 5-min global data set and higher resolution coverages where available. Another effort at NGDC has produced a corrected version of the global urban cover data set at 10-min resolution, which was released with the 1993 database.

In addition to the examples mentioned in this chapter, many other data sets have been contributed by users and are being processed for future CD-ROM releases.

Conclusion

This chapter has been concerned primarily with the issue of developing environmental characterization data from existing data sources to support global change research. The attempt to integrate data sets for broad distribution and use in the Global Ecosystems Database Project has revealed a general lack of consistency and scientific quality among existing data sources that seriously hinders their combined use (also see Slagle, this volume). We believe the solution to this problem is to encourage more commonly shared methods and peer review conventions within an overall context of data integration and environmental characterization.

Scientific standards emerge when an activity becomes formalized within a discipline. The growing concerns about rapid access to data, data quality, and uncertainty, metadata, interoperability of information systems and modelling support are indicative of such a need. The concept of environmental characterization as a primarily descriptive activity linked with information systems could form such a disciplinary focus. Given the current poor state of scientific data about earth systems, and the great attention being given to acquiring new remote sensing data and developing model simulations, characterization activities needed to link these two seem underemphasized and inadequately formalized. Without a unifying scientific purpose, data and information collection can result in poorly organized and unvalidated databases, with increasingly limited experimental relevance. The concept of global environmental characterization seems to encapsulate most of the important global change data and information issues, and is thus suggested as a unifying paradigm.

For both scientific and practical reasons, building characterization databases must remain an autonomous yet complementary activity to global change modelling. By recognizing the scientific value of descriptive work, especially in endeavours such as modelling that rely primarily on surrogate realities for both inputs and empirical testing, we can improve the quality, relevance and availability of key environmental information.

A characterization database is driven by the general requirements of models, and recognizes that many important questions are descriptive, requiring different methods and interests than questions about function. To provide an adequate representation of key environmental factors, one must be concerned primarily with the uncertainties and inadequacies of the information base. There is no substitute for experimentally validating each data set in the light of specific uses.

Meeting future information needs will depend on having accurate metadata and capabilities for data synthesis. Derivatives from existing data and combinations of data in the form of spatial data models are required to produce key variables and indices that may not exist in the observational database. Thus, data sets must be integrated in ways that support valid statistical comparisons and subsequent analysis. In this regard, analytical GIS may offer the structure and function needed to cross disciplines and provide a common ground for scientific review and validation.

The GEDP represents an approach to building a characterization database for global modelling, attempting to employ the philosophy presented in this chapter. It was a limited activity designed to explore methods and to provide a data publication outlet for ongoing research. The work to date has resulted in a useful compilation of available data and has stimulated parallel data set developments as well as improved data publication standards. In the continuation of the project, the primary emphasis will be on linking the database with models.

One of the questions addressed by the GEDP regards the role that data centres can play in global science. It is apparent that, aside from their traditional role as archives, data centres can be valuable partners in global research in several key ways. First, by establishing focused data integration activities that are linked to research and modelling and conducted under careful scientific review, data centres can play a key role in moving from raw data toward definitive environmental information. Second, by improving standards for data publication, data centres can play a key role in establishing the conventions and scientific exchanges needed for validating digital information. Third, by developing database structures in co-operation with developments in the field of GIS, data centres can provide a valuable link between information structure and analytical function that can offer users vastly improved resources.

We have also learned that improvements are needed at the sources of data. As a scientific community, we must consider the public distribution of digital information, through appropriate integration efforts and data centre activities, as a requirement of good science. Whatever piece of the puzzle we may be working on, our data should be designed and documented in such a way that they can contribute to experimentally sound multidisciplinary characterizations of the physical and ecological systems of the earth.

Notes

1. 'Design' here refers to the selection or development of key variables as well as their statistical representation and sampling designs.
2. A common use of the term characterization refers to individual numerical sets derived from spatial models, which represent a subset of the more encompassing environmental characterization cited here.

References

Committee on Earth Sciences (CES), 1989, *Our Changing Planet: The FY 1990 Research Plan, The US Global Change Research Program, A Report by the Committee on Earth Sciences,* July, Executive Office of the President, Office of Science and Technology Policy, Washington, DC.

Committee on Earth Sciences (CES), 1991, *Our Changing Planet: The FY 1992 Research Plan, The US Global Change Research Program, A Report by the Committee on Earth Sciences,* Executive Office of the President, Office of Science and Technology Policy, Washington, DC.

Committee on Earth Sciences (CES), 1992, *Our Changing Planet: The FY 1993 Research Plan, The US Global Change Research Program, A Report by the Committee on Earth Sciences,* Executive Office of the President, Office of Science and Technology Policy, Washington, DC.

Eastman, J. R. and Fulk, M., 1993a, Long sequence time series evaluation using standardized principal components, *Photogrammetric Engineering,* **59**(8) 1307–12.

Eastman, J. R., and Fulk, M., 1993b, Time series analysis of remotely sensed data using standardized principal components, in: Proceedings Vol. I 25th International Symposium; Remote Sensing and Global Environmental Change. Graz, Austria 4-8 April, pp. 1485–96.

Edwards, M. O., 1989, *Global Gridded Elevation and Bathymetry (ETOP05),* digital raster data on a 5-minute geographic (lat/long) 2160 × 4320 (centroid-registered) grid, 9-track tape, 18.6 mb, Boulder, Colorado: NOAA National Geophysical Data Center.

EROS Data Center (EDC) and National Geophysical Data Center (NGDC), 1992, *Monthly Global Vegetation Index from Gallo Bi-weekly Experimental Calibrated GVI (April 1985–December 1990),* digital raster data on a 10-minute geographic (lat/long) 1080 × 2160 grid, in *Experimental Calibrated Global Vegetation Index from NOAA AVHRR, 1985–1991,* 69 independent single-attribute spatial data layers on CD-ROM, 161 mb, Boulder, Colorado: National Geophysical Data Center.

Fleet Numeric Oceanographic Center (FNOC) (US Navy), 1985 (1981), *10-minute Global Elevation, Terrain, and Surface Characteristics* (reprocessed by the National Center for Atmospheric Research and the National Geophysical Data Center), digital raster data on a 10-minute geographic (lat/long) 1080 × 2160 grid, 9 files on 9-track tape or 2 floppy disks in compressed format, 28 mb, Boulder, Colorado: NOAA National Geophysical Data Center.

Gaston, G., Vinson, T., Jackson, P. and Kolchugina, Y., in press, Identification of carbon quantifiable regions in the former Soviet Union using unsupervised classification of AVHRR Global Vegetation Index, *International Journal of Remote Sensing.*

Kineman, J. J., 1992, *Global Ecosystems Database Version 1.0, User's Guide, Key to Geophysical Records Documentation No. 26,* Boulder, Colorado, USDOC/NOAA National Geophysical Data Center.

Kineman, J. J. and Ohrenschall, M. A., 1992, *Global Ecosystems Database Version 1.0: Disc A, Documentation Manual, Key to Geophysical Records Documentation No. 27,* Boulder, Colorado, USDOC/NOAA National Geophysical Data Center.

Leemans, R., 1989, *Global Holdridge Life Zone Classifications,* digital raster data on a 0.5 degree geographic (lat/long) 360 × 720 grid, multiple files on floppy disk, 0.26 mb, Laxenburg, Austria: IIASA.

Leemans, R. and Cramer, W. P., 1991, *The IIASA Database for Mean Monthly Values of Temperature, Precipitation, and Cloudiness on a Global Terrestrial Grid,* digital raster data on a 30-minute geographic (lat/long) 360 × 720 grid, 9-track tape, 10.3 mb, Laxenburg, Austria: IIASA.

Legates, D. R. and Willmott, C. J., 1989, *Monthly Average Surface Air Temperature and Precipitation,* digital raster data on a 0.5 degree geographic (lat/long) 361 × 721 grid (centroid registered on 0.5 degree meridians), 4 files on 9-track tape, 83 mb, Boulder, Colorado: National Center for Atmospheric Research.

Lerner, J., Matthews, E. and Fung, I., 1989, *Methane Emission from Animals: A Global High Resolution Database from the NASA Goddard Institute for Space Studies,* digital raster data on a 1 degree geographic (lat/long) 180 × 360 grid, 1 floppy disk, 1.3 mb, Boulder, Colorado: National Center for Atmospheric Research.

Matthews, E., 1983, *Global Vegetation, Land-Use, and Seasonal Albedo (NASA Goddard Institute for Space Studies),* digital raster data on a 1-degree geographic (lat/long) 180 × 360 grid, multiple files on 9-track tape, 0.8 mb, Boulder, Colorado: National Center for Atmospheric Research.

Matthews, E. and Fung, I., 1989, *Global Data Bases on Distribution, Characteristics and Methane Emission of Natural Wetlands from the NASA Goddard Institute for Space Studies,* digital raster data on a 1 degree geographic (lat/long) 180 × 360 grid, 1 floppy disk, 1.2 mb, Boulder, Colorado: National Center for Atmospheric Research.

NASA Committee on Earth Systems Sciences, 1988, *Earth System Science: A Closer View,* Report of the Earth System Sciences Committee, Washington, DC: NASA Advisory Council.

National Geophysical Data Center (NGDC), 1992, *Monthly Generalized Global Vegetation Index from NESDIS NOAA-9 Weekly GVI data (April 1985–December 1988),* digital raster data on a 10-minute geographic (lat/long) 1080 × 2160 grid, in *Global Ecosystems Database Version 1.0:* Disc A, 45 independent and 24 derived single-attribute spatial data layers on CD-ROM, 190 mb, Boulder, Colorado: NOAA National Geophysical Data Center.

NOAA-EPA Global Ecosystems Database Project (GEDP), 1992, *Global Ecosystems Database Version 1.0,* User's Guide, documentation, reprints, and digital data on CD-ROM, Boulder, Colorado: USDOC/NOAA National Geophysical Data Center.

Olson, J. S., 1992, *World Ecosystems (WE 1.3a and 1.4d),* digital raster data on a 10-minute geographic 1080 × 2160 grid, in *Global Ecosystems Database, Version 1.0:* Disc-A, three independent single-attribute spatial layers on CD-ROM, 5 mb, Boulder, Colorado: National Geophysical Data Center.

Pospeschil, F., 1988, *Micro World Databank II (MWDB-II): Coastlines, Country Boundaries, Islands, Lakes, and Rivers,* digital vector data at 1-minute resolution, compressed format on 1 floppy disk, 2.5 mb, Bellevue, Nebraska: MicroDoc.

Staub, B. and Rosenzweig, C., 1987, *Global Digital Datasets of Soil Type, Soil Texture, Surface Slope, and Other Properties,* digital raster data on a 1 degree geographic (lat/long) 180 × 360 grid, multiple files on floppy disk, 0.45 mb, Boulder, Colorado: National Center for Atmospheric Research.

Watson, J. F., 1978, Ecological characterization of the coastal ecosystems of the United States and its territories, in *Proceedings: Energy/Environment '78,* pp. 47–53, Los Angeles: Society of Petroleum Industry Biologists.

Webb, R. S., Rosenzweig, C. E. and Levine, E. R., 1991, *A Global Dataset of Soil Particle Size Properties,* digital raster data on a 1 degree geographic (lat/long) 180 × 360 grid, 0.51mb, New York: NASA Goddard Institute of Space Studies.

Wilson, M. F. and Henderson-Sellers, A., 1985, *A Global Archive of Land Cover and Soils Data for use in General Circulation Climate Models,* digital raster data on a 1 degree geographic (lat/long) 180 × 360 grid, multiple files on floppy disk, 0.8 mb, Boulder, Colorado: National Center for Atmospheric Research.

SECTION VI

Environmental modelling and geographic information systems

Understanding the changes in environmental patterns and processes and predicting the effects of those changes that are occurring at scales ranging from the ecosystem to the biosphere will necessarily require the incorporation of spatially explicit simulation modelling into our research. Steyaert and Goodchild initially examine the status of GIS as a technology to support atmospheric, hydrological and ecological simulation modelling. In addition to a comprehensive review of dynamic, physically based approaches to environmental modelling, they summarize current technological impediments and identify research challenges. For example, they highlight the need for developing more tightly linked approaches to integrated GIS and environmental simulation models, creating better graphical user interfaces that can overcome obstacles associated with current cumbersome interfaces, incorporating more sophisticated visualization and spatial analytic tools into GIS, and developing a more usable and functional modelling language. The remaining four chapters in this section provide some potential solutions to many of the recognized impediments.

Effective data management has traditionally been confined to development and maintenance of a particular database; it is the user's responsibility to 'manage' the data during and after the analysis. Kirchner argues that (1) data management should be considered a critical component of simulation modelling because of the wide variety of data required for parameter estimation, (2) model validation requires previously unused data, (3) simulation models can generate vast amounts of new data and (4) model code and metadata require special management considerations. On the basis of past experience, he further suggests several strategies for managing the data associated with and generated by simulation modelling.

As pointed out by Steyaert and Goodchild, true integration of GIS and environmental modelling will require more sophisticated spatial analytic tools and graphical user interfaces. Aspinall discusses issues associated with the quality of environmental data and the use of spatial analyses to convert data into information. He also describes two case studies: one in which GIS technology was used for modelling the effects of climate change on flora and fauna distributions and generating new hypotheses; and another in which principles of landscape ecology were employed to analyse relationships among spatial heterogeneity, scale and the ecology of different taxa. Coleman, Bearly, Burke and Lauenroth describe a new system that employs an innovative Graphical User Interface that tightly couples ecological simulation models with GIS.

Any scientist who has used GIS or simulation modelling to perform a complex series of analyses can remember the frustration of trying to reconstruct the maze of algorithms that were employed, identify which revisions of which data sets and GIS coverages were employed, and document which functions were performed and in what order. In the final chapter of this section, Lanter describes lineage metadata, essentially an abstract representation of maps and logic used within spatial analytic applications. He then demonstrates how comparative analysis of lineage metadata across different GIS application offers the possibility of automatically detecting and isolating patterns of spatial reasoning applied within GIS-based environmental models and discovering spatial analytic knowledge hidden in prior GIS applications.

21

Integrating geographic information systems and environmental simulation models: A status review

Louis T. Steyaert and Michael F. Goodchild

Geographic information systems (GIS) have emerged over the past decade as widely used software systems for input, storage, manipulation and output of geographically referenced data. This chapter examines the status of GIS as a technology to support environmental simulation modelling in the atmospheric, hydrologic and ecologic sciences. The status of GIS integration is reviewed, with emphasis on dynamic, physically based modelling approaches that integrate the effects of different environmental processes across multiple scales. GIS are currently used to assemble and manage large spatial databases, to perform spatial and statistical analyses and to produce effective visual representations of model results. The first steps have been taken toward more tightly linked approaches, including embedded modelling within the GIS. Integrated multidisciplinary modelling has created new opportunities for GIS, but has also exposed the weaknesses of the current generation of GIS technology, particularly in areas of functionality and data modelling. The final section of the chapter summarizes the current status of GIS in this field and suggests directions for future development.

Introduction

The 1980s saw extensive advances in geographic information systems (GIS), a technology now recognized as increasingly important to natural resource management and environmental assessment. GIS are widely used to assemble and manage large spatial databases, perform analyses and produce effective visual representations of management options.

Current GIS software packages range from expensive packages for scientific workstations to PC-based and public domain software. In general, they can be grouped into those that use a raster- or pixel-based representation, and those that use vector- or co-ordinate-based representation of points, lines and areas. Many vector-based GIS are constructed as hybrid systems connecting relational database management systems and graphical databases. Reviews of GIS technology can be found in introductory texts by Star and Estes (1990), Burrough (1986), and in the two-volume reference by Maguire *et al.* (1991).

Simulation models are sophisticated tools for characterizing and understanding environmental patterns and processes, and for estimating the effects of environmental change at local, regional and global scales. Recent attention has focused on developing land surface process modules for atmospheric models, applying distributed parameter approaches for water flow and contaminant transport modelling in surface and groundwater systems, and extending simulation modelling to complex ecological systems at broader scales. Advanced research concepts for integrating various dynamic simulation models across multiple time and space scales, such as coupled-systems modelling or land–atmosphere interactions modelling, are rooted in global change research programmes (ESSC, 1986, 1988; ICSU, 1986; IGBP, 1990; NRC, 1986, 1990).

The new generation of dynamic simulation models requires large amounts of diverse spatial data characterizing those dynamic land surface properties that exert fundamental controls over environmental processes (Steyaert, 1993). Some environmental processes exhibit strong spatial components such as dispersion, advection, slope runoff and spatial interactions within ecosystems. Research on spatial complexity and heterogeneity within landscapes is a priority within a wide range of disciplines. Thus, there is growing interest in the potential for integrating GIS technology and environmental simulation models in spatial database development, spatial processes, exploratory spatial analysis and model output visualization.

This chapter reviews the status of the integration of GIS technology and environmental simulation models in the atmospheric, hydrologic and ecological sciences (for a more extensive perspective, see Goodchild *et al.*, 1993). The main focus is on dynamic, physically based modelling of land–atmosphere interactions and water quality research. An attempt is made to illustrate how this interdisciplinary, multiscaled modelling approach has created new applications for GIS technology. The purpose of each type of dynamic simulation modelling application is described. Details on the modelling approach are followed by a discussion on the exact role of GIS in the modelling. In some cases, suggestions for enhancements to GIS technology are noted where the current capabilities fall short of what is needed.

Atmospheric Modelling and GIS

The integration of atmospheric modelling and GIS technology is just beginning. One promising area involves spatial data handling and analysis support for land surface parametrizations, the land process modules of atmospheric general circulation models (GCM) and mesoscale meteorological models. The use of GIS technology in regional air quality and tropospheric chemistry models is also under study (Novak and Dennis, 1993). Such simulation modelling is complemented by the use of remote sensing and GIS technologies to examine land surface processes related to the energy balance, soil moisture budget and radiative transfer. For example, process-orientated field experiments also provide the basis for the integration of atmospheric simulation models and remote sensing, with GIS technologies to understand environmental processes, develop algorithms and make regional extrapolations (NASA, 1991; Running, 1991b).

Land Surface Parametrizations for Atmospheric Modelling

Lee *et al.* (1993) reviewed the functions and data requirements for advanced land surface parametrizations such as the biosphere–atmosphere transfer scheme (BATS) of Dickinson *et al.* (1986), the Simple Biosphere model (SiB) of Sellers *et al.* (1986), and the Land–Ecosystem–Atmosphere–Feedback (LEAF) model (Lee, 1992). These land surface process models simulate interactions between the biosphere at the land surface and the free atmosphere and can include the exchanges of water and energy fluxes at time-steps of seconds to hours (NRC, 1990; Hayden, this volume).

The purpose of land surface parametrizations such as BATS, SiB and LEAF is to determine as realistically as possible the exchanges of radiative, momentum, and sensible and latent heat fluxes between the land surface and the lower atmosphere. In determining these exchanges within each model grid cell, the models take into account atmospheric forcing (solar radiation, temperature, humidity, wind and precipitation) and, in particular, the role of vegetation and other land surface properties. The models require detailed data on dynamic land surface characteristics such as albedo, surface roughness and stomatal conductance, all of which control the fluxes.

The aggregation or scaling-up of these data within variable sized grid cells of atmospheric models is also a research issue involving the proper representation and parametrization of subgrid processes. For example, Skelly *et al.* (1993) note that at least three distinct approaches to aggregation are used in simulation models: (1) determine the predominant land cover class within the model grid cell, (2) aggregate the land surface properties (roughness, etc.) by some weighted averaging approach, or (3) aggregate the fluxes after running the model.

Given these types of data issues, Skelly *et al.* (1993) suggested that GIS can assist in the manipulation of surface date for model input or for comparisons of model results. They see integration proceeding along two tracks: (1) utilizing the GIS as a data management, analysis and display tool that is linked to the land surface parametrization, and (2) fully integrating the land surface parametrization into the GIS. They call for a 'self-parametrizing' database in which the land surface data component is used to update the land surface parametrization within the atmospheric model automatically.

Lee (1992; personal communication), Smith *et al.* (1993) and Lakhtakia (personal communication) illustrate two approaches for spatial data handling and the tailoring of 1-km land cover characteristics data, such as described by Loveland *et al.* (1991), as input to the LEAF and BATS models, respectively. Lee (1992; personal communication) developed spatial data handling tools to ingest and process 1-km land cover data directly into the Colorado State University Regional Atmospheric Modeling System. These spatial data processing functions were used to change map projections, reclassify the land cover to the LEAF classification, aggregate the land cover to nested grids (10-km fine resolution and 40-km coarse resolution), and select the dominant land cover classes based on percentage composition within the grid cell (water, bare soil and predominant vegetation type).

In contrast, Smith *et al.* (1993) and Lakhtakia (personal communication) used similar spatial data handling tools within the land analysis system (LAS) to prepare land data for the Pennsylvania State University/National Center for Atmospheric Research mesoscale model version 4 (MM4). LAS software was also used to build land data for a soil moisture module used to initialize the MM4 system.

Regional Air Quality and Acid Deposition Modelling

Novak and Dennis (1993) and Birth *et al.* (1990) discuss the potential role of GIS technology as related to regional air quality and acid deposition models, including the Environmental Protection Agency's Regional Acid Deposition Model (RADM), Regional Oxidant Model (ROM), Urban Airshed Model (UAM) and the Regional Lagrangian Model of Air Pollution (RELMAP). These models are used to investigate physical and chemical processes associated with air pollution and to develop strategies for emissions control. The users need spatial analysis tools to evaluate and interpret model predictions, as well as capabilities to analyse collectively a variety of model results.

Novak and Dennis (1993) describe a co-operative pilot project in the EPA to determine the use of the ARC/INFO[1] GIS software for air pollution research and assessment activities. The goals were to enhance database management, provide users (researchers and policy analysts) with transparent access to spatial data for a wide variety of models (RADM, ROM, UAM and RELMAP), and take advantage of spatial display and analysis tools within the ARC/INFO GIS. They describe how a pilot interactive display and environmental analysis system in the ARC/INFO GIS was developed to permit end-users to study the relationships between measured and predicted air pollution, point- and area-source emissions, land use and health effects.

The pilot study concluded that a GIS is an effective tool to estimate emission inputs to air pollution models and to design emission control strategies. However, Novak and Dennis (1993) note that the GIS is limited in its ability to deal with temporal variability, finite difference calculations, three-dimensional analysis and efficient data conversions. The study recommended improved GIS functionality for animation of spatial fields, exploration of three-dimensional

---

1 Any use of trade, product, or firm names is for descriptive purposes only and does not imply endorsement by the US Government or the authors.

fields, enhanced data manipulation (sorts), transparent interfaces with other software packages and user friendliness.

Remote Sensing and GIS Technologies in Land Process Modelling

Although the future looks promising, GIS technology has made only modest inroads into studies involving remote sensing and land surface process research. In most cases, GIS-type functions in digital image processing systems or 'home-grown' generic GIS are used. However, there appears to be a growing role for GIS technology in major field experiments such as the International Satellite Land Surface Climatology Project (ISLSCP) (see for example NASA, 1991; Briggs and Su, this volume).

Shih and Jordan (1992) analysed Landsat thematic mapper (TM) mid-infrared (MIR) data and land use categories within a GIS to make qualitative estimates of surface soil moisture in central Florida. Image processing and GIS functions of the Earth Resources Laboratory Application Software were used in the study to classify the image (TM bands 2, 3 and 5), to georeference the classified image and the MIR band 7 image, to calibrate the MIR digital counts to in-band planetary albedo estimates, to aggregate land use classes according to band 7 albedo estimates, and to overlay the results for qualitative assessment of surface soil moisture categories.

Other examples include the use of remote sensing and GIS technologies to study land surface processes (Luval *et al.*, 1990; Goodin *et al.*, 1992). Luval *et al.* (1990) used the aircraft-based thermal infrared multispectral sensor with GIS-type functions to study canopy temperatures, thermal response numbers and evapotranspiration (ET) in the tropical forest regions of Costa Rica.

As part of a regional approach to estimate forest ET, Running *et al.* (1989) developed a simple GIS to integrate topographic, soils, vegetation and climate data for forested regions of western Montana. Key data input for each grid cell also included satellite-derived estimates of leaf area index, daily microclimatic data estimated from *in situ* and satellite data, soil water capacity and water equivalent snowpack data to estimate snowmelt runoff. Daily ET estimates were mapped for the region based on output from FOREST-BGC (biogeochemical cycles), a forest ecosystem simulation model. Further details on this coupled systems modelling approach are provided in the section on ecological modelling.

Hydrologic Modelling and GIS

There has been significant progress in the integration of GIS technology with hydrologic modelling in the fields of precipitation-runoff modelling for watersheds, including distributed parameter approaches, surface water quality modelling, and groundwater flow and contaminant transport modelling. Following some overview comments on the nature of hydrologic modelling, the status of GIS technology and model integration is summarized in terms of each of these three subject areas.

Maidment (1993) summarizes the basic elements of hydrologic modelling and examines the foundation for linking GIS technology and hyrologic modelling. He suggests that a GIS can add spatial specificity in the modelling and address regional and continental scale problems not yet resolved in hydrology. According to Maidment, a hydrologic model may be defined as the mathematical representation of the flow of water and its constituents on some part of the land surface or subsurface environment. He summarizes the challenge of modelling in three parts. First, the modelling of water flow is concerned with the disposition of rainfall in terms of runoff, infiltration, groundwater recharge, evaporation and water storage. Second, the hydraulics of flow are considered based on the discharge of water at a particular point. These include flow velocity and water surface elevation in a channel, or the Darcy flux and piezometric head in an aquifer. Third, transport issues related to water quality and pollutant flows are considered.

Maidment (1993) defines the spatial components of hydrologic modelling as watersheds (such as lumped models, linked-lumped models or distributed parameter models), pipes and stream channels concerned with channel roughness, aquifers (groundwater flow and transport), and lakes and estuaries. He categorizes models according to a taxonomy based on five sources of variation: time, three space dimensions and randomness. In contrast, Moore *et al.* (1993) define six model structures for surface–subsurface models using a topographic approach, including lumped models, hydrologic response units (HRU), grid-based models, triangulated irregular network (TIN)-based models, contour-based models and two- or three-dimensional groundwater models.

Maidment (1993) suggests that contemporary links between GIS and hydrologic modelling include hydrologic assessment, parameter determination, modelling in a GIS and linking a GIS with hydrologic models. These types of linkages are illustrated in subsequent discussions. For example, hydrologic assessment is illustrated by several types of water quality models. Watershed analysis tools and other approaches for hydrologic parametrization are noted. Dynamic models are linked to a GIS, mainly in terms of model input and output. There are now some excellent examples of tightly coupled GIS-model integration and there are realistic prospects for examples of hydrologic modelling within GIS. According to Maidment (1993) the new frontiers in hydrologic modelling with GIS include research on spatially distributed watershed properties, partial area flow, surface water–groundwater interaction, and regional and global hydrology.

Distributed Parameter Watershed Modelling

Traditional watershed modelling by the lumped-model approach is complemented by the distributed parameter approach. In lumped-systems modelling as described by Maidment (1993), the spatial properties of the watershed are averaged without consideration of the effects of location of land characteristics or topological relationships within the watershed or stream network. In contrast, the distributed parameter modelling approach uses detailed digital terrain data and, in some cases, soils, land cover and other spatial data in the modelling of hydrologic processes, and explicitly considers spatial location of each spatial unit within the watershed.

Rewerts (1992) and Engel *et al.* (1993) review the benefits of distributed parameter modelling relative to the simple lumped approach (in the context of surface flow and pollution modelling in a distributed parameter mode). They note that non-point-source pollution modelling is concerned with the movement of pesticides and nutrients, as well as soil erosion. Distributed parameter models incorporate the variability in landscape features that control hydrologic flow and transport processes (topography, soils and land use) and, therefore, are potentially a more realistic modelling approach. Engel *et al.* (1993) also note that distributed parameter modelling accounts for all parts of the watershed simultaneously and incorporates spatial and temporal variability due to land-scape attributes. They note that a third advantage of distributed modelling is that the results of 'plot-size' research can be extrapolated to the watershed scale.

Much of the distributed parameter modelling is now using GIS technology to perform watershed drainage system analysis (Band, 1986; Jenson and Domingue, 1988), to develop various types of HRU such as those defined by Leavesley and Stannard (1990), or to conduct research on basic runoff-infiltration processes such as in the concept of representative elemental areas described by Wood *et al.* (1990). For example, Band (1986) and Jenson and Domingue (1988) developed procedures to determine watershed drainage systems, including drainage boundaries, subbasins and stream network divides, based on the analysis of digital elevation models (DEM). Gandoy-Bernasconi and Palacois-Velez (1990) developed similar procedures for analysing DEM represented by a TIN. Vieux (1988, 1991) and Smith and Brilly (1992) review the use of these algorithms in hydrologic modelling.

A modified version of Band's algorithm is incorporated into the Geographic Resource Analysis Support System (GRASS) GIS (GRASS, 1988), while the Jenson and Domingue algorithm is the basis for such analysis tools within the ARC/INFO GIS (ESRI, 1991). Vieux and Kang (1990) describe a set of terrain analysis tools for the GRASS GIS called GRASS Waterworks. These GIS tools, coupled with standard GIS functions for boolean searches and overlay oper-ations, form the basis of much contemporary hydrologic modelling with GIS.

Hay *et al.* (1993) describe how a GIS is used in a coupled-systems modelling approach to assess the sensitivity of water resources of the Gunnison River Basin to climatic variability, including potential climatic change scenarios (see also Leavesley *et al.*, 1992). The output of nested general circulation and mesoscale models is linked to an orographic-precipitation model that provides the precipitation input for a precipitation-runoff modelling system (PRMS), a distributed parameter watershed model (Leavesley *et al.*, 1983). The output of

PRMS is used in a downstream flow accounting and routeing system to estimate total discharge of the pilot test basin. The GIS is used to manage, manipulate and analyse the topographic, soils, vegetation, climate and derived land surface characteristics data that help to determine the behaviour of major processes within this coupled systems approach. Such processes include the amount and spatial variability of precipitation through snow accumulation in the orographic-precipitation model, watershed processes (snow melt, evapotranspiration, infiltration and runoff), and stream flow routeing (Hay *et al.*, 1993). For example, the GIS is used to define elevation grids of various grid-cell sizes and then characterize terrain statistics in each cell for use in orographic-precipitation modelling. The GIS is also used to assess the effects of slope, aspect (wind orientation) and elevation as a function of scale on modelled precipitation. The gridded model precipitation estimates are then routed to either HRU or a second grid for PRMS modelling.

GIS is used by the PRMS modeller to define HRU boundaries, estimate parameters in HRU (or grid cells) and verify the model results. To illustrate, Leavesley and Stannard (1990) developed procedures for characterizing HRU based on terrain statistics (elevation, slope and aspect), soil and vegetation types and rainfall distribution. The watershed delineation tools of Jenson and Domingue (1988) and standard overlay capabilities of GIS are used to define HRU from these land data. GIS are also used to define basin characteristics in HRU for estimation of subsystem processes in the PRMS; for example, rainfall interception of vegetation based on vegetation type and density; infiltration based on soil type and slope; ET based on vegetation type, soil type, slope, aspect and elevation; surface runoff based on vegetation type and density, slope and soil type; and channel flow as a function of slope and stream length (Hay *et al.*, 1993). In the case of ET calculations, GIS is used to obtain TIN-based statistics of terrain, aspect and elevation within each model grid cell as a basis for area weighting solar radiation estimates for more realistic estimates of ET. Finally, PRMS estimates of snow cover are validated against satellite-derived estimates based on comparison in the GIS.

Gao *et al.* (1993) report on the integration of GRASS GIS with a physically based distributed parameter precipitation-runoff model for watershed analysis. The model consists of coupled modules for separate hydrologic processes, including rainfall intensity, vegetation canopy interception, infiltration, surface lateral flow given saturated soil conditions, channel flow and subsurface flow. A rectangular grid is used for horizontal discretization with vertical discretization for soil layers. One purpose of the modelling system is to examine the spatial and temporal variability of hydrologic processes, determine controlling factors and investigate scale effects. The GRASS GIS is used by the system to extract model input parameters by grid cell from the spatial database and to visualize model results. GRASS overlay capabilities are used with UNIX system programming macros to retrieve and edit basin characteristics within individual watershed boundaries. The real-time visualization of model results is accomplished by special UNIX shell programs that extract hydrographs and spatial output, such as overland flow intensity, as the model is executing. One UNIX shell runs the model and another shell counts the lines in the model output until the spatial domain is filled, and then pipes the results to GRASS for display. Gao *et al.* (1993) indicate that this dynamic linkage between the GIS and the numerical modelling system facilitates error analysis, sensitivity checks, model calibration and validation, and investigation of multiscale hydrologic processes. The GRASS GIS is also used for generalization of spatial data.

Other examples of distributed parameter modelling of precipitation-runoff processes in a watershed with some type of GIS interface are described by Smith and Brilly (1992), Johnson (1989), Neumann and Schultz (1989), Vieux (1991) and Drayton *et al.* (1992). In each of these, a GIS is used to define model parameters based on spatial analysis of terrain, soils and land cover data, with varied approaches to the integration of the model and GIS. In the work of Neumann and Schultz (1989), a prototype modelling system is described in which remote sensing and GIS technologies are simply used for model parameterization.

Surface Water Quality Modelling

Both qualitative and quantitative approaches address the non-point-source pollution modelling problem. The traditional qualitative approach with a GIS involves weighted compositing and overlay (Burrough, 1986). Quantitative approaches usually involve some type of distributed parameter modelling. Each of these approaches is illustrated in the following section.

In an example of the weighted compositing index approach, Blaszczynski (1992) described how the revised universal soil loss equation (RUSLE) is interfaced with a GIS to derive a soil erosion potential map for rangelands. The RUSLE is empirically based and is derived by the product of various factors related to rainfall and runoff, soil erodability, terrain slope length and steepness, land cover and management, and support practice. These factors are derived in a raster GIS based on the analysis of various spatial data input on terrain, soil survey and land use.

There is significant progress in the integration of GIS technology with physically based approaches for non-point-source pollution water quality modelling as evidenced by the reviews of Rewerts (1992), Engel *et al.* (1993) and Vieux (1991). Water quality and quantity modelling involving non-point sources of pollutants are dependent on spatially distributed attributes of a catchment or watershed. The heterogeneous spatial distributions of soil, land cover, land use and topography determine the type of contaminant, and affect key hydrologic processes such as runoff and infiltration that determine the fate of the contaminant during overland flow, infiltration and deep percolation into groundwater sources.

Engel *et al.* (1992) describe the integration of the GRASS GIS with two watershed non-point-source pollution models, the agricultural non-point-source pollution (AgNPS) model (Young *et al.*, 1987), and the areal non-point-source watershed environmental response simulation (ANSWERS) model (Beasley, 1977; Beasley and Huggins, 1982). Both AgNPS and ANSWERS are distributed parameter models designed to identify non-point-source pollution problems within agricultural watersheds. The models are typically implemented in the context of a decision support system to evaluate the effectiveness of erosion and chemical runoff movement techniques (Engel *et al.*, 1992). These models deal with runoff, erosion and chemical movement as individual processes. The models are run in a raster-based mode in which the watershed is divided into a set of regular grid cells. Although the data input requirements are quite large (for example, 21 input parameters for each grid cell in AgNPS), runoff characteristics and transport processes of sediments and nutrients are simulated for each cell and routed to its outlet. Thus, upland pollution sources can be identified.

Rewerts (1992) and Engel *et al.* (1993) discuss a modular toolbox approach consisting of various GIS tools to assist in data development and analysis requirements of the AgNPS and ANSWERS models. Separate modules are developed for hydrologic, erosion and chemical movement processes. They cite several advantages of this modular approach, including the ability to use modules with other distributed parameter models, the ease in updating or replacing modules, and the creation of a variety of modules based on different assumptions for process-level modelling.

Needham and Vieux (1989) and Donovan (1991) describe the use of ARC/ INFO to generate input files for the AgNPS model. The GIS is used to overlay the AgNPS grid cell structure with ARC/INFO coverages of topographic, soils, land use and other map-based data. The AgNPS input parameters are determined in the GIS by using an area-weighted analysis. The INFO files are exported as ASCII files for conversion to the AgNPS format.

Groundwater Flow and Contaminant Transport Modelling

An example of total GIS model integration is illustrated by the DRASTIC groundwater vulnerability model for state-wide assessment as described by Rundquist *et al.* (1991). This qualitative model requires the estimation of several hydrogeologic parameters that determine the name of the model: Depth to water, Recharge, Aquifer media, Soil media, Topography, Impact of vadose zone and hydraulic Conductivity. Each parameter is assigned an overall relative weight, and a rating for each possible categorical value of the parameter. For example, as illustrated by Rundquist *et al.* (1991), depth to water is assigned a weight of 5, and soil media has an overall weight of 2 in threat to groundwater vulnerability. The range of soils includes thin or absent (rating of 10) to non-aggregated clay (rating of 1). The DRASTIC model index is calculated as the summation of the products of weights and ratings for each parameter. Rundquist *et al.* (1991) describe how the ERDAS GIS was used to estimate each model parameter from map data inputs and then calculate the index as a weighted composite.

Harris *et al.* (1993) report on the integration of the ARC/INFO GIS with groundwater flow and contaminant transport models for the analysis of the San Gabriel basin groundwater system. In this tightly coupled approach, an integrated system of pre- and post-processing programs was developed to facilitate data flow between the GIS and the CFEST (coupled fluid, energy and solute transport) groundwater flow model. In general, the GIS was used to develop and calibrate the CFEST model, build and manage a large spatial database and perform comparative analysis of model output. Specifically, the initial runs were based on the US Geological Survey MODFLOW program (McDonald and Harbaugh, 1988), and the first task for the GIS was to convert MODFLOW input files to the CFEST format. GIS overlay capabilities were used to transfer model information from the orthogonal, regular MODFLOW grid to the irregular, finite element grid of the CFEST model. GIS query and spatial analysis tools were used in the conceptual model development of the CFEST implementation. The GIS was also used to update and manage the extensive input files during iterative calibration runs with the CFEST model to simulate historical conditions based on observations. The GIS managed a large and diverse spatial database, which included data on water quality, well characteristics, land use, topography, geology, institutional and political boundaries, and contaminant analysis (Harris *et al.*, 1993). The GIS also managed CFEST model input layers (grids, boundary conditions, and elemental and nodal properties) and model simulation output (water heads, contaminant concentrations and flow vectors).

In a second approach to integration, Hake and Cuhadaroglu (1993) describe the development of a menu-driven interface between the Intergraph GIS and the MODFLOW groundwater model. In this case, the interface was developed within the Intergraph MGE (Modular GIS Environment) Environmental Resource Management Applications (ERMA) software to simplify data input, analysis and display. The GIS-model integration generated the model grid, automatically transformed co-ordinates, extracted data from the database and provided overlay capabilities. The interface code features an impressive set of menus to invoke MODFLOW, define the grid, select various options to configure the simulation, define the output, run the model and display output for analysis.

Ecological Modelling and GIS

This section focuses on the growing use of GIS technology and dynamic simulation modelling in terrestrial ecological research. The focus is limited to examples pertaining to the land–atmosphere interactions modelling paradigm as formulated by the NRC (1990) and the IGBP (1990) and examples discussed by

Schimel *et al.* (1991) and Hall *et al.* (1988). Johnston (1993) reviews the use of GIS as a research tool within the fields of community and population ecological research, and Hunsaker *et al.* (1993) give a general review of GIS in the fields of freshwater and marine ecological modelling.

In general, recent developments in physiologically based ecosystem dynamics simulation models, forest succession models, landscape ecology and exploratory spatial analysis of disturbance patterns have created a promising foundation for GIS within these areas of ecological research. The following two sections summarize the status of GIS technology and ecological model integration in this context. The first section examines the role of a GIS as a tool to support dynamic simulation models operating at daily to monthly time steps for ecosystem dynamics, rangeland and crop simulations. The second section focuses on forest succession modelling, typically characterized by annual model time steps. The use of a GIS in disturbance modelling is also examined.

Ecosystem Dynamics, Rangeland and Crop Simulation Modelling

Examples of ecological models that fall in the land–atmosphere interactions paradigm include the regional hydro-ecological simulation system (RHESSys) recently described by Running *et al.* (1992), the CENTURY model discussed by Burke *et al.* (1991) and Schimel *et al.*, and various forest stand or ecological succession models (see for example Nisbet and Botkin, 1993). The first two examples and other physiologically based approaches for rangeland and crop simulation are described in this section. The ecological succession models illustrate processes operating over longer time scales and are discussed in the next section.

GIS technology is used to support the extrapolation of plot-level information to regional scales. For example, Running (1991a), Running *et al.* (1992) and Nemani *et al.* (1993) describe the RHESSys as an integrated modelling system in which a physiologically based forest ecosystem model (FOREST-BGC) (Running *et al.*, 1989) is used to extrapolate forest plot information to the regional and continental levels. The RHESSys is a coupled-system model that links hydrologic, ecological and microclimatological models with remote sensing, soils and climate data. The model is designed to simulate and map water, carbon and nutrient fluxes at various spatial scales. The simple or generic GIS approach discussed by Running *et al.* (1989) is replaced by a polygon-based approach for geographic information processing as discussed by Running (1991a) and Band *et al* (1991). Topographic analysis tools developed by Band (1986) define a template consisting of hillslopes, stream channels and subwatersheds that characterize the landscape structure. Polygons representing hillslopes and watersheds stratify leaf area index, microclimate data and soil water capacity into relatively homogeneous landscape units for model input. Spatial aggregation and scaling up are achieved by defining landscape templates of differing spatial resolutions.

The ARC/INFO GIS was coupled with the CENTURY ecosystem model to simulate the spatial variability of grasslands biogeochemistry (Burke *et al.*,

1990; Schimel and Burke, 1993). The GIS built regional climate and soils databases to drive the CENTURY model. Contoured maps of mean annual total precipitation and mean annual temperature were derived from monthly weather station data. The regional soil texture map was derived from county-level data. These three ARC/INFO coverages were overlaid to define unique polygons for the model simulation. The GIS was used to map the CENTURY output including soil organic carbon, annual above-ground net primary production, net nitrogen mineralization and nitrogen oxide fluxes (Burke *et al.*, 1990).

GIS technology is contributing to rangeland production and agricultural crop simulation research. Hanson and Baker (1993) illustrate a GIS as a database manager for the simulation, production and utilization of rangelands (SPUR) model. The GIS integrates climate, soils, hydrologic, plant, animal and terrain data for input, as appropriate, to the SPUR subsystem process modules for hydrology, plants, livestock, wildlife, insects and economics. The SPUR model is currently used in combination with GIS and remote sensing technologies to understand the spatial patterns and interactions caused by processes within rangeland ecosystems. Hanson and Baker (1993) suggest that a GIS can assist in the parametrization, execution and validation of models such as SPUR. This would contribute to the extrapolation of point data on soils, climate, plants and experimental plot results to larger scales; for example, pastures, ranches, landscapes and regions.

The potential role of remote sensing and GIS technologies to enhance crop simulations for small agricultural fields in the heterogeneous landscape environment of central South Carolina was demonstrated by Narumalani and Carbone (1993). Land cover was classified as to scrub/shrub, forest, wetland forest, wetland marsh, water, urban and agricultural based on analysis of SPOT multispectral images and National Aerial Photography Program photographs. Agricultural and non-agricultural land use regions were brought into an ARC/INFO GIS along with daily weather, soils and other data sets needed by a physiologically based soybean growth model. Although the analysis was constrained by available weather data and the accuracy of the image classification, the method does illustrate how spatial data in a GIS permitted aggregation of crop simulation results by running the model on the desired land use parcel and ignoring the non-agricultural lands. This type of study is relevant to the integration of simulation models with GIS and remote sensing technologies to routinely assess or monitor changing agricultural conditions throughout the growing season.

As an example of real-time monitoring functions, Kessell (1990) and Kessell and Beck (1991) describe the role of a generic GIS that was developed in Australia for forest fire modelling as part of an overall modelling system for natural resource management. The purpose is to assess fire hazard and potential, as well as provide a real-time fire monitoring tool. The software and hardware package has had many variations over the years and is called the geographical information and modelling system (GIMS). The system has individual modules for vegetation structure and floristics for each grid cell as a function of other GIMS data layers (such as elevation, slope, aspect and soil type), flammable fuel conditions, fire behaviour, and fire area and perimeter growth over time. There are also modules for natural resource management, including animal habitat, catchment protection and resource allocation.

Ecological Succession and Landscape Disturbance Modelling

Ecological succession modelling and the study of landscape structure and disturbance patterns are two promising areas for GIS technology (Kessell, 1990; Baker *et al.*, 1991; Polzer *et al.*, 1991; Silveira *et al.*, 1992; Nisbet and Botkin, 1993). The following examples illustrate the current status and likely directions of GIS in this type of research, particularly considering heterogeneous landscapes.

Nisbet and Botkin (1993) discuss how a GIS can display results of forest growth models such as JABOWA (Botkin *et al.*, 1972) and FORET (Shugart and West, 1977). However, they also suggest that an expanded role for GIS could result from inclusion of potential spatial interactions, which could affect forest regeneration, growth and mortality processes in the model. They suggested that conditions in adjacent cells, such as biomass levels or site heterogeneity, could be factors in the life-cycle process.

Smith and Urban (1988) used a roving window of influence to keep track of biomass and leaf area within adjacent cells in a JABOWA-like model. Similarly, Woodby (1991) used the JABOWA-II model to consider extinction probabilities of trees as a function of site heterogeneity and the proximity of trees to surrounding landscape. Model simulations of up to 1000 years showed a dependency on patch size, spatial variation of soil characteristics and occurrence of disturbances. The results suggested the importance of spatial interactions due to variability in soil quality, soil moisture availability, patch size and disturbance regime.

Based on these results, Nisbet and Botkin (1993) suggest that a suite of JABOWA-like forest growth models could be run in a GIS environment at the landscape scale to explore various hypotheses on the role of spatial interactions. They suggest that other complex spatial interactions could include the transport of materials (for example, by wind, water and biological vectors), windshed effects and pollution plume effects, and call for either an advanced modelling language to execute ecological models in a GIS or system calls to model subroutines from within the GIS.

Polzer *et al.* (1991) developed a software link between the PC ARC/INFO GIS and the ZELIG forest growth model (Urban, 1990). The GIS built and managed the spatial database and analysed and visualized model results. The link involved using the GIS to parametrize the model input based on the GIS data structure, converting this structure to the model input format, and reformatting the model output back to the GIS data structure following the model simulations. The link was accomplished through the dBASE programming language, although the conversion of multiple files was cumbersome. This case study demonstrates the need for transparent data conversion routines between the GIS and the external dynamic simulation software.

The role of disturbances as a factor in ecological succession is a major research topic in landscape ecology. Baker *et al.* (1991), Lowell (1991), and Silveira *et al.* (1992) illustrate three different approaches for modelling landscape disturbance, in each case with some GIS contribution. For example, Baker *et al.* (1991) describe the DISPATCH (DISturbance PATCH) simulation model that was developed as a theoretical framework to study disturbance and landscape interactions under global change scenarios. The model was designed to simulate repeated disturbances over time scales of hundreds of years, to simulate the effects of environmental variability or landscape structure on disturbance initiation and spread, and to provide information on quantitative changes in landscape structure. The components of the model include algorithms to generate climatic and disturbance regimes, five GIS map layers (vegetation type, patch age, elevation, slope and aspect), a disturbance probability map based on grid cell conditions, and spatial data tools for structural analysis of patches. As noted by the authors, however, the limitations of the model include the inability to validate long-term simulations and simulate disturbances (such as fires or floods) realistically and the absence of an extensive data set to calibrate the model.

Both Lowell (1991) and Silveira *et al.* (1992) used historical data sets with separate types of landscape disturbance models and a GIS for case study analyses of ecological succession. Lowell (1991) used discriminant function analysis (DFA) to analyse the spatial and temporal distributional changes associated with long-term ecological succession within a central Missouri wildlife research refuge. He concluded that the DFA does provide a useful framework for modelling spatiotemporal changes in the landscape patterns for this study area, which was farmed until 1937 when human intervention was halted (except for some grazing until the early 1960s) and the land was permitted to undergo natural ecological succession. He also suggests that the DFA is a parametric statistical tool for use in a GIS in ways that are entirely consistent with traditional overlay approaches that could have been used to analyse the central Missouri data. Specifically, the DFA provides a GIS with a means to develop temporal models that can be applied spatially.

Silveira *et al.* (1992) used qualitative modelling and analysis approaches entirely in an ERDAS GIS to investigate possible landscape and disturbance relationships in the Everglades ecosystem over time. The model was designed to simulate disturbance due to fire damage and postdisturbance vegetation succession. The modelling approach was rule-based, that is, the ERDAS GISMO module was provided with rules on fire behaviour, successional rates and other relationshps that were based on the literature.

Landscape ecologists and other researchers have made large strides in developing GIS techniques and exploratory spatial data analysis tools to investigate complex landscape structure including disturbance patterns (Walker and Belbin, 1990; Hall *et al.*, 1991; Baker and Cai, 1992; Johnston, 1992; Johnston, 1993; Michener *et al.*, 1992; Rossi *et al.*, 1992). For example, Johnston (1993) discusses the use of GIS techniques in combination with satellite image analysis to detect ecotones in vegetation patterns and to estimate forest successional rates over time based on the analysis of disturbance patterns within multitemporal satellite scenes. Baker and Cai (1992) developed GRASS GIS tools called r.le, which are designed to analyse more than 60 measures of landscape structure (for example, distance, size shape, diversity and texture). Rossi *et al.* (1992) discuss the application of advanced geostatistical tools for interpreting ecological spatial dependence.

Discussion and Directions for GIS

This chapter has reviewed a wide variety of GIS applications involving atmospheric, hydrologic and ecological sciences. Applications have been found to fall into several broad categories. First, GIS is used as a preprocessor of data, making use of its functions for projection change, resampling, generalization, windowing and other data extraction tasks, map digitizing and editing. Reformatting is often needed to meet the requirements of simulation packages. Second, GIS is used for storage of data since the database management facilities of GIS keep track automatically of many housekeeping functions, such as simple documentation, and provide a uniform mode of access. Third, GIS is used for statistical analysis of data and model results, often in combination with standard packages for statistics or geostatistics. Fourth, GIS is used for visualization, particularly for presenting results of simulations in map form, often in combination with other data. These functions are all available in the current range of GIS products and, in most of these areas, functions are currently available to meet most needs.

Current GIS differ widely in the methods used to represent geographic features and patterns, and in the functions supported by each package. The choice of an appropriate GIS to support simulation modelling is often complex, requiring substantial technical knowledge on the part of the user. Some modellers use commercial packages with proprietary data formats such as ARC/INFO, while others use public domain systems such as GRASS. In general, the current generation of GIS is considered difficult to use, in part because of the wide range and cumbersome nature of user interfaces.

GIS developers have been relatively slow to incorporate some of the more sophisticated forms of spatial analysis needed by environmental simulation modellers. For example, despite the importance of spatial interpolation to resampling of data, few packages offer the sophisticated methods of interpolation of geostatistics. Methods of spatial interpolation that allow the user to include a sophisticated knowledge of interpolated phenomena are also poorly developed. For example, it is not easy to find spatial interpolation algorithms in conventional GIS products that allow researchers to include discontinuities or weight observations by quality.

Many researchers have used GIS as a convenient source of functions for the display and mapping of data and results. Mapping software code can be complex, especially where functions must be included to mimic the skill of a trained cartographer in tasks such as positioning labels or where displays require the merger of many different types of information. Scientists with simple display requirements may find it more convenient to write their own code or make use of the increasing sophistication of scientific visualization packages like AVS (application visualization system) or Khoros rather than a GIS.

Goodchild *et al.* (1992) review the various forms of coupling that are possible when GIS technology and environmental models are used together. Loose coupling combines the capabilities of separate modules for GIS functions and environmental simulation by transferring simple files, often in ASCII format. Many GIS accept a wide range of input and output formats, but in many cases it is necessary to write a short piece of code as a bridge between the two modules. Loose coupling is effective for preprocessing data or displaying results, but may require the investment of substantial resources to build the necessary bridges. Tight coupling allows the two modules to run simultaneously and to share a common database. Because data formats and definitions must be shared, tight coupling is uncommon, although much more convenient for the researcher. Opportunities for tight coupling will increase as GIS data formats become more widely known, particularly through the use of open GIS toolboxes such as GRASS (1988), which allows the researcher access to source code. The development of the Spatial Data Transfer Standard (now adopted as Federal Information Processing Standard 173) will facilitate the development of tightly coupled GIS modelling systems.

Most of the applications reviewed in this chapter fall into one or more of the approaches for integration just described. Only a few use the GIS directly as the modelling environment, writing the simulation as a series of GIS commands rather than in a common source language. At present, such embedding of the model in a GIS is practical only in the case of finite difference models that can take advantage of the functionality of raster GIS for cell-based modelling, and the range of such raster capabilities falls short of the range needed for environmental simulation (Goodchild, 1993). Kemp (1992) argues that an ideal modelling language for environmental simulation would allow the modeller to ignore the details of the spatial representation of the model's variables. However, practical realization of this possibility is still in the future.

Several areas of successful integration of GIS technology and environmental simulation models have been described in this chapter, but advanced GIS tools are needed for further progress. Some of these capabilities such as efficient data conversion algorithms, exploratory spatial data analysis tools and a modelling language have been highlighted. The software development community is just beginning to address these needs (for an extensive review of GIS development directions see Maguire *et al.*, 1991), as well as other requirements such as temporal and three-dimensional data analysis capabilities that are typical of most dynamic simulation models. Hierarchical GIS are needed for regional extrapolations and other types of multiscale analyses. Object-orientated approaches are just beginning to appear in the literature, especially for hydrologic modelling. An understanding of the underlying environmental processes and modelling approaches is a prerequisite for the development of GIS tools in each of these areas.

Acknowledgement

The National Center for Geographic Information and Analysis is supported by the National Science Foundation, grant SES 88-10917.

References

Baker, W. L., Egbert, S. L. and Frazier, G. F., 1991, A spatial model for studying the effects of climatic change on the structure of landscapes subject to large disturbances, *Ecological Modeling*, **56**, 109–25.

Baker, W. L. and Cai, Y., 1992, The r.le programs for multiscale analysis of landscape structure using the GRASS geographical information system, *Landscape Ecology*, **7**, 291–302.

Band, L. E., 1986, Topographic partitioning of watersheds with digital elevation models, *Water Resources Research*, **22**, 15–24.

Band, L. E., Peterson, D.L., Running, S. W., Coughlan, J., Lammers, R., Dungan, J. and Nemani, R., 1991, Forest ecosystem processes at the watershed scale: Basis for distributed simulation, *Ecological Modeling*, **56**, 171–96.

Beasley, D. B., 1977, 'ANSWERS: a mathematical model for simulating the effects of land use and management on water quality', unpublished PhD dissertation, Purdue University, West Lafayette, Indiana.

Beasley, D. B. and Huggins, L. F., 1982, *ANSWERS — User's Manual*, EPA-905/9-82-001, Chicago, Illinois: US Environmental Protection Agency, Region 5.

Birth, T., Dessent, T., Milich, L. B., Beatty, J., Scheitlin, T. E., Coventry, D. H. and Novak, J. H., 1990, *A Report on the Implementation of a Pilot Geographic Information System (GIS) at EPA*, Research Triangle Park, North Carolina: US Environmental Protection Agency.

Blaszczynski, J., 1992, Regional soil loss prediction utilizing the RUSLE/GIS interface, in Johnson, A. I., Pettersson, C. B. and Fulton, J. L. (Eds) *Geographic Information Systems (GIS) and Mapping-Practices and Standards*, ASTM STP 1126, pp. 122–31, Philadelphia: American Society for Testing and Materials.

Botkin, D. B., Janak, J. F. and Wallis, J. R., 1972, Rationale, limitations, and assumptions of a northeastern forest growth simulator, *IBM Journal Research Division*, **16**, 101–16.

Burke, I. C., Schimel, D. S., Yonker, C. M., Parton, W. J., Joyce, L. A. and Lauenroth, W. K., 1990, Regional modeling of grassland biogeochemistry using GIS, *Landscape Ecology*, **4**, 45–54.

Burrough, P. A., 1986, *Principles of Geographical Information Systems for Land Resources Assessment*, Monographs on Soil and Resources Survey No. 12, New York: Oxford University Press.

Dickinson, R. E., Henderson-Sellers, A., Kennedy, P. J. and Wilson, M. F., 1986, *Biosphere–Atmosphere Transfer Scheme (BATS) for the NCAR Community Climate Model*, NCAR Technical Note NCAR/TN-275 + STR, Boulder, Colorado: National Center for Atmospheric Research.

Donovan, J. K., 1991, The US EPA's Arizona rangeland project: Integration of GIS and a non-point source pollution model, in *Proceedings of the Eleventh Annual ESRI User Conference*, pp. 261–72, Redlands, CA: Environmental Systems Research Institute.

Drayton, R. S., Wilde, B. M. and Harris, J. H. K., 1992, Geographical information system approach to distributed modelling, *Hydrological Processes*, **6**, 361–8.

Earth System Sciences Committee (ESSC), 1986, *Earth System Science Overview: A Program for Global Change*, Washington, DC: National Aeronautics and Space Administration.

Earth System Sciences Committee (ESSC), 1988, *Earth System Science: A Closer View*, Washington, DC: National Aeronautics and Space Administration.

Engel, B. A., Srinivasan, R. and Rewerts, C. C., 1992, Integrated watershed non-point source pollution models and geographic information systems, in *Managing Water Resources During Global Change*, AWRA 28th Annual Conference and Symposium, pp. 255–6, Bethesda: American Water Resources Association.

Engel, B. A., Srinivasan, R. and Rewerts, C. C., 1993, A spatial decision support system for modeling and managing agricultural non-point source pollution, in Goodchild, M. F., Parks, B. O. and Steyaert, L. T. (Eds) *Environmental Modeling with GIS*, pp. 231–7, New York: Oxford University Press.

Environmental Systems Research Institute (ESRI), 1991, *ARC-INFO GIS User's Guide*, Redlands, California: Environmental Systems Research Institute.

Gandoy-Bernasconi, W. and Palacois-Velez, O., 1990, Automatic cascade numbering of unit elements in distributed hydrologic models, *Journal of Hydrology*, **112**, 375–93.

Gao, X., Sorooshian, S. and Goodrich, D. C., 1993, GIS used for land phase hydrologic modeling, in Goodchild, M. F., Parks, B. O. and Steyaert, L. T. (Eds) *Environmental Modeling with GIS*, pp. 182–7, New York: Oxford University Press.

Geographical Resource Analysis Support System (GRASS), 1988, *Geographical Resource Analysis Support System Version 3.0*, Champaign, Illinois: US Army Corps of Engineers, Construction Engineering Research Laboratory.

Goodchild, M. F., 1993, The state of GIS for environmental modeling, in Goodchild, M. F., Parks, B. O. and Steyaert, L. T. (Eds) *Environmental Modeling with GIS*, pp. 8–15, New York: Oxford University Press.

Goodchild, M. F., Haining, R. P. and Wise, S., 1992, Integrating GIS and spatial analysis: Problems and possibilities, *International Journal of Geographical Information Systems*, **6**, 407–23.

Goodchild, M. F., Parks, B. O. and Steyaert, L. T. (Eds), 1993, *Environmental Modeling with GIS*, New York: Oxford University Press.

Goodin, D. G., Yang, L. and Yang, W., 1992, Integration of process models with remote sensing for monitoring surface processes, *ASPRS/ASCM/RT '92 Technical Papers*, pp. 213–21, Washington, DC.

Hake, R. and Cuhadaroglu, M. S., 1993, A finite difference groundwater model–GIS interface, in *ACSM/ASPRS Technical Papers*, pp. 213–22, 1993 ACSM/ASPRS Annual Convention, Bethesda: ASPRS/ACSM.

Hall, F. G., Botkin, D. B., Strebel, D. E., Woods, K. D. and Goeta, S. T., 1991, Large-scale patterns of forest succession as determined by remote sensing, *Ecology*, **72**, 628–40.

Hall, F. G., Strebel, D. E. and Sellers, P. J., 1988, Linking knowledge among spatial and temporal scales: Vegetation, atmosphere, climate and remote sensing, *Landscape Ecology*, **2**, 3–22.

Hanson, J. D. and Baker, B. B., 1993, Simulation of rangeland production: A case study in systems ecology, in Goodchild, M. F., Parks, B. O. and Steyaert, L. T. (Eds) *Environmental Modeling with GIS*, pp. 305–13, New York: Oxford University Press.

Harris, J., Gupta, S., Woodside, G. and Ziemba, N. , 1993, Integrated use of a GIS and a 3-dimensional, finite-element model; San Gabriel Basin groundwater flow analysis, in Goodchild, M. F., Parks, B. O. and Steyaert, L. T. (Eds) *Environmental Modeling with GIS*, pp. 168–72, New York: Oxford University Press.

Hay, L. E., Battaglin, W. A., Parker, R. S. and Leavesley, G. H., 1993, Modeling the effects of climate change on water resources in the Gunnison River Basin, Colorado, in Goodchild, M. F., Parks, B. O. and Steyaert, L. T. (Eds) *Environmental Modeling with GIS*, pp. 173–81, New York: Oxford University Press.

Hunsaker, C. T., Nisbet, R. A., Lam, D., Browder, J. A., Baker, W. L., Turner, M. G. and Botkin, D. B., 1993, Spatial models of ecological systems and processes: The role of GIS, in Goodchild, M. F., Parks, B. O. and Steyaert, L. T. (Eds) *Environmental Modeling with GIS*, pp. 248–64, New York: Oxford University Press.

International Council of Scientific Unions (ICSU), 1986, *The International Geosphere-Biosphere Program: A Study of Global Change, Report No. 1*, Final Report of the Ad Hoc Planning Group, ICSU Twenty-first General Assembly, 14–19 September 1986, Stockholm: International Council of Scientific Unions.

International Geosphere–Biosphere Programme (IGBP), 1990, *The International Geosphere-Biosphere Programme: A Study of Global Change, The Initial Core Projects, Report No. 12*, Stockholm: IGBP Secretariat.

Jenson, S. K. and Domingue, J. O., 1988, Extracting topographic structure from digital elevation data for geographic information system analysis, *Photogrammetric Engineering and Remote Sensing*, **54**, 1593–1600.

Johnson, L. E., 1989, MAPHYD — A digital map-based hydrologic modeling system, *Photogrammetric Engineering and Remote Sensing*, **55**, 911–17.

Johnston, C. A., 1993, Introduction to quantitative methods and modeling in community, population, and landscape ecology, in Goodchild, M. F., Parks, B. O. and Steyaert, L. T. (Eds) *Environmental Modeling with GIS*, pp. 276–83, New York: Oxford University Press.

Johnston, K. M., 1992, Using statistical regression analysis to build three prototype GIS wildlife models, in *Technical Proceedings GIS/LIS '92*, pp. 374–86. Bethesda: ASPRS/ACSM/AAG/URISA/AM-FM International.

Kemp, K. K., 1992, 'Environmental modeling with GIS: a strategy for dealing with spatial continuity', unpublished PhD dissertation, University of California, Santa Barbara.

Kessell, S. R., 1990, An Australian geographical information and modeling system for natural area management, *International Journal of Geographical Information Systems*, **4**, 333–62.

Kessell, S. R. and Beck, J. A., 1991, Development and implementation of forest fire modeling and decision support systems in Australia, in *Technical Proceedings GIS/LIS '91*, pp. 805–16. Bethesda: ASPRS/ACSM/AAG/URISA/AM-FM International.

Leavesley, G. H. and Stannard, L. G., 1990, Application of remotely sensed data in a distributed-parameter watershed model, in *Proceedings of Workshop on Applications of Remote Sensing in Hydrology*, Ottawa: Canada.

Leavesley, G. H., Branson, M. D. and Hay, L. E., 1992, Using coupled atmospheric and hydrologic models to investigate the effects of climate change in mountainous regions, in *Managing Water Resources During Global Change*, AWRA 28th Annual Conference and Symposium, Reno, Nevada. pp. 691–700.

Leavesley, G. H., Lichty, R. W., Troutman, B. M. and Saindon, L. G., 1983, *Precipitation-Runoff Modeling System — User's Manual*, US Geological Survey Water-Resources Investigations Report 83-4238. Reston: USGS.

Lee, T. J., 1992, 'The impact of vegetation on the atmospheric boundary layer and convective storms', unpublished PhD dissertation, Department of Atmospheric Science, Colorado State University, Fort Collins.

Lee, T. J., Pielke, R. A., Kittel, T. G. F. and Weaver, J. F., 1993, Atmospheric modeling and its spatial representation of land surface characteristics, in Goodchild, M. F., Parks, B. O. and Steyaert, L. T. (Eds) *Environmental Modeling with GIS*, pp. 108–22, New York: Oxford University Press.

Loveland, T. R., Merchant, J. W., Ohlen, D. O. and Brown, J. F., 1991, Development of a land-cover characteristics database for the conterminous US, *Photogrammetric Engineering and Remote Sensing*, **57**, 1453–63.

Lowell, K., 1991, Utilizing discriminant function analysis with a geographical information system to model ecological succession spatially, *International Journal of Geographical Information Systems*, **5**, 175–91.

Luval, J. C., Lieberman, D., Lieberman, M., Hartshorn, G. S. and Peralta, R., 1990, Estimation of tropical forest canopy temperatures, thermal response numbers, and evapotranspiration using an aircraft-based thermal sensor, *Photogrammetric Engineering and Remote Sensing*, **56**, 1393–401.

Maguire, D. J., Goodchild, M. F. and Rhind, D. W., 1991, *Geographical Information Systems: Principles and Applications*, London: Longman.

Maidment, D. R., 1993, GIS and hydrological modeling, in Goodchild, M. F., Parks, B. O. and Steyaert, L. T. (Eds) *Environmental Modeling with GIS*, pp. 147–67, New York: Oxford University Press.

McDonald, M. G. and Harbaugh, A. W., 1988, A modular three-dimensional finite difference groundwater flow model, *Techniques of Water Resources Investigations of the USGS*, Book 6, Chaper A1, Reston: USGS.

Michener, W. K., Jefferson, W. H., Karinshak, D. A. and Edwards, D., 1992, An integrated geographic information system, global positioning system, and spatio-statistical approach for analyzing ecological patterns at landscape scales, in *Technical Proceedings GIS/LIS '92*, pp. 64–76. Bethesda: ASPRS/ACSM/AAG/URISA/AM-FM International.

Moore, I. D., Turner, A. K., Wilson, J. P., Jenson, S. K. and Band, L. E., 1993, GIS and land surface-subsurface process modeling, in Goodchild, M. F., Parks, B. O. and Steyaert, L. T. (Eds) *Environmental Modeling with GIS*, pp. 196–230, New York: Oxford University Press.

Narumalani, S. and Carbone, G. J., 1993, The use of remote sensing for crop simulation studies, in *ACSM/ASPRS Technical Papers*, 1993 Annual Convention, pp. 230–9. Bethesda: ASPRS/ACSM.

National Aeronautics and Space Administration (NASA), 1991, *BOREAS (Boreal-Ecosystem-Atmosphere Study): Global Change and Biosphere-Atmosphere Interactions in the Boreal Forest Biome, Science Plan*, Greenbelt, Maryland: NASA.

National Research Council (NRC), 1986, *Global Change in the Geosphere-Biosphere, Initial Priorities for an IGBP*, US Committee for an International Geosphere-Biosphere Program, Washington, DC: National Academy Press.

National Research Council (NRC), 1990, *Research Strategies for the US Global Change Research Program*, Committee on Global Change (US National Committee for the IGBP), Washington, DC: National Academy Press.

Needham, S. and Vieux, B. E., 1989, *A GIS for AgNPS Parameter Input and Mapping Outputs*, ASAE Paper No. 89-2673, Winter Meeting, Chicago, Illinois.

Nemani, R. R., Running, S. W., Band, L. E. and Peterson, D. L., 1993, Regional hydro-ecological simulation system: An illustration of the integration of ecosystem models into a GIS, in Goodchild, M. F., Parks, B. O. and Steyaert, L. T. (Eds) *Environmental Modeling with GIS*, pp. 296–304, New York: Oxford University Press.

Neumann, P. and Schultz, G. A., 1989, Hydrological effects of catchment characteristics and land use changes determined by satellite imagery and GIS, in *Remote Sensing and Large-Scale Global Processes*, Proceedings of the IAHS Third International Assembly, IAHS Publ. No. 186, pp. 169–76, Baltimore, Maryland: IAHS.

Nisbet, R. A. and Botkin, D. B., 1993, Integrating a forest growth model with a geographic information system, in Goodchild, M. F., Parks, B. O. and Steyaert, L. T. (Eds) *Environmental Modeling with GIS*, pp. 265–9, New York: Oxford University Press.

Novak, J. H. and Dennis, R. L., 1993, Regional air quality and acid deposition modeling and the role for visualization, in Goodchild, M. F., Parks, B. O. and Steyaert, L. T. (Eds) *Environmental Modeling with GIS*, pp. 142–6, New York: Oxford University Press.

Polzer, P. L., Hartzell, B. J., Wynne, R. H., Harris, P. M. and MacKenzie, M. D., 1991, Linking GIS with predictive models: Case study in a southern Wisconsin Oak forest, *Conference Proceedings of GIS/LIS '91*, pp. 49–59. Bethesda: ASPRS/ACSM/ASG/URISA/AM-FM International.

Rewerts, C. C., 1992, 'ANSWERS on GRASS: integrating a watershed simulation with a geographic information system', unpublished PhD dissertation, Purdue University, West Lafayette, Indiana.

Rossi, R. E., Mulla, D. J., Journel, A. G. and Franz, E. H., 1992, Geostatistical tools for modeling and interpreting ecological spatial dependence, *Ecological Monographs*, **62**, 277–314.

Rundquist, D. C., Peters, A. J., Di, L., Rodekohr, D. A., Ehrman, R. L. and Murray, G., 1991, Statewide groundwater-vulnerability assessment in Nebraska using the DRASTIC/GIS model, *Geocarto International*, **2**, 51–8.

Running, S. W., 1991a, Computer simulation of regional evapotranspiration by integrating landscape biophysical attributes with satellite data, in Schmugge, T. J. and André, J. C. (Eds) *Land Surface Evaporation: Measurement and Parameterization*, pp. 359–69, New York: Springer.

Running, S. W. (Ed.), 1991b, 'Pre-BOREAS Ecological Modeling Workshop. Flathead Lake, MT, August 18–24, 1991', unpublished report, NASA Office of Space Science and Applications.

Running, S. W., Nemani, R. R. and Band, L. E., 1992, 'Scaling ecosystem processes from local to continental scales using a regional hydro-ecological simulation system', presentation at the JGR 1992 Fall Meeting.

Running, S. W., Nemani, R. R., Peterson, D. L., Band, L. E., Potts, D. F., Pierce, L. L. and Spanner, M. A., 1989, Mapping regional forest evapotranspiration and photosynthesis by coupling satellite data with ecosystem simulation, *Ecology*, **70**, 1090–1101.

Schimel, D. S. and Burke, I. C., 1993, Spatial interactive models of atmosphere-ecosystem coupling, in Goodchild, M. F., Parks, B. O. and Steyaert, L. T. (Eds) *Environmental Modeling with GIS*, pp. 284–9, New York: Oxford University Press.

Schimel, D. S., Kittel, T. G. F. and Parton, W. J., 1991, Terrestrial biogeochemical cycles, global interactions with the atmosphere and hydrology, *Tellus*, **43**, 188–203.

Sellers, P. J., Mintz, Y., Sud, Y. C. and Dalcher, A., 1986, A simple biosphere model (SiB) for use within general circulation models, *Journal of Atmospheric Science*, **43**, 505–31.

Shih, S. F. and Jordan, J. D., 1992, Landsat mid-infrared data and GIS in regional surface soil-moisture assessment, *Water Resources Bulletin*, **28**, 713–19.

Shugart, H. H. and West, D. C., 1977, Development of an Appalachian deciduous forest model and its application to assessment of the impact of the chestnut blight, *Journal of Environmental Management*, **5**, 161–79.

Silveira, J. E., Richardson, J. R. and Kitchens, W. M., 1992, A GIS model of fire, succession and landscape pattern in the Everglades, in *Technical Papers ASPRS/ACSM/RT '92*, Vol. 5: *Resource Technology '92*, pp. 88–97. Bethesda: ASPRS/ACSM.

Skelly, W. C., Henderson-Sellers, A. and Pitman, A. J., 1993, Land surface data: global climate modeling requirements, in Goodchild, M. F., Parks, B. O. and Steyaert, L. T. (Eds) *Environmental Modeling with GIS*, pp. 135–41, New York: Oxford University Press.

Smith, C. B., Lakhtakia, M. N., Capehart, W. J. and Carlson, T. N., 1993, Initialization of soil-water content for regional-scale atmospheric prediction models, Conference on Hydroclimatology, 73rd AMS Annual Meeting, 17–22 January, pp. 24–7. Boston: American Meteorological Society.

Smith, M. B. and Brilly, M., 1992, Automated grid element ordering for GIS-based overland flow modeling, *Photogrammetric Engineering and Remote Sensing*, **58**, 579–85.

Smith, T. M. and Urban, D. L., 1988, Scale and resolution of forest structural pattern, *Vegetatio*, **74**, 143–50.

Star, J. L. and Estes, J. E., 1990, *GIS: An Introduction*, Englewood Cliffs, New Jersey: Prentice-Hall.

Steyaert, L. T., 1993, A perspective on the state of environmental simulation modeling, in Goodchild, M. F., Parks, B. O. and Steyaert, L. T. (Eds) *Environmental Modeling with GIS*, pp.16–30, New York: Oxford University Press.

Urban, D. L., 1990, *A Versatile Model to Simulate Forest Pattern, A User's Guide to ZELIG Version 1.0*, Charlottesville, Virginia: Department of Environmental Sciences, University of Virginia.

Vieux, B. E., 1988, 'Finite element analysis of hydrological response areas using geographic information systems', unpublished PhD dissertation, Department of Agricultural Engineering, Michigan State University, East Lansing.

Vieux, B. E., 1991, Geographic information systems and non-point source water quality and quantity modelling, *Hydrological Processes*, **5**, 101–13.

Vieux, B. E. and Kang, Y., 1990, GRASS waterworks: A GIS toolbox for watershed hydrologic modeling, in *Proceedings of Application of Geographic Information Systems, Simulation Models, and Knowledge-Based Systems for Landuse Management*, Blacksburg, Virginia: Virginia Polytechnic Institute and State University.

Walker, P. A. and Belbin, L., 1990, The identification of spatial associations and their incorporation in geographic information systems, in *Proceedings of 4th International Symposium on Spatial Data Handling*, pp. 522–30, Columbus: International Geographical Union and Department of Geography, Ohio State University.

Wood, E. F., Sivapalan, M. and Beven, K., 1990, Similarity and scale in catchment storm response, *Reviews of Geophysics*, **28**, 1–18.

Woodby, D. A., 1991, 'An ecosystem model of forest tree persistence', unpublished PhD dissertation, University of California, Santa Barbara.

Young, R. A., Onstad, C. A., Bosch, D. D. and Anderson, W. P., 1987, *AgNPS: Agricultural Non-Point Source Pollution Model: A Watershed Analysis Tool*, US Department of Agriculture–Agriculture Research Service, Conservation Research Report 35, Washington: USDA-ARS.

22

Data management and simulation modelling

Thomas B. Kirchner

Data management is important in simulation modelling because a variety of data is needed for estimating parameters, validation of simulation models requires data not used to help construct the model, analysis of a model can produce large quantities of data, and model code represents data that require special management techniques. These issues are not typically encountered when managing data for empirical modelling. Simulation models require data about the states of the system at particular times and about the rate constants for processes that affect those states. Estimation of rate constants for a simulation model often requires data from many different short-term experiments that are not simply measurements of the states of interest. Validation of a simulation model is most efficacious when model results are compared with data that have not been used to derive equations or values for parameters used in the model. Such data are extremely valuable and usually rare resources that need to be managed carefully to derive the greatest benefit. Determining which data were used to derive parameter values is often problematic, particularly when parameters and validation data are obtained from the scientific literature. The volume of results obtained from a simulation model can create problems in data management. Model experiments may involve using data sets specific to a scenario being simulated or be tests of variants in the structure of the model. Experiments conducted with a simulation model, particularly experiments involving Monte Carlo methods, can produce large amounts of results relatively quickly. These results need to be saved for analysis and linked to the input data and model code for the experiments from which they were produced. Finally, models are subject to frequent revision and are often implemented in more than one version to enhance portability across platforms or to optimize the structure of the model for specific uses. Documenting differences between versions of the code and maintaining a history of changes is an important aspect of data management involving simulation modelling.

Introduction

Data management activities in ecology and database systems usually are concerned with the collection, storage, retrieval and analysis of field data.

Empirical models are used to help describe field data, identify relationships that exist among the data items and make interpolations within the domain of the data. Simulation models are frequently used to develop an understanding of the way systems function and to make extrapolations in time and space. The data management tasks associated with the development, use and analysis of simulation models are more extensive, and often more complex, than the tasks associated with the development of statistical or other empirical models. Management of data in simulation modelling can be a formidable task. Understanding the nature and extent of the data management tasks is an essential part of an effective simulation project.

Contrasting Empirical Models and Simulation Models

An empirical model is a mathematical representation of a set of data. The representation can be an arbitrary function, such as a polynomial, with coefficients chosen to provide a good fit to the data, or it can be a function based on laws or principles, such as exponential decay of radionuclides. Empirical models are based on the assumption that there exists a predictable relationship between one set of variables and another.

Functions can be fitted to data using direct numerical techniques, such as fitting a polynomial of order $n-1$ to n data points, or using iterative search techniques (Press *et al.,* 1988). In most statistical regression models, data are divided into sets of one or more independent variables that are used to predict one or more dependent variables. The models are usually linear equations that can be fitted using direct methodologies to minimize the sums of squared differences between the model predictions and the data. Non-linear models are usually parametrized using iterative search techniques. The units associated with the parameters and the distribution assumed to be exhibited by the parameters are implicitly defined by the data and the statistical methodology.

Time series models include time as an independent variable and are used to fit an equation to a set of measurements that were collected over a span of time. The equation typically consists of a weighted sum of sine or cosine functions that have coefficients chosen to provide a good fit to the data.

Empirical models are useful tools for describing data sets, particularly when interpolations are needed. Empirical models are not useful for extrapolation beyond the domain of the data used in their construction except when they are based on laws or principles (Zar, 1984). Time series models can be used for forecasting if one can assume that the process giving rise to the time series of data will continue to operate in the same way in the future. A strength of statistical models is that inference within the domain of the data is strong and a probabilistic measure of uncertainty can be derived.

Simulation models typically use mathematical expressions to describe the dominant processes that affect the behaviour of a system. In compartmental

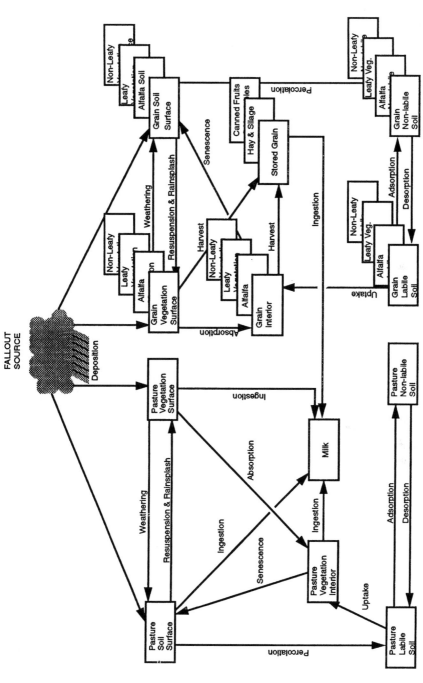

Figure 22.1 A compartmental simulation model represents the system of interest as a series of compartments and flows linking those compartments.

Note: The compartments contain material or energy. This diagram shows the PATHWAY radionuclide transport model.

simulation models, a system is visualized as a series of compartments linked by flows (Figure 22.1). The compartments are abstract or idealized containers of matter or energy representing one or more components of the real system, such as the biomass of plants on a square metre of land surface or the amount of contaminant in the tissues of a plant. The rate of flow between compartments is represented using differential equations, difference equations or discrete events. Simulation models may also use partial derivatives to represent the flow of material or energy in a continuous medium, such as the flow of water through soil. In any case the simulation requires the iterative solution of the rate equations or the simulation of events over the time span to be simulated. The length of each time step is determined by the type of equations being solved and, with differential equations, the accuracy with which the equations are to be solved.

Some simulation models incorporate stochastic processes into their structure. Stochastic processes are implemented by generating values for parameters periodically using pseudorandom number generators (Naylor *et al.,* 1968). The processes represent stochastic events such as precipitation events or the death of individual organisms. Models that make use of stochastic processes need to be run many times to generate many realizations of its predictions. Summaries, such as the mean and variance of a prediction, are used as the output variables of the model. However, analysis of the behaviour of the model often requires that the output from each realization, frequently at each time step, be saved for examination after the completion of the simulation.

Simulation models can be used successfully to make extrapolations in space and time because they are based on representing the dynamics of systems. However, simulation models, in general, do not have the inferential power of statistical models. The applicability of a model, its sensitivity to required input data, the quality of its predictions and the uncertainty surrounding its predictions must be based on extensive analyses. In the absence of such analyses the inference that one can draw from the model's predictions should be considered weak at best.

Data Management for Simulation Modelling

In simulation modelling it is useful to divide data requirements into three groups: data required for constructing the model, running the model and analysing the model. In addition, simulation models often produce large amounts of output, and these output data need to be managed.

Model Construction

Each process within a simulation model can be based on a unique set of data or, in some cases, a mechanism or process that is assumed in the absence of data.

Ecological simulation models are likely to have tens to hundreds of processes. The functions representing the processes can be based on theory or postulates or are determined empirically. Data used to parametrize a process may be in the form of replicated or repeated measurements from an experiment or a mean and an estimate of uncertainty using information derived from published literature.

In a simulation model a parameter is a component of a function, hence it is directly associated with the model's structure. A parameter is more than just a coefficient; a parameter will have explicitly defined units and should have an estimate of the uncertainty associated with it. The uncertainty could be expressed as a range of potential values, as a set of discrete values or as a distribution of values.

In a traditional structured approach to creating a simulation model, using languages such as Fortran or C, the structure of the model is represented in the code, and the code is designed to take a set of data as input, process that data and produce a set of data as output. In an object-orientated design the representation of a system is decomposed into a set of objects. The data and the methods or procedures associated with an object are encapsulated within the object, and objects can be created and destroyed as a simulation proceeds. The data used by an object are often made accessible only to procedures within the object. The differences between the object-orientated and structured-design paradigms have significant implications for data management.

The ability to create objects dynamically is particularly well suited to models representing the dynamics of individual-based models, to models that simulate the change in community structure during succession or due to disturbance, and to other models whose structure may change throughout simulated time or are based on scenario information for the system being simulated. As an example of a model with dynamic structure, consider the simulation of radionuclides through an environment. Some radionuclides, such as ^{131}I, undergo a simple decay into a stable form. Other radionuclides decay to form new radionuclides, which in turn decay. Such radioactive decay chains can be quite long and complex (Kocher, 1981). The object-orientated paradigm enables one to construct a basic framework for a simulation model of radionuclide transport that includes the capacity to add, at the time of execution, the structure necessary to simulate the daughter products (Figure 22.2). Thus much of the structure of the model can be determined at the time of execution. Information required to expand the structure can be specified as input to the model. The expansion, in turn, can affect what additional data need to be accessed by the model.

Errors in the database that define how expansions are to occur can greatly influence the results but nevertheless be difficult to recognize. For example, an error in the specification of a radioactive decay chain may cause several daughter products to be eliminated from consideration and produce dose estimates that are too low. An error of this type cannot be trapped as easily as, say, a parameter's value that is out of range or has inappropriate units. Therefore, careful verification of a database that influences the structure of a model should be undertaken.

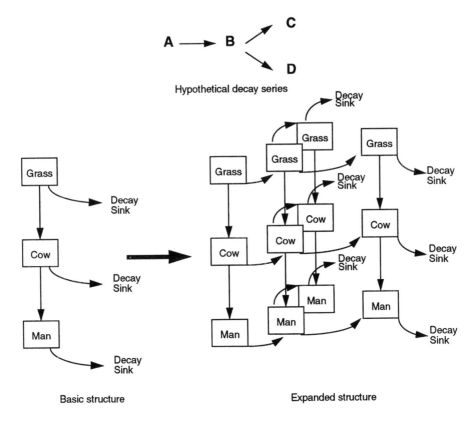

Figure 22.2 Object-oriented languages enable the construction of models which can expand their structure automatically during execution.

Note: For example, a radionuclide transport model could be designed as a basic template, as shown on the left. If the model is to simulate a radionuclide that has a chain of daughter products it would dynamically add the compartments and flows to produce the structure on the right. Data that describe how to expand the structure and parameter values for the flows would need to be obtained from a database.

Input Data

The input data for a simulation model can be classified as input parameters and driving data. Input parameters are used to specify coefficients for the model equations that are specific to the situation or scenario being simulated. For example, input parameters may define the kinds of species being simulated in a community or the kinds of contaminants to be considered in a model for risk analysis.

Driving data for a model are usually time-dependent values representing factors in the environment that affect the system and roughly correspond to the independent variables of an empirical model. Typical driving data for ecological simulation models are precipitation, temperature, wind rose information and other weather data; nutrient inputs; and data about management activities such as harvest rates and schedules. A basic assumption is that the behaviour of the system depends on the factors represented by the driving data, but that the driving data are at most only weakly influenced by the behaviour of the system. Driving data are often dependent on the site and scenario being considered. Unlike empirical models, obtaining new independent data for use with a simulation model does not necessarily require that the coefficients of the model be recomputed. Indeed, a simulation model that requires its parameters to be calibrated each time it is applied to a new situation may have limited utility.

The shift from a structured, procedural design of models to an object-orientated design can impose challenges for managing input data. In a traditional structural approach to building a simulation model, the model structure is static and defined completely prior to running the model. Thus, all of the data requirements for the model can be anticipated and specified. In addition, the input of data required for the simulation can often be centralized into one or few procedures.

An object-orientated design poses problems for accessing data required by the model and for storage of results. Typically the data required for each object need to be imported into the model as the objects are created or instantiated. Thus, the input of parameter values is highly distributed rather than centralized and can occur throughout the simulation as new objects are created. It is essential that the data obtained be accurate, that parameter values have the proper units and that appropriate action can be taken should an inconsistency or error be detected. These requirements suggest that object-orientated models are likely to require access to a structured database system rather than relying on simple input files, as is the case with most non-object-orientated models.

In designing an interface between a model and a set of data, the efficiency of access is an important consideration. Input and output (I/O) processing is relatively slow on most computers and thus has the potential for becoming a bottleneck in performance. The access of driving data may be particularly critical, especially in stochastic models or when Monte Carlo simulation for model analyses is being performed, because the number of accesses to the data can be high. The use of unformatted (binary) files should be considered because they avoid the overhead of translating character-based representations of numbers into the internal binary format required for use, and they can represent the full precision of any number with a few bytes of storage.

Efficiency is somewhat less important when accessing parameter values because access is often only required at the beginning of the simulation. However, it is critical that the units of parameters imported from a database match the units assumed for the variable in the expression where it is used. Although this may seem obvious, a mismatch in dimensions is a common source of error when running simulation models. Thus, one should consider implementing a scheme to verify that the dimensions on all variables are correct.

Model Analysis

The purpose of model analysis is to help one evaluate its predictive capabilities. Some authors prefer to speak of simulation models providing characterizations of systems rather than predictions, particularly when the inferential capability of a model is weak or the uncertainty of its estimates is high. In this chapter all model output will be referred to as constituting prediction of the model.

The analysis of simulation models is usually divided into four types of procedures: uncertainty analysis, sensitivity analysis, validation and verification. Each of these analyses is designed to answer different questions about the performance of a model.

Uncertainty Analyses

Parameters and inputs used in most ecological models have uncertainty associated with their estimates. The uncertainty may arise from a lack of knowledge about the parameter, from natural variability or from measurement errors. The process of propagating uncertainty in parameters and inputs through a model to estimate the uncertainty in the model predictions is called uncertainty analysis. The uncertainty in model predictions may be represented by parameters for a standard statistical distribution (such as the mean and variance or geometric mean and geometric standard deviation), by an empirical distribution or by a confidence interval. Uncertainty in model predictions may change through simulation time or across the domain of applicability for a model (Breshears *et al.*, 1989). The level of uncertainty in predictions may also change as the quality of the input data for a model changes.

There are three commonly used ways to propagate uncertainty through models: analytical methods using mathematical statistics, mathematical approximation techniques and Monte Carlo methods. Analytical and mathematical approximation techniques are usually only feasible for relatively small models of limited complexity. Discontinuities in model functions or the use of time-dependent driving variables make the analytical and delta methods difficult or impossible to use.

Monte Carlo methods can be applied to models of any size. However, Monte Carlo methods involve large numbers of simulations, and thus may require significant amounts of computer time. The Monte Carlo method of uncertainty analysis involves choosing parameter values using a stochastic selection scheme. The distributions can be sampled using simple random sampling or using a stratified design such as Latin Hypercube Sampling (McKay *et al.*, 1979). Typically, parameter values are selected prior to the start of each simulation. Numerous simulations are run, and the values of the output variables for each of those simulations are saved for later analysis. The values of the output variables may be saved at one or more times during the simulation.

The difference between the results obtained from the uncertainty analysis of a deterministic model and from a stochastic model is primarily one of interpretation. In a stochastic model the processes are envisioned as being stochastic and are simulated by sampling parameter values throughout the simulation. In uncertainty analysis the parameter values are assumed to be constant but not precisely known. The variation represented in model predictions from an uncertainty analysis represents the level of uncertainty associated with the predictions. In terms of data management either situation has the potential to produce a large quantity of output that must be managed for efficient storage and retrieval.

Sensitivity Analyses

In statistical models the importance of an independent variable is usually assessed by its explanatory power, that is, by how much of the total variation in the dependent variable is explained by the variation in the independent variable. Techniques such as regression and analysis of variance can be used to provide the statistics needed for ranking importance.

In simulation modelling, sensitivity analyses are used to establish the sensitivity of model predictions to the values of the parameters used in the model. Sensitivity analyses can be classed as global or local analyses. A local sensitivity analysis estimates the effect of a small perturbation in parameter values on the model prediction. The model is said to be sensitive to those variables in which a small perturbation produces a large change in the model predictions. Local sensitivity analyses are equivalent to determining the partial derivative of the response surface of the model at some point of interest in parameter space. Because of the complexity of most ecological models, local sensitivity analyses usually rely on numerical techniques using a factorial design (Steinhorst *et al.,* 1978; Rose, 1983).

Global sensitivity analyses are designed to partition the uncertainty in model predictions among the various inputs and parameters of a model. A parameter to which a model is sensitive accounts for a relatively large portion of the uncertainty in the predictions. Although it is theoretically possible to partition all of the uncertainty in model predictions using a factorial design, the number of simulations required usually prohibits such an approach. Instead, statistics such as partial correlations or partial rank correlations are often used to estimate the partitioning of the uncertainty (McKay *et al.,* 1976; Gardner *et al.,* 1982). This approach has the advantage that the Monte Carlo simulations used to estimate uncertainty can also be used to estimate the global sensitivity of the model predictions to the inputs and parameters.

Whether a factorial, stratified or random sampling scheme for sampling parameters is used, it is necessary that the values of the parameters used for a simulation be saved in such a way that they can be associated with the predictions of the model resulting from their use. In general, maintaining such a pairing of inputs and outputs is not a problem. However, it is not uncommon in performing model analyses to have selected a set of parameters that cause the model to terminate abnormally. This situation must be considered when setting up the modelling runs and appropriate steps taken to recover from the error, flag the error and maintain synchronization between inputs and outputs for the remainder of the simulations.

Validation

Validation is a term that is somewhat ambiguous among modellers. One definition of validation restricts the term's use to comparison of model results with observations from the system. Many consider a model 'invalid' unless it can be shown to predict results that compare well with observations. At the other extreme, validation is associated with the acceptability or usefulness of a model independent of whether its results compare favourably with observations; a model is 'valid' if it satisfies the objectives for which it was built. Forrester (1968) argues that all models are valid if for no other reason than they formalize and make explicit conceptual models in ways that facilitate critical evaluation.

Naylor and Finger (1967) recommend using a three-stage approach to validating simulation models. Three stages are associated with the methodologies of rationalism, empiricism and positive economics. Law and Kelton (1982) summarize the three-step approach as:

(1) establish the face validity for the model,
(2) test the assumptions used in the model and
(3) evaluate how well predictions used in the model correspond to observations from the system.

A model with high face validity is one that seems reasonable to people who are knowledgeable about the system being simulated. Face validity is established through the review by experts in the field of study of the structure of the model, the form of the equations used in the model and the values of parameters used in the model. Understanding the formulation of a model is necessary to evaluate the potential limitations of the model due to inherent constraints or assumptions.

Face validity is often achieved through publishing the model in refereed journals and by other peer review activities. To be effective, the review should include examination of the code as well as the mathematical notation for the equations. In addition, the source from which parameters were derived, units of the parameters, estimates of uncertainty on parameters and references related to functions used to represent processes should all be documented. Such documentation can be considered metadata for the model. Metadata are the information required to document a data set completely and properly. The metadata for the code for a model give the explanation or justification for the formulations and assumptions used in the model, the mathematical representation of the expressions, the appropriate units for parameters used in the expressions, citations of the literature used to develop the formulations, etc. The practical aspects of establishing face validity overlap with the verification step in model analysis, discussed in more detail below.

Models invariably contain assumptions. These assumptions should be made explicit in order to review the model fairly. When possible, the assumptions should be tested empirically. When empirical tests cannot be conducted it is frequently useful to examine the effects of substituting alternative mechanisms or structures. For example, it is often useful to demonstrate the effect of adding more or less structural complexity to a model.

Testing the assumptions used in a model often involves modifying the code for the model, the parameters used by the model, or both the code and parameters. To verify that the model experiments were set up and carried out correctly, it is essential that the changes be documented well and that the results of the experiments be associated explicitly with the resulting predictions.

The efficiency and quality of comparisons between models can be enhanced by using specialized sampling schemes. These sampling designs are designed to reduce variance in the results. Two common techniques are antithetic sampling and common random numbers. These sampling designs force correlation between input variables: negative correlation in the case of antithetic sampling and positive correlation in the case of common random numbers.

Antithetic sampling is a sampling scheme designed to reduce the variance in the estimate of the mean of an output from a model. The scheme requires that simulations be run in pairs and, for each pair, sampling is done to ensure that a high value of a parameter for the first simulation is paired with a low value from the second simulation. The outputs of the pair of simulations are averaged to produce a single estimate.

Common random numbers is a sampling scheme designed to reduce the variance in the difference between paired samples from simulations involving two models, or from one model running under two different sets of conditions. Random numbers are synchronized so that a random number used for a particular purpose in one system is used for the identical purpose in the other system.

Both of these sampling schemes require that attention be given to creating the input values and saving the results. This is sometimes problematic because of the nature of the pseudorandom number generators used to create the parameter values. Frequently all of the random number generators used in a model draw on a single uniform random number function. Whereas this function will always return the same sequence of values, the values assigned to a parameter in, say, the first simulation of one model may be different from the value assigned to that same parameter in the second model if the two models have different numbers of parameters being sampled.

A common method to ensure that the input values are paired correctly in the case of antithetic sampling, and that parameters are assigned the same sequence of values in the case of common random numbers, is to generate the parameter values prior to all simulations. The values are imported into the models as required.

The quantitative comparison of model predictions with observations from the system is most critical for models that are to be used as predictive tools. The types of quantitative tests that can be applied to the comparison of model results and empirical observations are dependent on the type and amount of data available (Naylor and Finger, 1967; Kirchner and Whicker, 1984). Tests that allow for statistical inference with regard to validation are not well developed for the kinds of data frequently available in ecology.

A fundamental principle for validation is that the data against which model predictions are tested must not have been used in the construction of the model. The independence of the validation data is necessary to avoid the criticism that the model should be expected to fit data to which it has been either explicitly or implicitly calibrated.

One of the important limitations in model evaluation is the frequent lack of data to use in quantitative model validation experiments. One reason that models are popular for the analysis of behaviour of systems is that experimental manipulation of many systems is impossible owing to the complexity of the system, the cost of obtaining data or regulations prohibiting experiments. For these same reasons, data that can be used in validation experiments are often uncommon. Data that do exist are often incomplete in the sense that only a limited subset of model outputs may be represented, and only some or none of the required input data may have been collected. Also, it can be difficult to ensure that the data that are available are truly independent of the data used to build the model. If the data are obtained from published values, one should verify that these data were not also used to provide estimates of parameters in the model. Tracking the exact sources of data used to provide estimates for parameter values can be a difficult undertaking. For this reason, the use of data collected after the completion of the model is often the best option.

In a well-designed validation experiment it is also most efficacious to sequester all data to be used for comparisons. Sometimes even seeing a particular pattern in a set of validation data is sufficient for a modeller to reconsider what values should be used for some parameters, or perhaps how a process should be modelled. In a 'blind' validation experiment, the validation data are identified, verified and protected by one or more people not directly associated with setting up the model for the test. These data archivists are also responsible for providing to the modellers the information necessary to run the model experiment, such as specific input data or scenario information.

In two series of such experiments with which I am familiar, the BIOMOVS program and the VAMP program, significant differences in the interpretation of the information about the inputs and scenarios led to widely different results being predicted by the models. After seeing the validation data several modellers were able to identify an error in their interpretation. A repetition of the experiment using the modified interpretation invariably reduced the amount of variation in the predictions across the models. However, it could no longer be said that the validation data were strictly independent of the results.

Verification

Verification is the process in which the code for the simulation model is checked to ensure that it functions as expected and does indeed accurately reflect the conceptual model. Verification of ecological models is frequently conducted only in a haphazard manner, if it is done at all. One common practice in modelling is to compare model results against data sets that were used to provide parameter estimates for the model. This 'feasibility analysis' simply assures that the model is capable of producing results similar to those observed and is a very weak demonstration of verification.

Of the four types of model analyses, verification is most likely to benefit most from changes in computer technology. The increase in power of computers has led to an increase in complexity in practically all types of computer application programs. The computer industry has responded with new tools to help solve the problems associated with the development and maintenance of large programs.

Computer Aided Software Engineering (CASE) tools are now available on most desktop and larger computer systems. These tools help design, generate code, maintain code and document large software applications. They are particularly helpful in situations where multiple participants are involved in the creation of an application, as is often the case with ecological simulation models. Some of these tools can help with the management of metadata about the model as well. However, they are expensive and are most useful only when the source code is larger than about 100 000 lines, which exceeds most simulation models.

Configuration management tools, such as SCSS and RCS under UNIX or similar commercial products available on DOS, provide a good intermediate level for maintaining and documenting changes to code. These systems provide mechanisms for storing, identifying and recovering versions of programs or other files. The files are checked in and out using a version number. Typically only the changes required to convert one version to the next are stored, so that the system has little impact on disk space requirements as compared with storing complete copies of all versions of the files. Managing the metadata for models is just as important as managing the code and has at least as many problems as managing the metadata for a scientific database (Stafford, 1993). Configuration management tools are usually insufficient for managing the metadata.

A common error in the development and use of simulation models is to create an expression that has unbalanced dimensions. Although a dimensional analysis of each line of code can locate such errors, it is a tedious task. Software for examining code has been used successfully for such analyses (Kirchner and Whicker, 1984). It is also possible to provide overloaded definitions for arithmetic and other functions in object-orientated languages such as C++ to analyse the dimensions of operands during the execution of the code. In all cases, the maintenance of a dictionary for the model variables that includes the units for the variables is essential. The identification of units on the values in all input files is also recommended.

Model Output

Frequently the goal of creating an empirical model is to establish the degree of relationship between items in a data set rather than interpolation or extrapolation. In other cases, such as models associated with GIS, the models may be used extensively to make predictions. In the former case there are few problems associated with managing the output from the model, whereas in the second case the management of output can pose significant problems.

Simulation models always produce results that must be analysed. The amount of output is usually large because simulation models produce a series of estimates throughout the time interval being simulated. It is not uncommon that all of the states estimated in the model be saved even though only a few may be of primary interest in order that analyses of the results can be made. In addition, characterization of the uncertainty on the estimates and of their sensitivity to the inputs can require that results from numerous simulations be saved.

Simulation models are typically used to make predictions rather than just identify relationships within a set of data. When considering even a few factors that affect predictions, the number of simulations required for a full factorial design becomes large. The application of simulation models to spatially explicit data sets also leads to the generation of large quantities of output.

To illustrate several of these problems consider the task of estimating doses from the ingestion of contaminated food to residents surrounding the Nevada Test Site from the nuclear tests conducted there (Friesen, 1985). The analysis required the construction of three simulation models that were linked serially (Figure 22.3). This design was implemented to increase the efficiency of doing the calculations by avoiding repeating calculations. For example, the same series of estimates of biomass could be used in all subsequent computations involving different radionuclides at the same location, so biomass was computed once and stored. In the second step, the **PATHWAY** model was used to provide estimates of concentrations in food products for each combination of events and agricultural scenarios. A deposition of $1\ \mu Ci\ m^{-3}$ was assumed for these calculations and the results adjusted for actual deposition in the third step, which brought in the site-specific factors that affect the dose calculations.

When computing just the estimates of concentrations in food products consumed by people, 86 nuclear tests were considered to be of potential significance. There were 21 radionuclides of concern in the fallout, and 15 sets of different agricultural practices were identified for the region. Thus, 27 090 unique cases had to be simulated. In order to provide estimates of uncertainty, the 86 tests were aggregated by date into 10 categories and 6.3×10^5 simulations were conducted. The third step made use of the estimated concentrations in food products and applied location-specific information about the food consumption rates of the people and the estimates of actual

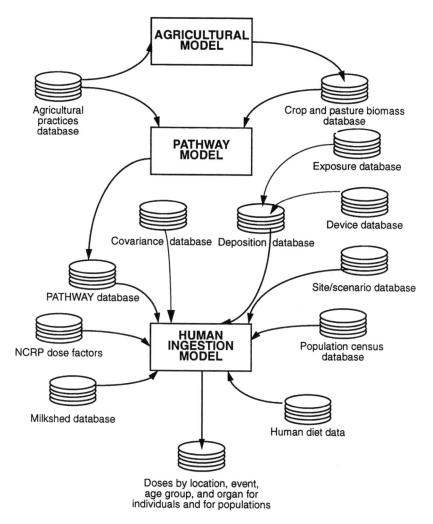

Figure 22.3 Estimating doses to residents surrounding the Nevada Test Site required consideration of agricultural practices across the region, radionuclide-specific transport information, and location-specific information about lifestyles and amounts of fallout.

Note: Three models were used to maximize computational efficiency.

deposition rates to give the sites specific-dose estimates. For the region, those estimates amounted to over 30 000 tables of organ doses by age class with an estimate of uncertainty. Over six million simulations were required to provide those estimates. The total number of data required or generated in this exercise amounted to about 200 megabytes. In a similar dose reconstruction exercise being conducted for the Hanford facility, the storage requirements are expected to be several gigabytes.

Efficiency should be considered when designing the output system for a model. Using unformatted or binary files typically provides the fastest, most precise and most compact way to write data, although the files are not, in general, portable across platforms. Object-oriented models can complicate saving output further because the data to be stored are likely to be distributed among objects, and the number of objects may be variable in some models. Large simulation experiments can produce thousands of sets of output that must be stored, and simply finding a way to identify a data set can be difficult. One can consider making use of the hierarchical file system available under UNIX, DOS and other operating systems if the output is stored in files and if the simulations can be stratified into logical groups.

Summary and Discussion

Simulation modelling poses many problems for data management. Input parameters must have both correct values and units that match those expected in the model. Driving data must also have the correct units and be accessible with a minimum of overhead. Validation data are scarce resources that must be carefully managed, especially when doing blind validation experiments. Such experiments require that careful attention be given to the specification of the input scenarios for the models. Ensuring proper sampling of input data for other model experiments, such as model comparison studies, uncertainty analyses and sensitivity analyses, often requires that input data for the simulations be created prior to running the model and that care be taken to handle properly instances where the model fails to complete an execution due to the particular set of data being used.

Managing the output data for a model can also be a significant challenge. Simulation models can produce large amounts of output, particularly when the model contains stochastic processes or is undergoing analysis involving Monte Carlo methods. Spatially explicit modelling or simulations conducted for a set of several factors can rapidly expand the number of cases to be simulated. Efficient storage of the results is required, and methodologies for identifying the many output files need to be implemented.

Finally, the code for the model is, itself, a form of data that must be carefully managed. Various versions of the model created during development and testing should be maintained for purposes of verification. The assumptions, sources of the equations and parameters, limitations and other commentary about the model are metadata that are essential for understanding the model and verifying that it is constructed properly.

Commercial database systems typically fall short in their ability to accommodate metadata owing to their complex and highly variable content. Using a DBMS to handle textual information, such as the source code for programs and the associated metadata for the model formulation, is also problematic. Configuration management systems fulfill part of the need to be able to store and recover various versions of programs as they evolve, but provide no direct support for linking metadata to the appropriate code. Hypertext systems can be used to provide linkages between code and metadata, but are ill equipped for providing configuration management. Integrated program development environments, such as Hewlett-Packard's SoftBench, link tools to provide solutions to many of the code maintenance problems.

There are two goals that often conflict when setting up a data management system for simulation modelling: providing rapid storage and access of data and maintaining integrity and security of the data. On one hand, efficiency of access is particularly critical for input of driving data and output of results when large numbers of simulations must be run, as in the case of model analysis experiments or large spatially explicit simulations. On the other hand, it is critical to ensure that the model is initialized using the correct values for parameters and state variables and that the correct driving data are used as input.

Traditional relational database systems are often suitable for handling the data management needs for empirical modelling, but fall short of providing the flexibility and the efficiency of retrieval and storage required for simulation models. Object-orientated database systems are becoming available commercially, and these systems have the potential to support some of the diverse input and output requirements of simulation modelling. The advantage of using a commercial system is that they are usually well designed for providing security of the data and for manipulating or extracting subsets of data for analysis. However, these systems usually do not provide rapid access to the large blocks of data such as are required for model input and output.

Considering data management during the design of the code for a model can greatly increase the efficiency of obtaining and saving data. To meet the goal of minimizing execution time of a model, one could consider writing and reading data using files rather than obtaining data directly from a DBMS, and then using a DBMS to manage those files. Using this scheme allows one to optimize the data flow into and out of the model using good programming strategies, and to optimize the management of the data using a DBMS. One method for minimizing the overhead associated with using a DBMS is to handle input and output of data within the model through binary files, and to manage those files as binary large objects (BLOBs). BLOBs are being accommodated increasingly in commercial DBMS in response to the need to handle graphical images but can be used equally well for data files for models.

Reading and writing data to a disk can be responsible for a large fraction of the execution time of a typical model. Data flow can be optimized by considering how to reduce the input/output overhead. The disk subsystems in most computers are most efficient when retrieving or writing data in large blocks, and often perform many functions without the need for direct interaction with the CPU. Thus one can sometimes reduce the overhead associated with data access and storage by moving data between the disk and memory in large blocks rather than one record at a time.

Some operating systems, such as UNIX and OS2, support multitasking or multithreaded processes. A multitasking environment allows several programs to execute simultaneously through time sharing of the CPU or the use of multiple CPUs. A multithreaded program allows two or more asynchronous threads, or tasks, to be executing within a single program. Using separate processes to deal with the storage and retrieval of data, or separate threads within a process for managing input and output of data, can allow the simulation to proceed with calculations. A good design to optimize efficiency is to have a thread or process loading into memory the data needed for a simulation while another thread or process performs other computations required to initialize the model. Likewise a separate thread or process can be used to store results in memory while the simulation proceeds. In a networked computer environment one can use a database server on another machine to extract and assemble data into files that are then made available to the model. When many simulations need to be performed, the model can run one simulation while the server is preparing data files for the next.

The diverse requirements for simulation modelling will necessitate a mixture of data management tools being employed. Such tools include integrated programming environments for program maintenance and documentation, database management systems for maintaining the different kinds of data associated with a simulation model, and traditional file creation and maintenance techniques. The trade-off between efficiency and the reliability of storage and retrieval of data needs to be considered early in large simulation efforts if maximum security for the data and optimal execution time for the model are to be achieved.

Acknowledgement

The research on which this chapter is based was supported in part by grant BSR-8612105 from the National Science Foundation, contract 9XA3-6821 J-1 from Los Alamos National Laboratory, and contract DE-AC08-86NV10503 from the US Department of Energy.

References

Breshears, D. D., Kirchner, T. B., Otis, M. D. and Whicker, F. W., 1989, Uncertainty in predictions of fallout radionuclides in foods and subsequent ingestion, *Health Physics*, **57**, 943–53.

Forrester, J. W., 1968, *Principles of Systems*, Cambridge, Massachusetts: Wright-Allen Press.

Friesen, H. N., 1985, *Offsite Radiation Exposure Review Project Fact Book*, NVO-295, Las Vegas: United States Department of Energy Nevada Operations Office.

Gardner, R. H., Cale, W. G. and O'Neill, R. V., 1982, Robust analysis of aggregation error, *Ecology,* **63,** 1771-9.

Kirchner, T. B. and Whicker, F. W., 1984, Validation of PATHWAY, a simulation model of the transport of radionuclides through agroecosystems, *Ecological Modelling,* **22,** 21-44.

Kocher, D. C., 1981, *Radioactive Decay Data Tables,* Springfield, Virginia: Technical Information Center, US Department of Energy.

Law, A. M. and Kelton, W. D., 1982, *Simulation Modelling and Analysis,* New York: McGraw-Hill.

McKay, M. D., Conover, W. J. and Beckman, R. J., 1979, A comparison of three methods for selecting values of input variables in the analysis of output from a computer code, *Technometrics,* **21,** 239-45.

McKay, M. D., Conover, W. J. and Whiteman, D. E., 1976, 'Report on the application of statistical techniques to the analysis of computer code', Informal Report LA-NUREG-G526-MS, Los Alamos Scientific Laboratory.

Naylor, T. M., Balintfry, J. H., Burdick, D. S. and Chu, K., 1968, *Computer Simulation Techniques,* New York: Wiley.

Naylor, T. H. and Finger, J. M., 1967, Verification of computer simulation models, *Management Science,* **14,** 92-101.

Press, W. H., Flannery, B. P., Teukolsky, S. A. and Vetterling, W. T., 1988, *Numerical Recipes in C,* New York: Cambridge University Press.

Rose, K. A., 1983, A simulation comparison and evaluation of parametric sensitivity methods applicable to large models, in Lauenroth, W. K., Skogerboe, G. V. and Flug, M. (Eds) *Analysis of Ecological Systems: State-of-the-Art in Ecological Modelling,* pp. 129-40, New York: Elsevier.

Stafford, Susan G., 1993, Data, data everywhere but not a byte to read: Managing monitoring information, *Environmental Monitoring and Assessment,* **26,** 125-41.

Steinhorst, R. K., Hunt, H. W., Innis, G. S. and Haydock, K. P., 1978, Sensitivity analysis of the ELM model, in Innis, G. S. (Ed.) *Grassland Simulation Model,* Ecological Studies 26, pp. 231-55, New York: Springer.

Zar, J. H., 1984, *Biostatistical Analysis,* Englewood Cliffs, New Jersey: Prentice-Hall.

23

GIS and spatial analysis for ecological modelling

Richard J. Aspinall

Ecological research has yet to take full advantage of opportunities provided by spatial analytical tools and modelling capabilities linked to geographic information systems (GIS). Some of the tools available (GIS, spatial analysis and ecological modelling) and the opportunities they present are reviewed from the perspective of the relevance and value of the outputs provided for a range of ecological research and environmental management applications. The tools allow questions concerning processes operating at different spatial scales to be addressed, and offer potential for gaining new insights into ecological relationships through detailed analysis of extensive databases stored and managed in GIS. This chapter presents issues associated with ecological and environmental data quality and the use of spatial analysis for converting data into information. Case studies are used to illustrate: (1) the use of GIS for process-based modelling of impacts of climate change on global, national and regional distribution of flora and fauna by generating spatial (geographic) and ecological hypotheses from automated investigation of databases, and (2) the application of landscape ecology principles in GIS for analysis of relationships among spatial heterogeneity, scale and the ecology of different taxa.

Introduction

Ecological research has yet to take full advantage of opportunities provided by spatial analytical tools and modelling capabilities linked to GIS. There are three main reasons for this situation. First, GIS is a relatively new technology that attempts to provide a set of generic tools for management and analysis of spatial data by the main users of GIS. Tools for data management are better developed than those for analysis; this reflects the dominant market for GIS (Goodchild, 1987). Methods are becoming available to link GIS and environmental modelling (Nyerges, 1992; Goodchild *et al.,* 1993) and GIS has been used to generate map outputs from models in ecological research (Grossman, 1991; Miller, 1992, this volume; Prentice *et al.,* 1992; Smith *et al.,* 1992). Notable omissions in GIS functionality include spatial analytical and process modelling capabilities needed for ecological research and related scientific applications (Goodchild,

1991a; Kemp, 1992). To capitalize on GIS data management capabilities and databases, new spatial analysis tools must be developed specifically for application in ecological research.

Second, spatial reasoning has received only limited attention in ecological research, where developments and experimentation have focused on elucidating ecological processes rather than analysis of pattern. This reflects interest in 'why' questions and mechanisms, investigation of the causes of observed patterns of distribution and abundance (Andrewartha and Birch, 1954) rather than analysis of spatial phenomena represented by the patterns of distribution. Dobson (1992) discusses the concept of spatial logic and its relationship with process studies in paleogeography, and has used analysis of both spatial and process logic in analysis of forest blowdown and lake chemistry in the Adirondack Mountains (Dobson *et al.,* 1990). Thus, although ecological research is relatively rich in theory and concepts that relate to processes and ecological functioning (for example, ecosystem science; niche theory, energy flow and trophic structure, and biogeochemical cycling; and population biology: inter- and intra-specific interactions, population regulation and life-history strategies), it is relatively poor in spatial concepts and spatial theory, which can underpin development of spatial analytical tools to enhance the functionality and relevance of GIS for use in ecological research. Levin (1992) has argued that the problem of pattern and scale is the central problem in ecology, providing a unifying concept and linking basic and applied ecology. Interaction of spatial and temporal scales with each other and with phenomena at different levels of ecological organization provides a framework for analysis and synthesis of many ecological problems, within which methodological and technical analytical tools can be identified and developed. GIS offer facilities for description and management of spatial environmental and ecological data, and, with appropriate tools, have the potential to be used to analyse and synthesize interactions and variability at different levels of spatial and ecological organization. For example, Walker and Walker (1991) have used a GIS to investigate questions related to energy development and climate change for the North Slope in Alaska. The GIS database is organized hierarchically, based on spatial and temporal scales of resolution and natural disturbance regimes (Delcourt *et al.,* 1983; Delcourt and Delcourt, 1988). The relationship between data sources and the appropriate scale and topic of investigation emerges as an important technical and methodological issue.

Third, the growth of interest in broad-scale (geographic) patterns and processes, global ecology (Southwick, 1983), and global geography (Calder, 1991) has taken place only within the past decade (Mooney, 1991). An explicit interest in the interaction of pattern and process for understanding global questions (Shugart, 1993) has highlighted the need to address problems associated with aggregation, simplification and scaling (Addicott *et al.,* 1987; Magnuson *et al.,* 1991; Levin, 1992; Simmons *et al.,* 1992). The opportunity to analyse remotely sensed data and increasingly large databases that are relevant to environmental resources (Skole *et al.,* 1993) and human activities allows

Table 23.1 Specification of classes of data quality and spatial analysis functionality required for use of GIS and spatial analysis in ecological research and modelling.

Class of Function	Purpose
Data Quality:	
Metadata description and analysis	Description and analysis of uncertainty and data quality
Error analysis	Management of error in analysis; error propagation; confidence limits for output
Scale, resolution, grain domain	Identify the appropriate application and scale domains for datasets
Spatial analysis:	
Data generalization and aggregation	Methods for generalizing spatial and object attributes of data
Data enhancement	Methods for generating synthetic data with finer basic spatial units
Pattern description	Methods for describing spatial pattern and spatial interaction in ecological distributions, processes; description of linkage between process and pattern
Pattern analysis	Methods for analysis of spatial patterns and spatial interaction effects; methods for generating spatial and ecological hypotheses about pattern–process relationships at a given spatial scale based on spatial and process logic
Pattern generation and hypothesis testing	Methods for generating spatial patterns from stochastic pattern-generating processes and comparison with observed distributions
Description and analysis of scale influences	Methods which translate between scales (scaling up and down) in data and models
Interfaces with models	Methods which allow location and spatial interaction effects to be interfaced with models of ecological process

questions relating to pattern and process at regional, national and global geographic scales to be addressed (Wallin *et al.*, 1992; Solomon and Shugart, 1993).

The interaction of pattern and process represents a theme which underpins the spatial analytical tools discussed in this chapter. Some of the characteristics of spatial analysis tools that are required for ecological research and modelling are presented in Table 23.1. Any specification of spatial analysis tools must also address issues associated with quality of spatial data (Goodchild and Gopal, 1989; Veregin, 1989; Chrisman, this volume), including propagation of uncertainty in data analysis (Openshaw, 1989), and identifying the appropriate scale for application of data given the level of abstraction that is frequently associated with data capture. The variable quality of spatial data can be an impediment to analysis since few methods are widely available that adequately manage the uncertainties present in spatial data throughout the analysis.

Spatial Analysis

Spatial analysis recognizes the location of objects and classes of objects. Goodchild (1987, 1991b) has identified six general classes of spatial analysis, although it is perhaps helpful to distinguish two categories of spatial analysis based on treatment of spatial effects: (1) statistical analysis of spatial data, and (2) spatial analysis.

Statistical analysis of spatial data recognizes location as a factor producing effects through spatial interaction, but the procedures developed control for these effects in application of analytical techniques and interpretation of results. This class treats the geography of phenomena as an external influence rather than allowing spatial relationships to have explanatory power (Openshaw *et al.,* 1987). Jeffers (1991) has identified a series of model families used in ecological research, many of which are based on statistical methods. Although none of these is explicitly spatial, specific implementations of any of the types may control for spatial effects and would therefore fall into this category.

Spatial analysis recognizes location as a property and context of data, and ascribes explanatory power to location and spatial interaction effects. Openshaw *et al.* (1987, 1990) have discussed the nature of spatial data and associated generic technical problems for spatial analysis in the context of database exploration and spatial hypothesis generation. The use of non-experimental spatial data (e.g., remotely sensed data or maps of environmental properties) presents fundamental problems for inference making and hypothesis testing (Leamer, 1978) and requires development of techniques tailored to the properties of spatial data and the purpose of analysis. Environmental databases used for research in global ecology are, however, generally suited to inductive methods for exploratory spatial data analysis in which pattern analysis is used to infer possible mechanisms and processes in the form of testable hypotheses (Openshaw *et al.,* 1987, 1990).

Case Study 1: Spatial Pattern Analysis to Generate Spatial and Ecological Hypotheses

A simple class of spatial analysis is the comparison of two maps (Unwin, 1981). This form of analysis has application in bioclimatic analysis, which has its basis in ecological theory describing the relationships between species distribution and climate (Woodward, 1987), and is of particular interest in the context of modelling impacts of global climatic changes. Bioclimatic analysis aims to characterize the climatic environment of a species to gain an understanding of possible causes of species distribution (why is a species where it is?), and predict the theoretical limits of distribution (where might a species occur?) (e.g., Lindemayer *et al.,* 191). It may be carried out by comparing a map of the distribution of a species with a map (or maps) of climatic variables to search for pattern matches. The two basic questions of bioclimatic analysis also define the outputs required: (1) quantified relationships between climatic variables and the response of the species being investigated (why is the species where it is?) and (2) maps indicating the likely distribution of the species or the climatic regime in which it occurs (where might it occur?). Spatial analysis can be used to address these questions, and a range of statistical and spatial methods have been used (Caughley *et al.,* 1987; Walker and Moore, 1988, Lindemayer *et al.,* 1991; Walker and Cocks, 1991). The data available for bioclimatic analysis and their application in modelling impacts of climate change place bioclimatic analysis firmly among topics requiring regional, national, continental and global scale study. The method is illustrated for Scots pine (*Pinus sylvestris*) in Scotland (Figures 23.1 and 23.2).

Data

Archive data on species distribution are an important resource for research on climate–species relationships and for predicting impacts of climate change (CCIRG, 1991). In Britain, grid cell data (10 km × 10 km; referenced to the UK National Grid; Figure 23.1 — inset) are available as a series of national atlases for a wide range of flora and fauna (e.g., Perring and Walters, 1962). Data describe presence, but not absence, within the grid square. Climate and other environmental data with a range of source scales are available as paper and digital maps, as well as from computer models of resource distribution. Monthly mean, maximum and minimum temperatures and rainfall records are available from meteorological stations in Scotland. From these records, baseline climatic data, as 30-yr means for the period 1951–80, are calculated for each month of the year and digitally mapped in a GIS. The approach combines trend surface analysis and point kriging with a digital elevation model to incorporate topographic effects (Aspinall and Miller, 1990; Aspinall and Matthews, in press). This produces climate data as continuously varying surfaces with a spatial resolution of 1 km. The approach also provides metadata describing the errors associated with the estimates of climate variables, these metadata providing an input to analysis of error propagation.

Analysis

The difference in resolution between the species data (10 km) and climate data (1 km) offers analytical opportunities (Openshaw, 1989), and is the basis for generating the hypotheses that describe bioclimatic relationships. Output has the

Figure 23.1 Model output of the climate regime of Scots pine in Scotland.

Note: Areas where Scots pine is more likely to occur are shown as darker tones. The map has a spatial resolution of 1 km. The small inset map shows the distribution of Scots pine from Perring and Walters (1962) with a 10 km spatial resolution on which the model is based.

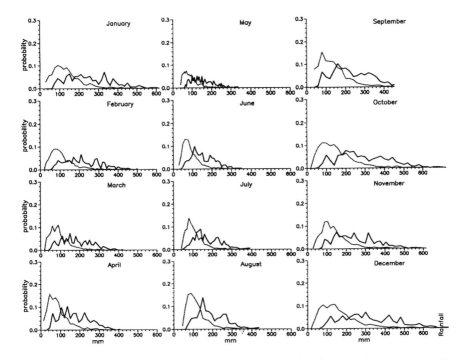

Figure 23.2 Graphs showing relationships between the distribution of Scots pine and rainfall (unbroken lines).

Note: The results for analysis of rainfall in each month are shown; these are spatial analyses, the presentation of results for each month indicating which months/seasons show significant agreement between patterns. For Scots pine, there is a pattern agreement for every month of the year. The dotted lines show the relationships with rainfall that would be expected if the species distribution was random. The relationship is measured as conditional probabilities.

spatial resolution of the fine data set (1 km^2), and the method can generate synthetic data of enhanced spatial resolution (Table 23.1). The analytical method used is based on Bayesian probability statistics and has been developed as a process for inductive exploratory spatial data analysis in GIS (Aspinall, 1992a). Inputs to a Bayesian model are calculated as conditional probabilities from an inductive learning process in which attributes of the data set to be modelled are compared with attributes of the predictor data sets. Random subsets of the data are used to generate error bounds for the analysis (Aspinall, 1992a).

Analysis uses the difference in resolution between the dependent (wildlife) and independent (climate) data. Coarse-resolution units in the dependent data allow multiple (stochastic) realizations of species presences to be generated at the finer spatial resolution of the independent data: 1 km^2 units within the 10 km^2 where the species is present are selected at random for comparison with the climate data.

Climatic conditions in the set of locations where the species is present are compared with the climatic conditions that would be expected if the species had a random distribution in Scotland; the differences are tested with chi-square. An advantage of this approach is that no prior knowledge of the expected probability distribution function is required. Since climate variables are represented in the database as surfaces of continuous variation they are simplified during analysis by applying class intervals. For each data set the class interval is increased gradually and the narrowest interval that produces a significant association is identified. This retains the maximum information content of the climate data. Wildlife–climate relationships that are significantly different from the random case are defined as conditional probabilities for presence, and are stored and graphed (Figure 23.2) to relate wildlife presence to gradients in the climatic data (Giller, 1984). These form hypotheses describing the quantified relationships between climatic variables and the response of the species being investigated. Multiple realizations allow uncertainty estimates to be generated for model parameters, and these are used in analysis of error propagation. A map indicating the likely distribution of the species (expressed as the climatic regime in which the species occurs) is produced by combining climate data sets using conditional probabilities for presence and the random distribution in Bayes theorem. This provides the second output required ('where might it occur?').

Error Analyses

Highlighting the need for adequate error analyses to accompany the main outputs are the following: the applied context, inductive approach, variability (and low frequencies for some parts of the variable range) of conditional probabilities for presence and randomness, mathematics of rescaling in Bayes theorem (error propagation), assumptions made in order to generate the multiple representation of conditional probability and uncertainty in the input data. Error analysis consequently is an important element of the method, and is integral to derivation, interpretation and use of the results of modelling. Two error analyses are performed. First, an iterative procedure in the model generation process is used to test the assumption of species 'presence' through-out the 10-km spatial unit of the atlas data set. Second, Monte Carlo simulation is used to determine sensitivity of the model to uncertainty associated with input data (climate) and model parameters (conditional probabilities) propagated through the analysis (Openshaw, 1989; Aspinall, 1992a).

The assumption of species presence in all 1 km^2 of each 10 km^2 is tested by generating conditional probabilities in a second run of the model but allocating 1 km^2 realizations of presence in 10 km^2 to either presence or random classes based on the results of the first model. Reallocation of the random class is not carried out. The rationale for this is that the 10 km^2 in which a species has been recorded may contain both suitable and unsuitable environmental conditions for that species — the species does not need to occur ubiquitously or uniformly throughout the square but only somewhere within it. The model is updated based on its first estimate. This indicates geographic areas in which the model is sensitive to assumptions made in the analysis.

The approach has been enhanced by adding a point pattern analysis procedure, the Geographical Analysis Machine (Openshaw *et al.,* 1987), which identifies spatial subsets in the species distribution data. These spatial subsets are analysed using the method already described, to identify any local (regional) differences in the species–climate relationships.

Outputs and Applications

The graphical and tabular (probability) results (Figure 23.2) represent hypotheses concerning ecological relationships between the species and its climatic regime. Maps are used to visualize this information (Figure 23.1) and the associated error sensitivities; computer displays allow greater flexibility than conventional paper maps for display and error visualization. These outputs describe and locate the climatic regime in which the species might be expected to occur and also define ecological relationships between species and climate. The approach makes effective use of biogeographic data and environmental data sets in GIS, these data being an important resource whose use and interpretation can provide valuable insights into climate–species relationships. The method presented is one way of gaining such insights, and the outputs generated provide useful information for a range of applications, including assessing impact of climate change and automated generation of testable hypotheses about process-based species–climate relationships. More generally, the approach has potential for hypothesis generation from spatial data and as a tool for inductive generation of synthetic spatial data (fine-resolution spatial data from coarse-resolution inputs) with known error tolerances.

Case Study 2: Spatial Analysis in Landscape Ecology

Spatial analysis can also be used to investigate spatial relationships within a single spatial data set. This offers particular potential for application in landscape ecology and for the study of species–environmental interactions at regional scales. In this case, spatial analysis is used to interpret and describe geometric properties of landscapes, leading to understanding of the complex interactions between different landscape components and linking description of environmental conditions with population biology (Levin, 1992). The approach is founded in the framework provided by landscape ecology (Forman and Godron, 1986), which provides theoretical and analytical approaches to integration of environmental and biological features across a range of spatial scales. The main principles are directed towards structural, functional and dynamic elements of landscape. Structural elements of landscape include patches and corridors of habitat and the environmental matrix in which patches and corridors are located. Measures of these elements describe their size, shape, number, type and configuration (contiguity and connectivity). Although structure is three-dimensional (McCoy and Bell, 1991; Flather *et al.,* 1992), few GIS can operate on three-dimensional data (Raper, 1989), and attributes describing vertical components are seldom attached to the horizontal components of data described in GIS. The analysis here deals only with structure in the horizontal plane, reflecting both a more mature development of GIS technology for 2-D analysis and applications than for 3-D, and the nature of spatial data collection, which is concentrated on the horizontal plane. Functional aspects of landscape ecology focus on movement and use of the structural elements by organisms. Since distribution of a species in a landscape is a product of both landscape structure and social behaviour, both environment and species ecology can be considered under functional factors. Temporal (dynamic) aspects of ecological systems include the dynamics of landscape use by organisms and changes in landscape structure through time. A range of temporal scales can be incorporated within the analytical and conceptual framework provided by landscape ecology although a static model is produced here.

These principles of landscape ecology are readily applied to the question of analysis of structure in habitat maps using spatial methods. The use of spatial analysis to describe and analyse landscape structure and spatial function is illustrated here for two species associated with upland moorland and grassland areas in Britain. These are curlew (*Numenius arquata*) and golden plover (*Pluvialis apricaria*), and the study area is about 30 km × 50 km centred on Balmoral, Morven and the Cabrach in Grampian Region, north-eastern Scotland.

Data

The distribution of these species was recorded from a bird survey reported on a 1 km^2 grid basis. One hundred and fifty-two 1 km^2 areas were visited between 8 April and 30 June 1988 and birds counted using transect survey. Survey counts are reported as counts of birds in each 1 km^2 of the GB National Grid; absence is recorded. Fledged young are not included in the counts. Curlew were recorded in 76 of the 152 sampled 1 km^2 during 1988, numbers varying between 1 and 13 in these squares. Curlew were absent from the remaining 76 squares at the time of the survey. Golden plover were reported from 91 squares and absent in 61. During 1989, a second survey was carried out in 68 randomly selected squares. In these, curlew were present in 14 and absent in 54 squares.

The habitat data are a subscene from a LANDSAT Thematic Mapper image from 17 April 1987; the subscene covers the moorland area of the bird survey. The image is geometrically corrected to the National Grid, and georeferenced with a digital elevation model and the bird survey data in an ERDAS image processing system. Since the bird survey data refer only to the upland grassland and moorland area, agricultural areas are masked on the image and excluded from analysis. The digital elevation model is derived from contours digitized from a 1:250 000 scale map; it has a 100-m pixel resolution with absolute vertical specification accuracy of ±30 m (Smith *et al.*, 1989).

Habitat Map

The generation of a habitat map for curlew (Figure 23.3) and a test of the output are described in Aspinall and Veitch (1993). The Bayesian method described for bioclimatic modelling is used to classify the imagery and DEM by recoding based on conditional probabilities calculated from the different resolution data sets. The image and DEM are used in their most detailed form, and all image bands are used. This analysis provides a map at the full spatial resolution of the imagery and contains a link between the coarse resolution of the bird survey data and the fine spatial resolution of the satellite image. It allows the bird data to drive the image classification with regard to the spectral properties of the image and altitude classes expressed in the DEM. The output represents both bird and habitat data. The map generated for curlew shows a strong association between high probabilities of curlew presence and grassland, dry-heath and grass-heath scrub vegetation types. Low probabilities of curlew presence (high probability of absence) are associated with woodland and subalpine vegetation types. An equivalent map was produced for golden plover (Figure 23.4) and identifies the mountain plateau and dwarf-shrub/coarse grassland vegetation types. It should be noted that although this example refers to interpretation of map output from a satellite image classification, any spatial representation of habitat data (for example a categorical map) can be analysed in a directly analogous manner.

Results: Measures of Spatial Structure and Function

Characterizing relationships of structure and function within these habitat data is an exercise in spatial analysis. For this, GIS can be configured to derive a wide variety of quantitative measures and indices: size, shape and quality can be calculated as properties of individual habitat patches (Aspinall, 1992b; Baker and Cai, 1992). Topological relations can be established between patches to model spatial functioning in the landscape as represented by contiguity or regional interactions. Figures 23.3 and 23.4 show a series of maps illustrating some of these measures in the habitat maps for curlew and golden plover. For curlew (Figure 23.3) the landscape is dominated by relatively few very large patches, with patch connections through valleys and along the hillsides. In contrast, the analysis for golden plover (Figure 23.4) shows no dominant patches, rather the habitat is constructed from a dense network of small patches associated with the upland plateau of the study area.

These analytical descriptions of landscape provide input to metapopulation studies and give an opportunity to develop detailed studies of biological and spatial interaction at a landscape scale. Further research is needed to examine the relevance of these spatial structural and functional measures critically.

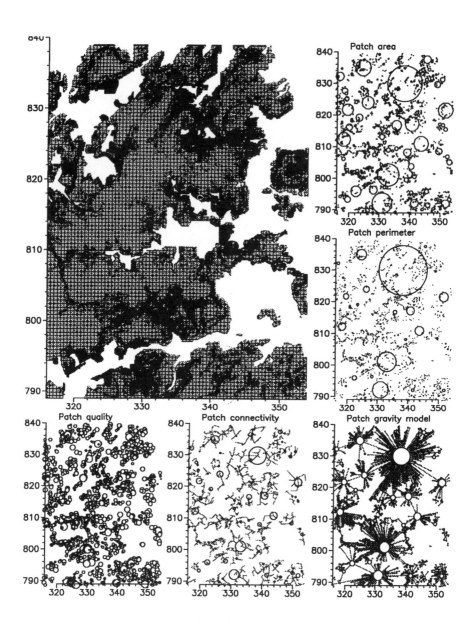

Figure 23.3 Curlew habitat maps.

Note: The data field (top left) is derived from spatial analysis of satellite imagery and a digital elevation using curlew presence and absence from a wildlife survey; the analytical procedure is the same as that used to generate bioclimatic models (Figures 23.1 and 23.2). Presence is more likely in the areas shown in darker shades. Spatial structure and functional linkages in habitat patches are characterized and illustrated. A gravity model is used to identify regional interaction between patches.

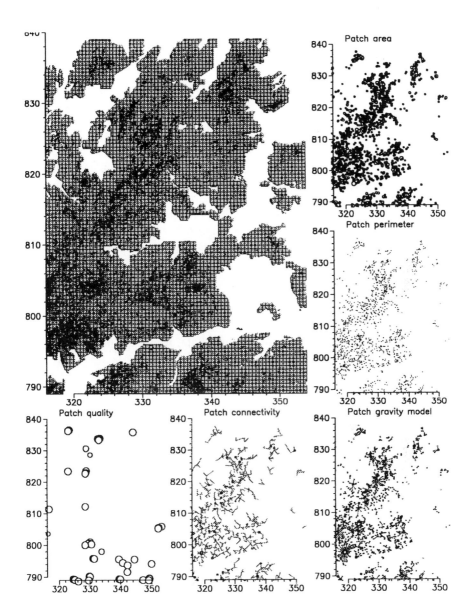

Figure 23.4 Golden plover habitat maps.

Note: The data field (top left) is derived from spatial analysis of satellite imagery and a digital elevation using golden plover presence and absence from a wildlife survey; the analytical procedure is the same as that used to generate bioclimatic models (Figures 23.1 and 23.2). Presence is more likely in the areas shown in darker shades. Spatial structure and functional linkages in habitat patches are characterized and illustrated. A gravity model is used to identify regional interaction between patches.

Potential applications include provision of new opportunities in ecological research, including investigation of scale effects in landscape dynamics; a tool for analysis of geographical aspects of designation in habitat and species conservation within regional landscapes; and a framework for analysis of off-site impacts of land use and land cover changes.

Data Source and Analysis

The source habitat data for this type of analysis can be maps or satellite imagery, the latter being preferred for two reasons. First, satellite imagery gives a wide coverage and regular updates are available. This gives it some advantages over field survey for gathering land surface data, although it has been relatively little used in ecological research (Simmons *et al.,* 1992). Second, and more importantly from the perspective of developing spatial analytical tools, maps usually provide habitat data as discrete, sharply bounded, internally homogenous polygons, while satellite imagery provides a surface of variation and can be used to investigate generalization and scale effects (Simmons *et al.,* 1992). Although the description of habitat patch structure presented here is based on reclassifying a data field (Goodchild, 1989) into a series of discrete polygons, the sensitivity of subsequent analysis to this threshold can be tested by changing the threshold used to identify the patches or by adjusting the resolution of the image. This is not possible with map source data, where the polygons are less easily modified and boundary uncertainty can, as yet, only be described using the perkal epsilon model (Blakemore, 1984) and fuzzy mathematical methods (Burrough, 1989). Further, these methods of uncertainty analysis apply to location of polygon boundaries rather than error associated with polygon heterogeneity (Goodchild, 1989), and are less relevant for the errors of greatest interest here: those associated with patch heterogeneity and areal properties of spatial structures. These aspects of data and data use in GIS can be investigated using a surface of variation in geographic space such as is presented by a (classified) satellite image. The use of satellite imagery, rather than categorical maps, in ecological studies in which space is an explicit component of the analysis will be fruitful in both ecological and spatial domains.

Discussion

This chapter presents some methods for spatial analysis that elucidate ecological hypotheses describing spatial distribution, and interpret spatial interactions in environmental conditions experienced by different species. Both methods support understanding and exploration of interactions between pattern and process. Questions of scale can be addressed using the methods presented here by coupling them with a structured analysis of ecological questions based in hierarchy theory (Allen and Star, 1983; O'Neill *et al.,* 1986; Urban *et al.,* 1987, O'Neill, 1989). Relationships between distribution and environmental variables can be investigated for environmental data of different source scale or spatial grain to characterize the spatial scales and environmental variability to which the species respond. Spatial scales which are too fine (fine grain, high resolution) appear as noise, while scales that are too coarse (large grain, low spatial resolution) appear as constants (Shugart *et al.*, 1991).

The two case studies presented are complementary, being based on comparison of different spatial data sets and analysis of structure and function within one spatial data set. Together they provide tools that can interpret a range of data at a variety of spatial scales, and potentially can link species distribution mapped at a broad national scale with climatic factors, which are important at the national level, and local habitat influences, which are important controls on local distribution and population size. In particular, the ability to construct measures of spatial association and interdependence for habitat patches using topological relationships allows patches of habitat to be assessed in their local and regional context. This can be a most useful tool for studies of animal movement in complex habitats and can lead to ecological understanding, with strong applications in landscape design and habitat management (Forman and Godron, 1986; Levin, 1992). The approach can also be used to integrate spatial elements of habitat with biological components drawn from study of species ecology. This is an obvious candidate for integrating ecological modelling, spatial analysis and GIS, and has been developed using an expert system interfaced with a GIS to model habitat use by deer (Folse *et al.*, 1989).

Spatial analysis in landscape ecology allows structural and functional aspects of landscape at a range of spatial scales to be addressed directly in an analytical sense. It also provides a framework for analysis of distribution–habitat relationships, which can be matched to the scales of both animal behaviour and habitat variation; biological functional units can be analysed against equivalent environmental units. Additionally, since the analytical framework provided by landscape ecology is at a scale compatible with organism behaviour and use of environmental resources, it has potential for application at a wide range of scales, including regional and subregional scales, and complements the bioclimatic modelling for analysis of spatial interaction and ecological processes.

The purpose of spatial analysis in ecological research is to support scientific investigation of ecological phenomena and provide tools that allow ecologists to manage and analyse interactions between scale in space and time and ecological organization (Levin, 1992). A toolbox approach is preferred for developing GIS, spatial analysis and ecological modelling since (although methods may be developed through particular applications) a generic set of spatial analysis methods will have applications beyond those in which they are developed. This is inevitable if tools properly address issues of scale, data quality, error propagation, and interaction of pattern and process. These are themes that cross levels of spatial, temporal and ecological organization and refer to qualities of the basic data and information resources of ecological research as well as the process of ecological science.

Conclusion

The spatial analytical methods presented in the two case studies provide generic tools for exploratory investigation of spatial data, potentially offering insights into processes operating at different scales of spatial and ecological organization. The first case study presents an approach that can automatically search for association between data in environmental databases using spatial logic, and suggests hypotheses about processes for further analysis. As environmental databases become more comprehensive and accessible, automated methods for analysis that suggest areas meriting further study will become increasingly useful and necessary. The second case study shows that spatial analysis can be used to describe and analyse spatial structure and function of habitats in complex landscapes, providing a generic method for quantifying this conceptualization of landscape. Further work is needed in the application of the measures of structure and function provided by this toolbox for landscape ecology, but the approach provides a link between description of pattern and scale provided by the methods of the first case study and the scale at which species respond to landscape variation.

The analytical methods presented also incorporate error analysis. This is necessary for environmental research that depends on spatial databases that are of variable quality. The information outputs from GIS analyses of environmental data too often have no estimate of reliability or uncertainty; new methods for analysis of these data must attend to description, management and analysis of data quality to encourage reliable interpretation of output.

Acknowledgement

Funding for the projects reported here is provided by the Scottish Office Agriculture and Fisheries Department. The bird survey was carried out by the Nature Conservancy Council (now Joint Committee for Nature Conservation), who funded the Macaulay Land Use Research Institute in an early investigation of the merits of satellite imagery for bird survey and modelling as part of their programme of research into nature conservation. I should like to thank Diane Pearson, Dr Peter Dennis, Dr R. V. Birnie and Prof T. J. Maxwell of the Macaulay Land Use Research Institute, three anonymous referees and the editors of this volume for their comments and suggestions.

References

Addicott, J. F., Aho, J. M., Antolin, M. F., Padilla, D. K., Richardson, J. S. and Soluk, D. A., 1987, Ecological neighborhoods: Scaling environmental patterns, *Oikos,* **49**, 340–6.
Allen, T. F. H. and Starr, T. B., 1983, *Hierarchy: Perspectives for Ecological Complexity,* Chicago, Illinois: Chicago University Press.

Andrewartha, H. G. and Birch, L. C., 1954, *The Distribution and Abundance of Animals,* Chicago, Illinois: University of Chicago Press.

Aspinall, R. J., 1992a, An inductive modeling procedure based on Bayes Theorem for analysis of pattern in spatial data, *International Journal of Geographical Information Systems,* **6**, 105–21.

Aspinall, R. J., 1992b, Spatial analysis of wildlife distribution and habitat in a GIS, in Bresnahan, P., Corwin, E. and Cowen, D. (Eds) *Proceedings of the 5th International Symposium on Spatial Data Handling,* Vol. 2, pp. 444–53, Columbia, South Carolina: International Geographical Union (IGU), Humanities and Social Sciences Computing Laboratory, University of South Carolina.

Aspinall, R. J. and Matthews, K., in press, Climate change impact on distribution and abundance of wildlife: An analytical approach using GIS, *Environmental Pollution.*

Aspinall, R. J. and Miller, D. R., 1990, Mixing climate change models with remotely-sensed data using raster based GIS, in Coulson, M. G. (Ed.) *Remote Sensing and Global Change,* Proceedings of the 16th Annual Conference of the Remote Sensing Society, pp. 1–11, Swansea.

Aspinall, R. J. and Veitch, N., 1993, Habitat mapping from satellite imagery and wildlife survey data using a Bayesian modeling procedure in a GIS, *Photogrammetric Engineering and Remote Sensing,* **59**(4), 537–43.

Baker, W. L. and Cai, Y., 1992, The role programmes for multiscale analysis of landscape structure using the GRASS geographical information system, *Landscape Ecology,* **7**(4), 291–302.

Blakemore, M., 1984, Generalisation and error in spatial data bases, *Cartographica,* **21**, 131–9.

Burrough, P. A., 1989, Fuzzy mathematical methods for soil survey and land evaluation, *Journal of Soil Science,* **40**, 477–92.

Calder, N., 1991, GIS/LIS and the new global geography, in *Proceedings of GIS/LIS '91,* **1**, xvii–xxxi. Bethesda, Maryland: ASCM/ASPRS.

Caughley, G., Short, J., Grigg, G. C. and Nix, H., 1987, Kangaroos and climate: An analysis of distribution, *Journal of Animal Ecology,* **56**, 751–61.

Climate Change Impacts Review Group (CCIRG), 1991, *The Potential Effects of Climate Change in the United Kingdom,* First Report, London: HMSO.

Delcourt, H. R. and Delcourt, P. A., 1988, Quaternary landscape ecology: Relevant scales in space and time, *Landscape Ecology,* **2**, 23–44.

Delcourt, H. R., Delcourt, P. A. and Webb III, T. A., 1983, Dynamic plant ecology: The spectrum of vegetation change in space and time, *Quaternary Science Reviews,* **1**, 153–75.

Dobson, J. E., 1992, Spatial logic in paleogeography and the explanation of continental drift, *Annals of the Association of American Geographers,* **82**(2), 187–206.

Dobson, J. E., Rush, R. M. and Peplies, R.W., 1990, Forest blowdown and lake acidification, *Annals of the Association of American Geographers,* **80**(3), 343–61.

Flather, C. H., Brady, S. J. and Inkley, D. B., 1992, Regional habitat appraisals of wildlife communities: A landscape-level evaluation of a resource planning model using avian distribution data, *Landscape Ecology,* **7**(2), 137–47.

Folse, L. J., Packard, J. M. and Grant, W. E., 1989, AI modeling of animal movements in a heterogeneous habitat, *Ecological Modeling,* **46**, 57–72.

Forman, R. T. T. and Godron, M., 1986, *Landscape Ecology,* New York: Wiley.

Giller, P. S., 1984, *Community Structure and the Niche,* London: Chapman and Hall.

Goodchild, M. F., 1987, A spatial analytical perspective on geographical information systems, *International Journal of Geographical Information Systems,* **1**(4), 327–34.

Goodchild, M. F., 1989, Modeling error in objects and fields, in Goodchild, M. F. and Gopal, S. (Eds) *Accuracy of Spatial Databases,* pp. 107–13, London: Taylor & Francis.

Goodchild, M. F., 1991a, Integrating GIS and environmental modeling at global scales, in *Proceedings of GIS/LIS 1991,* **1,** 117–27. Bethesda, Maryland: ASCM/ASPRS.

Goodchild, M. F., 1991b, Spatial analysis with GIS: Problems and prospects, in *Proceedings of GIS/LIS 1991,* **1,** 40–8. Bethesda, Maryland: ASCM/ASPRS.

Goodchild, M. F. and Gopal, S. (Eds), 1989, *Accuracy of Spatial Databases,* London: Taylor & Francis.

Goodchild, M. F., Parks, B. O. and Steyaert, L.T. (Eds), 1993, *Geographic Information Systems and Environmental Modeling,* Oxford: Oxford University Press.

Grossmann, W. D., 1991, Model- and strategy-driven geographical maps for ecological research and management, in Risser, P. G. (Ed.) *Long-term Ecological Research: An International Perspective,* SCOPE 47, pp. 241–56. New York: Wiley.

Hunsaker, C. T., Nisbet, R. T., Lam, D., Browder, J. A., Baker, W. L., Turner, M. G. and Botkin, D., 1993, Spatial models of ecological systems and processes: The role of GIS, in Goodchild, M. F., Parks, B. O. and Steyaert, L. T. (Eds) *Geographic Information Systems and Environmental Modeling,* Oxford: Oxford University Press.

Jeffers, J. N. R., 1991, From free-hand curves to chaos: Computer modeling in ecology, in Farmer, D. G. and Rycroft, M. J. (Eds) *Computer Modeling in the Environmental Sciences,* Institute of Mathematics and its Applications Conference Series No. 28, pp. 299–308, Oxford: Clarendon Press.

Kemp, K. K., 1992, Spatial models for environmental modeling with GIS, in Bresnahan, P., Corwin, E. and Cowen, D. (Eds) *Proceedings of the 5th International Symposium on Spatial Data Handling,* Vol. 2, pp. 524–33, Columbia, South Carolina: International Geographical Union (IGU), Humanities and Social Sciences Computing Laboratory, University of South Carolina.

Leamer, E. E., 1978, *Specification Searches: Ad Hoc Inference with Non-experimental Data,* New York: Wiley.

Levin, S. A., 1992, The problem of pattern and scale in ecology, *Ecology,* **73**(6), 1943–67.

Lindemayer, D. B., Nix, H. A., McMahon, J. P., Hutchinson, M. F. and Tanton, M. T., 1991, The conservation of Leadbeater's possum *Gymnobelideus leadbeateri* (McCoy): A case study of the use of bioclimatic modeling, *Journal of Biogeography,* **18,** 371–83.

Magnuson, J. J., Kratz, T. K., Frost, T. M., Bowser, C. J., Benson, B. J. and Nero, R., 1991, Expanding the temporal and spatial scales of ecological research and comparison of divergent ecosystems: Roles for LTER in the United States, in Risser, P. G. (Ed.) *Long-term Ecological Research: An International Perspective,* SCOPE 47, pp. 45–70, New York: Wiley.

McCoy, E. D. and Bell, S. S., 1991, Habitat structure: The evolution and diversification of a complex topic, in Bell, S. S., McCoy, E. D. and Mushinsky, H. R. (Eds) *Habitat Structure: The Physical Arrangement of Objects in Space,* pp. 3–27, London: Chapman and Hall.

Miller, D. R., 1992, Analysis of vegetation succession within an expert system, in Bresnahan, P., Corwin, E. and Cowen, D. (Eds) *Proceedings of the 5th International Symposium on Spatial Data Handling,* Vol. 2, pp. 381–90, Columbia, South Carolina: International Geographical Union (IGU), Humanities and Social Sciences Computing Laboratory, University of South Carolina.

Mooney, H. A., 1991, Emergence of the study of global ecology: Is terrestrial ecology an impediment to progress?, *Ecological Applications,* **1**(1), 2–5.

Nyerges, T. L., 1992, Coupling GIS and spatial analytical models, in Bresnahan, P., Corwin, E. and Cowen, D. (Eds) *Proceedings of the 5th International Symposium on Spatial Data Handling,* Vol. 2, pp. 534–43, Columbia, South Carolina: International Geographical Union (IGU), Humanities and Social Sciences Computing Laboratory, University of South Carolina.

O'Neill, R. V., 1989, Hierarchy theory and global change, in Rosswall, T., Woodmansee, R. G. and Risser, P. G. (Eds) *Scales and Global Change,* SCOPE 35, pp. 29–45, New York: Wiley.

O'Neill, R. V., DeAngelis, D. L., Waide, J. B. and Allen, T. F. H., 1986, *A Hierarchical Concept of Ecosystems,* Princeton, New Jersey: Princeton University Press.

Openshaw, S., 1989, Learning to live with errors in spatial databases, in Goodchild, M. F. and Gopal, S. (Eds) *Accuracy of Spatial Databases,* pp. 263–76, London: Taylor & Francis.

Openshaw, S., Charlton, M., Wymer, C. and Craft, A., 1987, A Mark 1 geographical analysis machine for the automated analysis of point data sets, *International Journal of Geographical Information Systems,* **1**, 335–58.

Openshaw, S., Cross, A. and Charlton, M., 1990, Building a prototype geographical correlates machine, *International Journal of Geographical Information Systems,* **4**, 297–312.

Perring, F. H. and Walters, S. M. (Eds), 1962, *Atlas of the British Flora,* London: BSBl, Thomas Nelson.

Prentice, I. C., Cramer, W., Harrison, S. P., Leemans, R., Monserud, R. A. and Solomon, A. M., 1992, A global biome model based on plant physiology and dominance, soil properties and climate, *Journal of Biogeography,* **19**, 117–34.

Raper, J. (Ed.), 1989, *Three Dimensional Applications in Geographical Information Systems,* London: Taylor & Francis.

Shugart, H. H., 1993, Global change, in Solomon, A. M. and Shugart, H. H. (Eds) *Vegetation Dynamics and Global Change,* pp. 3–21, London: Chapman and Hall.

Shugart, H. H., Bonan, G. B., Urban, D. L., Lauenroth, W. K., Parton, W. J. and Hornberger, G. M., 1991, Computer models and long-term ecological research, in Risser, P. G. (Ed.) *Long-term Ecological Research: An International Perspective,* SCOPE 47, pp. 211–39, New York: Wiley.

Simmons, M. A., Cullinan, V. I. and Thomas, J. M., 1992, Satellite imagery as a tool to evaluate ecological scale, *Landscape Ecology,* **7**(2), 77–85.

Skole, D. L., Moore, III B. and Chomentowski, W. H., 1993, Global Geographic Information Systems and databases for vegetation change studies, in Solomon, A. M. and Shugart, H. H. (Eds) *Vegetation Dynamics and Global Change,* London: Chapman and Hall.

Smith, J. M., Miller, D. R. and Morrice, J. G., 1989, An evaluation of a low-resolution DTM for use with satellite imagery for environmental mapping and analysis, in *Remote Sensing for Operational Applications,* Proceedings of the 15th Annual Conference of the Remote Sensing Society, pp. 393–8, Bristol, UK: Remote Sensing Society.

Smith, T. M., Shugart, H. H., Bonan, G. B. and Smith, J. B., 1992, Modeling the potential response of vegetation to global climate change, in Woodward, F. I. (Ed.) *The Ecological Consequences of Global Climate Change, Advances in Ecological Research,* **22**, 93–116.

Solomon, A. M. and Shugart, H. H. (Eds), 1993, *Vegetation Dynamics and Global Change,* London: Chapman and Hall.

Southwick, C. H. (Ed.), 1983, *Global Ecology,* Sunderland: Sinauer Associates.

Unwin, D., 1981, *Introductory Spatial Analysis,* London: Methuen.

Urban, D., O'Neill, R. V. and Shugart, H. H., 1987, Landscape ecology, *BioScience,* **37**, 119–27.

Veregin, H., 1989, Error modeling for the map overlay operation, in Goodchild, M. F. and Gopal, S. (Eds) *Accuracy of Spatial Databases,* pp. 3–18, London: Taylor & Francis.

Walker, D. A. and Walker, M. D., 1991, History and pattern of disturbance in Alaskan arctic terrestrial ecosystems: A hierarchical approach to analysing landscape change, *Journal of Applied Ecology,* **28**, 244–76.

Walker, P. A. and Cocks, A. D., 1991, HABITAT: A procedure for modeling a disjoint environmental envelope for a plant or animal species, *Global Ecology and Biogeography Letters,* **1**, 108–18.

Walker, P. A. and Moore, D. M., 1988, SIMPLE — An inductive modeling and mapping tool for spatially-oriented data, *International Journal of Geographical Information Systems,* **2**, 347–63.

Wallin, D. O., Elliott, C. C. H., Shugart, H. H., Tucker, C. J. and Wilhelmi, F., 1992, Satellite remote sensing of breeding habitat for an African weaver-bird, *Landscape Ecology,* **7**(2), 87–99.

Woodward, F. I., 1987, *Climate and Plant Distribution,* Cambridge: Cambridge University Press.

24

Linking ecological simulation models to geographic information systems: An automated solution

Martha B. Coleman, Tamara L. Bearly, Ingrid C. Burke and William K. Lauenroth

This chapter describes a system that links ecological simulation models to Geographic Information Systems (GIS). Ecological simulation models and GIS software are powerful tools by themselves, and when connected together form a vehicle that increases the rate and breadth of ecological modelling investigations. Selecting GIS data, connecting these data to ecological models, conducting model simulations and viewing results from a consistent graphical environment enables investigators more quickly and clearly to visualize the spatial dimension of ecological phenomena. The system described in this chapter, EcoVision, has the flexibility to incorporate multiple simulation models as well as different GIS packages that support a relational database structure. The predominant strength of our approach is the graphical user interface (GUI), which is not tied to any individual GIS but, instead, allows the user to access a GIS menu system when interactive work or data inspection is desirable. To conduct a model run, the scientist performs the following steps: (1) interactive selection of the analysis area, which allows the system to extract unique spatial units for the model run; (2) menu selection of model parameters and process schedule options, which initiate the model run; and (3) generation of output in a database format compatible with the selected GIS, which enables the scientist to visualize results graphically through the GIS as well as in tabular format. The ARC/INFO GIS and Century model are used for illustration of the system with reference to other supportable software and models.

Introduction

Ecological simulation modelling allows scientists to gain insight into processes and mechanisms that cause change in our environment. Until recently, ecological simulations have been conducted on small geographic data sets, and output (e.g., carbon loss, evapotranspiration rates or water storage) has been examined through the use of tables, graphs and plots (Parton *et al.*, 1987; Schimel *et al.*,

Input   Output
Map   Map

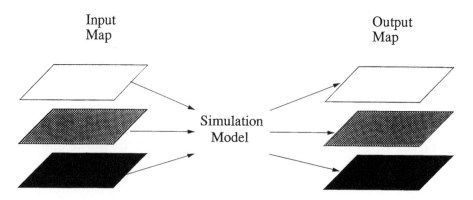

Simulation
Model

Figure 24.1 Conceptual flow of spatially derived data from the GIS to the ecological simulation model and back into the GIS.

1990, 1991). With emerging Geographic Information System (GIS) technologies and the continued development of digital spatial data, another dimension of analysis is now possible and practical. Using GIS to provide spatially explicit input to ecological simulation models and to spatially analyse the output has been a powerful method to investigate ecological problems (Running *et al.,* 1989; Burke *et al.,* 1990; Raich *et al.,* 1991). However, transfer of data between ecological models and GIS can be time consuming, and requires unique solutions for each model–GIS interaction. The need to standardize and automate this transfer of data and information between ecological models and GIS is the driving force behind the development of EcoVision.

EcoVision is a graphical user interface (GUI) that automates the connection between GIS data and ecological simulation models. Using a series of menus, EcoVision guides the user through the process of selecting GIS data, connecting these data to the ecological simulation model, conducting model simulations and viewing the results (Figure 24.1). Before further describing EcoVision, background information on ecological simulation models and GIS data will be presented. System design considerations and interface design principles will then be described, followed by a discussion of the EcoVision interface modules, and a case study.

Background

The primary requirement ecological models place on an interface system is the need for extreme flexibility. As knowledge and improved methodologies are generated, scientists refine and expand simulation models to accommodate new processes and options. This continued refinement is an essential part of the intellectual foundation for simulation modelling. For example, models such as

Century (Parton *et al.,* 1987; Parton *et al.,* 1988) undergo continual revision for the addition of more accurate or detailed algorithms. To explain more clearly the nature of the interface, brief descriptions of ecological simulation model characteristics and GIS data are provided below.

Ecological Models

In a generalized view, spatially explicit ecological simulation models require input of actual or potential site-specific data. Century was selected as the first model to be supported through EcoVision because it required several types of site-specific information. Century simulates change through time in the biogeo-chemistry of forests, grasslands and agricultural lands, and additionally can be used to model scenarios for herbivore interactions, crop rotations, fire events or climate change (Schimel *et al.,* 1991; Holland *et al.,* 1992). Century operates by reading a series of site-specific and fixed-input parameters organized through a predefined file structure. Model outputs are also structured through a fixed file format. Relationships within Century were derived from both site-level and regional data and, using GIS, the model has been applied to regional-scale databases in a number of recent studies (Burke *et al.,* 1990; Schimel *et al.,* 1990; Burke *et al.,* 1991).

Geographic Information Systems Data

GIS data format and content also play a role in the EcoVision interface function and design. With the current trends in computer technology, such as the decreasing cost of disk space and the continually increasing speed of central processing units (cpu), both vector and raster data are viable formats for the quantification of land-based resources (Burrough, 1986). The ARC/INFO (ESRI, Redlands, California) GIS software package supports both vector and raster data analysis. This software includes relational database capability for both formats.

Selection and use of specific GIS data layers are dependent on the scale and purpose of the ecological model simulation. Typical map data layers used for ecological simulations include precipitation, temperature, slope, aspect, land use, vegetation and various soil properties. The extent to which an ecological model can take advantage of the spatial relationships present in GIS data varies from model to model. Hydrologic models, for example, require neighbourhood relationships from GIS topology (spatial connection of graphic data elements) to simulate water movement through the landscape (Battaglin *et al.,* 1993; Krummel *et al.,* 1993). Plant and nutrient-based models such as Century utilize the vertical coincidence of data, such as the land use, soil properties and climate of specific locations, for the purpose of modelling stationary nutrient status and vertical movement (e.g., CO_2 release into the atmosphere). Lateral movement may be accounted for in the model with environment variables but is not driven by spatial site-specific data. Regional- and global-scale simulations rely heavily on GIS to determine vertical coincidence of data layers and to manage the associated area calculations (Burke *et al.*, 1991; Raich *et al.*, 1991).

System Design Considerations

In addition to the requirements that ecological models and GIS data place on EcoVision, two more system function and design issues were revealed during a user-needs assessment. The first of these addressed the need for longevity of the system: what type of system architecture would withstand the rate of change we experience in our hardware and software environments? The second issue considered the functions of the graphical user interface: how could we build a GUI to incorporate GIS data and a broad range of ecological models that also is easy to use?

Longevity

We identified two challenges that, if met, would help ensure longevity of the system. These challenges were the avoidance of unreliable software and adoption of a modular software design structure.

System longevity would be impaired if the interface relied on third-party software that became obsolete and unsupported. Completely avoiding future changes in software may not be possible, but the impact of such changes can be mitigated by selecting mature software with a large user base. ARC/INFO was selected as the initial GIS package because it supports both vector and raster format data analysis as well as relational database capability. This software has a large user base both locally and nationally, and supports a GUI toolkit that is visually and functionally compatible with several top-level menu systems such as XView (X Window-System-based Visual/Integrated Environment for Work-stations, Sun Microsystems, Inc.; Heller, 1991). XView is the window-building toolkit used to support the top level of EcoVision (Figure 24.2). XView was selected on the basis of the features, ease of maintainability and size of the local user base. Incorporating this type of window-building software removed total reliance on one ecological model or one GIS.

In addition to longevity considerations, flexibility in software design is critical. If the design is not flexible enough to easily incorporate changes that occur in the ecological simulation models and GIS software, EcoVision would cease to be useful or would require expensive maintenance. To maximize software flex-ibility, a modular design strategy was employed, which enables EcoVision to accommodate revision changes and expansion in the ecological models and

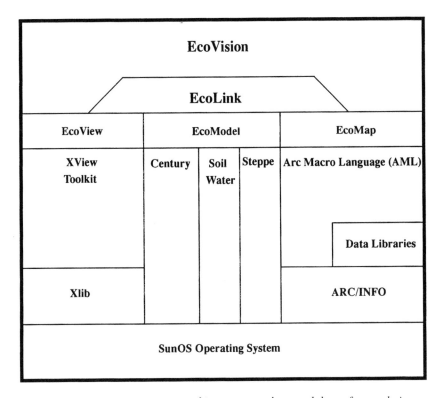

Figure 24.2 EcoVision system architecture reveals a modular software design.

Note: EcoView is the top-level graphical user interface that co-ordinates process flow and communications between the other modules. EcoModel contains the suite of simulation models available to the user. EcoMap manages the selection, viewing and use of GIS data, and EcoLink is the mechanism for process communication.

GIS software packages. This modular approach is facilitated by the XView software and follows a functional organization as illustrated in Figure 24.2.

Easy to Use

Consistency is the most important design element of a GUI that is easy to use (Gould, 1989; Marcus, 1990 a,b). Consistency is the characteristic of an interface that permits the user to learn quickly and to maintain familiarity with a system despite minor functional changes. This is important not only for ease in use and learning, but also to ensure accuracy in data handling.

Internal as well as external and real-world consistency must be considered during the design of a GUI. Internal consistency is the set of conventions within a particular application. In EcoVision, for example, there is a descriptive header at the top of each window (Figure 24.3). A *close* button is used for information windows and a *cancel* button is used in action windows. External consistency is

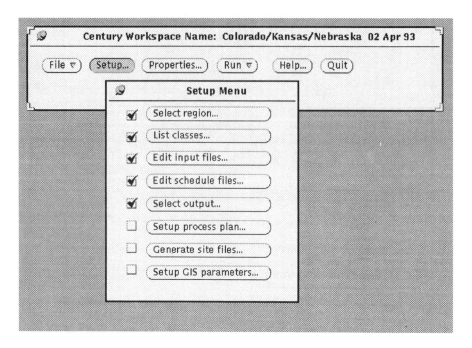

Figure 24.3 Interface consistency is illustrated in the EcoView ecological modelling set-up menu through the use of menu titles, placement of menu elements and button functionality.

the adherence to a set of industry-based standards (Marcus, 1990 a,b). There are no absolute standards at the current time, but there are some conventions that most X-based applications use. In EcoVision, external consistency is achieved by adopting some of the more universal conventions described by the OPEN LOOK GUI specifications (Sun Microsystems, 1990). Buttons that invoke a pull-down style menu, for example, have a label followed by an outlined triangle, whereas buttons that produce an independent window have a label followed by three dots (Figure 24.3). The EcoVision interface also displays a type of real-world consistency through the use of check boxes to mark off steps completed in a list (Figure 24.3.).

Consistency in data handling is more difficult to achieve because there are few pre-established standards among different ecological simulation models. A basic level of data consistency, therefore, is supported in EcoVision through the use of a digital file reference that contains the units of measure for each variable an ecological model accepts. During the selection of GIS data the information in this file is accessed and compared with the units supported by the selected GIS data layers. If inconsistencies are discovered, the user is notified and options are provided. If, for example, the selected model expects precipitation in centimeters and the selected GIS data layer provides precipitation in millimetres, EcoVision can make the simple data conversion. For more complex inconsistencies, however, more user interaction is required.

EcoVision System Design

The EcoVision interface is partitioned into four functional program modules: EcoView, EcoModel, EcoMap and EcoLink (Figure 24.2). EcoView is the application or *front-end* manager that, through EcoLink, co-ordinates activities between EcoModel and EcoMap. New models and GIS packages may easily be added into the existing modular software structure without changing the original software and by adding only minor changes to the top-level menu. Programming code and reference files required to support the incorporation of an ecological simulation model or GIS package are maintained in separate submodules.

EcoView

EcoView is the GUI that integrates the EcoVision system. By co-ordinating movement and communication between the ecological simulation models and the GIS (Figure 24.2), EcoView buffers the user from the details of linking data to and from the GIS. Built with the XView toolkit, EcoView operates as a conduit for accessing GIS data selection and model set-up tasks, and provides tools to manage the distribution of computing power for model processing and the allocation of disk space required for model output. Disk space requirements and allocation are reported to the user and, as the model simulation is processing, the status (reported as percentage complete) is available to the user from a menu selection.

EcoModel

EcoModel is the suite of simulation models available to the user. Currently, only Century is implemented, but the structure is present in EcoModel to store and maintain additional models. Incorporation of new models does not require that software internal to the ecological model be altered. Instead, as part of EcoLink, a data translator is written that handles data going into the model and data coming out of the model. This data handler takes the data and data descriptions sent by the GIS and reformats them into a form acceptable to the selected model. Support files containing data units and formats for each ecological model variable must be created for use by EcoLink. Each model is handled separately at present, but the incorporation of a standardized data interchange handler is being investigated.

Some models have separate submenu interfaces for the purpose of managing required input and output files. Century, for example, has two such menus currently under development. EcoMap can store and utilize many of these external model menus as they become available. This reduces the duplication of effort and allows faster incorporation of new model revisions and features. Using these external model menus sacrifices some overall consistency, but because EcoVision is designed with industry-based standards, many of these external model menus encountered in Century and several other models are visually consistent with EcoVision menu conventions.

EcoMap

EcoMap is the user interface that facilitates access to the GIS. The purpose of this module is to lead the user through the selection of the area of interest and the map layers to be used. Previously used study areas, such as state and county boundaries, and special interest areas, are stored in a data library for use by all models. Accessing predefined boundaries reduces duplication of effort as well as permitting separate model runs to be conducted on exactly the same geographic region. New areas are often needed, however, and, once created, are stored with the other study boundaries for future use.

Only one menu for each data layer type, such as land use or precipitation, is created. Each ecological simulation model that requires land use, for example, will use the same menu. Based on the model the user has chosen and other information provided through EcoView, EcoMap presents the user with only the data selection menus needed for the model run. With this menu design, addition of a new model may not require changes to EcoMap unless the model requires a data layer type not previously supported. EcoMap also provides the user with an environment for viewing model results in the form of maps, data reports and simple graphs.

EcoLink

A bidirectional exchange of data links ecological model processing with the GIS. The EcoLink module creates this connection between the model and the GIS. EcoLink is composed of UNIX network routines that implement the Transport Layer Interface (tli) to pass information between the two processes. These server-to-client programs are the mechanisms used to transfer individual data elements as well as data files.

Because of large processing requirements, the use of several networked machines is advantageous. As the geographic area of interest grows in size, the number of unique sites increases and consequently the number of simulation runs also increases. To reduce the time required to complete a full model run, EcoLink allows the user to start simulation runs on several different networked computers all invoked from a single machine via distributed processing.

Transferring model input parameters from the GIS to the model is accomplished by first having the client, EcoView, start up a server process on a machine that supports the GIS. After these two programs have established a protocol for communicating, EcoView begins the GIS program and the EcoMap interface. The EcoMap module is not seen by the user until GIS functions are required. When the user is ready to select GIS data, EcoMap is ready to directly access the GIS data library. Transfer files containing data and the associated formats are created, and these files are then passed back to the model via EcoLink. As the ecological model processes these data, output files are generated. Transferring the model output back into the GIS is done through a post-processing conversion program that writes the data in the specified GIS database file format.

Eco Vision Case Study

Although EcoVision is structured in a modular regime, movement between the modules is transparent to the user. The sequential steps in conducting a model session with EcoVision follow the data flow (Figure 24.4). In general, a session begins by choosing the ecological model, moves to the selection of GIS data and conversion of these data into model input, then through model processing, and finally to viewing and analysing model results. As the user initiates an EcoVision session, menus are presented that help guide the user through these modelling tasks.

For this case study, Weld County in northeast Colorado is the study area to be modelled through Century. Although much larger regions can be modelled through EcoVision, Weld County was selected because this procedure is easier to visualize on a small area. When a new Century model simulation is begun in EcoVision, the area of interest, in this case the Weld County boundary, must first be selected by the user. Once this area of interest is chosen all data layers required as model input must be selected (Figure 24.5).

In the EcoVision implementation of Century for grassland simulation modelling, land use, soil texture, precipitation and temperature are required. Other input variables such as percentage of rock may be used but are optional. For Weld County, the GIS data library accessed by EcoMap has two soil texture layers and three land user layers. Based on the modelling simulation goals, data descriptions and data browsing results, the user picks one of the soil map layers and one of the land use maps as input (Figure 24.6). Similar selections must be made for other required input such as precipitation and temperature.

Information from all selected map layers is combined and a list of codes describing each unique land area is created by EcoMap; this code is called a *class key*. As can be seen in Figure 24.6, class keys for this case study are developed from land use, precipitation, temperature and soil characteristics. The class key wf0300084015 is an object that represents one or more polygons and can be described as wheat–fallow, 30–40 cm precipitation, 8–10°C, 40 per cent sand,

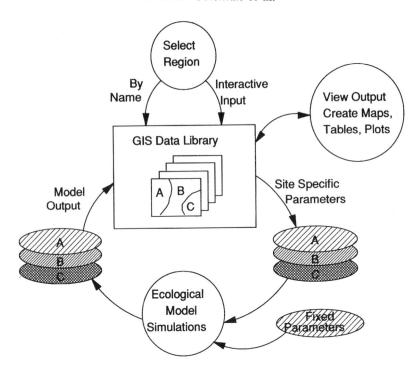

Figure 24.4 EcoVision data flow begins with the selection of a region or area of interest and GIS data layers.

Note: These data layers build the site-specific parameters that, along with fixed parameters, define the ecological scenario to be modelled. Model output are directed back into the GIS for viewing and potentially for further model processing.

and 15 per cent clay. The set of unique class keys in the same format as wf0300084015 is sent back to EcoView, via an EcoLink system file, interpreted and used to create site-specific input files for the Century model. Initial model runs for the establishment of steady-state conditions are run on this list of classes, with the assumption that all land use is grazing. Fixed input such as the beginning and ending simulation dates, crop varieties and crop rotations are defined through current Century submenus that are accessed and managed by EcoView (Figure 24.7). Likewise, the type of simulation model output variables must be selected.

At this point, computers to be used for processing the simulation are defined and processes are initiated. As the simulation is processing, the status can be assessed, or the user can set up a new modelling scenario. Model output variables previously chosen by the user are written directly into the GIS database along with the class key. Then, using the class key, these output data are attached to the map of the combined layers (Figure 24.8). This is made possible through the use of a relational database structure in the GIS.

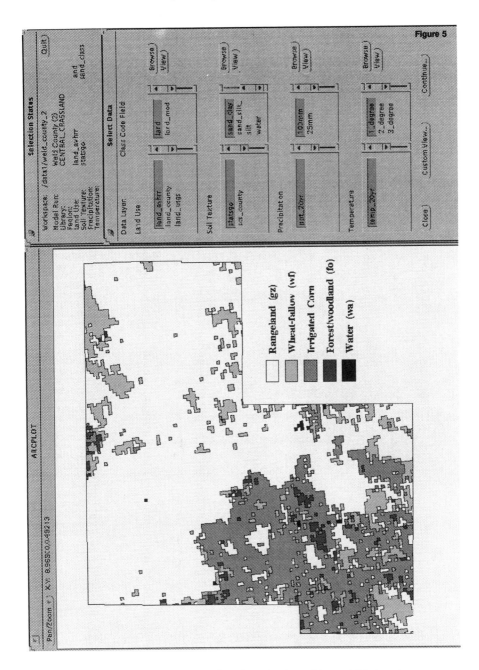

Figure 24.5 EcoMap user interface for selection of input GIS data layers.

Note: Land use for Weld County, Colorado, is the GIS library layer displayed. These data were adapted from a classification of Advanced Very High Resolution Radiometer (AVHRR) imagery at 1-km resolution (Loveland *et al.*, 1991).

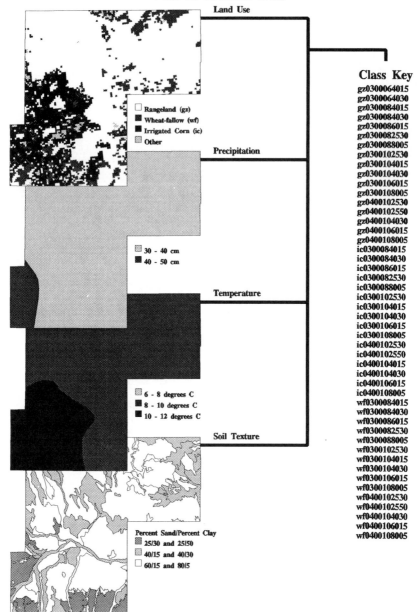

Class Key

gz0300064015
gz0300064030
gz0300084015
gz0300084030
gz0300086015
gz0300082530
gz0300088005
gz0300102530
gz0300104015
gz0300104030
gz0300106015
gz0300108005
gz0400102530
gz0400102550
gz0400104030
gz0400106015
gz0400108005
ic0300084015
ic0300084030
ic0300086015
ic0300082530
ic0300088005
ic0300102530
ic0300104015
ic0300104030
ic0300106015
ic0300108005
ic0400102530
ic0400102550
ic0400104015
ic0400104030
ic0400106015
ic0400108005
wf0300084015
wf0300084030
wf0300086015
wf0300082530
wf0300088005
wf0300102530
wf0300104015
wf0300104030
wf0300106015
wf0300108005
wf0400102530
wf0400102550
wf0400104030
wf0400106015
wf0400108005

Figure 24.6 Combination of selected GIS data layers for Weld County, Colorado form a list of codes describing each unique land area to be modelled through Century.

Note: Land use data were adapted from a classification of Advanced Very High Resolution Radiometer (AVHRR) imagery at 1-km resolution (Loveland *et al.*, 1991). Mean annual precipitation and temperature were generalized from national weather station sites (CLIMATEDATA, 1988). Soil texture was derived from the State Soil Geographic Data Base (STATSGO) for Colorado (USDA Soil Conservation Service, 1991).

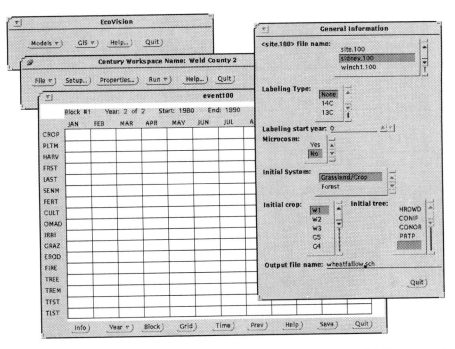

Figure 24.7 EcoView user interface linking Century ecosystem model (Parton et al., 1987) submenu system Event100 (Metherell et al., 1993).

Note: This particular menu is used to schedule the agricultural event sequence to be simulated by Century.

Conclusion

The role of EcoVision is crucial to the functionality that GIS and ecological models provide. Models and GIS software are powerful tools by themselves, but when connected together form a vehicle that increases the rate and breadth of ecological modelling investigations. Selecting GIS data, connecting these data to the ecological models, conducting model simulations and viewing results from a consistent graphical environment enables investigators more quickly and clearly to visualize the spatial dimension of ecological phenomenon. EcoVision is part of a new generation of software tools and promises to be a critical part of how modelling research is conducted in the future.

Acknowledgement

The authors thank Tom Kirchner and Elizabeth Olson for development of *Distributed Systems,* a tli-based toolkit. Support for this work was provided by

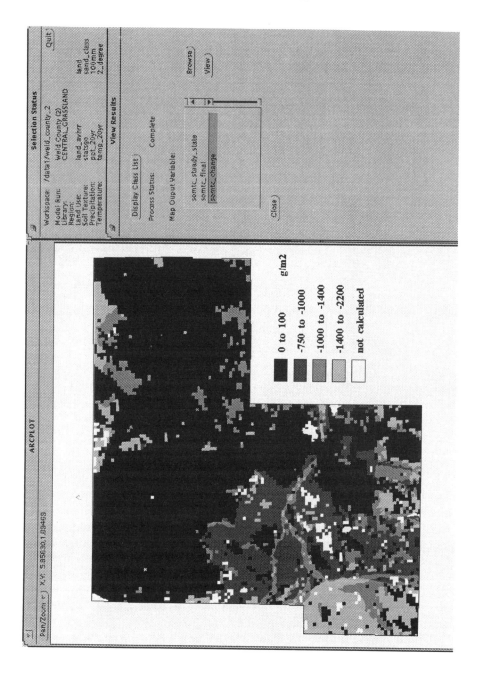

Figure 24.8 EcoMap user interface for viewing results of model simulations.

Note: Simulation data for soil carbon losses due to 50 years of historical cultivation for Weld County are displayed.

grants from the National Science Foundation (BSR 91-06183 and BSR 90-11659), the USDA Agricultural Research Service and the Colorado State Agricultural Experiment Station (1-50661).

References

Battaglin, W. A., Kuhn, G. and Parker, R., 1993, Using GIS to link digital spatial data and the precipitation-runoff modeling system, Gunnison River Basin, Colorado, *Proceedings, Second International Conference/Workshop on Integrating Geographic Information Systems and Environmental Modelling.*

Burke, I. C., Kittel, T. G. F., Lauenroth, W. K., Snook, P., Yonker, C. M. and Parton, W. J., 1991, Regional analysis of the Central Great Plains, *BioScience,* **14**(10), 685 92.

Burke, I. C., Schimel, D. S., Parton, W. J., Yonker, C. M., Joyce, L. A. and Lauenroth, W. K., 1990, Regional modeling of grassland biogeochemistry using GIS, *Landscape Ecology,* **4**, 45-54.

Burrough, P. A., 1986, *Principles of Geographical Information Systems for Land Resources Assessment,* Oxford: Clarendon Press.

CLIMATEDATA, 1988, *CLIMATEDATA,* Denver, Colorado: US West Optical Publishing.

Gould, M. D., 1989, Human factors research and its value to GIS user interface design, *Proceedings of GIS/LIS '89,* 542-50.

Heller, D., 1991, *XView Programming Manual,* Vol. 7, Sebastopol: O'Reilly and Associates.

Holland, E. A., Parton, W. J., Detling, J. K. and Coppock, D. L., 1992, Physiological responses of plant populations to herbivory and their consequences for ecosystem nutrient flow, *The American Naturalist,* **140**, 685-706.

Krummel, J. R., Dunn, C. P., Eckert, T. C. and Ayers, A. J., 1993, A technology to analyze spatiotemporal landscape dynamics: Application to Cadiz Township (Wisconsin), *Proceedings, Second International Conference/Workshop on Integrating Geographic Information Systems and Environmental Modeling.*

Loveland, T. R., Merchant, J. R., Ohlen, D. O. and Brown, J. F., 1991, Development of a land-cover characteristic database for the conterminous US, *Photogrammetric Engineering and Remote Sensing,* **57**, 1453-63.

Marcus, A., 1990a, Designing Graphical User Interfaces: Part I, *Unix World,* August, 107-15

Marcus, A., 1990b, Designing Graphical User Interfaces: Part II, *Unix World,* September, 121-7

Metherell, A. K., Harding, L. A., Cole, C. V. and Parton, W. J., 1993, *CENTURY Soil Organic Matter Model Environment Technical Documentation, Agroecosystem Version 4.0,* GPSR Technical Report No. 4, Fort Collins, Colorado: United States Department of Agriculture, Agricultural Research Service, Great Plains Research Unit, USDA-ARS.

Parton, W. J., Schimel, D. S., Cole, C. V. and Ojima, D. S., 1987, Analysis of factors controlling soil organic matter levels in the Great Plains grasslands, *Soil Science Society of America Journal,* **51**, 1173-9.

Parton, W. J., Stewart, J. W. B. and Cole, C. V., 1988, Dynamics of C, N, P and S in grassland soils: A model, *Biogeochemistry,* **5**, 109-31.

Raich, J. W., Rastetter, E. B., Melillo, J. M., Kicklighter, D. W., Steudler, P. A. and Peterson, B. J., 1991, Potential net primary productivity in South America: Application of a global model, *Ecological Applications,* **1**(4), 399-429.

Running, S. W., Ramakrishna, R. N., Peterson, D. L., Band, L. E., Potts, D. F., Pierce, L. L. and Spanner, M. A., 1989, Mapping regional forest evapotranspiration and photosynthesis by coupling satellite data with ecosystem simulation, *Ecology*, **70**, 1090–1101.

Schimel, D. S., Kittel, T. G. F. and Parton, W. J., 1991, Terrestrial biogeochemical cycles: Global interactions with the atmosphere and hydrology, *Tellus,* **43**(AB), 188–203.

Schimel, D. S., Parton, W. J., Kittel, T. G. F., Ojima, D. S. and Cole, C. V., 1990, Grassland biogeochemistry: Links to atmospheric processes, *Climate Change*, **17**, 13–25.

Sun Microsystems, 1990, *OPEN LOOK Graphical User Interface Application Style Guidelines*, Reading, Massachusetts: Addison-Wesley.

USDA Soil Conservation Service, 1989, *STATSGO Soil Maps*, Fort Worth, Texas: National Cartographic Center.

25

Comparison of spatial analytic applications of GIS

David P. Lanter

Lineage metadata are an abstract representation of maps and logic used within a spatial analytic application of a geographic information system (GIS). This chapter details the structure of such metadata and presents a series of comparison tests useful for identifying commonalities in logic applied within GIS applications to explicate spatial relationships between cartographic features encoded within a set of source maps. Comparative analysis of lineage metadata across different GIS applications offers the possibility of automatically detecting and isolating patterns of spatial reasoning applied within GIS-based environmental models.

Introduction

Much environmental information is based on assumptions that subsume the complexity of nature within a few summary variables. The common ancestry of earth's two million plus species results in patterns of environmental responses often reflected in spatial distributions. These result from organismal trade-offs made in response to environmental constraints such as resources or exposure to hazards (Tilman, 1989). Ecologists often seek to understand such behaviour in terms of concepts that are expandable across trophic levels. One such concept is the interaction between consumer(s) and resource(s). This interaction can be expressed in terms of spatial relations between members of the same species, members of different species, between species and resources, or species and hazards. Such spatial relations can be modelled with analytic geographic information system (GIS) functions applied to maps of such species, resources and hazards.

The lineage knowledge representation (Lanter, 1991) is an abstract representation of maps and logic applied within spatial analytic applications of GIS. Comparative analysis of such metadata offers the possibility of detecting patterns of spatial reasoning applied within environmental models that utilize GIS-derived data. Analysis of lineage metadata focuses on the role the GIS plays in explicitly encoding spatial relations implicit in a set of source maps. At this level, metadata attributes concerning cartographic scale and other qualities of source maps may be ignored with little consequence. Only the conceptual nature of feature classes within source map layers is relevant to identifying commonalities in spatial analytic derivation of explicit spatial relationships. Automatic comparison of abstract representations of spatial analytic applications focuses on spatial relations made explicit by algorithmic processing of geographic features encoded within source data files.

Spatial Analysis

Berry (1964) defined a geographic fact as an observation of a single characteristic observed at a single location at a particular point in time. Individuals possess cognitive images of such geographic facts making up their understanding of their environment (Hirtle and Heidorn, 1993). These mental maps are important because they are the basis for decisions concerning the environment (Muehrcke and Muehrcke, 1992). Decisions should be based on mental maps that are accurate and complete. When this is not the case, a cartographic map is required to upgrade the decision maker's mental map. Visual inspection of a cartographic map provides a synoptic view of an assemblage of geographic facts ('features') portraying both their spatial extent and interrelationships.

A map reader uses a thematic map when the need for information focuses on characteristics of a particular geographic distribution (Dent, 1993). Thematic maps are composed of a base map providing a locational frame of reference and thematic overlays consisting of qualitative or quantitative classes of cultural or scientific data. During thematic map reading the user visually integrates these base and thematic elements to understand the nature and extent of spatial associations. Nystuen (1968) identified three fundamental spatial associations: direction, distance and connectiveness. These, he argued, could be built up to include analysis of pattern, accessibility, neighbourhood and circulation. The task of spatial analysis, however, with traditional paper maps and map-based computer systems, is relegated to the visual apparatus of the map reader.

Spatial Analysis with GIS

Geographic information systems offer an automated alternative to visual inspection for spatial analysis using well-known software functions documented in the literature (Tomlinson and Boyle, 1981; Dangermond, 1983; Berry, 1987; Guptill *et al.*, 1988). These spatial analytic functions derive new cartographic themes consisting of features that explicitly encode spatial relations previously only inherent in a digital cartographic database. For example, visual inspection of the spatially registered thematic overlays of roads, streams and nests can be applied to identify nests within 30 m of the road and stream intersections. In this task, a reader would locate the road and stream intersections, locate all areas within 30 m of these intersections and identify the set of nests located within the areas: in this case, a GIS spatial analytic application based on overlaying road and stream maps to identify spatial coincidence, measure a distance zone of 30 m from the results of the previous step and overlay the distance zone with the nests to identify those that met the original criteria. The resulting functional application would result in algorithmic identification, data structure representation and database storage of the associated cartographic themes: road and stream intersections, areas within 30 m buffers of the intersections and nests within the areas. Note that nests identified in this way explicitly encode the desired spatial relationships that previously existed only implicitly within the original nests, road and stream overlays.

The ability of the analytic functions to derive new thematic features that encode spatial relations existing within source maps is what separates the GIS from computer mapping and computer-aided design systems (Cowen, 1988). Such computer mapping and computer-aided design systems are not equipped to identify the desired nests. The spatial analytic GIS functions do for spatial relationships what statistic functions do for measures of central tendency, and remotely sensed image classification functions do for thematic maps. They derive them from information implicit in the input data.

Abstracting Spatial Analytic Data Processing

Layer-based GIS typically collect thematically differentiated sets of features in spatially registered overlays called layers. During the life cycle of an operational GIS, there are three phases: locational inventory, spatial analysis and environmental management (Crain and MacDonald, 1983). During the location inventory phase, source thematic layers are acquired and entered into the GIS database. These may be the end result of the digital image processing of remotely sensed data, digitization of maps or *in situ* sampling. When spatial data are first loaded into a GIS database only source layers are available to be transformed into a derived layer, $n \geqslant 1$, $m = 0$ (Figure 25.1).

During spatial analysis, GIS applications process these data to explicate spatial relations implicit in the set of source maps comprising the location inventory. These efforts focus on identifying spatial relationships to test hypotheses or determine the nature and extent of geographic phenomena. Each step within such applications transforms a set of source or derived layers to derive a single new layer. Figure 25.1 illustrates the 'transformed into' relationship connecting source and derived input layers to the derived layer they were used to create. In the figure, cases where $(n = 1, m = 0) \vee (n = 0, m = 1)$ represent spatial analytic transformations of individual source or derived layers to create new layers. Such transformations include reclassification, distance measurement, connectivity, neighbourhood characterization and summarizing calculations. When $n + m > 1$, the transformation involves a set of input layers consisting of source, derived or both types of input data. These include arithmetic, statistical and logical overlay as well as drainage network and viewshed determinations. Each transformation derives a map that encodes a new spatial relationship used as a step toward explicating the desired spatial information.

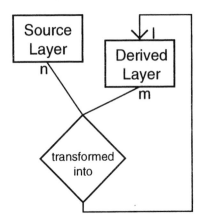

Figure 25.1 Model of data relationships in spatial analytic transformations.

Lineage Metadata

The lineage knowledge representation describes the transformations within a spatial analytic application as a hybrid semantic network/frame data structure. This structure represents each transformation connecting layers within a spatial analytic application. It consists of nodes representing spatially registered thematic layer types, L. Let L_i denote a layer type, $L= \{L_i : i = \text{source, derived}\}$. Furthermore, derived layers can be partitioned into intermediate and product layers, $L_{\text{derived}}= \{L_{\text{derived}.K} : K = \text{intermediate, product}\}$. Intermediate layers are derived as steps within a spatial analytic application connecting source layers and the application's product, which encodes explicit spatial relations implicit in the features of the source layers.

Semantic parent–child relationship types, R, link these layers to structure the steps that comprise an analytic GIS application (Figure 25.2), 'Parent' links connect input layers to the output layers they are directly used to create. 'Child' links connect the derived output layers to their inputs. Let R_j denote a relationship type, $R = \{R_j : j = \text{child, parent}\}$. Each R_j is defined as an ordered pair of input and output layers:

$$R_{\text{parent}}= (L_{\text{input}}, L_{\text{output}}), \quad R_{\text{child}}= (L_{\text{output}}, L_{\text{input}})$$

where

$$R_{\text{parent}}, R_{\text{child}} \; \varepsilon \; R$$

and

$$L_{\text{input}}, L_{\text{output}} \; \varepsilon \; L.$$

Each R_j is further specified by roles played by source and derived layers as data inputs and outputs,

$$R_{\text{parent}} \; \varepsilon \; \{(L_{\text{source}}, L_{\text{derived}}) \, (L_{\text{derived}}, L_{\text{derived}})\},$$

and

$$R_{\text{child}} \; \varepsilon \; \{(L_{\text{derived}}, L_{\text{derived}}) \, (L_{\text{derived}}, L_{\text{source}})\}.$$

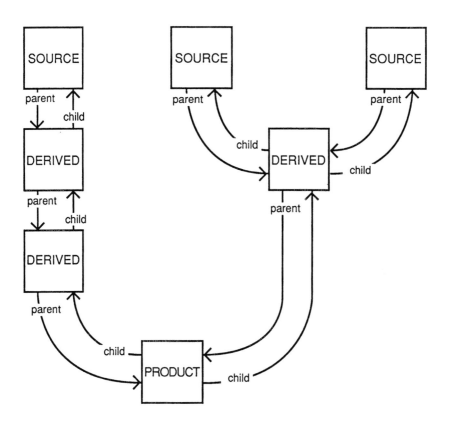

Figure 25.2 Semantic parent/child links between source and derived layers.

Traversing the links in a lineage knowledge representation of a GIS application is a basis for identifying the data dependencies created by analytic functions applied in the course of explicating spatial relationships between features within the source layers. Let l_i be an instance of a given layer type L_i, then a *parent*(l_{input}) operator can be defined to traverse the parent links of l_{input} to identify all of the instances of l_{output} directly derived from it. *Descendents* is a function that recursively traverses parent links to identify all layers derived from a particular source map within the lineage metadata representation of an application,

$$Descendents(l_{input}) = ()$$

if

$$parent(l_{input}) = (),$$

otherwise

$$parent(l_{input}) \cup (\cup_{\forall l_{output} \varepsilon parent(l_{input})} Descendents(l_{output}))$$

Conversely, the *child* (l_{output}) operator traverses child links of l_{output} to identify all instances of the input data, l_{input}, used in its derivation. The *Ancestors* follows these links to identify all the input layers used in its derivation,

$$Ancestors(l_{output}) = ()$$

if

$$child(l_{output}) = (),$$

otherwise

$$child(l_{output}) \cup (\cup_{\forall l_{input} \varepsilon child(l_{output})} Ancestors(l_{input}))$$

Child links connecting output layers to their inputs allow automatic deduction of which layers within a spatial database are sources and which are derived (Lanter, 1993). Derived layers are connected to their input layers by child links, sources are not. This deduction provides a basis for structuring additional metadata concerning the source data, spatial analytic logic applied to the source data and the resulting data product. Classifying layers into source, intermediate and product types serves as a basis for structuring lineage attributes.

The proposed National Spatial Data Transfer Standard (SDTS, 1992), recommends that digital cartographic source documentation include name, feature types, dates, responsible agency, scale, projection and accuracy attributes. Attributes of derived layers encode the transformation command and command parameters used in the layer's creation. Product layer attributes identify the particular analyst that created the spatial analytic application, the role played by the derived spatial data in the study, and the date associated with creation of the derived data. Each layer structured within the lineage knowledge representation is stored with its associated source description, command specification, and product-use attributes according to its type. Associated with L_i in L is an ordered list of attributes A_i, $A_i = \{A_{i1}, A_{i2}, \ldots A_{ik = f(i)}\}$. Specifically,

$$L_{source}, A_{source} = (\text{Name, Path, Features, Date, Scale,}$$
$$\text{Projection, Agency, Accuracy}),$$
$$L_{derived}, A_{derived} = (\text{Name, Path, Command, Parameters}),$$

and additionally

$$L_{derived.product}, A_{derived.product} = (A_{derived} \cup (\text{Analyst, Role, Release Date}, \ldots)).$$

Let α_{im} denote a value of A_{im}, then an instance l_i of a given layer type L_i, can be expressed as $l_i = (a_{i1}, a_{i2}, \ldots, a_{ik})$. Combining these metadata with the parent–child links described above results in a metalevel representation of digital cartographic feature data and transformational logic applied to them in the context of a spatial analytic application (Figure 25.3).

Spatial Analytic Data Equivalence

Lineage knowledge metadata can be compared to identify commonalities and differences in spatial reasoning. Comparison of lineage knowledge

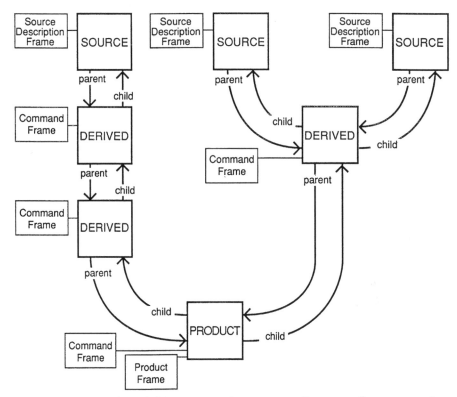

Figure 25.3 Lineage knowledge representation structures lineage attributes concerning source, derived and product layers and data dependency relationships between them.

representations of a set of spatial analytic applications provides a basis for determining the equivalence of synthesized feature data and logic used in their derivation. Data equivalence exists in any portions of two metadata structures that represent the same transformations applied to the same source data. The search for such equivalence focuses on commonality within two domains: (1) lineage attributes of layers, and (2) parent–child relationships between layers.

Source Layer Equivalence

Search for layers common to metadata representations of two applications must consider similarity of their source layers. Source layers within a GIS database are typically identified with a path and file name specifying the data's location within a hierarchically structured file system of a computer operating system. Such metadata, however, are only 'loosely' coupled to the data they represent. That is, two metadata representations of layers identified with the same path name may not point to the same data but to different data occupying the same file system location at different times. Therefore, additional evidence is necessary to conclude that two metalevel data representations refer to the same data.

Other source description attributes can be called on to provide this evidence. If, in addition to path names, the values for source description attributes such as feature content, date, agency, scale, date and time, etc. were equivalent then a conclusion could be made that the individual representations referred to the same source data. This can be denoted:

$$l_{source'} \equiv l_{source''}$$

iff

$$\forall A_{sourcek} \; \varepsilon \; A_{source} \wedge a_{source'k} = a_{source''k}$$

Relaxing the assumption that source layers must have the same path name to be considered in the search allows the use of other source description attributes to identify layers having different names or storage locations but having an equivalent content. That is, suppose

$$X_{source} = (A_{source\;feature}, A_{source\;date}, \ldots, A_{source\;accuracy}) \subset A_{source},$$

then,

$$l_{source'} \equiv l_{source''}$$

iff

$$\forall A_{sourcek} \; \varepsilon \; X_{source} \wedge a_{source'k} = a_{source''k}. \tag{25.1}$$

Derived Layer Equivalence

Layers derived from a shared source can be found by a software function that traverses the source's parent links in the lineage knowledge representation. Those found are analysed further to determine whether they represent the same spatial analytic information. Their equivalence is determined by comparing child links and command. That is, derived layers found to have an equivalent set of child links (i.e., input layers) are compared to see if the command and command parameters match as well. Layers found to have been created by the same command applied to the same input data are considered equivalent. This can be expressed by the following, if we let $l_{derived'}$ and $l_{derived''}$ denote occurrences of a layer type $L_{derived'}$, and suppose,

$$X_{derived} = (A_{derived\;command}, A_{derived\;parameters}) \subset A_{derived}$$

then,

$$l_{derived'} \equiv l_{derived''} \tag{25.2}$$

iff

$$(r_{child'} = r_{child''}) \wedge (\forall A_{derivedk} \; \varepsilon \; X_{derived} \wedge a_{derived'k} = a_{derived''k}).$$

Similarity Assessment

Comparative analysis of lineage metadata representations of spatial analytic applications offers the possibility of detecting patterns of spatial reasoning applied within environmental models. Analysis of such metadata focuses on the use of GIS derivation functions within cartographic models that explicitly encode spatial relations implicit in a set of source maps. At this level of conceptualization, metadata attributes for source layers used within equations (25.1) and (25.2) for scale, agency, projection and accuracy could be ignored with little consequence. Only the conceptual nature of the feature classes in the source layers is relevant to compare requirements for explication of spatial relationships.

Determination of the equivalence of source layer feature content, however, can be confounded by conflicts in names used to label feature classes. Such conflicts make it necessary to search for similarity in metadata representations of different realities and differences in metadata representations of the same reality (Batini *et al.,* 1992). For example, synonyms exist when the same feature classes are referred to by different names. Homonyms exist when the same name is used for different feature classes. The use of standard feature classification schemes provides a theoretical basis for ensuring that such synonym and homonym conflicts do not occur when comparing the thematic data content of source layers (see Jensen, 1986).

Examples of standardized feature classification systems include the US Geological Survey Land Use/Land Cover (Anderson *et al.,* 1976), Cowardin Wetland (Cowardin *et al.,* 1979), and California Wildlife-Habitat Relationships (Mayer and Laudenslayer, 1988) systems. Such systems, however, are each created with a specific emphasis. These are but a few of the different standardized (and non-standardized) classification systems used to label digital cartographic data by primary data producers. As a result, comparison of source materials used in spatial analytic applications must often be done across classification systems.

One possible solution to cross-classification system analysis is the classification 'crosswalk'. A classification crosswalk indexes similar categories of feature types found in alternative classification systems. For example, de Becker and Sweet (1988) established a correspondence between vegetation feature classes of the California Wildlife–Habitat Relationships (WHR) classification system with classes specified within nine alternative systems. This correspondence was based on a set of heuristic decision rules. A WHR classification type and another vegetation classification system's type were judged to correspond when their descriptions had a majority of geographic elements in common. This occurred in any of three ways: the WHR type was completely included in the other vegetation type, the other vegetation type was completely included in the WHR type or the WHR type and the vegetation type overlapped — but neither completely included the other. Development of crosswalks for other systems of classifying geographic feature types such as soils, hydrography, land use or land cover will help facilitate systematic comparison of source feature data across spatial analytic applications.

Source Similarity

Analogous criteria could be applied to extend the crosswalk concept to the determination of source layer similarity. For example, sources may be considered similar if feature classes in one include or overlap the feature classes in another. This can be expressed as

$$l_{source'} \approx l_{source''}$$

if   (25.3)

$$A_{source\ features} \ \varepsilon \ X_{source} \ \wedge \ (a_{source'\ features} \cap a_{source''features} \neq \varnothing).$$

This heuristic approach makes a source similarity determination based on the inclusion or overlap of the feature content of one source layer with those of another.

Another approach to source similarity determination is to use taxonomic association. This approach establishes associations of geographic feature classes through a shared connection at a higher level in a common taxonomy. For example, source materials used within a spatial analytic application for identifying croplands within a risk zone around industrial facilities could be viewed as similar to the source materials processed to identify orchards, groves and vineyards at risk from transported hazardous materials. Such similarity could be recognized by use of the US Geological Survey Land Use/Land Cover (LU/LC) Classification System. The croplands source layer in the first application can be identified as LU/LC Level II — class 21 (i.e., cropland and pasture) and the source layer containing orchards, groves and vineyards can be identified as LU/LC Level II — class 22 (orchards, groves, vineyards, nurseries and ornamental horticultural areas). The data within these two layers can be associated semantically via shared membership in class 2 at Level I (Agricultural land). In a similar fashion, industrial locations (Level II, class 13) can be associated with transportation routes (Level II, class 14) via shared membership in Level I, class 1 (urban or built-up land).

Derived Layer Similarity

Taxonomic association and feature overlap (25.3) tests can be used as a basis for modifying equation (25.2) to identify those derived layers resulting from the same processing applied to similar source materials:

$$l_{derived'} \approx l_{derived''}$$

if

$$(r_{child'} \approx r_{child''}) \wedge (\forall A_{derivedk} \ \varepsilon \ X_{derived} \wedge a_{derived'k} = a_{derived''k}).$$

Alternatively, derived layer similarity may focus on identifying similar processing of the same input data,

$$l_{derived'} \approx l_{derived''}$$

if   (25.4)

$$(r_{child'} = r_{child''}) \wedge (\forall A_{derivedk} \ \varepsilon \ X_{derived} \wedge a_{derived'k} \approx a_{derived''k}).$$

This heuristic approach could find cases where different parameters are used within the same transformation applied to the same source materials. The following would be identified as similar: a derived layer resulting from a buffer function applied with a 50 m parameter to a source layer containing lakes, and a derived layer resulting from a 100 m parameter for the buffer function applied to the lakes.

Equation 25.4 might also be applied to identify as similar data layers derived by similar functions applied within different GIS to the same feature data. The determination of similarity between the functions of different GIS requires that the functions of each be associated via generic spatial analytic functions specified within a standard taxonomy. This approach has been tested in a neural network based on interactive activation and competition (IAC) mechanisms (McClelland and Rumelhart, 1987). Lanter (1988) demonstrated a neural network trained with a simple GIS taxonomy to associate analytic commands of two GIS. Once trained, the software was able to translate commands between ERDAS and Map Analysis Package.

Additionally, it might be desirable to identify layers containing features representing similar spatial relations calculated with respect to similar feature class data:

$$l_{derived'} \approx l_{derived''}$$

if (25.5)

$$(r_{child'} \approx r_{child''}) \wedge (\forall A_{derivedk} \; \varepsilon \; X_{derived} \wedge a_{derived'k} \approx a_{derived''k}).$$

Equation 25.5 could be used to create a generalization of a pair of similar spatial analytic applications. For example, suppose one application identifies croplands (USGS LU/LC Level 1: 2, Level 2: 21) within a risk zone around industrial facilities (USGS LU/LC Level 1:1, Level 2:13), and another application identifies orchards, groves and vineyards (USGS LU/LC Level 1:2, Level 2:22) with access to transportation (USGS LU/LC Level 1:1, Level 2:14). This pair of applications could be both matched and generalized as agricultural land within a given distance of urban or built-up land.

Conclusion

This chapter has presented a formal approach for comparing spatial analytic applications of GIS. Lineage metadata have been shown as a basis for tests that determine the equivalence and similarity of source and derived maps. These tests can be used to detect patterns of data use and derivation. They are a basis for formulating generalizations of spatial analytic logic employed within prior applications of GIS. The techniques presented here result from integrating basic research into the structure of spatial thought with applied research in geographic tool design and use. The result is a new metadata management system capable of identifying patterns in prior analytic applications (Geographic Designs, 1993).

Lineage metadata analysis tools offer the environmental researcher capabilities for discovering spatial analytic knowledge hidden in prior GIS applications. This knowledge may take the form of answers to the following questions. Are there a finite number of spatial relationships studied within different environmental applications of GIS? Within such applications, are certain spatial relationships stressed more than others? Are common patterns of logic used to build up complex spatial relationships? Are particular spatial relationships consistently sought at different spatial, thematic and temporal scales? Are certain spatial relationships sought at one family of scales and not at others?

Acknowledgments

This research was conducted at Portugal's National Center of Geographic Information (CNIG), funded by a Fulbright Scholarship and supported by CNIG's President, Rui Goncalves Henriques. The author thanks CNIG's Armanda Rodrigas for her important contribution to the set notation used in this work.

References

Anderson, J. R., Hardy, E., Roach, J. and Witmer, R., 1976, *A Land Use and Land Cover Classification System for Use with Remote Sensor Data,* Professional Paper 964, Reston, Virginia: US Geological Survey.

Batini, C., Ceri, S. and Navathe, S. B., 1992, *Conceptual Database Design: An Entity-Relationship Approach,* Menlo Park: Benjamin/Cummings.

Berry, B., 1964, Approaches to regional geography: A synthesis, *Annals of the American Association of Geographers,* **54**, 2–11.

Berry, J. K., 1987, Fundamental operations in computer assisted map analysis, *International Journal of Geographical Information Systems,* **1**(2), 119–36.

Cowardin, L. M., Carter, V., Golet, F. C. and LaRoe, E. T., 1979, *Classification of Wetlands and Deepwater Habitats of the United States,* Washington, DC: US Fish and Wildlife Service.

Cowen, D. J., 1988, GIS versus CAD versus DBMS: What are the differences? *Photogrammetric Engineering and Remote Sensing,* **54**, 1551–5.

Crain, I. K. and MacDonald, C. L., 1983, From land inventory to land management: The evolution of an operational GIS, *Proceedings of Auto Carto VI,* **1**, 41–50.

Dangermond, J., 1983, A classification of software components commonly used in geographic information systems, in Peuquet, D. J. and O'Callaghan, J. (Eds) *Design and Implementation of Computer-Based Geographic Information Systems,* pp. 70–90, Amherst: IGU Commission on Geographical Data Sensing and Processing.

de Becker, S. and Sweet, A., 1988, Crosswalk between WHR and California vegetation classifications, in Mayer, K. E. and Laudenslayer, W. F. (Eds) *A Guide to Wildlife Habitats of California,* pp. 21–39, Sacramento: California Department of Forestry and Fire Protection.

Dent, B. D., 1993, *Cartography: Thematic Map Design,* Dubuque: Wm. C. Brown Geographic Designs, 1993, *GEOLINEUS Version 3.0 User Manual,* Santa Barbara, California: Geographic Designs.

Guptill, S. C., 1988, *A Process for Evaluating Geographic Information Systems,* Reston, Virginia: Federal Interagency Coordinating Committee on Digital Cartography–US Geological Survey.

Hirtle, S. C. and Heidorn, P. B., 1993, The structure of cognitive maps: Representations and processes, in Barling, T. and Golledge, R. B. (Eds) *Behavior and Environment: Psychological and Geographical Approaches,* pp. 170–92, London: Elsevie.

Jensen, J. R., 1986, *Introductory Digital Image Processing: A Remote Sensing Perspective,* Englewood Cliffs, New Jersey: Prentice-Hall.

Lanter, D. P., 1988, 'A neural network for GIS command language translation', unpublished research paper, University of South Carolina.

Lanter, D. P., 1991, Design of a lineage meta-database for GIS, *Cartography and Geographic Information Systems,* **18**(4), 255–61.

Lanter, D. P., 1993, A lineage meta-database approach towards spatial analytic database optimization, *Cartography and Geographic Information Systems,* **20**(2), 112–21.

Mayer, K. E. and Laudenslayer, W. F. (Eds), 1988, *A Guide to Wildlife Habitats of California,* Sacramento: California Department of Forestry and Fire Protection.

McClelland, J. L. and Rumelhart, D. E., 1987, *Explorations in Parallel Distributed Processing,* Cambridge, Massachusetts: MIT Press.

Muehrcke, P. C. and Muehrcke, J. O., 1992, *Map Use: Reading, Analysis and Interpretation,* Madison: JP Publications.

Nystuen, J. D., 1968, Identification of some fundamental spatial concepts, in Berry, B. and Marble, D. F. (Eds) *Spatial Analysis: A Reader in Statistical Geography,* pp. 35–41, Englewood Cliffs: Prentice-Hall.

Spatial Data Transfer Specification (SDTS), 1992, 'ASTM Section D18.01.05 Draft specification for metadata support in geographic information systems', presented at information exchange forum on spatial metadata, Federal Geographic Data Committee, Reston, Virginia: US Geological Survey.

Tilman, D., 1989, Population dynamics and species interactions, in Roughgarden, J., May, R. M. and Levin, S. A. (Eds) *Perspectives in Ecological Theory,* pp. 89–100, Princeton: Princeton University Press.

Tomlinson, R. F. and Boyle, A. R., 1981, The state of development systems for handling natural resources inventory data, *Cartographica,* **18**(4), 65–95.

SECTION VII

New analytical approaches

A central thesis of the first chapter (Stafford, Brunt and Michener) was that a new dynamic between science and technology has had and will continue to have a significant effect on the ways in which we manage and analyse environmental data and information. Sections II–VI dealt primarily with many of the issues most directly related to technology (scientific databases and information systems, QA/QC, needs for databases for broad-scale research, and environmental modelling and GIS) and organizations (data sharing and standardization). Although there is considerable overlap with the preceding sections, the six chapters in Section VII provide an indication of how the latest technological developments are being utilized to address broad-scale environmental issues. In many cases, the authors identify technological shortcomings and needs for data that mirror similar concerns raised in the previous sections. Thus, these six chapters not only serve as important case studies in their own right, but also serve to reinforce many of the conclusions reached earlier in the book.

Whereas many of the preceding chapters (especially in Section V) emphasized aspects of global change, studies on a regional scale will become increasingly important for characterizing environmental change and developing regional policies and management strategies for dealing with and possibly ameliorating the effects of such changes. Lathrop *et al.* discuss how a model for simulating forest ecosystems is being coupled with a GIS in order to use site-level research to estimate the impacts of atmospheric deposition and climatic change on the northeastern USA. As a part of this effort, they identify problems associated with existing data sets on regional soils and land cover, describing the condition as a 'tug-of-war between what input is needed vs what is available'. Further, they highlight the importance of understanding the spatial variability present in regional-scale data sets and how this information is critical for calibrating and validating broad-scale ecological models.

In Chapter 4, Jelinski, Goodchild and Steyaert discuss the need for developing better tools for detecting, characterizing and modelling changes in ecotones. Brown *et al.* describe two modelling approaches (empirical models of vegetation pattern that provide a regional- to landscape-scale perspective on the biophysical controls on pattern; and models of physiological processes in plants that characterize resource use and partitioning) and how they were used to assess the sensitivity of alpine treeline ecotone to climate change. They argue that the two approaches offer different perspectives that can be integrated for complementary, multiscale analyses linking ecotone patterns and processes, and propose a rule-based hierarchical model that integrates the two approaches.

As Hayden pointed out in Chapter 17, new research is needed for understanding the complex feedback loops between terrestrial ecosystems and the earth's climate (e.g., effects of land use on gas fluxes from major ecosystems). Cohen *et al.* illustrate how a GIS is being used to link both spatial and non-spatial data to climate and ecosystem models for quantifying the effects of logging on carbon fluxes from the forests of the Pacific Northwest region of the USA. Integral components of this research include remote sensing, GIS, complex schemes for database development and management, and a Graphical User Interface that facilitates ease of access to and visualization of the data and the modelling results.

In Section I, several specific challenges to broadening environmental research were identified in the four chapters: the need for better visualization and knowledge-discovery software for dealing with the immense volume of data being collected (Stafford *et al.,* and Gosz); incorporating the human dimension into ecological models and developing better mechanisms for integrating, synthesizing and modelling ecological knowledge (Brown); and developing new spatial techniques for analysing data at multiple scales (Jelinski *et al.*). The final three chapters in Section VII address these important issues from both an applied and a theoretical perspective. Miller presents a knowledge-based systems approach that synthesizes spatially explicit environmental data, with knowledge on vegetation dynamics and succession in order to assess and predict the rapidity and complexity of change in land cover of an area. The system provides guidance on GIS and model usage and offers decision makers the option of running models with different scenarios to gain insight into potential outcomes. Similarly, Mackay, Robinson and Band describe a knowledge-based simulation approach that provides fast, easy access to simulations and allows managers and decision makers to explore the results of various land management plans (e.g., relating reduced density of forest stands in a given area to runoff and erosion). Both approaches serve to insulate users from details about the data and the model, thereby facilitating the transformation of data into knowledge for decision making and policy development. In the final chapter of the section, Bradshaw and Garman examine the relationship between statistical and ecological measures of pattern and process by using simulated landscapes that were generated by various disturbance events. Their results indicate that statistical significance of pattern does not correspond systematically to ecological significance. Finally, they present avenues of research for reassessing the pattern-process paradigm.

26

GIS development to support regional simulation modelling of north-eastern (USA) forest ecosystems

Richard G. Lathrop, Jr, John D. Aber, John A. Bognar, Scott V. Ollinger, Stephane Casset and Jennifer M. Ellis

In an effort to model the impact of atmospheric deposition and climate change on north-eastern US forest ecosystems, we are coupling a forest ecosystem simulation model, PnET, with a grid cell-based geographic information system (GIS) to provide a powerful method of extrapolating site-level research for regional estimation of impacts and interactions. Working at the regional scale has required a simplification of the PnET model to rely on data that can be obtained through remote sensing and/or existing regional scale databases maintained by various US government agencies. This chapter examines the suitability of various digital data sets currently available through the US Geological Survey and the Soil Conservation Service for regional ecosystem modelling. Owing to high amounts of spatial variability, regional soils data are problematic. Coarse-resolution satellite imagery provides reasonable estimates of forest land cover but needs further refinement for other more specific land use/land cover categories.

Introduction

The north-eastern USA is among the most densely populated regions in the nation and also among the most densely forested. Not surprisingly, the region's forests have had a long history of human impact, of destruction and renewal. Largely deforested by the mid-1800s, the north-eastern region now supports extensive areas of northern hardwood, oak, pine and spruce/fir. These forests provide innumerable benefits to the region's population, ranging from timber production, watershed protection and wildlife habitat, to recreational opportunities. However, the region's forests may be entering a new phase in their history as high levels of atmospheric pollution and possible greenhouse-induced climate change threaten their integrity. The recent die-back of high altitude spruce/fir forests throughout the region is one symptom of this ongoing environmental change (Vogelmann and Rock, 1988).

Monitoring studies show a strong regional gradient of anthropogenic

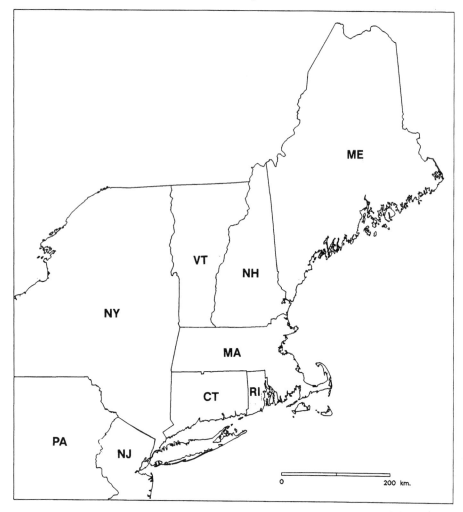

Figure 26.1 Map of north-eastern US forest regional change study area.

atmospheric deposition, with highest values in New York State and gradually decreasing values as one moves eastward into Maine (Ollinger *et al.*, 1993). Superimposed on this regional gradient are hotspots, which receive disproportionately higher levels of nitrogen and sulphur, located along the region's numerous mountain ranges (Lovett and Kinsman, 1990). The high inputs of nitrogen (N) and sulphur (S) to these normally nutrient-poor northern forest ecosystems is a cause of concern. There is some indication that excessive N loading (N saturation) may have adverse effects on both forest health and water quality (Aber *et al.*, 1989).

The desire to understand further the potential impacts of atmospheric deposition and climate change led to the initiation of a multi-investigator effort to understand and model regional environmental change in north-eastern forest ecosystems (Aber *et al.*, 1993). Figure 26.1 shows a map of the north-eastern USA forest change study area. The purpose of the north-eastern forest regional change modelling system is to provide a framework for synthesizing our present understanding of forest ecosystem processes within the region, and to use this information base to project mid- to long-term effects of different levels of climate change or atmospheric deposition on forest and water resources. This will be accomplished by linking observed patterns in ecosystem variables with spatially explicit mechanistic models to estimate forest productivity, water use and biogeochemical cycling at a regional level. The development of the geographic information system (GIS) to support this regional modelling effort is the focus of this chapter.

Regional Scale Modelling

Recently there have been several efforts to synthesize the available information on ecosystem-level processes such as net primary productivity, evapo-transpiration and biogeochemical cycling into mechanistic process-level models (Pastor and Post, 1986; Parton *et al.*, 1988; Running and Coughlan, 1988). A majority of previous modelling studies have investigated the dynamics of specific sites, for example at the level of individual forest stands. Extrapolating the methods and results of these site-specific studies is difficult owing to the temporal and spatial variability of major controlling factors. In attempting to 'scale up' to larger areas, a number of studies have adopted a watershed nutrient-balance approach as a way to integrate the local variability (e.g., see Likens *et al.*, 1977). More recently, investigators have started using a spatially distributed modelling approach to explicitly recognize and incorporate local-level variability, as well as link across scales (Hall *et al.*, 1988).

In the development of the north-eastern regional forest change modelling system, a modular approach has been adopted to provide flexibility in model development and continued refinement, as well as to facilitate integration of modules developed by different research groups (Aber *et al.*, 1993). The first component of the system is PnET, a monthly time step, lumped-parameter model of water and carbon fluxes designed to operate in a spatially distributed manner. PnET explicitly links carbon gain, transpiration and the development of canopy leaf area, and emphasizes the nature and degree of biological control over transpiration (Aber and Federer, 1992). The goal of PnET is to capture the essential variation between ecosystems and across seasons that is relevant to the monthly water and carbon balances of forest ecosystems. The central premise of the model is that aggregation (i.e., lumping of parameters) of climatic data to the monthly scale and biological data (such as foliar characteristics) to the biome level does not cause significant loss of information relative to long-term, mean ecosystem responses.

Table 26.1 PnET model input parameters and data source.

Input parameter	Source
Temperature — mean, max and min daily	NOAA Weather Station
Precipitation mean monthly	NOAA Weather Station/USGS DEM
Solar radiation — potential and actual	NOAA Weather Station/USGS DEM
Soil water holding capacity	SCS STATSGO
Vegetation canopy variables (depends on forest type) foliar retention time, foliar N content, leaf area index	AVHRR, USGS LU/LC

Linking Ecosystem Process Modelling to GIS

To handle the greater computational demands required to organize and retrieve information on any number of environmental parameters in a spatially explicit context, ecosystem modellers are increasingly using GIS. A GIS permits the integration of databases as multiple layers of driving variables that can be linked directly to ecosystem simulation models to estimate ecosystem process rates on a spatially distributed basis. Satellite remote sensing has also played an important role as a crucial source of data pertaining to the land surface characteristics that is readily input to a GIS. The coupling of an ecosystem process model to a GIS has been used to simulate the spatial variability in storage and fluxes of carbon (C) and nitrogen (N) for a regional analysis of the US Central Plains Grasslands (Burke *et al.*, 1990). Spatially distributed ecosystem modelling specifically designed to rely on satellite remote-sensing inputs has been used to simulate water and energy balances over mountainous terrain in Montana (Running *et al.*, 1989).

The PnET model was specifically designed to rely on landscape variables that can be obtained through remote sensing or existing regional scale databases (Table 26.1). Required ecosystem state and driving variables are either stored directly in the GIS or are calculated 'on-the-fly' in PnET from data stored in the GIS. For example, mean monthly precipitation is calculated internally for each grid cell based on the geographic location (latitude and longitude) and elevation (as contained in the GIS elevation file). Figure 26.2 illustrates the input GIS data required to run the PnET model and the output data produced.

In modelling ecosystem processes within the context of a GIS, there are two basic spatial data models to choose from: a raster or grid cell, and a vector or polygonal. A raster data cell is inherently compatible with environmental gradients continuous in space. A vector data model is inherently compatible with linear features/networks and discrete objects with distinct boundaries. The choice of GIS data model should be compatible with the modeller's conception

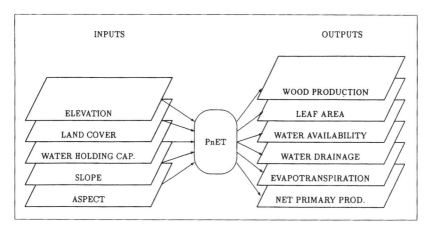

Figure 26.2 PnET input and output GIS data layers.

of the system in question. As we are conceptualizing the region as composing continuous environmental gradients rather than discrete objects, we adopted a grid-cell-based approach. A grid cell data model is particularly suited to integrating and manipulating multiple ecosystem input variables, each stored as a separate thematic layer. In addition, many of the regional scale spatial data sets to be used are either already in or are readily transformed to a grid cell or raster format.

Most of the spatial data processing has been undertaken using the GRASS (*G*eographic *R*esources *A*nalysis *S*upport *S*ystem) GIS and Arc/Info software systems. GRASS is a public domain, grid-cell-based (but with some vector capability) software system developed and maintained by the US Army Construction Engineering Research Laboratory (USACERL, Champaign, Illinois). Arc/Info (ESRI, Redlands, California) is a commercial GIS software system that contains polygonal and, more recently, grid cell (GRID) capabilities. Depending on the data type (e.g., grid cell or polygonal) and the spatial analysis functions required, the most appropriate software package was used. Once processed, the spatial data can be interchanged between any number of grid-cell-based GIS. The final data storage and linkage to the PnET model has been accomplished using both GRASS and IDRISI, a grid-cell-based GIS (Clark University, Worcester, Massachusetts).

Presently the spatial database is stored at two levels of spatial resolution: (1) 100-m grid cells referenced in the Universal Transverse Mercator (UTM) co-ordinate system (zones 18 and 19); and (2) 30-arc-seconds referenced in the geographic (latitude, longitude) co-ordinate system of the World Geodetic System 1972 Datum (WGS 72). The coarser resolution of 30-arc-seconds (approximately 1-km grid cells) was chosen to match the resolution of the Advanced Very High Resolution Radiometer (AVHRR) satellite image data.

Spatial Database Development: Scale and Accuracy

The question of what constitutes an appropriate scale at which to model various ecosystem processes remains open. The choice of the above scales is somewhat arbitrary in that they have no inherent ecological meaning but were determined by the scale of the available data. Ideally, one should be working at the same scale at which the ecological phenomenon or process of interest is operating (Allen and Starr, 1982). This ideal is difficult to put into practice because the major driving variables controlling forest ecosystem processes generally have different patterns and scales of variation. The broad-scale regional gradients of climate and atmospheric deposition are superimposed over landscape-level variations in topography, soil and vegetation. A further difficulty that must be addressed in regional modelling is how to handle the spatial heterogeneity in land surface characteristics. To enhance computing efficiency, there must be some form of data reduction through spatial aggregation, but the distributional complexities of different land surface characteristics must also be addressed adequately. Our objective is to run the PnET model at the coarsest scale possible without seriously sacrificing predictive capability.

To examine the effect of spatial scale on the behaviour of the PnET model, we ran the model at various grid-cell resolutions (30, 100, 300, 450, 600, 1000 and 2000 m) for a 350 km^2 area of predominantly deciduous forest in northern New Jersey. Input data originally mapped at 30 m were successively aggregated to coarser scales using mean or majority rule, depending on the parameter of interest. The PnET outputs were summed over the entire study area and weighted by the amount of forest area (which varied with changing spatial resolution) to give a single output number at each resolution for each output parameter (Figures 26.3a and 26.3b). As spatial resolution increases, the model output parameters tend to increase slightly, then start to level off at greater than 500 m. The percentage change from 30 m to 2000 m was on the order of only 5 per cent. These preliminary studies indicate that the PnET model behaviour is not particularly sensitive to varying spatial resolution, at least across the range of scales investigated.

Spatial modelling efforts within the context of a GIS generally combine a variety of data sources mapped at different scales. Depending on the level of error inherent in the source data and the error produced through data capture and manipulation, the products of GIS analyses, and especially map overlay analyses, may possess significant amounts of error (Walsh *et al.*, 1987). We have recognized this problem and have attempted to analyse the validity of our individual spatial data sets. We have yet to evaluate the errors generated when individual data layers are combined in the context of the spatial ecosystem simulation model.

The remainder of this chapter discusses the development of the spatial database to support the regional forest modelling effort. The suitability of some of the various regional digital data sets available for regional level ecosystem

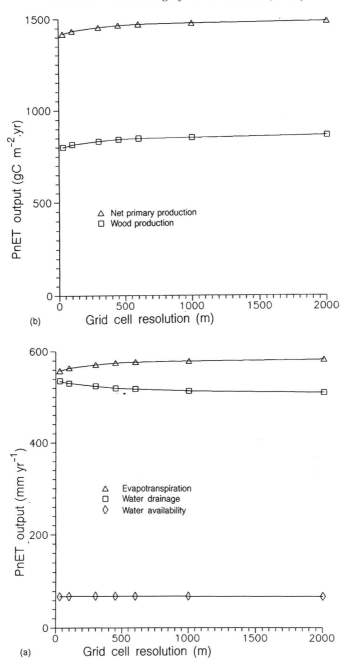

Figure 26.3(a) Plot of PnET outputs of water drainage (mm yr⁻¹), evapotranspiration (mm yr⁻¹) and water availability (mm) vs varying grid cell resolution (m); (b) Plot of PnET outputs of net primary production (gC m⁻².yr) and wood production (gC m⁻².yr) vs varying grid cell resolution (m).

modelling are addressed. In particular, the utility of regional soils and land cover data sets in the context of ecosystem simulation modelling are analysed by comparing them against other 'independent' data sets.

Regional Watershed and State/County Boundaries

The US Geological Survey (USGS) is the primary US distributor of spatial data in both a traditional analogue form and, increasingly, in a digital form. The USGS has made a concerted attempt to improve the quality and availability of digital data by compiling and digitizing maps at consistent scales and according to national standards (US Geological Survey, 1983). Regional scale 'wall-to-wall' digital data sets are generally available at a scale of 1:250 000 (100-m grid cells). Additional data sets are available at a scale of 1:1 000 000 or 1:2 000 000. Presently, higher resolution data (e.g., 1:24 000 or 1:100 000) are available only for limited areas. There will be slight locational discrepancies between the various scaled data sets owing to different map sources and the effect of generalization.

The 1:2 000 000-Scale Digital Line Graph (DLG) data, which are available for the entire country, provide a means to reference visually the regional GIS database. Political boundaries, major transportation routes, hydrological networks (e.g., rivers, coastlines and watershed boundaries) are available. These data were used as a template for determining watershed- or county-level averages (e.g., estimates of total N deposition by watershed or forest cover type by county).

Terrain Data

Land surface elevation and topography exert a major physical control on a number of environmental factors, such as temperature, solar radiation and precipitation. Fortunately, digital elevation data, generally called Digital Elevation Model (DEM) data, are readily available for most of the United States at a scale appropriate for regional modelling efforts. The USGS distributes 1° DEM data as 1° × 1° blocks with 3 × 3 arc-second data spacing derived from 1:250 000 scale source maps. Higher resolution 7.5-minute DEM data (with 30-m grid cell resolution and a 1:24 000 scale) are presently incomplete for the north-eastern states region. Fifty separate 1° DEM coverages were required to cover eastern New York and all the New England states. These individual data sets were processed to create a single coverage. This data set was then processed to create regional maps of slope and aspect that were used in deriving estimates of solar radiation. In mountainous areas, the derivation of slope and aspect from DEM's generalized to coarser resolution (e.g., to 30-arc-second) becomes questionable owing to the smoothing of the spatially heterogeneous terrain.

Climate and Atmospheric Deposition Data

Spatial patterns of precipitation, maximum and minimum temperature, solar radiation and atmospheric deposition across the north-eastern states were evaluated and summarized by Ollinger *et al.* (1993) in a simple spatial model as a function of topographic and geographic position within the region (see also, Ollinger *et al.*, in press). All variables were derived from long-term (30-year) records for 307 National Oceanic and Atmospheric Administration (NOAA) weather stations across New England and New York State. Since higher elevation sites were poorly represented in the NOAA data set, precipitation data from several other high-elevation sites were included, though these records were generally of shorter time duration (7–11 years). Wet deposition of major ions was derived from National Atmospheric Deposition Program/National Trends Network (NADP/NTN) data. Annual means were used to compute long-term, volume-weighted mean concentrations for 26 locations, each containing from three to 11 years of data (mean = 6.7 years).

Ollinger *et al.* (1993) used multiple regression analyses to relate mean monthly precipitation and mean annual deposition to station latitude, longitude and elevation. The resulting regression models were incorporated within the PnET model to calculate the precipitation and wet deposition fields according to the appropriate equations for latitude, longitude and elevation (from the 1:250 000-scale elevation data).

Regional Soils Data

One of the input variables needed by PnET is the soil water-holding capacity (WHC), which provides a measure of the soil moisture potentially available for plant growth. The availability of detailed soils information has been a major stumbling block for regional scale ecosystem modelling. While much of the conterminous USA has been mapped as part of the US Soil Conservation Service (SCS) county-level soil surveys, there are gaps, especially in areas that have historically had low agricultural potential. For example, large sections of northern New England's extensive forested mountain areas have not been mapped in any great detail. In addition, most of the county soil surveys are not presently in digital form or, worse yet, have not been mapped to a rectified base (thus, are not directly suitable for input to a GIS). Recognizing this critical gap, the SCS has been actively trying to create three digital geographic databases: the Soil Survey Geographic Data Base (SSURGO); the State Soil Geographic Data Base (STATSGO); and the National Soil Geographic Data Base (NATSGO) (Reybold and Teselle, 1989).

The SCS has recently completed STATSGO for the north-eastern states. At a scale of 1:250 000, STATSGO was designed to be used primarily for regional resource planning, management and monitoring. STATSGO soil maps are compiled by generalizing more detailed soil maps such as county soil surveys. Data for STATSGO are collected and archived in 1° × 2° topographic quadrangle units, and adjoining 1° × 2° units are matched both within and between states. The digital STATSGO data are distributed as complete state-wide coverages in a polygonal format.

Generalized soil maps are derived from detailed soil maps by combining their delineations to form units that are more extensive but less homogeneous. In a detailed SSURGO soil map, each map unit is usually represented by a single soil component, typically a soil series phase. In contrast, on a STATSGO map, each map unit contains up to 21 components. While the percentage of map unit area composed by each component is known, there is no visible distinction as to the spatial location of these components within the map unit delineated. Each individual soil component is further broken down into different layers corresponding to the different horizons in a soil profile.

For our simulation modelling purposes, we desired a single attribute value (e.g., a single WHC value) for each map unit, in this case each grid cell. To do this, first for every map unit component, the low and high WHC values reported for each soil layer were averaged, multiplied by the thickness (depth) of the layer, and then summed to compute the value for each individual map unit component. The individual WHC values for each component were then weighted by the percentage area of that component to derive a weighted average (by area) for the entire map unit.

The weighted averaging that was done to derive a single WHC value for each STATSGO map unit had a 'smoothing' effect, dampening local variability to highlight regional trends. For example, at the regional level the major trends in WHC are evident with lower values in the sandy glacial deposits of Long Island and Cape Cod (WHC<7 cm), intermediate values (WHC = 12–18 cm) for many of the stony, glacial till soils of New England, and higher values (WHC = 23–28 cm) in the prime agricultural regions of the Caribou region of northern Maine and the St Lawrence lowlands of upper New York State.

For highlighting regional trends, the above STATSGO-derived WHC maps are adequate; however, caution is warranted when using the same maps for modelling at a landscape level because of the inherent generalized nature of the STATSGO data. As one means of assessing the validity of the above STATSGO-derived WHC maps for our modelling purposes, we compared STATSGO data with SSURGO data for two county subareas: (1) Grafton County, New Hampshire; and (2) Hunterdon County, New Jersey. The higher spatial resolution SSURGO data were scaled up to calculate a mean WHC for each STATSGO map unit (at a 1:250 000 scale). The variability of scaled up SSURGO data within a STATSGO class was quite large, with a coefficient of variation of 25 per cent. Regression analysis of the scaled up SSURGO vs the STATSGO data showed little to no relationship (Grafton County: $n = 83$, $R^2 = 0.11$; Hunterdon County: $n = 43$, $R^2 = 0.03$).

This comparison highlights the potential problems in the use of STATSGO to create single attribute maps for landscape or even regional scale modelling. The underlying assumption in our linked GIS-simulation model is that each grid cell is relatively homogeneous. While this assumption is reasonably valid for the finer resolution SSURGO data (though even at finer scales, soil attributes are notoriously spatially heterogeneous and difficult to map), clearly this is not the case for STATSGO data. The STATSGO within-unit variability for WHC is quite large, with the mean coefficient of variation for the various states ranging from 50 per cent to 66 per cent. The STATSGO within-unit variability is often greater than the between-unit variability. In areas of complex terrain where there is a greater heterogeneity of WHC values over short spatial distances, the within-unit variability of STATSGO is exacerbated.

The SCS recognizes fully the limitations of STATSGO and does not recommend its use in a 'traditional' 1-to-1 mapping (i.e., 1 attribute value to 1 map unit) as used in the foregoing (Soil Conservation Service, 1991). Alternatively, the SCS suggests that to present information on a single attribute such as WHC, a series of maps should be produced that show the percentage of the map unit that meets a specified criterion (Bliss and Reybold, 1989), for example, map 1 of WHC < 5.0 cm; map 2: WHC 5–9.9 cm, and map 3: WHC > 10.0 cm. Within each map, there might be four classes of map units: (1) 0–25 per cent of map unit meets criterion (e.g., WHC < 5.0 cm); (2) 26–50 per cent; (3) 51–75 per cent; and (4) 76–100 per cent. This approach holds promise but adds to the complexity of the modelling process by requiring additional model runs to capture the range of variability.

Regional Land Use/Land Cover Data

Land use and land cover are among the most important and widely used environmental data sets. Though often used interchangeably, there is a subtle distinction between the terms land use and land cover. Land cover consists of the biophysical materials covering the earth's surface and land use is the human use to which the land is put. The availability of up-to-date land use/land cover data for the USA is an issue of concern shared by many resource management agencies at all levels of government (FGDC, 1992). As part of the regional change modelling effort, we investigated several sources of land use/land cover data.

Owing to its varied topography and the impact of extensive human development, the north-eastern USA is characterized by a diversity of land use/land cover types. However, our primary focus was on mapping the major forest community or cover types: northern hardwoods, oak/hickory, mixed pine and spruce/fir. The forest vegetation canopy variables of foliar retention time, foliar N content and leaf area index are based on forest cover type. The primary distinction that is important to the PnET model is the distinction between coniferous evergreens and deciduous broad-leaves. Thus, the two major criteria used in evaluating the various possible sources of land cover data were (1) the

Table 26.2 Land use/land cover classification categories.

USGS LU/LC		AVHRR-LU/LC	CONUS
Urban residential	⇒	Urban/residential	Urban/residential
Cropland	⇒	Cropland	Cropland
Rangeland	⇒	Rangeland	Cropland/woodland
Forest			
deciduous		Northern hardwoods	Northern hardwoods
		Southern NE forest/non-forest	Woodland/croplands
coniferous		Spruce/fir	Spruce/fir
mixed		Northern NE mixed	Northern NE mixed
		Central NE mixed forest/non-forest	
Water	⇒	Water	Water
Wetlands	⇒	Wetlands	
(non-forest)			
Tundra	⇒	Tundra	
Barren	⇒	Barren	

accuracy in distinguishing forest from non-forest, and (2) the accuracy in distinguishing conifer from deciduous forests.

Digital land use/land cover data at a scale of 1:250 000 are available for a majority of the USA. The USGS Land Use and Land Cover (LU/LC) maps are compiled to portray the Level I–II categories of the land use/land cover classification system documented by Anderson *et al.* (1976). These data were produced by manual interpretation of high-altitude, colour-infrared aerial photographs (usually at scales smaller than 1:60 000) and compiled to a standard $1° \times 2°$ 1:250 000 topographic base map. All features are delineated in a polygonal format, though the data are available in a grid cell format (called Composite Theme Grid or CTG) at 200-m grid cell resolution. Twenty-five $1° \times 2°$ CTG data sets were acquired and processed for the study region.

The use of manual interpretation in developing the USGS LU/LC data set has pros and cons. Skilled photointerpreters can incorporate a variety of contextual information and thus generally have high accuracy in delineating non-forest land cover classes such as cropland, urban and residential (Table 26.2). A major flaw in the USGS LU/LC data for our purposes was the discrepancy in interpretation of deciduous vs coniferous vs mixed forests across the entire region. Individual photointerpreters were responsible for individual $1° \times 2°$ map sheets. Discrepancies in the interpretation of the three forest cover types (e.g., mixed vs deciduous forest) are readily apparent where map sheets join. To a certain degree, the USGS LU/LC data are obsolete as they are now 10–20 years out of date. For example, LU/LC classes may be incorrect in a rapidly developing suburban fringe but still correct in a comparatively stable agricultural or forest area.

Remote sensing provides a ready means of characterizing forest cover type/ status over large regions of the earth's surface. The spatial and spectral resolution of Landsat Thematic Mapper (TM) is generally considered to be sufficient to classify and map general forest cover types (Iverson *et al.,* 1989b). However, use of TM-derived forest classification for all of the north-eastern USA was not considered because of the prohibitive cost (within the framework of our grant support) associated with the purchase and time required to process approximately 18 scenes.

NOAA Advanced Very High Resolution Radiometer (AVHRR) satellite imagery is another possible source of regional scale land cover data. Due to the high temporal frequency, ready availability and low cost, the use of AVHRR to map major regional land cover type and monitor change is an inviting prospect (Tucker *et al.,* 1985). AVHRR's comparatively coarse resolution (approximately 1-km grid cells) does present some problems because the accuracy of the resulting land cover maps is difficult to validate owing to the lack of suitable 'ground truth' at an appropriate scale (Townshend and Tucker, 1984; Gervin *et al.,* 1985).

Recently there has been an effort by investigators at the USGS EROS Data Center to develop a land cover characteristics database for the conterminous USA based on AVHRR data (dubbed CONUS) (Loveland *et al.,* 1991). Database development involved a stratification of vegetated and barren land, an unsupervised classification of multitemporal AVHRR imagery collected as 19 biweekly data sets from March to October 1990, and post-classification stratification of classes into homogeneous land cover regions using ancillary data. This CONUS data set was acquired and evaluated for the north-eastern US region. The original 167 different land cover class designations were evaluated for their suitability for this particular region. For instance, the CONUS data set does not distinguish urban land cover classes, thus the New York City metropolitan area was originally mapped as desert! The CONUS classes were reassigned as appropriate; as in the foregoing example, all desert categories were reclassed as urban (Table 26.2).

To investigate further the utility of AVHRR for regional land cover mapping, the classification of a single image (NOAA 11 Orbit 05589, 31 October 1989) was undertaken. A late autumn image, as compared with a midsummer image, was used in an attempt to delineate primarily coniferous (with leaves on) vs deciduous (with leaves off) forested regions. Red, near infrared and thermal infrared wavebands (bands 1, 2, 3) were used in a hybrid classification that combined unsupervised clustering and supervised training set delineation. The thermal infrared waveband was useful in distinguishing urban areas with a higher thermal signature. This traditional 'top-down' rather than the 'bottom-up' or spectral mixing model approach advocated by Iverson *et al.* (1989a) was undertaken because the goal was to distinguish the different regional forest cover types, not just forest vs non-forest areas.

Owing to AVHRR's rather coarse spatial resolution and limited spectral wavebands, the delineation of non-forest categories (e.g., urban/residential, cropland/rangeland, water, non-forested wetlands, tundra, barren) is often difficult. To improve the accuracy of non-forest classification, the non-forest Level I categories were extracted from the USGS LU/LC data, coarsened (using majority rule) and merged into the single-date AVHRR classification. For our purposes, the somewhat dated nature of the USGS LU/LC data was not deemed critical. The major forest or mixed forest/non-forest cover types in the merged data set (AVHRR-LU/LC) were derived solely from the AVHRR classification (Table 26.2).

In an attempt to validate the two AVHRR-derived land cover data sets, CONUS and AVHRR-LU/LC were compared with an independent regional land cover data set. The US Forest Service, through the Forest Inventory and Analysis (FIA) programme, conducts periodic inventories of all states to provide up-to-date information on forest resources. The FIA data are based on a sampling procedure utilizing measurements from both aerial photography and ground plots, and provides estimates of forest land area and major forest cover types by county (Powell and Dickson, 1984). The major forest cover types are classified as follows: softwood (white/red pine, spruce/fir and hard pine), oak (oak/pine and oak/hickory), northern hardwood (elm/ash/maple, northern hardwoods and aspen/birch). FIA data that were obtained for each of the seven states in the region were based on inventories from the early to mid-1980s. Area statistics of total forest and major cover type by county were extracted from the CONUS and AVHRR-LU/LC data set and compared with FIA estimates using regression analysis. The regression coefficient of determination, R^2, and the degree to which the data approximate a 1:1 line were used as a basis for comparing the data sets. The FIA comparison does not provide a test of the accuracy of these land cover data sets on a per pixel or grid cell basis, but rather on a per county basis.

For estimating total forest area (per FIA), the CONUS and AVHRR-LU/LC data sets gave similar results. If only 'pure' forest categories were included, the two data sets tended to underestimate forest area (per FIA) by approximately 65 000 hectares. By including the 'mixed' woodland/cropland categories (but not CONUS crop/woodland), both data sets more closely matched the FIA estimates in a nearly one-to-one relationship (Figures 26.4a and 26.4b). The regression of CONUS and AVHRR-LU/LC on FIA total forest area showed little bias (Figures 26.4a and 26.4b) and high coefficient of determination (CONUS $R^2 = 0.88$; AVHRR-LU/LC $R^2 = 0.98$).

The delineation of deciduous vs. coniferous forest type is not as satisfactory. For the coniferous forest cover type, there was little relationship between the CONUS 'pure' conifer class and the FIA data ($R^2 = 0.017$). Combining the CONUS conifer and mixed coniferous/deciduous classes strengthened the relationship ($R^2 = 0.759$), but there was still not a good one-to-one relationship (Figure 26.5a). The relationship between the FIA data and the AVHRR-LU/LC 'pure' and the combined 'pure' and mixed conifer class had a high coefficient of determination ($R^2 = 0.71$ and $R^2 = 0.94$, respectively) but had a slight displacement from a 1:1 relationship (Figure 26.5b). For the deciduous forest cover type (combined oak and northern hardwood types), the combined CONUS deciduous 'pure' forest and mixed forest/non-forest classes tended to overestimate the FIA data ($R^2 = 0.63$) (Figure 26.6a). The combined AVHRR-LU/LC

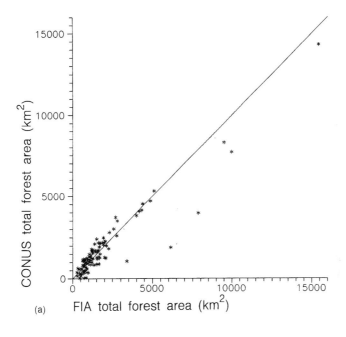

(a)

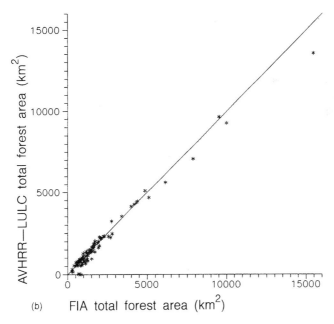

(b)

Figure 26.4(a) Plot of CONUS vs FIA total forest area (km²) by county (R² = 0.88) with one-to-one line; (b) plot of AVHRR-LU/LC (predicted) vs FIA (observed) total forest area (km²) by county (R² = 0.98) with one-to-one line.

deciduous 'pure' forest and mixed forest/non-forest classes gave a reasonable one-to-one relationship (Figure 26.6b) but with a relatively low coefficient of determination ($R^2 = 0.44$).

Overall, the coarse-resolution AVHRR data gave good regional estimates of total forest cover. Merging the higher resolution USGS LU/LC data appeared to enhance further the delineation of non-forest area. The use of the higher resolution data (e.g., manually interpreted aerial photography as in the case of the USGS LU/LC) provides useful information as to the nature of the non-forest categories that may be potentially important in regional modelling applications and difficult to distinguish from spectral classification of satellite data alone (e.g., to differentiate urban/suburban from agricultural land uses).

The CONUS approach of biweekly data sets spanning the entire growing season is appealing; however, it appears to lose some of the discrimination between conifers and deciduous forests that is afforded by the leaf-off imagery used in the single-date classification. The use of leaf-off imagery, either with the biweekly growing season data or as a separate pre- or post-clustering stratification, would enhance future CONUS-type continental-scale, land cover characterization analyses.

Summary and Conclusion

The first component of the north-eastern forest regional change modelling system, PnET, has provided a ready means of codifying our present understanding of forest ecosystem processes, specifically water and carbon fluxes (Aber *et al.,* 1993). To overcome the constraints of data availability at the regional scale, the PnET model was specifically designed such that the construction of models of individual processes, or sets of processes, is as simple as possible, requiring only a few widely available parameters as input (e.g., through remote sensing or existing regional databases). By necessity, the development of PnET and the spatial database have been strongly linked through the tug-of-war between what input is needed vs what is available at the regional scale.

The use of GIS has greatly facilitated the creation of the spatial database needed to support the regional forest change modelling effort. At present, the GIS is primarily serving as a front- and back-end to the simulation model, utilizing the data storage and display capabilities that the GIS software provides. The GIS acts as a data integration tool to assemble the various ecosystem state parameters or driving variables at the same scale in the same georectified coordinate system for input to the simulation model. In this approach, each individual grid cell is treated independently; landscape-level linkages such as spatial configuration or contiguity are not taken into account (e.g., Costanza *et al.,* 1990). At our presently limited state of knowledge of the impact of landscape-level linkages on forest ecosystem processes at a regional scale, a more comprehensive landscape ecology approach is not yet possible.

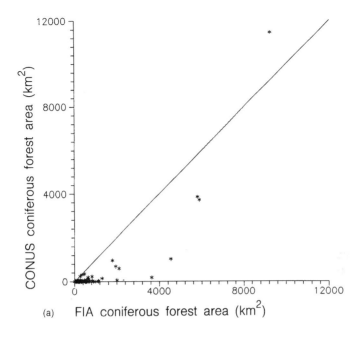

(a)

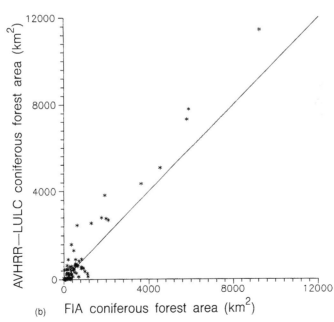

(b)

Figure 26.5(a) Plot of CONUS vs FIA coniferous forest area (km^2) by county (R^2 = 0.76) with one-to-one line; (b) plot of AVHRR-LU/LC vs. FIA coniferous forest area (km^2) by county (R^2 = 0.94) with one-to-one line.

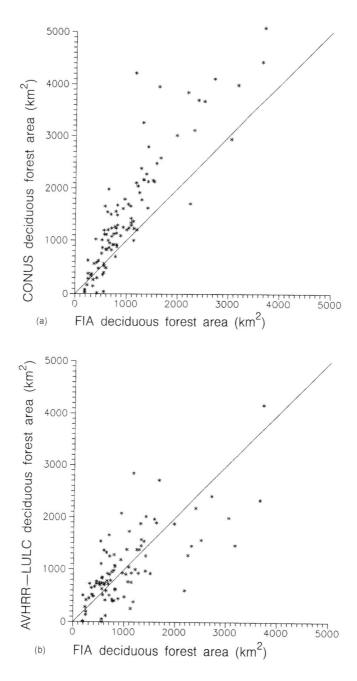

Figure 26.6(a) Plot of CONUS vs FIA deciduous forest area (km²) by county (R² = 0.63) with one-to-one line; (b) plot of AVHRR-LU/LC vs FIA deciduous forest area (km²) by county (R² = 0.44) with one-to-one line.

Understanding the spatial variation of key driving variables is crucial to both calibrate and validate efforts to model ecosystem processes at the larger landscape to regional scales. The development of the regional soils and land cover database illustrates the problems encountered. The rather coarse nature of regional scale data makes an assessment of their accuracy difficult. Where other regional data sets are available, such as the FIA data, some type of validation, or at least comparison, is possible. While accounting for the proportional distribution of classes (e.g., land cover types or soil components) within a map unit or grid cell is feasible, most ecosystem simulation models assume a homogeneous grid cell (1 cell:1 attribute value) and are not yet configured to handle a mixed grid cell situation. To deal with varying degrees of spatial heterogeneity, we envision two alternatives. The first approach works within the constraints of the 1 cell:1 attribute mode but has a nested data structure with subareas of higher spatial resolution within a coarser grained regional matrix. This would provide the ability to increase the model resolution in areas of greater spatial heterogeneity. The second approach does not assume a homogeneous grid cell but attempts to handle spatial complexity by including summary statistics of variability or the frequency distribution of classes in each grid cell as additional input to the model.

At this stage of the project, the development of the GIS database to support our regional ecosystem analysis is complete. The next step is to run the PnET model for several intensive study sites such as the Hubbard Brook Long-Term Ecological Research area to assess the model's validity on a watershed or landscape level. Additional study sites that span the regional atmospheric deposition gradient will be examined to assess the validity of the model for regional scale analysis. The crucial step of properly scaling and validating these regional level models remains a challenging area of research. Regional scale investigations will be increasingly important for determining the regional impacts of environmental change and developing policies to deal with or ameliorate the change.

Acknowledgement

We thank Jesslyn Brown and colleagues at the EROS Data Center in providing access to the CONUS data set. The assistance of Darlene Monds and Russell Kelsea of the SCS in obtaining the STATSGO and SSURGO data is greatly appreciated. We thank several anonymous reviewers for their useful comments which greatly enhanced the final manuscript. This work was supported by the National Aeronautics and Space Administration under grant NASA-NAGW-1825 and the National Science Foundation under grant BSR-8910988.

References

Aber, J. D., Driscoll, C., Federer, C. A., Lathrop, R. G., Lovett, G., Mellillo, J. M., Steudler, P. and Vogelmann, P., 1993, A strategy for the regional analysis of the effects of physical and chemical climate change on biogeochemical cycles in northeastern US forest, *Ecological Modelling,* **67**, 37–47.

Aber, J. D. and Federer, C. A., 1992, A generalized, lumped-parameter model of photosynthesis, evapotranspiration and net primary production in temperate and boreal forest ecosystems, *Oecologia,* **92**, 463–74.

Aber, J. D., Nadelhoffer, K. J., Steudler, P. and Melillo, J. M., 1989, Nitrogen saturation in northern forest ecosystems — hypotheses and implications, *BioScience,* **39**, 378–86.

Allen, T. F. H. and Starr, T. B., 1982, *Hierarchy,* Chicago: University of Chicago Press.

Anderson, J. R., Hardy, E. E., Roach, J. T. and Witmer, R. E., 1976, *A Land Use and Land Cover Classification System for Use with Remote Sensor Data,* Professional Paper 964, Reston, Virginia: US Geological Survey.

Bliss, N. B. and Reybold, W. U., 1989, Small-scale digital soil maps for interpreting natural resources, *Journal of Soil and Water Conservation,* **44**, 30–4.

Burke, I. C., Schimel, D. S., Yonker, C. M., Parton, W. J. and Joyce, L. A., 1990, Regional modeling of grassland biogeochemistry using GIS, *Landscape Ecology,* **4**, 45–54.

Costanza, R., Sklar, F. H. and White, M. L., 1990, Modeling coastal landscape dynamics, *BioScience,* **40**, 91–107.

Federal Geographic Data Committee (FGDC), 1992, *Forum on Land Use and Land Cover: Summary Report,* Reston, Virginia: US Geological Survey.

Gervin, J. C., Kerber, A. G., Witt, R. G., Lu, Y. C. and Sekhon, R., 1985, Comparison of level 1 land cover classification accuracy for MSS and AVHRR data, *International Journal of Remote Sensing,* **6**, 47–57.

Hall, F. G., Strebel, D. E. and Sellers, P. J., 1988, Linking knowledge among spatial and temporal scales: Vegetation, atmosphere, climate and remote sensing, *Landscape Ecology,* **2**, 3–22.

Iverson, L. R., Cook, E. A. and Graham, R. L., 1989a, A technique for extrapolating and validating forest cover across large regions: Calibrating AVHRR data with TM data, *International Journal of Remote Sensing,* **10**, 1805–12.

Iverson, L. R., Graham, R. L. and Cook, E. A., 1989b, Applications of satellite remote sensing to forested ecosystems, *Landscape Ecology,* **3**, 131–43.

Likens, G. E., Bormann, F. H., Pierce, R. S., Eaton, J. S. and Johnson, N. M., 1977, *Biogeochemistry of a Forested Ecosystem,* New York: Springer.

Loveland, T. R., Merchant, J. W., Ohlen, D. O. and Brown, J. F., 1991, Development of a land-cover characteristics database for the conterminous US, *Photogrammetric Engineering and Remote Sensing,* **57**, 1453–63.

Lovett, G. M. and Kinsman, J. D., 1990, Atmospheric pollutant deposition to high elevation ecosystems, *Atmospheric Environment,* **24**, 2767–86.

Ollinger, S. V., Aber, J. D., Federer, C. A., Lovett, G. M. and Ellis, J. M., in press, *Modeling Physical and Chemical Climate Variables across the Northeastern US for a Geographic Information System,* USDA Forest Service Northeastern Station Technical Report, Broomall, Pennsylvania: USDA Forest Service.

Ollinger, S. V., Aber, J. D., Lovett, G. M., Lathrop, R. G. and Ellis, J. M., 1993, A spatial model of atmospheric deposition for the northeastern US, *Ecological Applications,* **3**, 459–72.

Parton, W. J., Mosier, A. R. and Schimel, D. S., 1988, Rates and pathways of nitrous oxide production in a shortgrass steppe, *Biogeochemistry,* **6**, 45–58.

Pastor, J. and Post, W. M., 1986, Influence of climate, soil moisture, and succession on forest carbon and nitrogen cycles, *Biogeochemistry,* **2**, 3–27.

Powell, D. S. and Dickson, D. R., 1984, *Forest Statistics for Maine 1971 and 1982*, USDA Forest Service Northeastern Station Resource Bulletin NE-81, Broomall, Pennsylvania: USDA Forest Service.

Reybold, W. U. and TeSelle, G. W., 1989, Soil geographic data bases, *Journal of Soil and Water Conservation*, **44**, 28–9.

Running, S. W. and Coughlan, J. C., 1988, A general model of forest ecosystem processes for regional applications, I: Hydrological balance, canopy gas exchange, and primary production processes, *Ecological Modeling*, **42**, 125–44.

Running, S. W., Nemani, R. R., Peterson, D. L., Band, L. E., Potts, D. F., Pierce, L. L. and Spanner, M. A., 1989, Mapping regional forest evapotranspiration and photosynthesis by coupling satellite data with ecosystem simulation, *Ecology*, **70**, 1090–1101.

Soil Conservation Service (SCS), 1991, *State Soil Geographic Database (STATSGO) Date Users Guide*, US Soil Conservation Service Miscellaneous Publication 1492, Washington, DC: US Soil Conservation Service.

Townshend, J. R. G. and Tucker, C. J., 1984, Objective assessment of Advanced Very High Resolution Radiometer data for land cover mapping, *International Journal of Remote Sensing*, **5**, 497–504.

Tucker, C. J., Townshend, J. R. G. and Goff, T. E., 1985, African land-cover classification using satellite data, *Science*, **227**, 369–75.

US Geological Survey, 1983, *USGS Digital Cartographic Data Standards: Overview and USGS Activities*, Circular 895A, Reston, Virginia: US Geological Survey.

Vogelmann, J. E. and Rock, B. N., 1988, Assessing forest damage in high-elevation coniferous forests in Vermont and New Hampshire using Thematic Mapper data, *Remote Sensing of Environment*, **24**, 227–46.

Walsh, S. J., Lightfoot, D. R. and Butler, D. R., 1987, Recognition and assessment of error in geographic information systems, *Photogrammetric Engineering and Remote Sensing*, **53**, 1423–30.

27

Remote sensing and GIS techniques for spatial and biophysical analyses of alpine treeline through process and empirical models

Daniel G. Brown, David M. Cairns, George P. Malanson, Stephen J. Walsh and David R. Butler

Two modelling approaches are used to assess the sensitivity of alpine treeline to climate change. Each of the models is limited to a range of spatial scales. First, empirical models of vegetation pattern provide a regional- to landscape-scale perspective on the biophysical controls of the pattern. Generalized linear and additive models are presented as techniques for diagnostic and predictive modelling of spatial pattern. Second, plant physiological process models characterize resource use and partitioning, particularly net carbon gain, at the plant and stand scales. Regression residuals from empirical spatial models were analysed to define regional factors — including geomorphology and geology, drainage basin configurations, summer seasonal solar radiation, snow potential and soil moisture potential — affecting model performance. Experiments with a process model provided insights about limits of available models for use at treeline. Alpine treeline is often found at lower elevations than site-specific resources would indicate because of local and regional factors. Important processes, like edge-related evapotranspiration, and density-independent factors controlled at the regional scale (e.g., wind), are not included in the model. The two modelling approaches offer different perspectives on vegetation pattern that can be integrated for complementary, multiscale analyses linking pattern and process at the alpine treeline ecotone. A rule-based hierarchical model is suggested for integrating the approaches.

Introduction

Studies of treeline dynamics indicate that many northern and alpine treelines have advanced within the last 100 years (Innes, 1991). In the mountainous west of North America, increased tree growth at the alpine treeline during at least part of the twentieth century has been reported from Colorado (Hansen-Bristow *et al.*, 1988), the Pacific Northwest (Graumlich and Brubaker, 1986), the Sierra

Nevada (Scuderi, 1987), and the Canadian Rockies (Kearney, 1982; Luckman, 1986; Colenutt and Luckman, 1991); however, see Graumlich (1991) for caveats. Causes of treeline advance and increased tree growth are uncertain, but many studies cite climate change as the most likely explanation (cf., Hansen-Bristow, 1986; Kullman, 1988, 1993; Graumlich, 1991; Innes, 1991, and references therein).

Although individual trees may be sensitive to climatic change, some studies show that the geographical position of the alpine treeline ecotone (ATE) is remarkably stable. Ives and Hansen-Bristow (1983) and Hansen-Bristow and Ives (1984, 1985) examined the ATE in the Front Range of Colorado, and found it to be positionally static over the course of the last several decades. Butler *et al.* (in press) used repeat photography over the most recent 20-year period to illustrate similarly the stability of upper treeline in the Lewis Range of eastern Glacier National Park, Montana.

It is clear, then, that although climatic conditions play a prominent role in the geographical positioning of the ecotone, other biophysical factors are also, at least locally, important in determining where treeline will be located. Climatic factors affecting tree growth at alpine treeline include short growing seasons, low minimum and seasonally-averaged air temperatures, frozen soils, drought stress, high levels of solar radiation, irregular snow accumulation and strong winds (Hansen-Bristow, 1986). Local-scale controlling variables include topography, plant physiology, geology and substrate, geomorphology, fire, other natural and human-induced disturbances, and the form of the transition from forest to tundra (Hansen-Bristow, 1986; Walsh *et al.*, 1992; Walsh, 1993).

Payette *et al.* (1989) noted that the response of treeline to climate change may be strongly influenced by the structure of the transition (or spatial/biophysical sequence of component types) from trees through krummholz to tundra. Krummholz is a growth form of conifers in which stems are deformed and the overall appearance is that of a low-lying (approximately 0.5–1 m in height) mat. The forest–alpine tundra ecotone is 'characterized by a cessation of the true, closed-canopy forest, an increase in the degree of deformation of the trees, a gradual reduction in their size, and a reduction in the surface area occupied by them so that islands of trees separated by alpine tundra communities of shrubs and herbaceous perennials become more characteristic with increasing altitude' (Hansen-Bristow and Ives, 1984) (also see Figure 27.1). In this chapter, we explore the limitations of two modelling approaches (spatial-empirical and process-mechanistic models) for assessing the sensitivity of the ATE to climate change at different spatial scales. The models are examined for the insights they provide concerning processes controlling the location and pattern of alpine treeline. We suggest some processes or factors that may be lacking in each of the modelling approaches. Our study focuses on the alpine treeline–tundra transition in the central part of the Lewis Range in eastern Glacier National Park, Montana, USA.

We suggest a hierarchical modelling approach for understanding the multiscale controls of the ATE (O'Neill *et al.,* 1988). Figure 27.2 depicts the

Figure 27.1 The forest–alpine tundra ecotone, with several krummholz islands or patches visible (arrows), eastern Glacier National Park.

Note the relatively rapid transition in a narrow elevational band from forest (left) to alpine tundra (lower, right).

patterns and processes affecting treeline at a variety of scales. Hierarchy theory suggests that, for a given scale of investigation, broader scale patterns and processes act as constants or constraints on the scale of observation, and finer scale patterns and processes operate at frequencies too high to be detected (Allen and Starr, 1982). By nesting two modelling approaches, we are able to understand and characterize a span of several levels in the hierarchy. The models are implemented within the context of an Integrated Geographic Information System (IGIS; Davis *et al.,* 1991). All relevant data sets, both primary and secondary, are geographically registered to a common co-ordinate system to facilitate overlay and analysis of multiple layers.

Empirically Based Spatial Models

The initial phase of our study involves the observation and modelling of vegetation–environment relationships at the landscale scale (30 m × 30 m resolution), through the integration of satellite digital data (Landsat Thematic Mapper) with terrain and biophysical disturbance data within the IGIS (Brown,

Landscape (100–10 000 ha)

Components = vegetation communities

Influences on treeline = mesoscale climatology,
topography, disturbance

Modelling approach = spatial empirical models
(TEMTREE)

Stand (0.1–1 ha)

Components = plants

Influences on treeline = snow, elevation, soil,
slope, insolation,
moisture, disturbance

Modelling approach = stand models (BOFORS)

Plant (<0.1 ha)

Components = leaves, roots

Influences on treeline = situation
(nutrients, energy
and water fluxes)

Modelling approach = physiological models
(FOREST-BGC)

Figure 27.2 Scale hierarchy of patterns and processes affecting treeline.

1992). Gradient analysis was presented by Whittaker (1967) as a means of studying the relationships between vegetation patterns on the landscape and environmental conditions. Statistical relationships between values of an environmental variable (e.g., elevation) and vegetation (usually characterized by species importance values) are examined for the deduction of information about a species' realized niche. Gradient models, the equations which represent the vegetation–environment relationships, can be used to understand overlapping species niches (Austin *et al.,* 1990) and to produce potential species range distributions (Davis and Goetz, 1990; Yee and Mitchell, 1991). Whereas traditional gradient analyses are based on species ranges, we use structurally defined vegetation communities that can be identified through the classification of satellite digital data. Given the abruptness of the ATE, spatial models of entire vegetation communities can provide insight into the relative importance of controls on the ecotone.

Characterizations of topographic and environmental gradients used in gradient models have often been qualitative in nature and/or based on observer judgement (Oberbauer and Billings, 1981; Allen *et al.,* 1991). Topographic descriptors, including windward slope, wet meadow, lee slope, snow drift, windblown and minimal snow cover (Oberbauer and Billings, 1981; Burns and Tonkin, 1982), have been used to characterize soil moisture and wind exposure gradients. Topoclimatic potential models, based on gridded elevation data, were developed and/or adapted to provide more quantitative and objective measures of the potential environmental impact on vegetation patterns (Brown, 1991).

Topographical empirical models of treeline (TEMTREE) were developed to capture the gradient relationships at the ATE. Collectively, the TEMTREE models provide (1) a heuristic device, in and of itself, for understanding and exploring processes operating at landscape scales and larger; (2) a predictive device for estimating potential new climax vegetation under different climatic gradient conditions; and (3) a framework within which to structure runs of the physiological process models. For each purpose, the geographical analysis routines available within a geographic information system (GIS) can be used to analyse and map expected and observed spatial patterns of vegetation. Specifically, a database management system (DBMS) linked with statistical and spatial analysis routines, digital terrain modelling functions and the standard suite of GIS functions (e.g, buffer, overlay and distance calculations) provides an environment for exploring the reasons for and implications of a given set of geographical patterns.

Hypotheses regarding the processes that have affected vegetation patterns can be generated through map analysis and comparison. Additionally, comparison of observed and expected levels of vegetation development can be used to identify anomalous sites, where vegetation is overdeveloped or underdeveloped compared with the expected, for simulation and explanation using process models. The predictive ability of the model depends on the variables that are used in the model. Currently, our model is limited by the use of topographical surrogates for climatic variables. Future versions of the models will include optimal interpolations (kriging and co-kriging) of archived climatic data so that specific predictions can be made based on specific changes in averaged climatic parameters.

Vegetation Stand Models

At scales of 0.1–1.0 ha the interactions among individual plants become important processes in the functioning of the ecosystem (Shugart and Prentice,

1992). In studies in plant community ecology, apparent emergent properties are seen at this scale. These properties seem to be the result of both species interactions and of non-equilibrium conditions brought about by time lags. Models at this level of analysis have been widely used in studying potential responses to climatic change and to disturbance, and have been reviewed recently (Shugart *et al.*, 1992; Malanson, 1993). Most forest models are scaled to <1 ha, but other models are not restricted (e.g., Malanson *et al.*, 1992). Malanson (1993) noted that this level of modelling can serve a useful place in a hierarchy of spatially scaled models. They can use the output of finer scaled physiological models for parametrization, and they can include spatially explicit landscape structure and non-equilibrium responses to temporal dynamics.

Many of the models at this scale have developed from the JABOWA–FORET line of simulations (Botkin *et al.*, 1972; Shugart and West, 1977). One of the more recent developments is the BOFORS model, which is modified to take into account the canopy structure of coniferous trees and the resulting gap structure (Bonan, 1992). This model also includes soil temperature, frost and some effects of fire. Although we will not discuss stand modelling in this chapter, we believe that the BOFORS model can potentially be used in a hierarchy between the empirical spatial models described above and the physiological process models described below.

Plant Physiological Process Models

Within the physiological modelling paradigm, models are available for a variety of plant species in both agroecosystems and non-farm ecosystems. In contrast with the gradient models discussed earlier, physiological models represent mechanistically the growth processes of the system of interest. Such plant growth processes operate at a wide range of spatial scales and, as a consequence, the spatial scale of physiological models varies considerably. The goals of process models range from simulating the gas exchange of a single leaf (e.g., Norman, 1993) to those that predict the functioning of entire biomes (Running and Hunt, 1993). With regard to ecosystem processes, such as those of interest at the ATE, models operating at the scale of individual trees are of particular note. Models at this spatial scale strive to simulate elements of forest growth, water balance, nutrient cycling and canopy absorption (e.g., Running and Coughlan, 1988; McMurtrie *et al.*, 1990; Wang and Jarvis, 1990; Aber and Federer, 1992; Weinstein, 1992). However, few of these models account for such variables as wind and snowpack, which are important at the ATE.

The mechanistic representation of plant growth provided by process models allows for the investigation of the interaction of environmental stresses encountered by a single tree or community at the ATE. The ability to represent the interactive nature of stresses placed on trees at the ATE is important because the ATE is hypothesized to represent the location on the landscape where the carbon balance of a tree is zero (Stevens and Fox, 1991). Carbon balance is calculated as the difference between photosynthetic inputs and reductions in the carbon pool due to respiration and tissue death (equation 27.1)

$$\text{Carbon balance} = \Sigma \text{ Photosynthesis} - \Sigma \text{ Respiration} - \Sigma \text{ Senescence} \quad (27.1)$$

The key processes represented in the carbon balance are influenced by a suite of environmental factors, acting both singly and in concert. Therefore, using a physiological model, it should be possible to determine the position along an elevational gradient at which the ATE occurs.

We will discuss the usefulness and difficulties associated with using an existing physiological model (FOREST-BGC) at treeline. FOREST-BGC is a physiologically mechanistic model that simulates processes involved in the carbon, water and nutrient cycles of forest ecosystems. The model specifically calculates processes related to evaporation, transpiration, photosynthesis, respiration, carbon allocation, litterfall and decomposition. Evaporation, transpiration, photosynthesis and respiration are treated in a daily time step within the model, while carbon allocation, litterfall and decomposition are treated in a yearly time step. Although the majority of work done with FOREST-BGC has been at the regional scale (Running and Coughlan, 1988; Running and Nemani, 1991), the 'model treats fluxes only in the vertical dimension, so that horizontal homogeneity is assumed for any defined area' (Running and Coughlan, 1988). The ability to define the operative spatial scale of the model, based on user-specified parameters, facilitates the use of FOREST-BGC for modelling growth at the ATE.

FOREST-BGC requires site-specific climate (e.g., daily precipitation, and minimum and maximum temperatures) and vegetation structure data. These data are easily obtained through the use of remote sensing techniques or, in the case of the climate data, from meteorological stations. The single most important driving variable, with regard to vegetation structure, is total leaf area index (LAI). The model uses LAI to calculate canopy interception, transpiration, respiration, photosynthesis, carbon allocation and litterfall. LAI can be determined either on the ground or by remote sensing techniques (cf. Running *et al.*, 1986).

Agren *et al.* (1991) note that one of the major weaknesses of physiologically based models is their oversimplified consideration of carbon allocation. The most recently published version of FOREST-BGC has overcome this problem by incorporating a dynamic allocation of carbon to leaves, roots and stems based on water and nitrogen limitations (Running and Gower, 1991).

FOREST-BGC has successfully been used to simulate the hydrologic balance and primary production for a forest near Missoula, Montana (Running and Coughlan, 1988). In addition, it has been used to simulate the effects of climatic change on the regional carbon balance of a forest in Montana (Running and Nemani, 1991). Scuderi (1993) used the model at treeline sites in the Sierra

Nevada to identify the relative importance of precipitation and temperature to tree growth. Although FOREST-BGC has been used at ATE locations, it consistently overestimates carbon balance at those locations (L. Band, personal communication), probably because FOREST-BGC was designed to simulate forest interior conditions. Conditions at the ATE are significantly different from those at interior locations. In particular, processes that negatively influence the carbon balance, such as frost desiccation and cuticular abrasion due to blowing snow, are not included in the model.

The most obvious environmental limiting factor at treeline locations is temperature. Early researchers of alpine treeline stressed the importance of temperature in determining the location of the ecotone. In particular, it was noted that the location of treeline corresponds to the point on a mountainside where the mean temperature of the warmest month of the year is 10°C (Zotov, 1938; Daubenmire, 1954). This criterion proved to be a useful estimate of the location of the ecotone; however, it does not provide much insight into the actual mechanism (or mechanisms) by which temperature acts to limit the growth of trees with respect to altitude. A variety of mechanisms have been suggested to explain the effect of temperature on determining the elevation of treeline. Examples of such mechanisms include intracellular freezing of water at extreme low temperatures (Becwar *et al.*, 1981; Becwar and Burke, 1982) and shortening of the growing season (Tranquillini, 1979).

Study Area

Glacier National Park (GNP) is a United Nations-designated International Biosphere Reserve comprising approximately 0.4 million hectares astride the Continental Divide in north-west Montana (Figure 27.3). The Continental Divide exerts a great influence on the climate of the park; west of the Divide, influences of Pacific maritime air dominate, whereas east of the Divide the climate is harsher, with lower temperatures, stronger winds and less precipitation (Walsh *et al.*, 1992).

The ATE in the park is complex, and dependent on conditions of aspect, slope shape and steepness, orientation to prevailing winds, shading from surrounding slopes and disturbance history (Butler and Malanson, 1989; Walsh and Kelly, 1990; Kelly, 1991; Brown, 1992; Bian and Walsh, 1993). In a study of alpine treeline in the Logan Pass area in central GNP, Kelly (1991) used a digital elevation model and interpretation of Landsat Thematic Mapper imagery to derive mean elevation and slope angle values for krummholz. Mean krummholz elevation was 1722 m, and mean slope angle was 22°. Steeper slopes were likely to encounter disturbances precluding tree establishment, in the form of snow avalanches (Butler, 1979), unstable talus and debris flows.

To reduce the complexity of reactions to climate in our study area, we restricted our examination of the ATE to the drier eastern portion of GNP east of the Continental Divide (Figure 27.3). There, upper treeline is dominated by

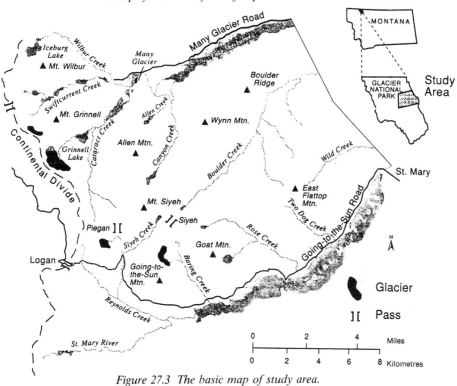

Figure 27.3 The basic map of study area.

subalpine fir (*Abies lasiocarpa*) and five-needle pines (*Pinus flexilis* and *Pinus albicaulis*), with locally some Engelmann spruce (*Picea engelmannii*) and alpine larch (*Larix lyallii*). At many locations east of the Divide, the drier climate precluded extensive Pleistocene glacial scouring of higher elevations, leaving broad, gently sloping uplands at and above treeline (Butler and Malanson, 1989). At such sites, there appear to be no landscape disturbance factors (such as snow avalanches or debris flows) precluding upward movement of treeline in response to climatic change. However, wind may be an important factor at these sites. In other locations, steep glaciated slopes are conducive to mass movements which create a geomorphically controlled treeline (Griggs, 1938; Butler *et al.*, in press). Although fire has apparently affected treeline in some locations in GNP (Habeck, 1969), its influence in our study area is minimal at the elevations of treeline (Brown, 1992).

Methods

TEMTREE

Landsat Thematic Mapper digital data, which had been normalized for the effects of topography, were classified through a hybrid of supervised and unsupervised procedures. The Thematic Mapper data were acquired on 6 August 1986. Two visible channels, the near-infrared channel, and a visible:near-infrared ratio were used in the ISODATA clustering algorithm to identify 50 spectrally homogeneous clusters. Ground control data, collected through a transect-based field sampling approach, were used to group the spectral clusters by structural characteristics of the vegetation canopy: percentage cover of deciduous and coniferous species, meadows and brush. Four treeline components were identified: closed-canopy forest, open-canopy forest, meadow/krummholz and tundra/barren/snow. The vegetation classes were defined to take advantage of the spectral information derived from the Thematic Mapper sensor. Given the broad class definitions, we were able to attain approximately 85 per cent overall accuracy in the classification, relative to ground control site classifications.

Topoclimatic potential models were developed and/or adapted to provide quantitative and objective measures of the potential environmental impact on vegetation patterns (Brown, 1991). In addition, because the models are calculated from 7.5-minute digital elevation models, broad coverage of the resulting spatial variables is available for prediction. Models of topographic site water potential (Beven and Kirkby, 1979; Burt and Butcher, 1985; O'Loughlin, 1986), solar radiation potential (Bonan, 1989; Dubayah *et al.*, 1990), and snow accumulation potential (Frank, 1988) were developed and/or adapted as hypotheses for explaining the patterns of vegetation communities at treeline (in addition to elevation). The topographic site water potential index characterizes the downslope movement of water and its collection in areas of low slope (Wolock *et al.*, 1989). The snow potential index includes topographic curvature, slope aspect and elevation in its calculation. High values indicate areas on leeward, concave slopes, which should receive high snow accumulations (Brown, 1992). The solar radiation potential program (SOLARPOT) is based on the relative exposure of a site to the sun's direct rays throughout a day, and incorporates an algorithm to test for shadows.

Factor analysis was used to reduce the redundancy and multicollinearity in the topographic predictor variables. Two orthogonal factors were extracted from the four variables: elevation, solar potential, topographic moisture potential and snow accumulation potential. Factor 1 represented an exposure gradient, where higher values represented sites that were more exposed to wind and solar radiation. Factor 2 was a ridge–valley gradient, related to both elevation and site moisture potential. Higher values of Factor 2 were at higher elevations and tended to be on ridges with lower site moisture potential.

Geomorphic and biophysical disturbances are active in the Rocky Mountains (Peet, 1988; Butler and Walsh, 1990). Spatial patterns of disturbance can account for much of the potential deviation between potential (climax) vegetation types and actual patterns of vegetation. The primary disturbances affecting the ATE within the study area appear to be snow avalanches, debris flows, snow accumulation and ablation patterns, and historical fires. The IGIS (Integrated Geographic Information System) has been developed to examine the spatial pattern of the disturbances and to explore relationships between topographical, biophysical and topological controls of feature distributions and characteristics (Butler and Walsh, 1990; Walsh *et al.,* 1990; Walsh and Butler, 1991). The database includes information on the location of 121 snow avalanche paths, 157 debris flows and 15 historical fires for GNP since the early 1900s. Locational and attribute information for each of the disturbance types is organized within the Arc/Info software of the IGIS. Sets of morphometric variables are related to each path and flow location. Date of the fire serves as the attribute identifier for each of the historical fires. Snow accumulation and ablation patterns are represented through the snow potential index, and through multitemporal Landsat multispectral scanner mapping of season-sequential snow patterns and the use of Markov analysis (Allen and Walsh 1993). The multitemporal satellite analysis of accumulation and ablation patterns was achieved through radiometric and geometric alignment of seasonal satellite images and the use of Markov transition probabilities for identifying 'from' and 'to' states of classified snowcover conditions. The Markov probabilities were derived from four satellite images representing a set of time periods extending from early spring to early summer for a typical snow-year. Markov analyses were calculated for the entire study area, watershed subdivisions and for the six elevation zones ranging from 1600 m to 2350 m.

Generalized additive models (GAM) were used to identify the form of relationships between topographical factors, disturbance patterns and vegetation classes (Yee and Mitchell, 1991). Logistic regression equations (Periera and Itami, 1991) were then constructed to characterize those relationships for the four treeline components. (Table 27.1 lists the forms of the equations but, for simplicity, not the estimated coefficients.) Each of the models was derived using a systematic spatial sample (at a 20-pixel lag) of pixels that were between 1600 m and 2350 m elevation. This lag was chosen as a trade-off between decreasing spatial autocorrelation at greater lags and increasing sample size at smaller lags.

The variables that were significant for explaining the locations of each of the treeline components are listed in Table 27.1. The Kappa statistic is a measure of agreement between the predicted pattern of the treeline component and the observed pattern for a systematic spatial sample that was spatially offset (by 10 pixels) from the data set with which the models were calibrated. Kappa is a measure of agreement for categorical maps and ranges between 0 and 1.0. The statistic characterizes the actual proportion of sites that is classified the same on two maps less the amount of agreement expected in a random classification (Bishop *et al.,* 1975). The estimate of chance agreement is based on the marginal totals in the category 'confusion matrix'. Using the four derived logistic equations, expected patterns of vegetation were mapped from the predictive variables listed in Table 27.1. Based on the expected vegetation map, which resulted from the assignment of all sites with the 1600 m to 2350 m elevation range to one of the four treeline components, the level of fit between the expected and observed patterns was 0.376 (Kappa).

Table 27.1 Significant variables included in the logistic regression models for each of the four treeline components. [a,b,c]

Closed forest	Open forest	Meadow	Tundra/barren
+ Factor 1	+ Factor 2	+ Factor 1	− Factor 1
− Factor 1^2	− Factor 2^2	+ Factor 2	+ Factor 2
− Factor 2	− Geomorph	− Factor 2^2	+ Geomorph
− Geomorph			
+ Fire			

Kappa statistic[d]

0.461	0.157	0.100	0.614

Notes:

[a]The sign represents direction of the effect (positive or negative).

[b]Superscripts represent exponentiation of variable for quadratic function estimation.

[c]Factor 1 represents relative degree of exposure to sun and wind. Factor 2 is positively related to elevation and negatively related to the topographic moisture index.

[d]Kappa is the predictive accuracy of the individual models, calculated with the sample data offset 10 pixels (in the x and y directions) from the data used to calibrate the models. All Kappa values are significantly different from zero at 0.95 level.

Quantitative comparisons of the satellite-derived (observed) treeline component map and the expected pattern from TEMTREE by drainage basin revealed regional scale controls of treeline location. Maps, such as Figure 27.4, were derived to assess visually the degree of deviation between the predicted treeline component patterns (individually and as a group) and the observed pattern. We hypothesized that processes operating at the scale of the drainage basins were affecting treeline patterns, and that those processes could be characterized by quantifying the effects of terrain. Here, basin-averaged solar radiation, as distinct from site-specific solar radiation receipts, is suggested as one potential regional scale control. Bian and Walsh (1993) indicated that the prevailing scale for terrain variability was directly related to the average basin width of between 2.25 km and 3 km based on 13 drainage basins that ranged in size from 375 ha to 3000 ha.

The measures we use for comparison of expected and observed patterns are based on the 'confusion matrix' (van Genderen, 1977). The Kappa statistic is used to measure the degree of agreement between the expected and observed maps. An index, the Symmetry Index, was also constructed to represent the symmetry of the confusion matrix (Figure 27.5). This index does not character-ize how well two thematic maps agree. But, assuming that classes can be ranked and converted from nominal to ordinal scales (from low to high or vice versa), the index represents the nature of the disagreement (in one direction or the other). A confusion matrix in which sites predicted to be closed-canopy forest are actually open-canopy forests or meadow is qualitatively different from the situation in which predicted meadow sites are actually open- or closed-canopy

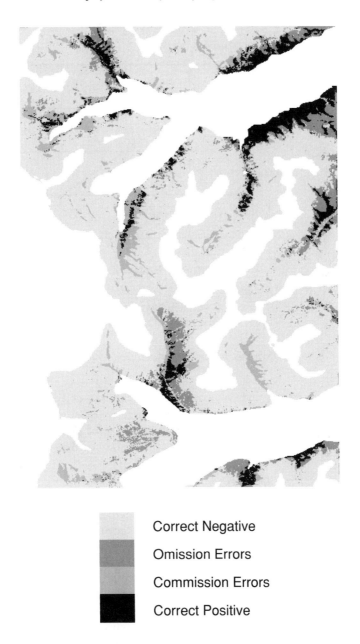

Correct Negative

Omission Errors

Commission Errors

Correct Positive

Figure 27.4 Residual pattern for the closed-canopy forest TEMTREE model.

Note: Correct Positive and Correct Negative indicate correspondence of estimated presence and absence (respectively) of closed-canopy forest between the logistic regression model and the digital classification. Omission errors were incorrectly estimated by the model as absence, whereas commission errors were incorrectly estimated as presence, of closed-canopy forest.

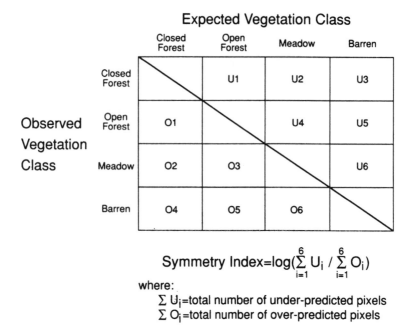

Figure 27.5 Symmetry index calculated from confusion matrix.

forests. The symmetry index quantifies such differences between confusion matrices. Values of the symmetry index greater than zero indicate that vegetation tends to be better established than predicted. Values less than zero indicate that vegetation tends not to be as well established as expected.

Landscape Pattern Indices

Landscape pattern at the ATE, observed through satellite data or modelled through TEMTREE, can be quantified through the derivation of spatial indices within the IGIS environment (Turner, 1990). The following indices may be useful in assessing the influence of landscape pattern on the sensitivity of treeline to climatic changes. The Dominance Index measures the deviation from the maximum possible landscape diversity or heterogeneity. Values near zero indicate that the landscape has many land cover types represented in approximately equal proportions. Values near one indicate that the landscape is dominated by one or a few land cover types. The Contagion Index measures the adjacency, defined as horizontal or vertical joins, of land cover types and indicates whether or not a clustered pattern is present on the landscape. A value of zero indicates a landscape with a single land cover type where all possible adjacencies occur with equal probability. Values near one indicate a landscape with a clustered pattern of land cover types.

The manner of the transition from forested to bare surfaces and its corresponding spatial pattern at the ecotone reflect the biophysical factors that are affecting or have affected the ATE. Payette *et al.* (1989) reported that treelines may respond differently to climate changes depending on the width and patterning of the transition zone from subalpine forests to tundra and barren surfaces. The spatial extent and character of the treeline transition relates to the sensitivity of treeline for vegetation establishment and growth.

The ATE within the study area was stratified horizontally by selected watersheds and the Dominance and Contagion indices were calculated for land cover types at the ecotone (derived from classified Landsat Thematic Mapper digital data) and elevation, slope angle and slope aspect surfaces (derived from digital elevation models) (DEMs; Table 27.2). The study area was also stratified vertically into six evenly distributed elevation classes ranging from 1600 m to 2350 m. Dominance and Contagion indices were calculated for land cover, elevation, slope angle, slope aspect, surficial geology and substrate geology from thematic coverages organized within an IGIS environment by elevation class. Surficial geology and substrate geology (lithology) were digitized from US Geological Survey (USGS) map products (Ross, 1959; Carrara, 1990). Figure 27.6 illustrates how the indices varied for each variable across the six elevation zones. A lower dominance value indicates a more even distribution or a less fragmented landscape, while a high dominance value indicates a more fragmented landscape dominated by a lower number of class types. Surficial geology and land cover show the sharpest increase in dominance values of the variables evaluated. Substrate geology increases at a less pronounced rate. All three variables showed higher rates of dominance value change (increases) beyond an elevation class of 1800 m. A lower contagion value indicates a more dissected distribution of class types, while a higher contagion value indicates a more clumped distribution of class types. Land cover and surficial geology were the only variables that showed an increase in contagion values. Substrate geology showed little deviation with elevation variation, whereas slope angle and slope aspect showed decreases in contagion values beginning at elevation class 3 and class 4 respectively. The dominance and contagion analyses suggest the importance of topography when operating at the landscape level and the variability in landscape features both horizontally and vertically within the study area.

FOREST-BGC

To investigate the effect of temperature on the carbon balance of trees predicted by a single-year run of FOREST-BGC, we used 1984 climate data from Missoula, Montana. This enabled a comparison with most previous runs of the model, which used those data (Running and Coughlan, 1988; Running and Hunt, 1993). We reduced daily maximum and minimum temperature data by

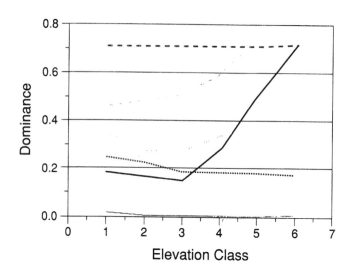

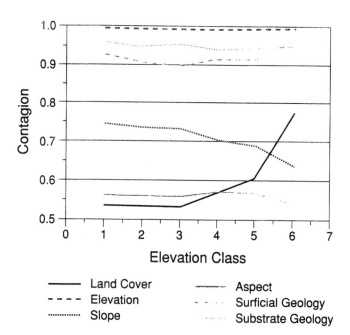

Figure 27.6 Dominance and contagion index values for land cover, elevation, slope angle, slope aspect, surficial geology and substrate geology by elevation zone.

Note: Class 1 ranges from 1600 m to 1725 m and Class 6 from 2226 m to 2350 m; all elevation values are equally distributed between the six classes.

Table 27.2 Dominance and contagion index values for land cover, elevation, slope angle and slope aspect, by drainage basin.

Basin	Land cover Dom.	Cont.	Elevation Dom.	Cont.	Slope angle Dom.	Cont.	Slope aspect Dom.	Cont.
Pyramid Pk	0.271	0.662	0.254	0.926	0.196	0.792	0.057	0.511
Boulder Ck	0.278	0.630	0.253	0.914	0.302	0.780	0.015	0.520
Belly River	0.146	0.614	0.280	0.925	0.200	0.771	0.029	0.505
Kennedy Ck	0.207	0.608	0.278	0.933	0.264	0.794	0.062	0.528
Lk. Sherburne	0.160	0.623	0.257	0.921	0.282	0.766	0.047	0.512
St Marys Lk	0.251	0.641	0.254	0.909	0.282	0.756	0.038	0.518
Reynolds Ck	0.215	0.634	0.237	0.918	0.230	0.751	0.037	0.515
Gunsight Lk	0.306	0.724	0.249	0.903	0.213	0.781	0.070	0.526
Red Eagle Ck	0.388	0.625	0.197	0.903	0.233	0.720	0.030	0.525
Grinnell Ck	0.220	0.633	0.236	0.913	0.212	0.756	0.055	0.488

amounts between 5°C and 30°C at 5°C increments to determine the magnitude of reduction necessary to produce a zero carbon balance. FOREST-BGC is designed to be sensitive to leaf area index (LAI), therefore, at each temperature reduction, we ran the model using four LAI values (3, 6, 9 and 14). In Glacier National Park, trees growing at the ATE exhibit a variety of growth forms ranging from upright trees to matted krummholz. Stands of upright coniferous forest generally have LAI between 3 and 9, and krummholz subalpine fir have an LAI near 14 (Hadley and Smith, 1987).

Results and Discussion

Landscape-Scale Controls

Individual logistic regression models were constructed for each treeline component in separate elevation zones. Significant variables explaining the patterns of each treeline component in each elevation zone are given in Table 27.3. Only those models that were significant (p-value < 0.05) are listed. This analysis confirmed the importance of disturbances, particularly those resulting from geomorphic processes, for controlling vegetation patterns at the ATE. Certain geomorphic processes and conditions (avalanche paths, talus slopes and slopes steeper than 34°) are likely to favour the occurrence of the barren class and preclude the occurrence of forests. These effects are particularly strong between 1900 and 2200 m. Also affecting vegetation patterns within this zone is the relative degree of exposure to solar radiation and wind. Higher exposure seems to favour open-canopy forests while being negatively related to barren surfaces. This may be related to the influence of snow cover on reducing the growing season on sheltered sites. Forests, therefore, tend to ascend to higher elevations on sites that are snow free for a longer period.

Table 27.3 *Significant variables for each treeline component by elevation zone, with Kappa values.*[a]

	Elevation zone (in metres)				
	1600–1750	1750–1900	1900–2050	2050–2200	2200–2350
Closed forest	– Geomorph (0.24)[b]	– Geomorph (0.11)[b]	– Geomorph (0.11)[b]	–	–
Open forest	– Fire	– Fire	– Geomorph	+ Factor 1 – Geomorph – Fire	–
	(0.25)[b]	(0.12)[b]	(0.07)	(0.30)[b]	
Meadow	– (0.20)	–	+ Factor1	–	–
Tundra/barren	–	–	– Factor 1 + Geomorph (0.44)[b]	– Factor 12 + Geomorph (0.39)[b]	–

[a]See notes for Table 27.1.
[b]Kappa statistic value is significant at the 0.95 level. Others not significant.

Kappa values for individual basins ranged from 0.10 to 0.45, and nearly all basin pairs had significantly different Kappa values. Symmetry values varied from – 1.0 to + 1.0. The range of Kappa and Symmetry values indicated that basins were different in terms of the model fit. The basins do not relate directly to the watersheds listed in Table 27.2. Spatially cumulative solar radiation within a basin is a potential basin-scale influence that results from overall basin orientation. The solar radiation model used in the TEMTREE model did not include re-radiation by adjacent terrain, which can increase the effective amount of radiation achieved at a site. Even if re-radiation by adjacent terrain had been included in the model, thermal advection within intensely radiated basins is likely to redistribute energy in these basins more than in those that receive lower average insolation.

There seems to be a general trend in which vegetation in a basin becomes better developed, relative to the model expectations, as basins are increasingly warmed by higher basin average solar radiation (Figure 27.7). Such a relationship is consistent with the hypothesis of temperature control at treeline. At the highest insolation levels, however, increasing basin average radiation results in less successful establishment and growth of certain plant communities (particularly forests) than expected. The two basins with highest average insolation are south-east facing and open directly on to one of the two large lakes within the study area. The existence of flagged trees (trees with branches growing only on one side, indicating high winds or other disturbance) at the openings of those two basins indicates that strong winds may be hindering the development of dense forests in otherwise amenable environments. Although topographical channelling of regional scale flow is likely to be important in the study area, the pattern of vegetation growth by basin suggests that thermal winds may also be

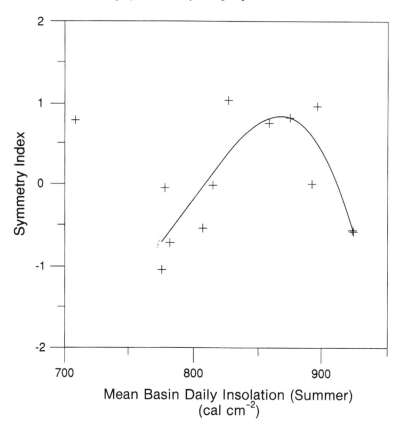

Figure 27.7 Relationship between mean basin solar radiation and symmetry index (R² = 0.70).

important in some basins. Whereas increased temperature resulting from higher insolation has a positive effect on vegetation development in most basins, strong temperature gradients, like that set up between the two basins with the highest insolation levels and the lakes, will tend to produce strong wind currents (Pielke and Segal, 1986). There may be a compounding of desiccating influences in basins that experience the highest levels of insolation. Higher air temperatures and thermally induced flows would both contribute to plant desiccation within those basins.

Basin form may account for additional basin-scale control over treeline. The one basin that did not fit the trend in Figure 27.7, Allen Creek basin, has a broad, gently sloping, upland area that is at high elevations relative to the study area (>2800 m). The gentle nature of this upland area makes it suitable for tundra and krummholz development at an elevation that, in the majority of the study area, is usually barren. Also, basin and mountain forms affect the magnitude and direction of channelized wind flows.

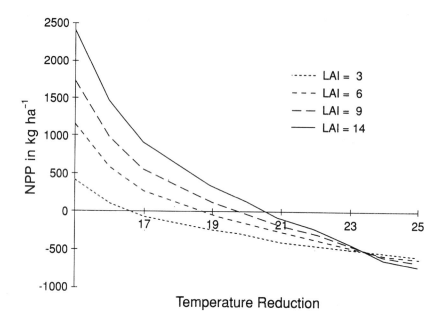

Figure 27.8 Changes in net primary productivity (NPP) with decreasing temperature predicted by FOREST-BGC.

The implications of disturbances and regional scale controls (i.e., thermally induced mesoscale air flow patterns and basin form) need to be understood in order to characterize potential changes in ecosystems due to climatic changes. All disturbance regimes are subject to variation in a changing climate (Overpeck *et al.,* 1990). Changes in the patterns, if not the frequencies and magnitudes, of disturbance resulting from snow avalanches, debris flows, fires, wind and snow accumulation must be included in ecosystem response models of ATE if such models are to be effective predictors of future ecosystem patterns. DISPATCH, a prototype disturbance frequency/magnitude model, may provide a starting point for such research (Baker *et al.,* 1991).

Plant-Scale Controls

To produce a zero carbon balance from FOREST-BGC for GNP, simulated reductions in Missoula temperature values between 16°C and 21°C were necessary (Figure 27.8). Due to its high photosynthetic capacity (LAI = 14), krummholz vegetation required the largest temperature reductions to produce a zero carbon balance, although LAI values of 6, 9 and 14 all produced a zero carbon balance between temperature reductions of 18–21°C.

To determine the elevation increase corresponding to the temperature decrease necessary to produce a zero carbon balance, we used the observed environmental lapse rate. Finklin (1986) notes that, on average, tempertaure decreases 7.8°C per 1000 m in Glacier National Park. A temperature reduction of 20°C would, therefore, correspond to an increase of approximately 2500 m in elevation. The upper limits of the ATE in Glacier National Park occur at approximately 2100 m asl. The 2500 m increase in elevation above the location of the Missoula base station (~ 1000 m asl) would place the ATE at 3500 m asl, approximately 1400 m higher than it is actually located.

These results lead us to conclude that the position of the ATE cannot be adequately modelled based entirely on temperature decreases with altitude. We hypothesize that additional sources of carbon loss account for the discrepancy between our modelled results and the true elevation of the ATE. Potential sources of carbon loss include lethal desiccation, due to both frost drought and increased advective heat exchange. Additionally, the abrading action of high–velocity winds may play an important role in reducing the carbon balance of trees growing at the ATE. In the future, it will be necessary to incorporate other environmental factors, including frost drought, abrasion by wind and advective heat exchange, into physiologically mechanistic growth models in order realistically to model tree growth at the ATE.

Hierarchical Modelling Approach

To scale up a hierarchy from leaf to ecosystem, it is desirable to develop a hierarchical model of ecosystem processes (Malanson, 1993). This model might best be conceived as a metamodel in which models at different hierarchical levels are connected and controlled. Ecosystem-level processes are the sum of processes of the component organisms and the abiotic realm, but they cannot be understood by naively adding components. Hierarchy theory provides an epistemological basis for the study of distinct levels of ecological organization. Allen and Starr (1982) argued that a hierarchical approach combines reductionist and holist research strategies in a valuable way. We propose to use finer scaled models to calibrate coarser scale models in such a modelling approach. We propose that a rule-based approach can be used for this hierarchy, as it has been for physiologically based transfer models (Neilson *et al.*, 1992).

Crossing a range of model levels involves changing both temporal and spatial scales. Models running at fine spatial scales ·(i.e., a leaf) usually simulate processes with fast rates and so iterate at seconds/minutes over the course of hours. For individual plants, models run at intermediate rates and iterate at minutes/hours over days. Stand models iterate at days/years over multiple years. The problem of scaling up has been addressed in the literature. Acock and Reynolds (1990) argued that it is important to scale up from individual tree models to forests, but they propose extrapolating from individual tree models to the whole forest by aggregating information in similar cells based on GIS analysis. Acock and Reynolds (1990) also state that coarse-scale models should not be considered because they represent approaches that are too empirical. More recent analyses have recognized the conceptual and computational problems with a fine-scale mechanistic approach (e.g., Reynolds *et al.,* 1993). For example, simple geographical aggregation is likely to produce error. Rastetter *et al.* (1992) discussed the aggregation of fine-scale data in coarse-scale models. They explained four approaches for addressing the problem: partial transformations of fine-scale equations to coarse-scale process descriptions; a similar but less rigorous approach using a moment expansion; partitioning of aggregation into classes; and calibration of fine-scale equations using coarse-scale data. Combination of the approaches may provide the best means of addressing the problem realistically.

For the multiple controls of the position and pattern of alpine treeline, we propose a hierarchical system that assigns pieces of the landscape to different models for evaluation and projection. The hierarchy is based on pixels and groups of pixels so that remote sensing and GIS techniques can be integrated. The rules will be developed iteratively as we see how models actually perform in specific situations.

In the first step, we will assign treeline area pixels that are controlled by processes that would be slow and intermediate in their response to climatic change. These include geological, geomorphological and some disturbance factors. Geological controls such as cliffs of exposed bedrock can easily be determined from remote sensing and GIS sources. Geomorphological controls can be determined from the same sources but require interpretation by a geomorphologist. Disturbances such as avalanche paths and fires can also be similarly determined (Butler and Walsh, 1990). For the present, these sections of treeline will be stored and not analysed further.

In the second step, we will examine those treeline areas that may respond more quickly to climatic change, although the time lags may still be in the order of decades (e.g., Butler *et al.,* in press). Here, more specific rules are used to assign areas to particular models. In areas of low exposure and moderate relief, the empirical TEMTREE model performs well. In areas of greater relief but with moderate exposure, a version of BOFORS, modified to take into account specific edge effects (cf. Kupfer and Malanson, 1993), will be used. In areas of higher exposure, where wind is a major factor, we will use a substantially modified version of FOREST-BGC. This modification will take into account frost drought, advective heat exchange, and loss of carbon due to abrasion by wind. The detail of this model is appropriate to sites where the carbon balance calculation becomes more complicated.

In a third step, pixels from the study area will be aggregated into classes within each model. The classes will depend on the variables to which each model is sensitive. By grouping pixels into classes each pixel need not be simulated separately, but simulations can be run for a case representing all pixels in a given class. For validation, this procedure will run for treeline areas in Glacier National Park that were not part of the original field studies. If the model works well, climatic change scenarios will be simulated to elucidate the rate of change in treeline position, especially in eastern areas of the park where extensive tundra areas are not far above present treeline.

A recognized problem in ecological research is making projections at higher levels of organization based on processes studied at lower levels (e.g., Field and Ehleringer, 1993; Levin, 1993). For practical purposes at least, it appears that ecosystems contain emergent properties that cannot be reproduced by simple extrapolation from smaller scale processes, yet cannot be understood without reference to these processes. To understand the effects of climatic change on ecosystems and landscapes, it is necessary, therefore, to conduct studies at these levels that are scaled up from a foundation of studies at the plant and leaf level. Such a scaling up allows us to integrate processes in a more mechanistic way than would an additive or extrapolative approach. Importantly, scaling up allows the incorporation of feedbacks into our understanding of the ecosystem. These feedbacks might involve important secondary effects in addition to the direct responses to moisture, temperature and CO_2, such as disturbances. The organization of integrated processes and feedbacks at the ecosystem scale can determine the existence of homeostasis or the degree of resistance and resilience in the ecosystem. One of the important products of this research will be heuristic. First, it will identify those factors that must be understood at the leaf and plant levels in order to predict ecosystem responses, and will set a stage for future experimental physiological ecology. Second, the modelling, in and of itself, will address theoretical developments on procedures for scaling information within a hierarchical system.

Conclusion

The analyses presented in this chapter represent the collaborative efforts of the authors to develop spatially explicit analytical techniques for understanding and modelling the alpine treeline ecotone. We have presented findings from initial phases of our work that have sought to describe the important processes operating at the ecotone. For this purpose, we employed techniques that addressed both spatial patterns and processes. Examinations of pattern and process are important and interrelated components of a complete description of alpine treeline, and other ecosystems or ecotones. We have proposed a hierarchical, rule-based model of vegetation change at alpine treeline that combines both spatial and process knowledge. The model represents an effort that seeks to take advantage of multiple-scale representations within a single framework.

Although some of our work is in progress, a number of specific conclusions may be drawn at this time. In particular, a number of observations may be made regarding the nature of the alpine treeline ecotone that have relevance to our ability to model its potential response to global climate change:

- geomorphic and biophysical disturbances exert significant control on vegetation patterns at the alpine treeline ecotone;

- basin-scale processes, such as basin-averaged insolation, thermally induced winds and drainage basin form, may be important, at least locally;
- the role of the spatial organization of the components of the alpine treeline ecotone needs to be better understood and characterized;
- simple reductions in temperature are insufficient to represent mechanistically how tree growth processes differ at treeline; and
- a hierarchy of models that represent processes at various scales is needed to understand the likely responses of the alpine treeline ecotone to climatic change and to best use the variety of data available.

Acknowledgements

The research described in this chapter was funded in part by grants from the National Science Foundation to DRB, GPM and SJW; and from the National Aeronautics and Space Administration to SJW and DGB. We thank officials of Glacier National Park for housing support during the summers of 1991 and 1992, as well as for necessary co-operation, permits and logistical support. Assistance in the field was provided by Ms Katherine Schipke and Mr Hampton Hager in 1991, and by Mr Bill Welsh (1992). Laboratory analyses in support of the project have been provided by Tom Allen, Tom Evans and Sean McKnight.

References

Aber, J. D. and Federer, C. A., 1992, A generalized, lumped-parameter model of photosynthesis, evapotranspiration and net primary production in temperate and boreal forest ecosystems, *Oecologia,* **92**, 463–74.
Acock, B. and Reynolds, J. F., 1990, Model structure and data base development, in Dixon, R. K., Meldahl, R. S., Ruark, G. A. and Warren, W. G. (Eds) *Process Modelling of Forest Growth Responses to Environmental Stress,* pp. 169–79, Portland: Timber Press.
Agren, G. I., McMurtrie, R. E., Parton, W. J., Pastor, J. and Shugart, H. H., 1991, State-of-the-art of models of production–decomposition linkages in conifer and grassland ecosystems, *Ecological Applications,* **1**, 118–38.
Allen, R. B., Peet, R. K. and Baker, W. L., 1991, Gradient analysis of latitudinal variation in Southern Rocky Mountain forests, *Journal of Biogeography,* **18**, 123–40.
Allen, T. H. F. and Starr, T. B., 1982, *Hierarchy, Perspectives for Ecological Complexity,* Chicago: University of Chicago Press.
Allen, T. R. and Walsh, S. J., 1993, Characterizing multitemporal alpine snowmelt patterns for ecological inference. *Photogrammetric Engineering and Remote Sensings,* **59**, 1521–1529.
Austin, M. P., Nicholls, A. O. and Margules, C. R., 1990, Measurement of the realized qualitative niche: Environmental niches of five *Eucalyptus* species, *Ecological Monographs,* **60**, 161–177.
Baker, W. L., Egbert, S. L. and Frazier, G. F., 1991, A spatial model for studying the effects of climatic change on the structure of landscapes subject to large disturbances, *Ecological Modelling,* **56**, 109–25.

Becwar, M. R. and Burke, M. J., 1982, Winter hardiness limitations and physiography of woody timberline flora, in Li, P. H. and Sakai, A. (Eds) *Plant Cold Hardiness and Freezing Stress: Mechanisms and Crop Implications*, Vol. 2, pp. 307–32, New York: Academic Press.

Becwar, M. R., Rajashekar, C., Hansen Bristow, K. J. and Burke, M. J., 1981, Deep undercooling of tissue water and winter hardiness limitations in timberline flora, *Plant Physiology*, **68**, 111–14.

Beven, K. J. and Kirkby, M. J., 1979, A physically-based variable contributing area model of basin hydrology, *Hydrological Sciences Bulletin*, **24**, 43–69.

Bian, L. and Walsh, S. J., 1993, Scale dependencies of vegetation and topography in a mountainous environment in Montana, *Professional Geographer*, **45**, 1–11.

Bishop, Y. M. M., Fienberg, S. E. and Holland, P. W., 1975, *Discrete Multivariate Analysis: Theory and Practice*, Cambridge, Massachusetts: MIT Press.

Bonan, G. B., 1989, A computer model of the solar radiation, soil moisture, and soil thermal regimes in boreal forests, *Ecological Modelling*, **45**, 275–306.

Bonan, G. B., 1992, A simulation analysis of environmental factors and ecological processes in North American boreal forests, in Shugart, H. H., Leemans, R. and Bonan, G. B. (Eds) *A Systems Analysis of the Global Boreal Forest*, pp. 404–27, Cambridge: Cambridge University Press.

Botkin, D. B., Janak, J. F. and Wallis, J. R., 1972, Rationale, limitations and assumptions of a northeastern forest growth simulator, *IBM Journal of Research and Development*, **16**, 101–16.

Brown, D. G., 1991, Topoclimatic models of an alpine environment using digital elevation models within a GIS, *Proceedings GIS/LIS '91 Conference*, Atlanta, GA, 2, 835–44, Bethesda, MD: American Society for Photogrammetry and Remote Sensing.

Brown, D. G., 1992, 'Topographic and biophysical modeling of vegetation patterns at alpine treeline', unpublished PhD dissertation, Department of Geography, University of North Carolina at Chapel Hill.

Burns, S. F. and Tonkin, P. J., 1982, Soil-geomorphic models and the spatial distribution and development of alpine soils, in Thorn, C. E. (Ed.) *Space and Time in Geomorphology*, pp. 25–43, Boston: George Allen and Unwin.

Burt, T. P. and Butcher, D. P., 1985, Topographic controls of soil moisture distributions, *Journal of Soil Science*, **36**, 469–86.

Butler, D. R., 1979, Snow avalanche path terrain and vegetation, Glacier Park, Montana, *Arctic and Alpine Research*, **11**, 17–32.

Butler, D. R. and Malanson, G. P., 1989, Periglacial patterned ground, Waterton-Glacier International Peace Park, Canada and USA, *Zeitschrift für Geomorphologie*, **33**, 43–57.

Butler, D. R., Malanson, G. P. and Cairns, D. M., in press, Stability of alpine treeline in northern Montana, USA, as revealed by repeat photography, *Phytocoenologia*.

Butler, D. R. and Walsh, S. J., 1990, Lithologic, structural, and topographic influences on snow avalanche path location, eastern Glacier National Park, Montana, *Annals of the Association of American Geographers*, **80**, 362–78.

Carrara, P. E., 1990, *Surficial Geology, Glacier National Park, Montana*, USGS Miscellaneous Investigations Series, MAP I-1508-D, Reston, Virginia: US Geological Survey.

Colenutt, M. E. and Luckman, B. H., 1991, Dendrochronological investigation of *Larix lyallii* at Larch Valley, Alberta, *Canadian Journal of Forest Research*, **21**, 1222–33.

Daubenmire, R., 1954, Alpine timberlines in the Americas and their interpretation, *Butler University Botanical Studies*, **11**, 119–36.

Davis, F. W. and Goetz, S., 1990, Modeling vegetation pattern using digital terrain data, *Landscape Ecology*, **4**, 69–80.

Davis, F. W., Quattrochi, D. A., Ridd, M. K., Lam, N. S.-N., Walsh, S. J., Michaelsen, J. C., Franklin, J., Stow, D. A., Johannsen, C. J. and Johnston, C. A., 1991, Environmental analysis using GIS and remotely sensed data: Some research needs and priorities, *Photogrammetric Engineering and Remote Sensing*, **57**, 689–97.

Dubayah, R., Dozier, J. and Davis, F. W., 1990, Topographic distribution of clear-sky radiation over the Konza Prairie, Kansas, *Water Resources Research*, **26**, 629–90.

Field, C. B. and Ehleringer, J. R., 1993, Introduction: Questions of scale, in Ehleringer, J. R. and Field, C. B. (Eds) *Scaling Physiological Processes*, pp. 1–4, San Diego: Academic Press.

Finklin, A. I., 1986, *A Climatic Handbook for Glacier National Park — with Data for Waterton Lakes National Park*, General Technical Report INT-204, Ogden, Utah: USDA, Forest Service, Intermountain Research Station.

Frank, T. D., 1988, Mapping dominant vegetation communities in the Colorado Rocky Mountain Front Range with Landsat Thematic Mapper and digital terrain data, *Photogrammetric Engineering and Remote Sensing*, **54**, 1727–34.

Graumlich, L. J., 1991, Subalpine tree growth, climate, and increasing CO_2: An assessment of recent growth trends, *Ecology*, **72**, 1–11.

Graumlich, L. J., and Brubaker, L. B., 1986, Reconstruction of annual temperature (1590–1979) for Longmire, Washington, derived from tree rings, *Quaternary Research*, **25**, 223–34.

Griggs, R. F., 1938, Timberlines in the northern Rocky Mountains, *Ecology*, **19**, 548–64.

Habeck, J. R., 1969, A gradient analysis of a timberline zone at Logan Pass, Glacier Park, Montana, *Northwest Science*, **43**, 65–73.

Hadley, J. L. and Smith, W. K., 1987, Influence of krummholz mat microclimate on needle physiology and survival, *Oecologia*, **73**, 82–90.

Hansen-Bristow, K. J., 1986, Influence of increasing elevation on growth characteristics at timberline, *Canadian Journal of Botany*, **64**, 2517–23.

Hansen-Bristow, K. J. and Ives, J. D., 1984, Changes in the forest–alpine tundra ecotone: Colorado Front Range, *Physical Geography*, **5**, 186–97.

Hansen-Bristow, K. J. and Ives, J. D., 1985, Composition, form and distribution of the forest–alpine tundra ecotone, Indian Peaks, Colorado, USA, *Erdkunde*, **39**, 286–95.

Hansen-Bristow, K. J., Ives, J. D. and Wilson, J. P., 1988, Climatic variability and tree response within the forest–alpine tundra ecotone, *Annals of the Association of American Geographers*, **78**, 505–19.

Innes, J. L., 1991, High-altitude and high-latitude tree growth in relation to past, present and future global climate change, *The Holocene*, **1**, 168–73.

Ives, J. D. and Hansen-Bristow, K. J., 1983, Stability and instability of natural and modified upper timberline landscapes in the Colorado Rocky Mountains, USA, *Mountain Research and Develoment*, **3**, 149–55.

Kearney, M. S., 1982, Recent seedling establishment at timberline in Jasper National Park, Atla., *Canadian Journal of Botany*, **60**, 2283–7.

Kelly, N. M., 1991, 'Role of topography in the establishment and maintenance of treeline ecosystem components in Glacier National Park, Montana: An integration of remote sensing methods and digital elevation models', unpublished master's thesis, Department of Geography, University of North Carolina at Chapel Hill.

Kullman, L., 1988, Short-term dynamic approach to tree-limit and thermal climate: Evidence from *Pinus sylvestris* in the Swedish Scandes, *Annales Botanici Fennici*, **25**, 219–27.

Kullman, L., 1993, Pine (*Pinus sylvestris* L.) tree-limit surveillance during recent decades, Central Sweden, *Arctic and Alpine Research*, **25**, 24–31.

Kupfer, J. A. and Malanson, G. P., 1993, Observed and modeled directional change in riparian forest composition at an eroding cutbank, *Landscape Ecology*, **8**, 185–200.

Levin, S. A., 1993, Concepts of scale at the local level, in Ehleringer, J. R. and Field, C. B. (Eds) *Scaling Physiological Processes*, pp. 7–19, San Diego: Academic Press.

Luckman, B. H., 1986, Reconstruction of Little Ice Age events in the Canadian Rocky Mountains, *Géographie physique et Quaternaire*, **40**, 17–28.

Malanson, G. P., 1993, Comment on modeling ecological response to climatic change, *Climatic Change*, **23**, 95–109.

Malanson, G. P., Westman, W. E. and Yan, Y. L., 1992, Realized versus fundamental niche functions in a model of chaparral response to climatic change, *Ecological Modelling*, **64**, 261–77.

McMurtrie, R. E., Rook, D. A. and Kelliher, F. M., 1990, Modelling the yield of *Pinus radiata* on a site limited by water and nitrogen, *Forest Ecology and Management*, **30**, 381–413.

Neilson, R. P., King, G. A. and Koerper, G., 1992, Toward a rule-based biome model, *Landscape Ecology*, **7**, 27–43.

Norman, J. M., 1993, Scaling processes between leaf and canopy levels, in Ehleringer, J. R. and Field, C. B. (Eds) *Scaling Physiological Processes*, pp. 41–76, San Diego: Academic Press.

Oberbauer, S. F. and Billings, W. D., 1981, Drought tolerance and water use by plants along an alpine topographic gradient, *Oecologia*, **50**, 325–31.

O'Loughlin, E. M., 1986, Prediction of surface saturation zones in natural catchments by topographic analysis, *Water Resources Research*, **22**, 794–804.

O'Neill, R. V., DeAngelis, D. L., Waide, J. B. and Allen, T. F. H., 1988, *A Hierarchical Concept of Ecosystems*, Princeton, New Jersey: Princeton University Press.

Overpeck, J. T., Rind, D. and Goldberg, R., 1990, Climate-induced changes in forest disturbance and vegetation, *Nature (London)*, **343**, 51–6.

Payette, S., Filion, L., Delwaide, A. and Begin, C., 1989, Reconstruction of tree vegetation response to long-term climatic change, *Nature (London)*, **341**, 429–32.

Peet, R. K., 1988, Forests of the Rocky Mountains, in Barbour, M. G. and Billings, W. D. (Eds) *North American Terrestrial Vegetation*, pp. 64–101, New York: Cambridge University Press.

Periera, J. M. C. and Itami, R. M., 1991, GIS-based habitat modeling using logistic multiple regression: A study of the Mt. Graham red squirrel, *Photogrammetric Engineering and Remote Sensing*, **57**, 1475–86.

Pielke, R. A. and Segal, M., 1986, Mesoscale circulation forced by differential terrain heating, in Ray, P. S. (Ed.) *Mesoscale Meteorology and Forecasting*, pp. 516–48, Boston: American Meteorological Society.

Rastetter, E. B., King, A. W., Cosby, B. J., Hornberger, G. M., O'Neill, R. V. and Hobbie, J. E., 1992, Aggregating fine-scale ecological knowledge to model coarser scale attributes of ecosystems, *Ecological Applications*, **2**, 55–70.

Reynolds, J. F., Hilbert, D. W. and Kemp, P. R., 1993, Scaling ecophysiology from the plant to the ecosystem: A conceptual framework, in Ehleringer, J. R. and Field, C. B. (Eds) *Scaling Physiological Processes*, pp. 127–40, San Diego: Academic Press.

Ross, C. P., 1959, *Geology of Glacier National Park and the Flathead Region, Northwestern Montana*, USGS Professional Paper 296, Washington: Government Printing Office.

Running, S. W. and Coughlan, J. C., 1988, A general model of forest ecosystem processes for regional applications, I: Hydrologic balance, canopy gas exchange and primary production processes, *Ecological Modelling*, **42**, 125–54.

Running, S. W. and Gower, S. T., 1991, FOREST-BGC, a general model of forest ecosystem processes for regional applications, II: Dynamic carbon allocation and nitrogen budgets, *Tree Physiology*, **9**, 147–60.

Running, S. W. and Hunt, E. R., Jr, 1993, Generalization of a forest ecosystem process model for other biomes, BIOME-BGC, and an application for global-scale models, in Ehleringer, J. R., and Field, C. B. (Eds) *Scaling Physiological Processes: Leaf to Globe*, pp. 141–58, San Diego: Academic Press.

Running, S. W. and Nemani, R. R., 1991, Regional hydrologic and carbon balance responses of forests resulting from potential climate change, *Climatic Change,* **19**, 349–68.

Running, S. W., Peterson, D. L., Spanner, M. A. and Teuber, K. B., 1986, Remote sensing of coniferous forest leaf area, *Ecology,* **67**, 273–6.

Scuderi, L. A., 1987, Late Holocene upper timberline variation in the southern Sierra Nevada, *Nature (London),* **325**, 242–4.

Scuderi, L. A., 1993, A 2000-year tree ring record of annual temperatures in the Sierra Nevada Mountains, *Science,* **259**, 1433–36.

Shugart, H. H. and Prentice, I. C., 1992, Individual-tree-based models of forest dynamics and their applications in global change research, in Shugart, H. H., Leemans, R. and Bonan, G. B. (Eds) *A Systems Analysis of the Global Boreal Forest,* pp. 313–33, Cambridge: Cambridge University Press.

Shugart, H. H., Smith, T. M. and Post, W. M., 1992, The potential for application of individual-based models for assessing the effects of global change, *Annual Review of Ecology and Systematics,* **23**, 15–38.

Shugart, H. H. and West, D. C., 1977, Development of an Appalachian deciduous forest model and its application to assessment of the impact of the chestnut blight, *Journal of Environmental Management,* **5**, 161–79.

Stevens, G. C. and Fox, J. F., 1991, The causes of treeline, *Annual Review of Ecology and Systematics,* **22**, 177–91.

Tranquillini, W., 1979, *Physiological Ecology of the Alpine Timberline,* Berlin: Springer-Verlag.

Turner, M. G., 1990, Landscape change in nine rural counties in Georgia, *Photogrammetric Engineering and Remote Sensing,* **56**, 379–86.

van Genderen, J. L., 1977, Testing land-use map accuracy, *Photogrammetric Engineering and Remote Sensing,* **43**, 1135–7.

Walsh, S. J., 1993, Spatial and biophysical analysis of alpine vegetation through Landsat TM and Spot MX/PN data, in *ACSM/ASPRS Annual Convention and Exposition Technical Papers,* pp. 426–37, New Orleans: American Congress on Surveying and Mapping/American Society for Photogrammetry and Remote Sensing.

Walsh, S. J. and Butler, D. R., 1991, Biophysical impacts on the morphological components of snow-avalanche paths, Glacier National Park, Montana, *Proceedings GIS/LIS '91 Conference,* Atlanta, Georgia, **1**, 133–43, Bethesda, MD: American Society for Photogrammetry and Remote Sensing.

Walsh, S. J., Butler, D. R., Brown D. G. and Bian, L., 1990, Cartographic modeling of snow-avalanche path location within Glacier National Park, Montana, *Photogrammetric Engineering and Remote Sensing,* **56**, 615–21.

Walsh, S. J. and Kelly, N. M., 1990, Treeline migration and terrain variability: Integration of remote sensing digital enhancements and digital elevation models, in Frazier, J. W., Epstein, B. J., Schoolmaster III, F. A. and Lord, J. D. (Eds) *Papers and Proceedings of Applied Geography,* pp. 24–32, Binghamton, NY: John W. Frazier.

Walsh, S. J., Malanson, G. P. and Butler, D. R., 1992, Alpine treeline in Glacier National Park, Montana, in Janelle, D. G. (Ed.) *Geographical Snapshots of North America,* pp. 167–71, New York: Guilford Press.

Wang, Y. P. and Jarvis, P. G., 1990, Description and validation of an array model — MAESTRO, *Agricultural and Forest Meteorology,* **51**, 257–80.

Weinstein, D. A., 1992, *The Response of Plants to Interacting Stresses: TREGRO Version 1.74. Description and Parameter Requirements,* Palo Alto, CA: Electric Power Research Institute.

Whittaker, R. H., 1967, Gradient analysis of vegetation, *Biological Review,* **42**, 207–64.

Wolock, D. M., Hornberger, G. M., Beven, K. J. and Campbell, W. G., 1989, The relationship of catchment topography and soil hydraulic characteristics to lake alkalinity in the Northeastern United States, *Water Resources Research,* **25**, 829–37.

Yee, T. W., and Mitchell, N. D., 1991, Generalized additive models in plant ecology, *Journal of Vegetation Science,* **2**, 587–602.

Zotov, V. D., 1938, Some correlations between vegetation and climate in New Zealand, *New Zealand Journal of Science and Technology,* **19**, 474–87.

28

Using a GIS to model the effects of land use on carbon storage in the forests of the Pacific Northwest, USA

Warren B. Cohen, Phillip Sollins, Peter Homann, William K. Ferrell, Mark E. Harmon, David O. Wallin and Maria Fiorella

Linking ground, reference data sets to climate and other simulation models within a Geographic Information System (GIS) is an important step towards solving problems that have global change implications. One such problem is the effects of land use on gas fluxes from major ecosystems of the world. This chapter illustrates by example how a GIS is being used to link both spatial and non-spatial reference data to climate and ecosystem models, and thereby quantify the effects of logging on carbon fluxes from the forests of the Pacific Northwest (PNW) region of the USA. Images from the Landsat archive are available for the study period, 1972-92. The satellite data are used to map forest conditions and detect logging over time. Climate models are used to develop maps of monthly temperature and precipitation regimes over the study area. A carbon storage model, developed for the PNW region, simulates the amount of carbon stored in living and detrital pools as a function of climate and vegetation conditions within a forest stand. By quantifying changes in carbon storage in the region and accounting for the flow of wood harvested from forest stands through the forest products sector, fluxes of carbon from the PNW region over the 20-year study period are determined.

Introduction

Ameliorating the effects of global climate change on natural and intensively managed ecosystems will require considerably more knowledge than we currently possess about the responses of ecosystems to changes in temperature, moisture and natural and human-caused disturbance. A central question yet to be answered is the role of the terrestrial ecosystem in regulating atmospheric concentrations of greenhouse gases (Houghton *et al.*, 1987; Keeling *et al.*, 1989; Tans *et al.*, 1990). Answering this question largely depends on our ability to assess how carbon storage in the terrestrial ecosystem will change. Two primary factors can directly alter terrestrial carbon stores: (1) effects of changing climate and atmospheric CO_2 concentrations on ecological processes, and (2) natural

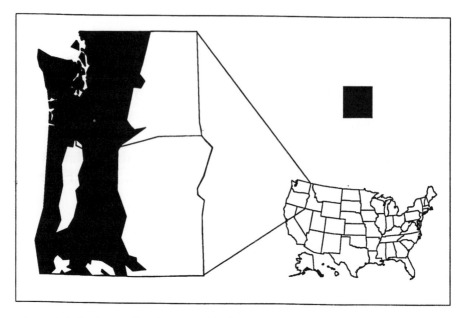

Figure 28.1 Study area for the Pacific Northwest region carbon project, which includes forested portions of western Oregon and Washington.

disturbances and land use. This study focuses on land use effects, a critical topic globally, and one that has been identified as a priority issue in research initiatives at national and international levels (ISSC, 1990).

Our research examines the forested portions of the PNW region of the USA (Figure 28.1), where land use change (in the form of forest clearing) over the last several decades has been extensive (Spies *et al.*, in press). This study requires vast amounts of data that must be acquired, screened, sorted, stored, linked, processed and intelligently managed. These data include nearly 100 Landsat Multispectral Scanner (MSS) and Thematic Mapper (TM) satellite images, digital elevation model (DEM) data, digitized land ownership maps, precipitation and temperature raster images derived from climate simulation models, look-up tables (LUT) derived from a carbon storage model, derived forest land cover and carbon storage maps, and ancillary ground and statistical data. The focus of this chapter is to illustrate how a geographic information system (GIS) is used in our study to link the required spatial and non-spatial data to map carbon stores in, and determine carbon fluxes from, the forests of the PNW region. This chapter does not present results, as only preliminary results are available; these are the subject of a forthcoming paper.

A conceptual model for this research (Figure 28.2 a, b) shows that several major steps are involved. First, three separate and independent procedures must be completed. These are (1) the development of forest age maps from MSS and TM data for five different time periods, from 1972 to 1992; (2) the development

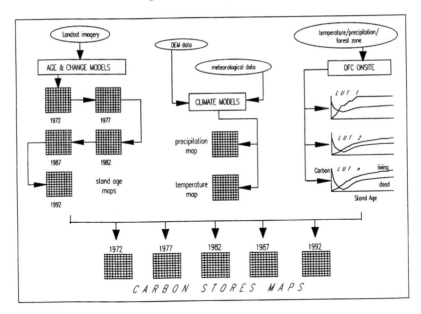

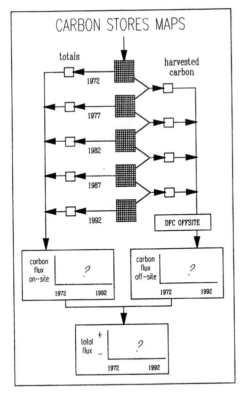

Figure 28.2 Conceptual models for (a) the creation of carbon stores maps for five different years between 1972 and 1992, and (b) calculation of carbon fluxes both within the forest ecosystem (on site) and within the forest products sector (off site).

of precipitation and temperature regime maps; and (3) the creation of carbon
stores look-up tables from the ONSITE module of our carbon model,
Disturbed Forest Carbon (DFC) (Harmon *et al.,* 1990). Subsequently, these
three steps are combined to create carbon stores maps for each of the five time
periods, which are then overlaid to determine changes in carbon stores in the
forest ecosystem. Carbon removed from the site (i.e., harvested carbon) is then
processed using the OFFSITE module of DFC. Finally, carbon fluxes are
determined.

Mapping Land Cover Type and Forest Age Through Time

Satellite imagery was used to develop maps of land cover and forest age for the
study area for each of the five different time periods. In this section, general
procedures for mapping forest conditions are described, followed by more
specific details.

General Procedure

The satellite images used include MSS data, mostly from the years 1972, 1977
and 1982 and TM data from the years 1988 and 1992. From these data, a
baseline year was selected for an initial land cover and forest age map. Ideally,
we would have used the 1972 imagery as a baseline, creating a cover type and age
map for that year. Then stand age could be incremented while stepping through
the 20-year sequence. At each interval a change detection algorithm would be
used to locate harvested stands, which would be set back to age zero. This
idealized strategy was not chosen for one very practical reason: significantly
better mapping accuracy can be expected using the TM images. This is because
of the improved spatial and spectral characteristics of TM images relative to
MSS images. Fortunately, most of the ancillary data available to assist in
mapping are from the period between 1985 and 1988, further strengthening the
probability that more accurate maps can be produced with the TM data. Given
these circumstances, 1988 was chosen as the primary baseline year, and
procedures to track changes through time were altered as follows.

First, the 1992 images are spatially registered to those of the 1988 images, and
the idealized procedure outlined above is followed for that interval; that is, four
years are added to the age of all stands not harvested and the harvested stands
are set back to age zero. For the earlier dates of imagery, the appropriate
number of years can be subtracted from the baseline 1988 age map. The only
problem is that, for stands logged between 1972 and 1988, it is necessary to
determine the age of the stand that existed prior to harvesting. This problem

requires that an additional baseline year be established using the 1972 MSS images. Then, when needed, the 1972 baseline is used.

The expected lower accuracy associated with 1972 baseline MSS age map should not be a significant problem for this study. This is because the overwhelming majority of stands harvested between 1972 and 1988 were over 200 years of age (unpublished data), and forest stands of that age should be relatively easy to identify in MSS imagery (Cohen *et al.*, 1992). Furthermore, although the amount of carbon stored in forest stands changes dramatically over the first 200 years of age, changes in carbon stores after that age are small (Harmon *et al.*, 1990).

Baseline Years: Land Cover Type and Forest Age

Our approach to mapping land cover type and forest age for the baseline years of 1972 and 1988 involves a combination of unsupervised classification and regression analysis. The unsupervised classification depends on the use of ancillary ground data and aerial photographs as reference sources, both of which are available for the full study area. Rather than mosaicking all images of a baseline year and classifying the whole study region at once, which would be computationally prohibitive, the region is first stratified into ecoregions. Stratification involves mosaicking several spatially contiguous satellite images together, then segmenting the mosaic into units that have a greater degree of spectral and ecological homogeneity than those that would result if image boundaries were used as units. The digitized ecoregion map layer of Omernik and Gallant (1986) was used for stratification. This map also has a class that identifies the major non-forested valley floors of the study area; these were excluded from our analyses.

The unsupervised classification resulted in the following classes: water; high alpine snow and ice; urban and other non-forest; open, sparsely vegetated forest; partially vegetated forest; fully vegetated mixed deciduous/coniferous forest; and fully vegetated coniferous forest. Within the forest classes, regression relationships were developed between forest stand age (obtained from the ancillary ground data) and a variety of vegetation indices (derived from the image data and topographic measures calculated from DEM data that had been spatially registered to the satellite imagery). In developing the regression models, locations of the forest stands are digitized in the imagery and DEM data, summaries of the spectral and topographic data are extracted from each forest stand, the summaries are exported to a statistical analysis software package containing the ground-based age data, and regression analysis procedures are performed (Cohen and Spies, 1992). The regression models were then applied to the imagery to create the forest age maps. Roughly, one-half of the ground data were used to develop the regression models, the other half were reserved to quantify the error of the resultant age map. Models are developed and evaluated iteratively until desirable accuracy is reached. After each ecoregion stratum is processed as outlined, the resulting maps are 'stitched' together into the baseline map.

Non-baseline Years: Changes Through Time

Change detection, when done in a spatially explicit manner, involves spatial overlay of two or more images from different dates. Several different change detection algorithms can be used, such as subtracting one image from another, or dividing one image by another. For this study, the simple subtraction method is used.

To locate forest stands that have been harvested between any two successive dates, a vegetation index image of each date is used. The vegetation index chosen here is Tasseled Cap brightness (Kauth and Thomas, 1976; Crist *et al.,* 1986). For the 1988 to 1992 interval (forward direction) the 1988 brightness image is subtracted from the 1992 brightness image (Figure 28.3a). Forest stands that have been harvested will have the highest digital numbers (DN) in the resultant change image. Then, a threshold DN must be found, below which change due to more natural successional forces is the cause. This results in a harvest mask for the 1988 to 1992 interval. If four years are added to the age of forest stands that have not been harvested, and harvested stands are set back to age equals zero, the result is an age map for the year 1992.

Going back in time (backward direction) involves a slightly different approach (Figure 28.3b). For example, consider what happens when the 1982 brightness image is subtracted from the 1988 brightness image. Already known is the age of all forest stands in 1988. Thus, those stands that have not been harvested are six years younger, and this amount can be subtracted from their age. However, the age of the stand that existed prior to harvesting (i.e., 1982) must be determined. In the forward direction the age of stands prior to harvesting was already known. In the backward direction the age determination is made by referring to the 1972 baseline map, and adding 10 years to the age of the forest stands that were not harvested. Like the procedure for going forward in time, this also results in a harvest mask for the 1982 to 1988 time interval and an age map for the year 1982. The backward direction procedure is followed for each of the other two earlier time periods of this study, 1972 to 1977 and 1977 to 1982.

There is a substantial benefit to the use of 1988 as the primary baseline year that has, thus far, not been discussed. When going back in time, stands less than 16 years of age (1988–72) will have been clearcut during the study period. This gives us a means of determining the accuracy of our age models for the open, sparsely vegetated and partially vegetated forest classes. If we find that our age models for these classes are inaccurate, we can adjust them accordingly.

Radiometric Normalization and Geometric Registration

As already noted, adjacent images must be mosaicked prior to using the ecoregion overlay for stratification and images must be spatially overlaid for

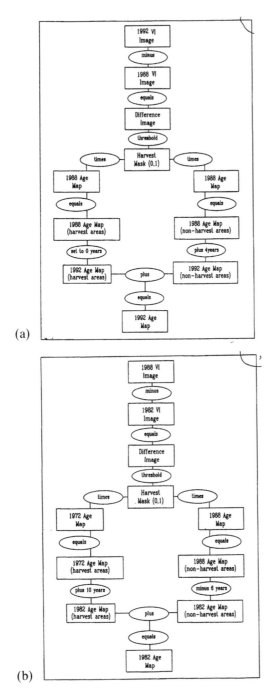

Figure 28.3 Schemes for detection of changes in forest age in (a) the forward direction, and (b) the backward direction.

detection of forest harvesting. These processes require consideration of radiometric normalization and geometric registration issues, so that associated errors are minimized.

Landsat scenes collected on different dates will have varying sun angles and atmospheric conditions, thus their radiometric properties may be considerably different, even for non-harvested forest stands. This could seriously affect the classification of the imagery. Radiometric normalization may partially alleviate this problem, since it involves the use of techniques to 'correct' multiple satellite images for spatial and temporal differences caused by differential atmospheric conditions and illumination and viewing geometries. We use a normalization method developed by Hall *et al.* (1991) that 'matches' digital numbers of each band of a subject image to those of a reference image. First, dark and bright radiometric control sets are selected from the reference and subject images in the Kauth and Thomas (1976) two-dimensional brightness–greenness space. The control sets are selected interactively by viewing images and highlighting potential control set pixels in the images. Then, raw band-to-band linear transformations are developed from regression relationships between control sets of the subject and reference images. The resultant linear transformations are used to normalize separately each band of the subject image.

To normalize TM images radiometrically, a single 1988 TM scene was chosen as the reference image, and all other TM images (from both 1988 and 1992) are radiometrically matched to that image. The same procedure is used for the MSS images. Radiometric matching among these different types of imagery is not feasible because of the different spectral sensitivities of the MSS and TM sensors. Thus, for change detection, the transition from TM to MSS (i.e., the 1982 to 1988 interval) is most likely to yield errors.

Geometric, or spatial, image registration for any given year (i.e., mosaicking) follows radiometric normalization, and is relatively simple. To accomplish this, we take advantage of the overlap among images, by simply 'stitching' them together. Using the 1988 mosaicked set, each cell of which has been previously geocoded, a geometric rectification algorithm (available in a commercial image processing software program) is used to geometrically register mosaics from the other years to the 1988 mosaic.

The primary problem with respect to geometric registration concerns change detection, as even small degrees of misregistration can cause erroneous results around 'sharp' edges such as clearcut and forest boundaries (Townshend *et al.*, 1992). To minimize these effects, all change images are smoothed using a standard smoothing filter, after the proper harvest detection threshold has been determined, as in Figures 28.3a and 23.8b.

Precipitation and Temperature Simulation

To generate the precipitation data layer, the PRISM model of Daly *et al.* (in press) is used. PRISM applies a unique combination of physical and statistical concepts to the analysis of precipitation. The model uses DEM to estimate the elevations of precipitation stations at a 'proper' orographic scale, and a windowing technique and DEM to group stations into individual topographic facets. PRISM then estimates precipitation at a DEM grid cell through a regression of precipitation on orographic elevation (elevation determined from a smoothed DEM).

To generate the temperature data layer, a standardized climatic database for North America is used (Marks, personal communication). This database contains long-term mean monthly temperatures from meteorological stations of the Historic Climate Network. Temperatures from the stations are then interpolated across a 10-km grid cell.

Carbon Storage Computations

Harmon *et al.* (1990) developed the Disturbed Forest Carbon (DFC) model to assess the effect of logging old-growth coniferous forests on carbon flux to the atmosphere. The model's ONSITE module tracks carbon stores in a forest stand in living and detrital pools as a function of time since major disturbance (ie., stand age). Living carbon pools include foliage, branch, bole, coarse roots and fine roots, and detrital pools include fine litter, fine woody debris, coarse woody debris and labile soil carbon. Major driving variables for ONSITE are forest cover type and mean monthly temperature and precipitation. Using these driving variables, ONSITE is run to create several carbon look-up tables (LUT). The model is run once for each cover type, temperature and precipitation combination desired. The LUT are ASCII files in the form of an n-by-m matrix, where n is the number of time intervals (commonly 0–500 years in five-year age increments), and m is the number of carbon pools, plus one for stand age. Using available ground data, estimates of carbon values in the various pools in the LUT at different ages are checked for their accuracy.

For forest soils, the ONSITE module of DFC models only changes in soil carbon that occur over the succession of a forest stand from clearcut to old growth (the labile fraction). The 'stable' fraction of soil carbon varies markedly across the Pacific Northwest. Values range from lows of 30–40 mg ha^{-1} on some high-elevation pumice soils (Boone *et al.*, 1988) to over 250 mg ha^{-1} at some forested sites along the Oregon coast, where the amount of carbon in the soil can equal or even exceed that in vegetation and litter (Binkley *et al.*, 1992). There is virtually no correlation between the stable fraction of soil carbon levels and age or biomass of the forest. Other factors of soil formation, such as topography, climate and parent material, appear to control the stable fraction of soil carbon more than does vegetation. Depth of volcanic ash may be an especially important datum because the ash-derived soils of the PNW appear to have accumulated unusually large amounts of carbon.

To estimate the stable fraction of soil carbon the Oregon STATSGO soil database is used (Soil Conservation Service, 1991). The Oregon database, which has recently become available, consists of 217 spatially explicit map units for the state of Oregon, of which 97 are totally or partially contained in the study area. Each map unit is subdivided into as many as 21 components, and each component is described by minimum and maximum values of several soil characteristics. The minimum and maximum values of organic matter concentration, bulk density and rock content were averaged and used to calculate mineral soil carbon content (mg ha^{-1}) to a depth of 20 cm. Mineral soil carbon content could not be determined to a greater depth because organic matter data were generally limited to the surface mineral soil horizon. The adequacy of this approach to estimate soil C over large regions is being tested by comparing results with measurements made at 400 locations in western Oregon. If the validity of this approach is confirmed, the Washington State STATSGO data will also be used to estimate the stable soil C for the study area.

Carbon Storage Mapping

The DFC model estimates carbon stores in forest stands. Extrapolation of DFC to a regional scale is a multistep process, as shown in Figure 28.2. Thus far, development of the first three separate and independent procedures has been described. Now, these steps must be combined for creation of the regional carbon stores maps. To accomplish this we use several spatially registered data layers, and the carbon stores LUT. The spatial data include the maps of forest age, Omernik and Gallant (1986) ecoregions, temperature and precipitation. Starting at cell 1,1, the cover type, temperature and precipitation map values are referenced. These values dictate which carbon storage LUT to use, and stand age, as referenced in the age map, dictates where in the LUT to find the amount of carbon stored in each living and detrital pool for that cell. These values are then written to new, carbon stores maps, with one map constructed for each pool.

This crucial step in the process, where the age maps, climate maps and carbon LUTs are combined is accomplished within the KHOROS programming environment, and more specifically, the KHOROS CANTATA graphical user interface (Rasure *et al.,* 1990). KHOROS is a software development environment for data processing, graphics and visualization. CANTATA is a graphically expressed data-flow-orientated language that provides a visual programming environment. This environment permits a high level of user interaction at a visual level, enabling one to link models and data (both spatial and non-spatial) and evaluate results, while much of the computer code for the linkages is automatically created.

Carbon in Forest Products

The OFFSITE module of DFC simulates the disposition of trees during and after harvest, taking into account breakage, the various products made from tree

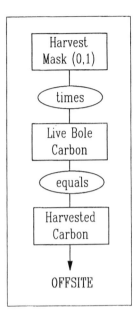

Figure 28.4 Calculation of harvested carbon for input to the OFFSITE module of DFC.

boles, and rates of decay and destruction. Within OFFSITE, harvested boles are converted to various forest products including plywood, lumber, paper, fuel and waste products. Processed logs have a parameter to define ratio of raw materials converted to products and waste. Waste products are added to the short-term pool, assuming the major use was as paper or fuel.

To determine how much carbon is harvested for a given time interval, the harvest mask for that time interval and the bole carbon map for the earlier date of that interval are used (Figure 28.4). For example, to calculate the amount of carbon to process with the OFFSITE module for the first time interval, 1972 to 1977, the bole carbon map of 1972 is overlaid by the harvest mask of that time interval. Then, using a masking algorithm, the total amount of carbon harvested is determined and that value is sent to the OFFSITE module for processing.

Carbon Fluxes

The net regional flux of carbon between the forest and the atmosphere is determined by the change in both on-site and off-site carbon stores over a specific interval. To derive the stores of carbon in a given pool on site in the region for a particular year the values in all cells of that carbon stores map are summed (Figure 28.2b). Subsequently, a temporal profile of on-site carbon stores by pool (live, detrital and total) for the entire region is created. Likewise, the total carbon stores off site for each year are determined. The temporal profile of regional carbon exchange between the forest and the atmosphere (carbon flux) is then computed by combining data on carbon stores on site and off site.

Conclusion

Understanding the complex feedback mechanisms and the multiple interactions between the earth's terrestrial ecosystems and its climate is a dominant focus of current scientific research. This focus is multifaceted, with emphases on development and validation of both predictive and observational models, collection and utilization of an expansive array of data, formation of sophisticated computing interfaces, and development and utilization of emerging analytical tools and techniques. Integral components of the current research environment are remote sensing, GIS, complex database development and management schemes, methods to facilitate ease of access to and visualization of the various types of data, and results of model output.

This chapter illustrates research that takes advantage of many of the important tools and techniques required to address large-scale ecosystem–climate interactions that affect global change. Specifically, an approach for determining carbon stores and fluxes from the forests of the Pacific Northwest (PNW) region of the USA has been described. This approach requires an array of both spatial and non-spatial data derived from ground surveys, remote sensing and models. Furthermore, these data are brought together and processed in a GIS, which uses a graphical user interface to facilitate visualization of outputs.

Acknowledgement

This research was funded in part by the Ecology, Biology, and Atmospheric Chemistry Branch, Terrestrial Ecology Program, of NASA (grant no. W-18,020), and by the HJ Andrews LTER Program, the Global Change Research Program and the Inventory and Economics Program of the PNW Research Station, USDA Forest Service.

References

Binkley, D., Sollins, P., Bell, R., Sachs, D. and Myrold, D., 1992, Biogeochemistry of adjacent conifer and alder/conifer stands, *Ecology*, **73**, 2022–33.

Boone, R. D., Sollins, P. and Cromack, K., 1988, Soil carbon and nitrogen patterns along a mountain hemlock death and regrowth sequence, *Ecology*, **69**, 714–22.

Cohen, W. B., and Spies, T. A., 1992, Estimating structural attributes of Douglas fir/ western hemlock forest stands from LANDSAT and SPOT imagery, *Remote Sensing of the Environment*, **41**, 1–17.

Cohen, W. B., Wallin, D. O., Harmon, M. E., Sollins, P., Daly, C. and Ferrell, W. K., 1992, Modeling the effect of land use on carbon storage in the forests of the Pacific Northwest, in *International Geosciences and Remote Sensing Symposium,* Vol. 2, pp. 1023–26, New York: IEEE Publishing Services.

Crist, E. P., Laurin, R. and Cicone, R. C., 1986, Vegetation and soils information contained in transformed Thematic Mapper data, in *Proceedings, IGARSS '86 Symposium,* pp. 1465–70, Tempe, Arizona: Ecological Society of America Publishing Division.

Daly, C., Neilson, R. P. and Philips, D. L., in press, PRISM: A digital topographic model for distributing precipitation over mountainous terrain, *Journal of Applied Meteorology.*

Hall, F. G., Strebel, D. E., Nickeson, J. E. and Goetz, S. J., 1991, Radiometric rectification: Toward a common radiometric response among multidate, multisensor images, *Remote Sensing of the Environment,* **35**, 11–27.

Harmon, M. E., Ferrell, W. K. and Franklin, J. F., 1990, Effects on carbon storage of conversion of old-growth forests to young forests, *Science,* **247**, 699–702.

Houghton, R. A., Boone, R. D., Fruci, J. R., Hobbie, J. E., Melillo, J. M., Palm, C. A., Peterson, B. J., Shaver, G. R. and Woodwell, G. M., 1987, The flux of carbon from terrestrial ecosystems to the atmosphere in 1980 due to changes in land use: Geographic distribution of the global flux, *Tellus,* **39**B, 122–39.

ISSC (International Social Science Council), 1990, *A Framework for Research on the Human Dimensions of Global Environmental Change,* Series 3, Paris: ISSC/ UNESCO.

Kauth, R. J. and Thomas, G. S., 1976, The tasseled cap — a graphic description of the spectral-temporal development of agricultural crops as seen by Landsat, in *Proceedings, Symposium on Machine Processing of Remotely Sensed Data,* 4b, pp. 41–51, West Lafayette, Indiana: Purdue University.

Keeling, C. D., Bacastow, R. B., Carter, A. F., Piper, S. C., Whorf, T. P., Heimannn, M., Mook, W. G. and Roeloffzen, H. J., 1989, A three-dimensional model of atmospheric CO_2 transport based on observed winds, I: Analysis of observational data, *Geophysical Monograph,* **55**, 165–236.

Omernik, J. M. and Gallant, A. L., 1986, *Ecoregions of the Pacific Northwest,* USEPA/ 600/3-86/033, Corvallis, Oregon: Environmental Research Laboratory, US Environmental Protection Agency.

Rasure, J., Williams, C., Argiro, D. and Sauer, T., 1990, A visual language and software development environment for image processing, *International Journal of Imaging Systems and Technology,* **2**, 183–99.

Soil Conservation Service, 1991, *State Soil Geographic Data Base (STATSGO), Data Users Guide,* Miscellaneous Publication Number 1492, Soil Conservation Service, United States Department of Agriculture.

Spies, T. A., Ripple, W. J. and Bradshaw, G. A., in press, The dynamics and pattern of a managed coniferous landscape, *Ecological Applications.*

Tans, P. P., Fung, I. Y. and Takahashi, T., 1990, Observational constraints on the global atmospheric CO_2 budget, *Science,* **247**, 1431–8.

Townshend, J. R. G., Justice, C. O., Gurney, C. and McManus, J., 1992, The impact of misregistration on change detection, *IEEE Transactions in Geosciences and Remote Sensing,* **30**, 1054–60.

29

Coupling of process-based vegetation models to GIS and knowledge-based systems for analysis of vegetation change

David Miller

Empirical and process-based models have been developed for studying plant succession and growth at scales varying from the individual species to the region. A knowledge-based system (KBS) approach is presented that synthesizes spatially explicit environmental data with knowledge on vegetation dynamics and succession. The KBS controls the organization and selection of GIS functions for running the succession model. Metadata on spatial data and procedural knowledge are maintained as inputs to system guidance and system auditing. Vegetation survey data, compiled and represented at resolutions compatible with ground, aerial photograph and satellite image platforms, are synthesized for input to process-based models of vegetation change. Contextual measures of the distribution of semi-natural vegetation with respect to the ecological environment are used within the assessment of likely changes in its boundaries. The adjacency of land cover types and the rate of transition between vegetation types at the boundary are inputs to a simulation of the succession model. From this, an assessment of the rapidity of change and the possible complexity of changes in land cover can be made as well as a prediction of future land cover for an area.

Introduction

The aim of this chapter is to demonstrate how diverse fields of expertise may be linked within a knowledge-based system to map current, and to predict future, distribution of semi-natural vegetation. That is, vegetation in areas where there is no intensive agricultural cropping or recently sown pastures; these areas often are dominated by heather moorland, bent-fescue grasslands and non-plantation woodland (Taylor, 1986).

Coupling independent environmental models and GIS using knowledge-based techniques may be achieved simply by exchanging files between GIS and an environmental model where the GIS capabilities are only used for the input of data and the display of output. Close coupling requires a sufficiently open GIS architecture to provide the necessary linkages and interfacing, possibly in the form of a toolbox. The system described in this chapter permits both loose and tight coupling (Goodchild *et al.*, 1993), according to the specific objectives.

Models of vegetation change may be based on knowledge of past states at a location. A synthesis of spatial data processing techniques with reference to a process-based understanding of vegetation dynamics may provide for the development of a strategy for *what* methodology and *where* to observe for monitoring those semi-natural classes that have been least successfully mapped. If enough knowledge about the environment is available, the coupling of non-remotely sensed data and remotely sensed data may be used to characterize features identified from the imagery.

Bracken (*Pteridium aquilinum* L. Kuhn) was chosen as the vegetation type to be used in building the prototype system. Almost uniquely in the areas of the UK not intensively farmed, bracken can rapidly spread and dominate the land. It can pose a threat to animal health if eaten, causing cancer of the throat and stomach, and perhaps to human health owing to the carcinogens that are believed to leach into the water supply from the rhizomes (Taylor, 1986). Research into predicting the land most susceptible to bracken spread can permit more efficient targeting of financial resources towards control measures (chemical or non-chemical) into high-risk areas.

Extensive research has been carried out on bracken ecology, its interaction with other plant species, and its geographical environment (Watt, 1947). It spreads (reproduction is by rhizomes or spores) at rates that lead to a loss of agricultural land, the reclamation of which can be costly (Taylor, 1986). In this study we concentrate on rhizomal spread. Using bracken as the subject of the study means that techniques developed are applicable to other plant species for which models of growth are available, particularly those which are described as 'competitive' in their growth strategy (Page, 1989).

The extent of the bracken distribution is known aproximately, but has proved difficult to map accurately, particularly across large areas. Visual description of the plant varies through the year, making detection problematic. Variable canopy dimension and structure (Watt, 1947) result in variability in photographic recording (Howard, 1970) and in reflectance when using satellite imagery. Matching spatial and spectral resolutions to the detection and measurement of bracken may be difficult from remotely sensed courses alone (Birnie and Miller, 1987). A probability mapping approach using satellite imagery and additional map data can be used to predict the distribution of core areas of bracken.

Implementation of the Knowledge-Based System

Overview

An expert system shell, SBS (Baldock *et al.*, 1987), is being used as a framework within which to develop the GIS coupled to the vegetation succession model. Separation of types of knowledge into different groups or 'sources' helps modularity but complicates control and communication within the system. Using a 'blackboard system' (Engel *et al.*, 1990), each source may have its own type of knowledge and processing (Smith *et al.*, 1987). The output of that source is recorded on the blackboard, which acts as a common database. A scheduler is used to select the knowledge source appropriate to the current task. This shell uses the language POP-11 (Barrett *et al.*, 1986) for the scheduling of EXPERT routines, which retain system-wide information (e.g., current estimates of accumulated errors). Information specific to individual EXPERTs (e.g., soil type details) is held in a frame structure, which manipulates the object-referenced data but not with an object-oriented code. Priorities assigned to each EXPERT may be changed during a system run. The EXPERTs operate 'opportunistically' on the current task according to their goal, priorities and prerequisites. Access to the raster-format, spatial data sets is by means of FORTRAN routines initiated by EXPERT routines.

The system executes its goals by means of 'matching' information between database entries and current problem details. This requires that the information have a defined structure and type. The information library within the system, accessible to all EXPERTs, is the POP-11 database. This is a mechanism containing a list of data items that may in itself be a list. Simple statements such as those in Figure 29.1 form the basis of the database content; the information may be quantitative or qualitative. Accompanying this information is some estimate of confidence in the relationship. Further, error assessments for one class will have consequential impacts on error assessments of classes which it is adjacent to or confused with (Veregin, 1989).

'Frame' Representation of Knowledge

Frame-based representations of objects allow *class–subclass, is-a* and *has-a* relations (Towers, 1987; Lynn Usery *et al.*, 1988). Attributes are assigned to these objects which, if changed, may have a 'trigger-system' for recalculating other associated attribute values. For example, changes in the area of one object will necessitate a change in the area of neighbouring objects. The frame is used to represent a group of entities with attendant facts (Towers, 1987; Lynn Usery *et al.*, 1988). For example, a frame may contain slots for each type of soil map unit represented in the spatial data set. These slots are, in turn, frames containing details of individual soil map units. These can be dynamically built and temporarily retained. Transfer of information from frame to GIS function parameter requires that an instance of the frame be interpreted by the code within an EXPERT routine (Lynn Usery *et al.*, 1988).

Knowledge Base and Metadata

Sources of ecological knowledge were sought to provide the basis for predicting and explaining the distribution of vegetation and its interactions. Principal

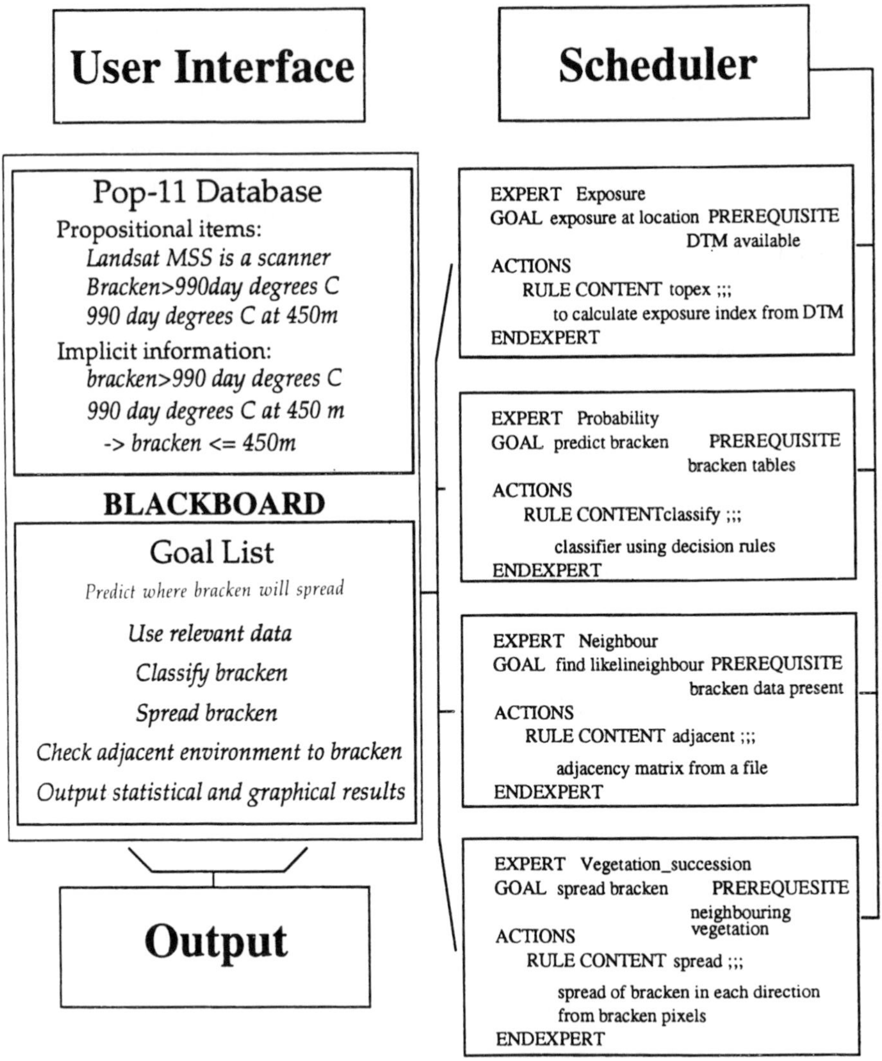

Figure 29.1 Schematic diagram of the SBS expert system shell with example system contents.

Source: Based on Baldock *et al.* (1987).

sources were published literature of vegetation in Great Britain is general, and Scotland in particular (McVean and Ratcliffe, 1962), journal and conference publications on vegetation distribution and change, and botanical details of bracken (Page, 1989), heather (*Calluna vulgaris*), and Nardus or Molinia grasslands (McVean and Ratcliffe, 1962).

Work by Townshend (1983) and Birnie and Miller (1987) can be used for selection of the appropriate sensor platform for spatial resolution, timeliness of remotely sensed data and suitability of sensor platform to monitoring changes in vegetation. Models, as published by these authors, are examples of metadata. From them, deductions can be made and inferences drawn with respect to ground observations. All characteristics of the model output (for example, uncertainties and constraints) are then carried through the remainder of the processing.

To manage, share and use spatial data effectively, it is important to have information about the data (often resulting from complex processing steps, specific parameters, and judgements and objectives of the previous user) and their structure (Clancey, 1992). Information such as data statistics can be computed directly from the basic spatial data and may or may not be stored.

A data history model for both quantitative and qualitative spatial metadata was designed and implemented. The design covers the minimum content and processing requirements for the spatial metadata relevant to this project. Access is required to the metadata for query and update and, where appropriate, for use in data processing procedures. Any operations on a data set preserve the data set's metadata. The metadata accompany a data transfer automatically, including transfer of the characteristics of a data set prior to processing and an explanation of why the processing step was performed.

The content of the meta-database includes the contents of the header information associated with each data file. The scale of maximum use follows a guideline of 2.5 times the source scale. The nature of the data source is described as classified (from satellite imagery) as opposed to satellite (raw data) or digitized. Data currency relates to the estimated relevant length of time of the data. Data processing lineage is also recorded in terms of processing routines and parameters.

Coupling GIS and Environmental Models

The links between the environmental, remote sensing and GIS models, and the GIS functions must be designed in the data model at the outset of the project. The models need to supply the GIS and receive in return the following sorts of information.

The models have to supply questions regarding spatial data (for example, look for vegetation types adjacent to bracken); error tolerance levels (e.g., accuracy of satellite classification in terms of omission and commission); information on data to use; and explanation prompts. The GIS has to supply output from functions for spatial analysis (e.g., adjacent environmental conditions for land cover and soil types); image processing functions; and uncertainty measures. The expert system scheduler provides the GIS function to call, altering EXPERT routine priorities; conflict resolution of goals matching and priority levels of EXPERTs being the same; and system auditing and information management (e.g., to respond to explanation prompts). The scheduler receives information on the data sets that are available; the functions available; and error assessments after each process.

The EXPERT routines and the POP-11 database provide the vehicles for transferring information between the goal list and the GIS functions in the form of subgoals, function parameters and variable values. The environmental models may be process-based understanding of phenomena, where sufficient data and information can be expressed as knowledge of the subject and the depth of knowledge may reflect the understanding of the processes involved but does not depend on it. Analysis of pattern may generate hypotheses on process (Openshaw, 1993; Aspinall, this volume).

Operational Enquiry

Overview

Figure 29.1 shows a schematic overview of the system. The goals listed are not all present at once; they are a summary of stages in the example system run. Other subgoals may be added by EXPERT routines to attain the original goal. The goal is checked against a history file of previously undertaken tasks, which includes those undertaken as part of a previous system run. The first task is carried out by routines that operate at a level of metaknowledge (Buchanan and Shortliffe, 1977), which is knowledge encoded about how to undertake a task, the principles that should be adhered to and in what order tasks should be undertaken.

Mapping Bracken Distribution

In response to a system-prompted enquiry that requests the estimates of acceptable accuracy, percentage errors for bracken identification and area of bracken spread are entered. A look-up table approach is used to match error in area with resolution and thus to sensor type and scales of available data sets.

Spatial data are available for 10 experimental sites in Scotland covering a range of conditions in the physical environment and representing various land use types. Vegetation observations were made at a hierarchy of scales. Soils, topographic and selected land use (forestry, built-up areas and arable/improved grassland) data sets, and satellite imagery were used in analysis of each area.

On the first system run, simple overlay operations are initiated by rules embedded in the EXPERT routines. Statistical summaries of the results are entered into frame slots. Comparisons between bracken presence and selected environmental factors in the test area were used to assist classification in an enlarged area. A maximum-likelihood classification algorithm was modified to incorporate non-continuous data sets like soil types (Kittler, 1979). Existing algorithms based on mean vectors and calculating equi-probability levels imply that any interpolated values in an image (satellite, slope, aspect or altitude) have some meaning. However, in a digital representation of a soil map, no meaning can be attributed to intermediate values. The facility to alter the prior probability of each class was also incorporated, using Bayes's Theorem as the decision-rule framework within the likelihood classifier. The measured preferences of bracken for certain topographic or soil conditions were used as weighted variables within the classification.

Once separated into the most likely class at every location, the likelihood surfaces are usually treated as contiguous classes (Wood and Foody, 1989). All levels of the likelihood surface may be processed using the same topological spatial relation. The second and third likelihood levels are also retained. The difference between likelihood levels for each pixel provides a guide to the confidence with which each class can be used. For example, comparison of the distribution of the second or third likelihood levels of bracken with aerial photographic and ground observations showed a correspondence with subdominant bracken presence (Wood and Foody, 1989; Goodchild *et al.*, 1993).

Predicting Bracken Spread

Vegetation succession is a widespread and readily observed ecological phenomenon (Miles, 1988). Currently, primary succession is subtle, slow in occurrence and limited in scale in Great Britain. Secondary succession between semi-natural vegetation types is caused mainly by changes in grazing pressures or burning regimes. When these pressures are constant or non-existent, the reproduction processes of the vegetation are vegetative or sporophitic. However, the nature of these trends differs for different types of vegetation (Miles, 1988).

Ecological knowledge of the bracken spread is summarized as follows: establishment of prothali is rare; main spread is by rhizomes; an invading front may advance up to 0.5 m yr^{-1} and in 'ideal' conditions up to 1.5 m yr^{-1}; many existing transitions to *Pteridium*-dominated stands are from existing, low-density frond populations rather than by invasion; and bracken does not grow much above 45 m, suggesting that it is intolerant to cold or exposure. These 'statements' include quantitative and qualitative information. Both statement (simplified) and a reliability level are encoded into a frame. Similar frames are built for grassland, moorland, forestry and agricultural grasses.

Figure 29.2 shows a conceptual model of successional trends in semi-natural vegetation in upland Scotland. In any sequence there is 'continuous' replacement of species — a continuum through time. This continuum has to be divided into a

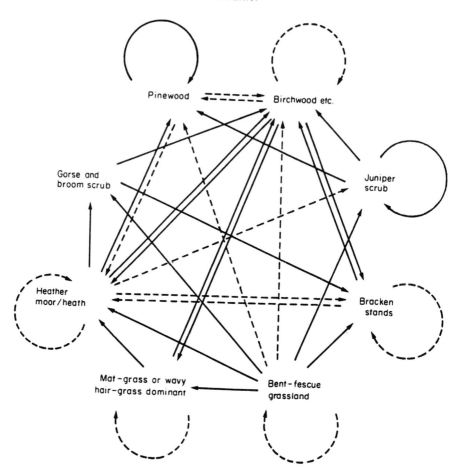

Figure 29.2 Successional transitions between common grazing types of semi-natural vegetation with low grazing pressures.

Source: Based on Miles (1988).

small number of discrete states where one species may be dominant or co-dominant with others. Quantification of these transitions comes from the field observations made at each experimental site, which are then standardized against time.

Spatial Context of Bracken Distribution

Contextual information of the vegetation cover may be lost when interpreted as polygonal information without retaining information on the nature of the boundary. The mature hinterland bracken may locally show signs of degeneration, which may be exhibited as low growth or discoloration. Regression in bracken may be due to patch ageing, pollution, climatic factors or local soil conditions. These changes become obvious when bracken cover reveals gaps.

Similarly, spatial patterns may denote the extent and location of changes in a vegetation stand. The role of 'nucleation' (a core patch) was studied by Watt (1947), who identified its role in determining spatial pattern and temporal variability of the community. This concept combined the expanding spatial front of a building phase and the existence of a persistent clonal patch. Therefore, the prediction of bracken spread requires some representation and analysis of bracken as a set of discrete patches.

Assessing the likelihood of spread involves the following: (1) checking bracken heterogeneity as measured by the number of bracken polygons in an area minus the number of 'holes' in those polygons (the Euler number). This is calculated from different probability levels to give a measure of the susceptibility to bracken spread according to the rules in the POP-11 database; (2) checking the environment adjacent to that occupied by bracken. The altitude, slopes, aspect, soil types, exposure levels and land cover types adjacent to existing bracken are summarized and each bracken pixel is assigned a 'suitability' score of its neighbouring environment to bracken spread; (3) assesssing the type of changes which are being observed to identify active and static edges in the bracken stands; (4) running the vegetation succession model (Miles, 1988), which allocates likelihoods of spread according to the neighbouring vegetation type (Figure 29.2); (5) applying a classification matrix as a summary of the conditions at each location around the boundary.

Output and Explanation of Results

Output is both numerical and graphical. The areas that have undergone changes in vegetation cover are summarized. A further enquiry from the user will elicit an explanation of the procedures undertaken during the system run. Graphical output is of raster data sets showing where changes are expected to occur and a reliability index of those predicted changes.

Accuracy assessments of change predictions are made by overlaying predictions with historical data of bracken distribution. A feedback loop involving the calculation of transition matrices based on the incremental inclusion of measured changes is not yet complete. This provides a means of 'learning' what changes to expect and their spatial extent. Currently, the success of predicting where changes will occur (75 per cent) is higher than that of predicting how much change will occur (55 per cent). If there is evidence of a 'hole' in a well-established bracken stand, the active edge can be progressed as a bracken front retreating rather than spreading. The model relies on the identification of the vegetation types occupying such a hole and some information on their dynamics.

The 'explanation' facilities of the KBS compiled descriptions of both *which* rules and *what* knowledge were used by each EXPERT routine and (if requested) at which times decisions were made and the degree of certainty associated with each; and *why* bracken may have spread or may spread in the future at particular locations, using GIS functions to assess context.

The first form of description is a listing of the system routines used and the decisions made by the scheduler. The second description uses data stored in the frames of *what* can be inferred from frames or EXPERT routines being accessed and *why* that has implications for changes in bracken distribution. This is the value of the KBS approach (Buchanan and Shortliffe, 1977). The extent of the explanation depends on the depth of the knowledge base: the credibility depends on the sources of knowledge and the reliability on the inferences.

Knowledge Transfer

Transfer formats for data exchange between systems are necessary for effective use of different computer packages. A considerable investment is being made to standardize transfer formats between nations and reappraise the minimum appropriate information to be transferred with digital map data (Federal Geographic Data Committee, 1992). Metadata about the models' constraints and assumptions are as important as those relating to spatial data sets and should be transferable with other details of models (Kirchner, this volume).

If there is a role for KBS and use of spatial data within the study then there is likely to be a need for transfer of knowledge bases between different systems. The nature of the transferral of a knowledge base depends on the nature of the representation of knowledge within a KBS. The value of the transfer is to retain and build on the expert knowledge that has been formalized such that it does not remain isolated as a single routine or subject-specific system. For example, much of the KBS work done in medical imaging has relevance and application in satellite image processing and GIS.

One approach is to treat packages or routines as subroutines, with single calls to initiate them. This has all the advantages of modularity of system development and tractability of system implementation. However, in intradisciplinary work, which is often the nature of GIS applications, building up knowledge bases across subject domains at equivalent reference levels may offer greater potential for an interdisciplinary facility to undertake applications using spatial analyses.

One difficulty in designing and implementing a knowledge-transfer protocol is the relatively small number of opportunities available for testing transfer success. A successful, simple test with a system developed for medical image analyses has been undertaken with the EXPERTs and frame structures presented in this chapter. The transfer included frame-encoded knowledge of what and why functions were called. If such knowledge transfer can be linked into the metadata initiatives, there may be greater scope for making GIS facilities genuinely interdisciplinary tools.

Conclusion

The potential for non-expert users to run models with different scenarios provides decision makers with an insight to what the outcomes may be. The user who has an interest in understanding the scale of ecological phenomena in time and space should benefit from a tool that provides some guidance on spatial analyses. Description and explanation of pattern and process and their linkages can be (gradually) entertained by the system. The knowledge encoded into the KBS should never constrain the user to pre-ordained solutions or explanations. Its role is to help facilitate guidance on GIS usage and coupling of process-based models for more informed implementations and to manage information flow from raw data to final processing and outputs. Greater value would be obtained by the user who has an appreciation of the scale and magnitude of temporal changes and can evolve the basis for coupling dynamic models and spatial data through relationships learned by the knowledge-based system.

The assumptions made in using the succession model require refinement in two ways: (1) to assess sensitivity of model predictions at medium or small scales against changes in grazing intensity; (2) to input measures of animal uptake in grazed areas to assess grazing pressure. These changes (which may be significant at a farm unit level) would make the system outlined more applicable to decisions on land management such as bracken clearance or control measures.

The extent to which expert system tools can readily be built into or on GIS applications depends largely on the nature of future users. If the backgrounds of users in environmental applications become more diverse, then the degree to which the technology and the user expertise can be formally coupled becomes a more valuable area of research. The importance of maintaining records of data lineage throughout the processing stages will also grow as digital data sets are made more widely available (Lanter, this volume). If a common remit of users develops more narrowly along the lines of specialization of, for example, cartography, image processing or forestry and agriculture, then the expert system coupling may be less useful. In either case, the need for theoretical and developmental work into the nature and form of data models will continue to increase in importance.

Acknowledgement

The author would like to thank The Scottish Office Agriculture and Fisheries Department for their funding of the work from which this chapter was derived. Acknowledgement is also due to Richard Aspinall for detailed discussion on context and objectives of this approach, Jane Morrice for compilation of data associated with the prototype KBS and Matt Wells for advice on the running of Poplog and SBS. Thanks also to the Medical Research Council and Population Cytogenetics Unit at the University of Edinburgh for use of the SBS expert system shell.

References

Baldock, R. A., Ireland, J. and Towers, S. J., 1987, *SBS User Guide*, Edinburgh: Medical Research Council and Population Cytogenetics Unit.

Barrett, R., Ramsay, A. and Sloman, A., 1986, *POP-11: A Practical Language for Artificial Intelligence*, Chichester: Ellis Horwood.

Birnie, R. V. and Miller, D. R., 1987, Lessons from the bracken survey of Scotland: The development of an objective methodology for applying remote sensing techniques to countryside mapping, in Adams, W. and Budd, J. (Eds) *Proceedings of Monitoring Countryside Change*, Chichester: Silsoe College, Packard Publishing.

Buchanan, B. G. and Shortliffe, E. H., 1977, *Rule-Based Expert Systems, the MYCIN Experiments of the Standard Heuristic Programming Project*, Reading, Massachusetts: Addison-Wesley.

Clancey, W. J., 1992, Model construction operators, *Artificial Intelligence*, **53**(1), 1–115.

Engel, B. A., Beasley, D. B. and Barrett, J. R., 1990, Integrating expert systems with conventional problem solving techniques using blackboards, *Computers and Electronics in Agriculture*, **4**, 287–301.

Federal Geographic Data Committee, 1992, *Information Exchange Forum on Spatial Meta-Data*, Reston, Virginia: US Geological Survey.

Goodchild, M. F., Perks, B. O. and Steyaert, L. T. (Eds), 1993, *Environmental Modeling with GIS*, Oxford, UK: Oxford University Press.

Howard, J. A., 1970, *Aerial Photo-Ecology*, pp. 321, London: Faber and Faber.

Kittler, J., 1979, Image processing for remote sensing, *Philosophical Transactions of the Royal Society of London, A*, **309**, 323–35.

Lynn Usery, E., Altheide, P., Deister, R. R. P. and Barr, D. J., 1988, Knowledge-based GIS techniques applied to geological engineering, *Photogrammetric Engineering and Remote Sensing*, **54**(11), 1623–8.

McVean, D. N. and Ratcliffe, D. A., 1962, *Plant Communities of the Scottish Highlands: A Study of the Scottish Mountain, Moorland and Forest Vegetation*, Monographs of the Nature Conservancy, No. 1, London: HMSO.

Miles, J., 1988, Vegetation and soil change in the uplands, in Usher, M. B. and Thompson, D. B. A. (Eds) *Ecological Change in the Uplands*, 7, pp. 57–74, Special publications series of the British Ecological Society.

Openshaw, S., 1993, A concepts rich approach to spatial analysis for GIS, in *Proceedings of GIS Research UK*, Keele: University of Keele.

Page, C. N., 1989, Taxonomic evaluation of the fern genus *Pteridium* and its active evolutionary state, in Thomson, J. A. and Smith, R. T. (Eds) *Bracken '89: Bracken Biology and Management*, pp. 23–34, Australia: Institute of Agricultural Science Occasional Publication No. 40.

Smith, T., Peuquet, D., Menon, S. and Agarwal, P., 1987, KBGIS-II, A knowledge-based geographical information system, *International Journal of Geographical Information Systems*, **1**(2), 149–72.

Taylor, J. A., 1986, The bracken problem: a local hazard and global issue, in Smith, R. T. and Taylor, J. A. (Eds) *Proceedings of Bracken '85, Bracken Ecology, Land Use and Control Technology*, pp. 21–42, Leeds.

Towers, S. J., 1987, 'Frames as a data structure for SBS', internal report, Medical Research Council and Population Cytogenetics Unit, University of Edinburgh.

Townshend, J. R. G., 1983, Effects of spatial resolution on the classification of land cover type, in *Ecological Mapping from Ground, Air and Space*, pp. 101–12, Natural Environment Research Council, Institute of Terrestrial Ecology.

Veregin, H., 1989, *A Taxonomy of Error in Spatial Databases*, National Center for Geographical Information and Analysis Technical Paper No. 89–12, Santa Barbara, California: National Center for Geographical Information and Analysis.

Watt, A. S., 1947, Pattern and process in the plant community, *Journal of Ecology*, **35**, 1–22.

Wood, T. F. and Foody, G. M., 1989, Analysis and representation of vegetation continua from Landsat Thematic Mapper data for lowland heaths, *International Journal of Remote Sensing*, **10**(1), 181–92.

30

A knowledge-based approach to the
management of geographic information
systems for simulation of forested
ecosystems

D. Scott Mackay, Vincent B. Robinson and Lawrence E. Band

Forested ecosystems in mountainous terrain show significant heterogeneity in atmospheric and land-surface interactions. Forest ecosystem simulation modelling requires management of geographically based information detailing these complex interactions. We present a knowledge-based approach for a geographically based information system designed to infer and simulate objects using simple attributes. Higher order objects are recursively defined using query models and aggregation. They are organized and accessible as distinct, identifiable landscape units such as hillslopes and forest stands. Object-orientated and knowledge-based tools simulate ecosystem processes on the basis of higher order geographic features. Ecological simulations are organized around a model of terrain based on a hierarchy of nested watersheds, hillslopes and streams. We organize spatial relations into a graph of topological links around which is built a graphical query interface. The interface translates user requests into database queries using a query model that maps graphical objects to database objects. We show how the query system integrates structural, logical and procedural knowledge to provide answers to complex queries based on simulation results. We demonstrate that this system insulates the user from implementation details of the simulation system, and that it can manage the ecological/spatial objects in a transparent manner throughout simulation experiments.

Introduction

This chapter describes a knowledge-based geographic information systems (GIS) approach to organizing both data and simulation models on the basis of distinct, identifiable landscape units. The system integrates spatial object management with simulation models. Within the forest ecosystem modelling community there is a growing interest in developing the capability of simulating carbon, water and energy flux processes at regional scales (e.g., BOREAS project; also see Brown *et al.,* this volume; Cohen *et al.,* this volume; Lathrop *et al.,* this volume). Current modelling capability is at the watershed scale (e.g., RHESSys, Band *et al.,* 1991, 1993; TOPOG, O'Loughlan, 1990). Realistic simulation of larger, spatially heterogeneous regions requires substantial amounts of data. The volume and variability of data can hinder the reliable management of both data and simulations. In particular, extremely variable topography, canopy cover and soil water availability requires breaking a region into a large number of unique landscape components, or units. An appropriate breakdown of a region might be in terms of watersheds and hillslopes. Data and simulation models then need to be structured on an individual basis for each of the landscape components. Once organized in this manner, there is still a need to update the information base to reflect management scenarios such as thinning, which would be prescribed in spatially explicit terms. This presents a tedious and error-prone task for regional scale modelling. Furthermore, simulation data requirements depend on the amount of physical realism required. A trade-off exists between providing a detailed representation of land surface processes and efficiently managing the information required for these detailed representations. At one extreme, physically based models that simulate processes with high levels of spatial detail (e.g., forest plots) are impractical for ecosystem simulations of hillslopes, watersheds and regions of watersheds, as these demand quantities of data that are generally unavailable. At the other extreme, lumped models aggregate individually observed properties into mean or calibrated parameters without addressing the variance and covariance of the parameters. Spatial heterogeneity, implicit in variance/covariance of joint-parameter distributions, is lost in the aggregation. Without explicit representation of space, interactions between individuals or aggregates of individuals in space or time cannot be monitored. To manage large-scale ecosystem simulations, a compromise between these extreme modelling approaches has to be found. The compromise should retain enough physical basis to allow spatial variability of processes, while allowing for direct measurement of parameters such as leaf area index (LAI) from remote sensing imagery, or soil wetness from topography and soil transmissivity. Such compromises allow for the representation of spatial and temporal (or spatiotemporal) variability in aggregate responses of regional scale ecosystems while keeping the simulations computationally tractable.

Watershed evapotranspiration (ET) and canopy net photosynthesis (PSN) are sensitive to the spatial distribution of soil moisture and canopy properties (Band *et al.,* 1993). Factors controlling the spatial distribution of soil water include saturated hydraulic conductivity and thickness of the hydrologically active zone, local slope, catchment area, aspect, annual variation in canopy cover, net rainfall, radiation and temperature (Moore *et al.,* 1991). These data are difficult to obtain at a scale, in time and space, compatible with the scale of representation in the simulation model. In particular, land surface information acquired from high-resolution remote sensing imagery, digital elevation data and soil maps may have different levels of spatial accuracy and scale. All of these data have to be georeferenced to a common system of geographic objects and managed over time.

While we make no claim to solve the problems of accuracy and scale of data, we do manage the information system within these limitations.

The ability to portray spatiotemporal heterogeneity of surface properties and processes plays an important role in the management of ecosystem models. For example, FOREST-BGC (Running and Coughlan, 1988) simulates net primary productivity (NPP), a measure of change in biomass and detritus produced during a year, and evapotranspiration (ET) of water from the forest canopy and litter. Both NPP and ET are correlated with maximum sustainable leaf area index (LAI) (Waring and Schlesinger, 1985); LAI is correlated with stand density and maturity. As a stand matures, growth efficiency lessens owing to competition between individual leaves for available light and to increased respiration costs. Once a stand reaches its maximum sustainable LAI, further growth of individuals within the stand requires a reduction in the number of surviving individuals, resulting in natural, long-term changes in the structure of the stand. Terrestrial ecosystems survive by adapting to availability of water, light, nutrients and temperature through changes in species composition, morphology and physiology changes within a species, and community density (Larcher, 1980). Thus, changes in LAI reflect the spatiotemporal dynamics of structural, environmental and functional aspects of a forest stand. Furthermore, disturbances such as fire and disease may result in a near-instantaneous reduction in LAI in part, or all, of a stand. This shifts the allocation of biomass from living tissue to detritus, increases soil water and nutrient supplies, allows more radiation through the canopy, increases precipitation reaching the ground, and increases soil temperature. Failure to accommodate new inputs as a simulated forest stand evolves results in bias and the accumulation of error, particularly for long simulation periods. Managing spatiotemporal land surface patches as they split and merge in simulated time is a long-term goal of ecological modelling, but the remote sensing tools needed for detecting changes in terrestrial vegetation are becoming available. For example, Peterson *et al.* (1987) show that combinations of red and near-infrared reflectances measured by satellite sensors are strongly related to vegetation density. Furthermore, remotely sensed imagery has been successfully used for vegetation classification on local to global scales (Tucker *et al.,* 1985). The increasing availability of remotely sensed imagery at high spatial and temporal resolutions should allow for detection of changes in LAI at spatiotemporal scales needed for modelling terrestrial ecosystems over large regions and long time spans.

Spatiotemporal variability of soil wetness is an extremely important factor in determining maximum sustainable LAI, particularly in water-limited environments. Models of water and nutrient movement through soil use parameters such as hydraulic conductivity. The decrease in hydraulic conductivity with depth, through the soil profile (or soil transmissivity), is due to reduced soil capillary sizes with depth which leads to lateral movement of water. When combined with topographic convergence, lateral movement of water through the soil results in increased soil moisture levels on lower parts of slopes and within hillslope hollows (O'Loughlan, 1986). Simulation of water movement through soil is typically described using mechanistic models that describe the movement from point to point. For watershed and regional forest ecosystem modelling, the mechanistic approach is a computational burden, requiring data that are generally unavailable. At larger scales it may be more important to represent the spatial variability of land surface process than to model accurately what happens at any one point. Current modelling approaches seek indices that simplify the representation of physical systems by capturing their essential functional properties (Moore *et al.,* 1991). A soil wetness index functionally incorporates contributing drainage area per unit contour width, soil transmissivity and local gradient. A difficulty exists with obtaining soil information at spatial resolutions or accuracies needed to model adequately the spatial variability of soil transmissivity. Soil is generally given as a map layer of polygons representing soil series, from which soil parameters are inferred, but spatial variations in the parameters are limited to the level of detail provided by the polygons. Furthermore, soil series maps are generated for a particular purpose, which may have different data requirements from ecological modelling, and are generally not adequate for modelling within-object (hillslope or stand) spatial variability (e.g., within-hillslope variability).

The representation of topography is of fundamental importance in any automated approach to modelling of forested ecosystems in mountainous regions. The preceding discussion illustrates its importance concerning soil wetness. Moore *et al.* (1991) list a number of attributes that are derived from topography and are important for the following:

- Elevation for climate, as orographic effects influence temperature and precipitation amounts and hence limit the types of vegetation communities that can be supported;
- slope azimuth (aspect) for modelling solar irradiation, which influences temperature, evapotranspiration and photosynthetic activity;
- gradient (slope) for computing overland flow and subsurface flow routeing, water velocity and runoff rates;
- mean slope of dispersal area for controlling the rate of soil drainage;
- upslope contributing area per unit contour width for estimating runoff volume and steady-state runoff rates; and
- planform curvature for controlling convergence and divergence of flow, and hence soil water availability.

With wide availability of digital elevation data and automated tools for analysing them, topography is perhaps the least problematic data domain needed for ecosystem modelling. Digital elevation data are widely available at high resolution, and research in automated partitioning of watersheds on digital elevation data has matured for grid elevation data (Band, 1986, 1989), triangular irregular networks (Jones *et al.,* 1990), and digitized contours (Moore *et al.,* 1988).

The preceding discussion illustrates the potentially large number of complex landscape units and relations that must be managed while preserving the integrity of the database and simulation model. This chapter describes a knowledge-based GIS approach to organizing both data and simulation models on the basis of distinct, identifiable landscape units. The system integrates spatial object management with simulation models. This requires some consideration of the data requirements of both GIS and simulation models, which means that system integration is not straightforward. In particular, it is important that the GIS organize data so that they are compatible with the data requirements of simulation modelling (Moore *et al.,* 1991). GIS and simulation models organize their information in fundamentally different ways. A GIS manages large quantities of generic spatial objects registered to a common system of geographic co-ordinates, adjacency between spatial objects and object attributes. Objects in a GIS are typically in the form of primitives such as points, lines, polygons or layers. However, objects in a simulation model are defined in terms of system state, mass and energy flux, and interaction and dynamics of species or individuals (Kircher, this volume). Thus, simulation objects have to be managed in time. There has recently been interest in managing temporal information in GIS (Langran, 1992), and temporal reasoning is an integral part of knowledge-based management systems such as Telos (Mylopoulos *et al.,* 1990). However, in this chapter it is more significant that simulation objects are at a conceptually higher level than the generic spatial objects, as they incorporate physical laws, functional relationships and constraints, in addition to spatial relations. To bridge the conceptual gap between GIS objects and simulation objects, we have designed a hybrid knowledge-based management system that transforms simple observations and relations in collected data into higher order concepts. The hybrid system integrates structural, deductive and procedural knowledge representation. Structural knowledge forms hierarchical, topological and functional linkages between spatial entities, as well as common registration of multisource data. Deductive knowledge infers new facts from initial observations, and provides decision support for designing modelling strategies. Finally, procedural knowledge integrates existing GIS and simulation tools so that the system can be adapted to changing needs.

The next section gives a discussion of our methodology in designing the system. This is followed by a description of a system implementation and testing for a small, mountainous watershed. We finish with some conclusions relating our results to environmental information management and analysis.

System Design Methodology

Knowledge-based Management

Terrain analysis and simulation modelling tasks require specialized tools that are generally separate from the information management facilities provided by a GIS. We assume that information systems are knowledge-intensive, as are the specialized tools needed to perform tasks such as terrain analysis and ecosystem simulation. As such, all components of the system are viewed as types of

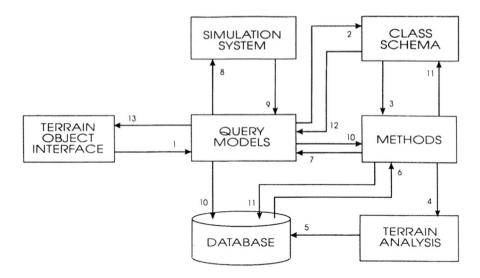

Figure 30.1 KBLIMS is managed by query models which integrate structural schemes, database facts and methods, terrain analysis tools and simulation models.

Note: Query models provide an integrated view of information to the terrain object interface with which users interact with the system. Labels on arrows indicate relative position of information flow during a typical query requiring the use of all KBLIMS modules. Transactions begin and end at the user interace, and information flow is co-ordinated by the query models.

knowledge that can be incorporated into a hybrid knowledge-based system (also see Miller, this volume). Figure 30.1 shows the structural components and the connections between components of the Knowledge-based Land Information Manager and Simulator (KBLIMS). Connections are labelled to show the sequence of operations that form a transaction to answer a query requiring all components of the system. Transactions begin and end with human–computer interaction through the Terrain Object Interface.

The KBLIMS knowledge base is organized around a query model manager (Query Model), which distinguishes among three types of knowledge: (1) extensional and intensional predicates, (2) structural descriptions, and (3) procedures. Extensional (fact) and intensional (rule) predicates are respectively incorporated into Database and Methods modules (Mackay *et al.*, 1991, 1992b). The database contains explicitly known facts, while methods stores implicit facts in the form of deductive rules. In the discussion below, we provide some examples using a Prolog-like syntax. In these examples, facts have the form $p(a,b)$, which means that proposition (e.g., a spatial relation) p holds between individuals a and b. A rule has the form $p(a,b)$:- $q(a,c),r(c,b)$, which means that p is true if both propositions q and r are true.

Aggregation and generalization are important for modelling the real world in an information system. Mylopoulos *et al.* (1990) argue that abstraction mechanisms are important in modelling all aspects of information systems development and maintenance. Generalization enables a class of individual objects to be thought of generically as a single-named object. For example, generalization can be used to express the idea that hillslopes, divides and watersheds are all kinds of topographic objects. Aggregation refers to an abstraction in which a relationship between objects is regarded as a higher level object. For example, the relationship between a patch of land, a stream and the movement of water from the patch of land to the stream forms an aggregate, or higher order object called a hillslope. The concept of a hillslope embodies the notion of a functional relationship between a patch object represented by a polygon GIS object, and a stream represented by a line GIS object.

Terrain analysis and simulation modelling tools are viewed as procedural knowledge, as they embody specialized knowledge from their respective domains in the form of programs. Query models build concrete objects by combining predicates, procedures and abstract structural descriptions. Concrete objects are instantiated counterparts of their respective class descriptions. The transformation from object class descriptions to instantiated objects and the isA relation are defined as follows:

> *Definition 1 (Class description).* A class description is a three-tuple $< \Phi, \Gamma, P >$ in which Φ is the class name, Γ is a set of generalized classes $\{ \Phi'_1, \Phi'_2, ..., \Phi'_N \} \in \Gamma$ such that the relation isA(Φ, Φ'_i) holds for all $i \in \Phi'$, and P is a set of attribute descriptions $(\rho, \Phi)_j \in P$, each consisting of an attribute name, p_j, and a reference class, Φ_j.
>
> *Definition 2 (isA relation).* The isA relation is recursively defined such that isA (c_0, c_0) holds, and (isA (c_0, c_1) and isA (c_1, c_2) implies isA (c_0, c_2)) holds.

Class descriptions provide statements about the organization and attributes of classes of objects with similar properties. They do this by defining the position of an object class within its generalization, or class–subclass hierarchy defined by the isA relation, and by aggregating related attributes into a single structure. Since class descriptions relate simple data types to higher order objects, we refer to them as knowledge about objects. Knowledge describes how low-level data and procedures should be organized and integrated.

Associated with the isA relation is the notion of attribute inheritance. The knowledge base supports multiple inheritance of attributes from generalized classes, Φ', to specialized classes, Φ, defined by the relation isA (Φ, Φ'). Inherited descriptions are used by query models to instantiate object structures, but specialized attribute descriptions always take precedence over generalized attribute descriptions when an object is instantiated.

> *Definition 3 (Query).* A query σ on knowledge base II is defined as II $\times \rho(\Phi, v, \phi, f(\Phi, \phi)(v))$ where p is a 4-ary predicate defining a relation between object class Φ and reference class ϕ, v is an object identifier, and function $f(\Phi, \phi)(v)$ computes all instances of class ϕ satisfying σ.

Queries are matched against object methods or explicit facts to obtain results.

For example, to find transpiration for a patch of land the system would match query:

$$transpiration(patch, 1, mm, \textit{Transpiration})? \qquad (30.1)$$

to fact:

$$transpiration(patch, 1, mm, 0.5520) \qquad (30.2)$$

or to a method that computes hillslope area using procedural knowledge. Since transpiration may be computed by simulating water balance for patch 1, but may not be explicitly defined in a database, the system needs to recognize that running the simulation is a precondition to answering the query for transpiration. However, the structure of the query is the same regardless of whether it is matched against an explicit fact or against methods that define facts implicitly.

It is sometimes useful to retrieve aggregate totals for queries on a number of objects of the same class. For example, we might want to aggregate the attributes of a set of patches prior to running any simulations if all of the patches in the set have similar properties. Alternatively, we might want to aggregate the results obtained by running simulations on a number of patches and computing total transpiration (m^3) for the patches. To perform aggregation in queries, we distinguish between two types of identifiers: atomic identifiers and complex identifiers. Atomic identifiers are single integers taken from the topological encoding within the spatial database. Each instantiated object of a given class has a unique identifier. Complex identifiers (e.g., (1,2)), are made up of unique atomic identifiers from some class C, and denote aggregates of objects in class C. Results of queries on objects with complex identifiers are obtained by applying the query over each of the objects represented by the atomic identifiers within the complex. For example, the following query asks for total transpiration for patches 1 and 2:

$$transpiration(patch, (1,2), mm, \textit{Transpiration})? \qquad (30.3)$$

Since it is unlikely that our knowledge base will, or should, have explicit facts on total transpiration for patches 1 and 2, we need to be able to express this query in terms of the facts we have or those we can compute. Query (30.3) can be written out in long format using a rule that relates transpiration of the complex object to the transpiration facts of the atomic objects:

transpiration(patch, (1,2), mm, *Transpiration*) :–
transpiration(patch, 1, mm, T_1), area(patch, 1, mm, A_1),
transpiration(patch, 2, mm, T_2), area(patch, 2, ha, A_2),
Transpiration is DIVIDE(SUM(T_1*A_1, T_2*A_2), SUM(A_1, A_2)). (30.4)

which breaks the aggregate query conjunction of atomic queries of the same object class Φ (patch) and reference class ϕ (mm), returns individual transpirations T_1 and T_2, and then builds *Transpiration* by computing the weighted average of T_1 and T_2 using the relative patch sizes.

Object methods provide access to the database but hide procedural aspects of queries. System-defined methods include models for asserting attributes, retracting attributes, retrieving attributes and checking constraints on objects. Each method has an external interface that provides declarative semantics for queries. The system takes a description for an object attribute and automatically generates an implementation query. This strategy forms a distinction between an implementation layer and class description layer. Layering permits changes in low-level procedures without changing the semantics of descriptions defined at a higher level.

Deductive rules are specialized kinds of methods that infer new facts. Deduction refers to the fact that inferences made during a previous transaction are available as explicit facts for subsequent transactions. A motivation for having deductive rules is that we generally cannot foresee all application scenarios. Furthermore, an end-user may select, or browse, a set of objects without knowing specifically what kinds of questions to ask about these objects. The system needs to be able to retain the retrieved objects and assists the end-user in asking more specific questions about these objects. In such cases, the system infers new facts based on these specific queries. For example, if we want to compute total evapotranspiration for a patch of land we might define the following deductive rule:

evapotranspiration(patch, *Identity,* mm, *Evapotranspiration*) :–
evaporation(patch, *Identity,* mm, *Evaporation*),
transpiration(patch, *Identity,* mm, *Transpiration*),
Evapotranspiration := SUM(*Evaporation, Transpiration*).   (30.5)

which would be called by a query to retrieve evapotranspiration. If the rule succeeds, then a fact for evapotranspiration is added, or asserted, to the database and is available, for a given time (e.g., time-stamp) or set of conditions of simulation, for use in subsequent queries.

Definition 4 (Query model). A query model Σ is an ordered set of individual queries $\{ \sigma_1, \sigma_2, \ldots, \sigma_N \}$ such that $\sigma_i \neq \sigma_j$ for all i,j.

A query model specifies a set of queries, similar to a script that act as a single transaction on the knowledge base. It follows from Definition 3 that inter-transaction data are retained, as queries are deductive. Results of a query within a transaction add information to the knowledge base, which is then available for use by subsequent queries in the same transaction. This implies an intratransaction information-sharing property of query models. Transaction atomicity is maintained by viewing each query model as a sequence of short transactions, each one query in length.

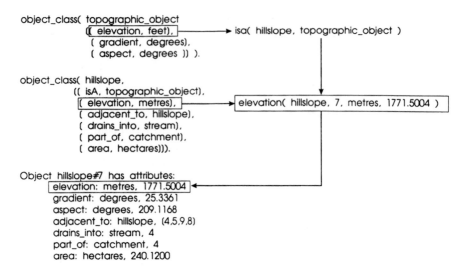

Figure 30.2 Topographic_object and hillslope descriptions are used in creating an object for Hillslope No. 7.

Note: Hillslope No. 7 inherits attributes from topographic_object, but uses the elevation attribute given in the more specialized hillslope description.

Definition 5 (Object structure). An object structure is a three-tuple $< \Phi, v, \Sigma >$ for class name Φ in class description $< \Phi, \Gamma, P >$, v is the object identity, and Σ is a query model such that query $\sigma_k \in \Sigma$ is isomorphic to attribute description $(\rho, \phi)_k \in P$ for all k.

An object structure reflects the description given in the class schema. Figure 30.2 illustrates how a particular class is defined, and what the corresponding object structure is for a particular instantiation. Figure 30.2 illustrates how a particular attribute, elevation, is inherited from the generalization, topographic_object, to the subclass, hillslope, then instantiated with value type, metres, specified in hillslope rather than value type, feet, defined in topographic_object. The target object structure is instantiated by (1) retrieving all attribute descriptions from the isA graph, such as the attributes given in the topographic_object and hillslo·e class descriptions, (2) resolving any attribute cancellations such as replacing the attribute (elevation, feet), with the attribute (elevation, metres), (3) generating a query model from the retrieved descriptions, and (4) populating the object structure with the result of applying the query model to the knowledge base.

Terrain Analysis

Information used for deriving terrain objects in KBLIMS is layer-based (e.g., grid elevation, remotely sensed imagery). The Terrain Analysis system translates this layer-based geographic information into object-based geographic information. The layer-based model of geographic information represents spatially distributed attributes as a set of data layers each of which defines a distribution of a single attribute over a defined space. For example, a continuous surface of elevation values in a grid matrix is a mapping of geographic space on to a set of elevation values. A layer of soil polygons expresses an arbitrary partitioning of a geographic space on to attributes from some soil domain (e.g., soil series). The layer-based approach is efficient for performing locally based spatial operations, including point-based operations (e.g., gradient at point p computed from finite difference of elevations within a neighbourhood of p), and zoning operations (e.g., given soil polygons in one layer and gradient in another layer, compute mean gradient for each soil polygon).

The object-based model defines structures representing unique entities in the modelled reality, along with topological, geometrical and other attributes associated with those entities. The object-based model combines generic spatial objects (points, lines, polygons), explicit spatial relations between objects (lines with left and right polygons) and attributes (e.g., area) into a single, unified data model. By having a unified data model the object-based model of GIS avoids having to maintain a dual database model (Bracken and Webster, 1989). In a dual database, updates to a spatial database are made as separate operations from updates to the relational database supporting spatial entity attributes. Independent updates increase the likelihood of corrupting the database (Robinson and Mackay, 1993). One of the major and immediate sources of corruption is difficulty in maintaining referential integrity (Halustchak, 1993). For this reason, a unified object-based model 'is particularly important in developing spatial decision-support systems (Armstrong and Densham, 1990), and expert systems (Goodenough *et al.,* 1987). Spatial objects are manipulated by set-based operations (e.g., a patch represented by a point-set), metric operations (e.g., computing patch area), Euclidean operations (e.g., measuring junction angles) and topological operations (e.g., solving path-based queries). A major advantage of the object model is the ease with which topological operations are performed on objects.

The layer and object-based models are not mutually exclusive as it is possible to translate from layer to object and back. Worboys and Deen (1991) suggest that a suitable model of transformation between the layer and object models is point-set topology. The goals of our terrain knowledge base mainly require a layer-to-object transformation, but the other direction may be used to generate maps showing snapshots of simulation results. Since we are working mainly with raster data (e.g., DEM and satellite imagery) our point-set topology is approximated by pixels and their neighbourhoods. Transformation from image data to symbolic elements follows a three-step process (Figure 30.3): (1) extraction of drainage basins and their segmentation into hillslopes and stream links, (2) analysis of the hillslope and stream links, and (3) generation of an object-based database. A drainage area transform (DAT), which is a tree representing contributing drainage area for each cell in a grid digital elevation

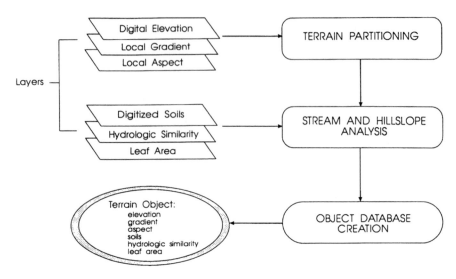

Figure 30.3 The terrain analysis tools transform layer-based image data into object-based symbolic data for use within the knowledge base.

model, is recursively computed. By setting a threshold contributing drainage area the DAT is pruned into a stream network (Band, 1989), and then each stream link is topologically encoded and assigned left and right hillslope polygons (Band, 1986). This approach adaptively scales terrain partitioning from a few large hillslopes to many small hillslopes for any given drainage basin, depending on the amount of pruning that takes place. It is this scale flexibility that allows layer-to-object transformation without significant loss of essential land surface heterogeneity that is so important for realistic ecosystem simulations. The scaling tool retains enough physical basis so that processes vary spatially while also allowing for direct measurement of land surface and environmental parameters. Analysis of hillslopes and streams extends the work of Lammers and Band (1990) to topology such as stream link connections, stream link to hillslope connections, and hillslope to hillslope connections, and computation of hillslope distributional information for soil hydraulic properties and leaf area index. This information is incorporated into an object database (Mackay *et al.*, 1991, 1992a).

Once the terrain information is stored as objects it can be used by rules to build new objects and relations. For example, if we know that a hillslope is part of one catchment basin and is topologically connected to other hillslopes, then adjacency between catchments can be inferred as follows:

$$\text{adjacent\_to(catchment, } C_1\text{, catchment, } C_2\text{) :-}$$
$$\text{part\_of(hillslope, } H_1\text{, catchment, } C_1\text{),}$$
$$\text{adjacent\_to(hillslope, } H_1\text{, hillslope, } H_2\text{),}$$
$$\text{part\_of(hillslope, } H_2\text{, catchment, } C_2\text{).} \qquad (30.6)$$

Note that there are two relations for expressing adjacency between objects. Since the system is object-orientated it does not limit relations to be homomorphisms (Maier, 1983). While this requires a little more overhead to maintain internally, it reduces the burden of defining object classes as there is no limitation on how many different object descriptions can share a relation. Since it depends on the object class, the behaviour of a method associated with the relation is polymorphic (Atkinson *et al.*, 1989).

Simulation System

The Simulation System is based on the model components of RHESSys (Band *et al.*, 1991, 1993), including MTCLIM, FOREST-BGC (Running and Coughlan, 1988), and TOPMODEL (Beven and Kirkby, 1979), designed to simulate the distribution of forest carbon and water flux, and storage processes over a mountainous landscape. MTCLIM extrapolates daily meteorological observations from base stations using adiabatic lapse rates, orographic gradients and topographic correction of solar radiation. FOREST-BGC is a stand-level model of forest carbon, water and nutrient budgets. It computes daily photosynthesis, respiration, evaporation, snowmelt and runoff production using site conditions such as current stand characteristics, observed or computed daily meteorological data and soil properties. In addition, daily and seasonal hydrologic flux processes such as snowmelt, interception, runoff and evapotranspiration are computed.

LAI is important for monitoring seasonal and annual changes in the spatial distribution of vegetation. LAI is estimated from remote sensing imagery, such as normalized difference vegetation indices (NDVI) using a red (0.6–0.7 μm) chlorophyll absorption band, and near-infrared (0.7–0.8 μm) high-water-reflectance band (Peterson *et al.*, 1987). LAI estimates are calibrated for coniferous trees, so a correction for canopy depth to control for higher NDVI gained from deciduous scrub is made using a water absorption band in the near-to-middle infrared (1.55–1.75μ) region of the spectrum (Band *et al.*, 1993; Nemani *et al.*, in press). Since remotely sensed imagery can be acquired at regular intervals from airborne and satellite sensors, estimates of LAI can be made at high spatial and temporal resolutions.

Within-stand distribution of soil wetness is a controlling factor on the distribution of vegetation in water-limited, mountainous environments. Differences in soil transmissivity and local topography result in variability over space and time in soil water availability. FOREST-BGC alone does not consider this within-stand variability of soil moisture so RHESSys incorporates TOPMODEL, a quasi-distributed model of watershed runoff production and soil moisture accounting. TOPMODEL computes spatiotemporal variation in soil moisture, using a soil wetness, or hydrologic similarity, index that scales functional behaviour of hillslope and watershed hydrology from areal mean values, using local variations in topography and soil hydraulic properties. Hydrologic similarity χ at point i is computed as

$$\chi_i = \ln[\frac{a}{\tan\beta}]_i + [\ln(T_e) - \ln(T_i)] \qquad (30.7)$$

where a is upstream contributing area per unit contour width, derived from the drainage area transform, T_i is soil transmissivity at point i, T_e is an areally integrated soil transmissivity for a given patch, and β is the topographic gradient at i. TOPMODEL uses the similarity index to scale system behaviour by assuming that locations with similar index values have the same soil moisture.

Metaknowledge is used to integrate the Simulation System and the knowledge base. Metaknowledge provides a means of describing how the knowledge base interfaces with the Simulation System, including data structuring, procedural calls and retrieval of results. Current versions of the Simulation System are designed to operate both as a stand-alone system and as an integral part of the knowledge base. System-defined metaknowledge provides a seamless view of integration of system components to the end-user, hence it is not visible at the user interface level. The Simulation System operates on simulation objects that are specialized kinds of objects, defined within the knowledge base to integrate information from topography, remote sensing and soils domains. This is shown conceptually in Figure 30.4, in which classification (e.g., hillslope is an instance of topography), generalization (e.g., simulation object is a subclass of hillslope and vegetation density and soil hydraulic properties), and aggregation (e.g., simulation object comprising distributions of spatially varying attributes) contribute to the definition of a simulation object class. With the integration of a knowledge base and a simulation system, there is a need to manage simulation objects in time. Three classes of time are used: (1) valid time for objects, (2) transaction time for objects, and (3) simulation-defined time. Valid time refers to the time at which an object is true in the reality being modelled by the knowledge base. This might be the time at which simulation results were generated by a simulation system. Transaction time refers to the time when an object is stored in the knowledge base. Simulation-defined time is a kind of user-defined time (Jenson *et al.*, 1992) where the user of the knowledge base is the simulation system. Any objects created using the simulation results need to include some identification of simulation time. For example, a simulation object at location X created from simulation results for day 180 of year 1988 is different from a simulation object created from results of day 365 at the same location X. As such, object identity is defined, in part, by the simulation time.

Implementation and Testing

Ideas presented above were implemented in the Knowledge-Based Land Information Manager and Simulator (KBLIMS), and tested on Soup Creek, a 15 km^2 watershed in the Swan Range of the Rocky Mountains of north-western Montana. Soup Creek is predominantly covered by conifer forest composed of lodgepole pine (*Pinus contorta*), Douglas fir (*Pseudotsuga menziesii*), larch (*Larix occidentalis*), red cedar (*Thuja plicata*), subalpine fir (*Abies lasiocarpa*), and Englemann spruce (*Picea englemannii*). The topography is steep, with an elevation range of about 1000 m, resulting in significant microclimatic differences

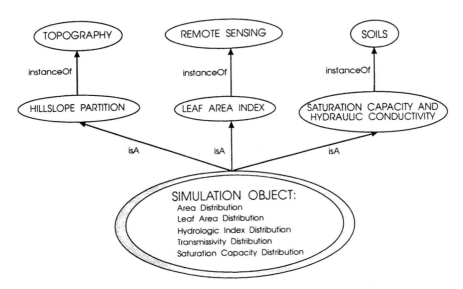

Figure 30.4 A simulation object is the integration of a hillslope partition, leaf area index and soil saturation capacity and hydraulic conductivity.

Note: Attributes inherited from the topography, remote sensing and soils domains may be represented as distributional information within simulation objects.

between hillslopes. This is illustrated in Table 30.1, which shows mean values of the hillslope attributes computed by the terrain analysis tools, including hillslope mean soil water-holding capacities (SWC). Topographic and soil variability result in variable mean hydrological similarity (Lambda), and highly variable mean LAI. Figure 30.5 shows LAI draped over the surface topography of Soup Creek. The view is from the basin outlet looking up the valley. The circular, dark patches of low LAI on the valley bottom are clearcuts. The effects of clear cutting on the hydrologically active zone near streams cannot be addressed if only a mean LAI is used for each hillslope. Within-hillslope variability of LAI as a function of hydrological similarity is shown in Figure 30.6 for four microclimatic regimes. The four microclimatic regimes are low-elevation south-facing, high-elevation south-facing, low-elevation north-facing, and high-elevation north-facing hillslopes. Within a given microclimatic regime in Soup Creek, LAI generally increases with increasing soil wetness. LAI is highest on south-facing hillslopes at low elevations, and lowest on north-facing hillslopes at higher elevations. The distributions of LAI are identified as attributes of simulation objects to account for clearcuts and other disturbances.

Topological relations derived from the terrain analysis are used to support graphical queries. Derivation of topological relations from drainage basin objects is illustrated in Figure 30.7. Links between objects, such as hillslopes 1 and 3, are physically stored as binary relations but are recognized as graphical objects having start and end window co-ordinates in a graphical window. Figure

Table 30.1 Mean surface property attributes generated by the terrain analysis system, for each hillslope object of Soup Creek watershed.

Object Id	Aspect (degrees)	Elev (m)	Gradient (degrees)	LAI	SWC	Area (m)	Lambda (ha)
1	177.26	1600.49	30.49	10.67	0.1082	129.33	5.70
2	330.06	1627.64	30.94	7.70	0.1224	101.34	5.44
3	167.26	1847.78	30.76	9.10	0.1017	23.67	5.33
4	236.69	1885.87	33.69	5.99	0.1024	35.28	5.26
5	206.74	1519.03	17.29	10.13	0.1629	18.18	5.47
6	7.51	1640.07	28.51	7.96	0.1346	38.43	5.83
7	209.94	1771.50	26.89	6.59	0.1316	238.32	5.98
8	23.36	1692.34	27.25	7.14	0.1452	101.43	5.61
9	184.81	1834.96	22.24	8.15	0.1163	14.67	5.92
10	245.60	1923.53	29.18	6.31	0.1051	57.60	5.96
11	242.93	1597.21	10.90	4.04	0.1425	8.73	5.77
12	45.31	1711.43	22.45	6.61	0.1638	25.56	5.97
13	225.19	1623.83	13.68	6.10	0.1500	3.24	5.97
15	259.13	1968.24	28.49	4.96	0.1077	172.17	6.17
16	13.79	1842.99	23.90	6.07	0.1241	83.16	6.07
17	334.18	1883.00	24.94	6.13	0.1151	43.47	6.00
18	58.28	1925.02	26.95	5.32	0.1086	65.07	6.04
19	25.15	1941.57	30.65	4.95	0.1101	25.11	5.52
20	94.85	1932.33	27.41	5.71	0.1138	16.56	5.75
21	308.53	1838.30	33.74	6.92	0.1056	16.38	4.81
22	31.64	1847.57	34.72	8.03	0.1078	38.25	5.43

Note: Leaf area index (LAI) is derived from remotely sensed imagery, soil water-holding capacity (SWC) is derived from soil series, and mean hydrologic similarity index (Lambda) is derived from topographic and soils data.

30.8 shows the main components of KBLIMS performing a query, in which an object is selected (in the window labelled 2), a local neighbourhood is retrieved in (3), and simulation results are shown in (4). Each graphical object is of a named type, and is organized around a small number of binary topological relations providing access to all attributes and spatial relations while giving the end-user a simple, uncluttered graphical query tool. Figure 30.9 illustrates the use of recursion and aggregation, in which the initial query in window (1) produces a hypergraph (2) asking for a spatial relation. The result of the transitive closure of the relation drains_from is shown in window (3), and the aggregated result for the retrieved objects is computed in (4).

Conclusion

Forest ecosystems exhibit large spatial and temporal variations in land-surface processes and properties. Topographic, soils and vegetation data required for simulating regional scale forest ecosystems are difficult to obtain at scales

Figure 30.5 Leaf area index (LAI) of Soup Creek, looking upstream from the basin outlet.

Note: In general, LAI decreases with distance from the main channel, with the exception of a set of clearcuts (dark patches) near the valley bottom. Disturbances introduce significant bias in correlations between topography and LAI requiring the use of distributional information.

commensurate with assumptions regarding inputs to simulation models. For this reason data integration and data management requirements for the accurate portrayal of processes in simulations hinder ecosystem studies of large areas. In this chapter we have shown how we manage simulations within these limitations by managing information on both simulations and data, rather than just collecting data. This shifts the focus towards an information systems approach that integrates information from multiple data domains, including topography, remote sensing and soils. These data are typically collected and stored as layers within a GIS. However, ecosystem simulations are made on the basis of higher order objects, such as hillslopes or stands, as they are more meaningful to an ecologist than layers of pixels and polygons. Higher order objects provide a compromise in spatial resolution and detail for watershed and regional scale

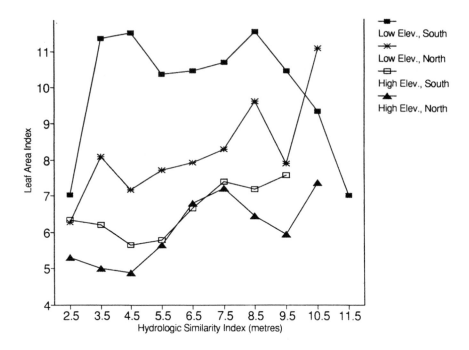

Figure 30.6 Leaf area index is correlated with hydrologic similarity to provide a spatial distribution within simulation objects.

Note: In this figure are shown distributions for low elevation south-facing (hillslope 1) and north-facing (hillslope 2), and high elevation south-facing (hillslope 10) and north-facing (hillslope 17) objects of Soup Creek basin. Differences within curves are related to available water, while between-curve variability results from radiation and temperature differences between hillslopes.

ecosystem simulations, by defining explicitly the physical relationships between well-defined objects. A parsimonious coupling of GIS and simulations is made using object-orientated techniques that transform simple observations and relations into higher order concepts based on structural, deductive and procedural knowledge. Layer-based information is transformed into object-based information, which is defined at a conceptual level in a knowledge base. The knowledge base stratifies its knowledge into layers to hide implementation details from conceptual descriptions. Layering allows changes in low-level tools such as simulation models and terrain analysis tools without changing the semantics of hillslope and forest stand objects.

Transformation of layer-based data to object-based information requires analytical techniques to create aggregates that retain enough spatial variability to provide a meaningful distribution of processes in space, and make simulation of large regions feasible. This analysis is typically a preprocessor to further analysis, but there are also merits to having aggregation as a post-simulation

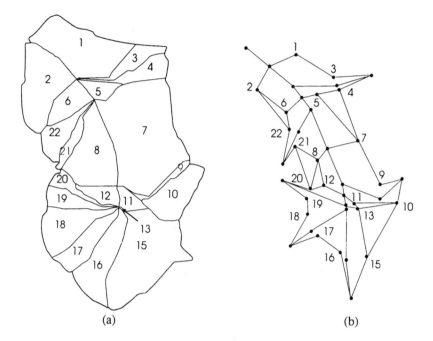

Figure 30.7 A layer of hillslopes and streams: (a) is transformed into a graph of topological links, (b) which are stored as binary topological relations.

Note: A graphical view of relations captures some structure of the domain and provides efficient browsing of the database.

process. KBLIMS supports post-simulation aggregation using common aggregate operators, such as count, average and minimum, which are found in many query languages (Maier, 1983). If a query requires different criteria for forming aggregates than are predefined by the aggregation preprocessor, then the two forms of aggregation are made at cross-purposes. For example, the terrain analysis system designed to retain variable slope and aspect may produce a large number of hillslopes with similar properties but at different locations in space. A query asking for evapotranspiration on south-facing slopes has to aggregate simulation results for all south-facing hillslopes. We have found that post-simulation aggregation becomes extremely inefficient for larger numbers of hillslopes. However, *a priori* aggregation introduces significant bias in simulations (Band, in press). A reasonable approach to solving this dilemma is to make the aggregation query dependent, so that objects are identified and aggregated to satisfy the requirements of the query. This approach would benefit from flexible reasoning techniques to classify objects on the basis of form and function (Mackay *et al.,* 1992a).

Another important issue is the ease with which information can be accessed. This is relevant to humans interacting with the information system, and other

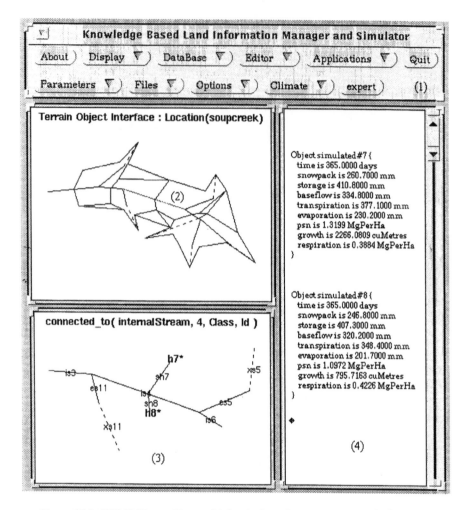

Figure 30.8 KBLIMS provides multiple windows for interaction with the system.

Note: The top level of the system is accessed via a toolbar (1). Hypermedia tools allow for both coarse and fine-grained search within the spatial database. The sequence of operations shows a portion of the Terrain Object Interface graph is selected (2) and its local neighbourhood shown in (3). This spatial query result is then used to ask for simulation attributes associated with the local neighbourhood (4).

programs accessing information in a distributed computing environment. KBLIMS provides a simple language based on query models that execute user-defined or program-defined queries. The purpose of the language is to provide a canonical model for communication between programs, such as an information system communicating with a simulation system. Queries are deductive so a user, human or program, may define a simulation experiment by first identifying a set of objects as a spatial query, then specifying some action to be performed

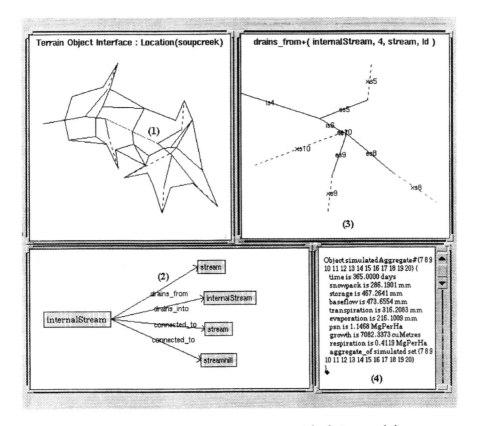

Figure 30.9 KBLIMS will do recursive queries on spatial relations, and then aggregate results of simulations for objects retrieved in recursive search of space.

Note: In this case a position in the stream network is selected (1), a recursive spatial relation is selected from a hypergraph (2), the transitive closure of the spatial relation is computed, and (4) results are aggregated.

on these objects, such as a combined simulation query and aggregation query. In order to ease the burden of the human user in specifying queries, a user interface provides graphical objects with which the user interacts. Graphical objects are visualizations that correspond to relations in the knowledge base. Relations attach themselves to object-class schemes to support telescoping of object views, from broad overviews of many objects down to a focused view of specific objects. The objects have knowledge on how explicitly to parametrize and run simulation models. Typical use of the simulation system is managed by the knowledge base using its metaknowledge. This allows for integrating systems designed to operate both as integrated or stand-alone programs while maintaining a seamless view at the user interface level.

We foresee considerable benefit of a KBLIMS approach to simulation for resource management and research. The potential exists for a forest manager to sit down with a graphical interface and explore various scenarios, for example, asking the system to reduce stand density in spatially explicit areas within a watershed and report on any changes in runoff and erosion. The system would retrieve the relevant spatial objects, reduce LAI, run the simulation models needed and report back to the interface. Researchers can also use a KBLIMS approach in developing and testing their models by using the information management tool to maintain a record of results tagged by a set of constraints for simulation. The information management tool provides fast, easy access to simulation results without the need for end-users explicitly to create and maintain input files for the simulation programs.

Acknowledgement

This work was partially supported by an operating grant from the Natural Sciences and Engineering Research Council (NSERC) of Canada, research grant NAGW-952 from the National Aeronautics and Space Administration (NASA) and an NSERC Doctoral Fellowship. Support of the computing facilities of the Institute for Land Information Management at the University of Toronto is also gratefully acknowledged.

References

Armstrong, M. P. and Densham, P. J., 1990, Database organization for spatial decision support systems, *International Journal of Geographical Information Systems,* **4**, 3–20.

Atkinson, M., Bancilhon, F., DeWitt, D., Dittrich, K., Maier, D. and Zdonik, S., 1989, The object-oriented database system manifesto, in *Proceedings of the International Conference on Deductive and Object-Oriented Databases,* pp. 40–7, Amsterdam: Elsevier Science Publishing Co.

Band, L. E., 1986, Analysis and representation of drainage basin structure with digital elevation data, in *Proceedings of the 2nd International Symposium on Spatial Data Handling,* pp. 437–50, Williamsville, New York: International Geographical Union.

Band, L. E., 1989, A terrain-based watershed information system, *Hydrological Processes,* **3**, 151–62.

Band, L. E., 1993, Effect of land surface representation on forest water and carbon budgets, *Hydrology,* **150**, 759–72, Amsterdam: Elsevier Science Publishing Co.

Band, L. E., Patterson, P., Nemani, R. and Running, S. W., 1993, Forest ecosystem processes at the watershed scale: Incorporating hillslope hydrology, *Agricultural and Forest Meteorology,* **63**, 93–126.

Band, L. E., Peterson, D. L., Running, S. W., Coughlan, J. C., Lammers, R. B., Dungan J. and Nemani, R., 1991, Forest ecosystem processes at the watershed scale: Basis for distributed simulation, *Ecological Modeling,* **56**, 151–76.

Beven, K. J. and Kirkby, M. J., 1979, A physically based, variable contributing model of basin hydrology, *Hydrological Sciences Bulletin,* **24**, 43–69.

BIM, 1990, *Prolog by BIM,* Belgium: BIM, Kwikstraat, 4-3078 Everberg.

Bracken, I. and Webster, C., 1989, Towards a typology of geographical information systems, *International Journal of Geographical Information Systems,* **3**, 137–52.

Goodenough, D. G., Goldberg, M., Plunkett, G. and Zelek, J., 1987, An expert system for remote sensing, *IEEE Transactions on Geoscience and Remote Sensing,* **25**, 349–59.

Halustchak, O., 1993, Proposed spatial data handling extensions to SQL, in Robinson, V.B. and Tom, H. (Eds) *Towards SQL Database Extensions for Geographic Information Systems,* pp. 69–84, Gaithersburg, Maryland: National Institute of Standards and Technology.

Jenson, C. S., Clifford, J., Gadia, S. K., Segev, A. and Snodgrass, R. T., 1992, A glossary of temporal database concepts, *ACM Sigmod Record,* **21**(3), 35–43.

Jones, N. L., Wright, S. G. and Maidment, D. R., 1990, Watershed delineation with triangle-based terrain models, *Journal of Hydraulic Engineering,* **116**, 1232–51.

Lammers, R. B. and Band, L. E., 1990, Automating object representation of drainage basins, *Computers and Geosciences,* **16**, 787–810.

Langran, G., 1992, *Time in Geographic Information Systems,* London: Taylor & Francis.

Larcher, W., 1980, *Physiological Plant Ecology,* 2nd Edn, Berlin: Springer.

Mackay, D. S., Band, L. E. and Robinson, V. B., 1991, An object-oriented system for the organization and representation of terrain knowledge for forested ecosystems, in *GIS/LIS '91 Proceedings,* pp. 617–26, Bethesda, Maryland: American Society for Photogrammetry and Remote Sensing.

Mackay, D. S., Robinson, V. B. and Band, L. E., 1992a Classification of higher order topographic objects on digital terrain data, *Computers, Environment, and Urban Systems,* **16**, 473–96.

Mackay, D. S., Robinson, V. B. and Band, L. E., 1992b, Development of an integrated knowledge-based system for managing spatiotemporal ecological simulations, in *GIS/LIS '92 Proceedings,* pp. 494–503, Bethesda, Maryland: American Society for Photogrammetry and Remote Sensing.

Maier, D., 1983, *The Theory of Relational Databases,* Rockville, Maryland: Computer Science Press.

Moore, I. D., Grayson, R. B. and Ladson, A. R., 1991, Digital terrain modeling: A review of hydrological, geomorphological and biological applications, *Hydrological Processes,* **5**, 3–30.

Moore, I. D., O'Loughlan, E. M. and Burch, G. J., 1988, A contour-based topographic model for hydrological and ecological applications, *Earth Surface Processes and Landforms,* **13**, 305–20.

Mylopoulos, J., Borgida, A., Jarke, M. and Koubarakis, M., 1990, Telos: Representing knowledge about information systems, *ACM Transactions on Information Systems,* **8**, 325–62.

Nemani, R., Pierce, L. L. and Band, L. E., in press, Remote sensing of leaf area index, *International Journal of Remote Sensing.*

O'Loughlan, E. M., 1986, Prediction of surface saturation zones in natural catchments by topographic analysis, *Water Resources Research,* **22**, 794–804.

O'Loughlan, E. M., 1990, Modeling soil water status in complex terrain, *Agricultural and Forest Meteorology,* **50**, 23–38.

Peterson, D. L., Spanner, M. A., Running, S. W. and Tueber, B., 1987, Relationship of thematic mapper simulator data to leaf area index of temperature coniferous forests, *Remote Sensing of the Environment,* **22**, 323–41.

Robinson, V.B. and Mackay, D. S., 1993, On heterogeneous geographic information systems, architectures, spatial data models, transactions, and database languages, in Robinson, V. B. and Tor, H. (Eds) *Towards SQL Database Language Extensions for Geographic Information Systems,* pp. 1–35, Gaithersburg, Maryland: National Institute of Standards and Technology.

Running, S. W. and Coughlan, J. C., 1988, A general model of forest ecosystem processes for regional applications, I: Hydrologic balance, canopy gas exchange and primary production processes, *Ecological Modeling,* **42**, 125–54.

Tucker, C. J., Townshend, J. R. G. and Goff, T. E., 1985, African land cover classification using satellite data, *Science,* **227**, 369–74.

Waring, R. H. and Schlesinger, W. H., 1985, *Forest Ecosystems: Concepts and Management,* Orlando, Florida: Academic Press.

Worboys, M. F. and Deen, S. M., 1991, Semantic heterogeneity in distributed geographic databases, *ACM SIGMOD Record,* **20**(4), 30–4.

31

Detecting fine-scale disturbance in forested ecosystems as measured by large-scale landscape patterns

G. A. Bradshaw and Steven L. Garman

As GIS-based and satellite data have become increasingly accessible, it is possible to integrate empirical and simulation approaches to pattern analysis and to translate knowledge of ecosystem processes at the stand level to landscape and regional scales. As a result, there has become an increased reliance on pattern to provide insight into understanding ecological processes. Because different processes may produce similar patterns, there is a critical need to understand what information regarding ecosystem processes (e.g., disturbance) is retained and detectable from quantitative measures of spatial pattern. The relationship between statistical and ecological measures of pattern and process in the Pacific Northwest, USA, is explored using simulated landscapes generated by varying disturbance events. Simulation results indicate that statistical significance of pattern does not correspond systematically to ecological significance. A predictable correspondence between process (i.e., fine-scale disturbance) and pattern (i.e., large-scale landscape structure) only occurred consistently under the restricted conditions of intense or multiple-event disturbances.

Introduction

The past few years have been marked by an unprecedented burgeoning of geographic information systems (GIS) and remote sensing applications in natural resource sciences and ecology. These applications span physical scales ranging from the microscopic (e.g., soil structure profiles) to regional and global scales. Regionalized variable analysis (e.g., spatial statistics) is undergoing a similar revival in the environmental sciences (Ford and Renshaw, 1984; Bradshaw, 1991; Turner *et al.*, 1991). With a parallel growth in computer technology, an increased reliance on these techniques is anticipated (Stafford *et al.*, this volume).

The new directions in database quantity and quality have shifted the research focus from fine-scale sampling to landscape and higher order scales where the spatial and temporal patterns of data are considered explicitly. Ecosystem scientists are working not only with multivariate data sets but also with multi-scale datasets, and hence the errors associated with each data layer. With increased usage of and reliance on raster- or vector-based databases such as satellite imagery and digital maps, it is essential to establish an understanding of the information content of landscape patterns in both a rigorous and quantitative manner. As a result of the growing technology, we are confronted with new problems that are both ecological and statistical in nature. One of these issues deals with pattern explicitly: what is the ecological and statistical significance of pattern?

A plethora of landscape metrics, statistics, time-series techniques and geostatistical methods are employed for the sole purpose of identifying pattern (e.g., Milne, 1988; Legendre and Fortin, 1989; Cohen *et al.,* 1990; Turner *et al.,* 1991; Bradshaw and Spies, 1992). Based on these methods, comparisons between landscapes and inferences relating to these statistical measures are drawn. Now that the initial excitement of the discovery of spatial pattern has passed, scientists are left with the less glamorous task of determining what the real significance of these numbers is relative to the ecosystem processes responsible for pattern generation; namely, determination of ecological and statistical significance of spatial pattern. Ecological significance does not necessarily imply statistical significance and vice versa.

For example, if a landscape has sustained two types of disturbance (e.g., fire and bark beetle outbreak) over a given time period, it is ecologically significant if one disturbance consistently occurs prior to the second disturbance event. In the case of a beetle outbreak, a consistent ordering of events may imply cause and effect, that is, the presence of fire may be required to precondition the susceptibility of the forest to beetle outbreak. However, measures of landscape pattern in the two distinct cases may not be distinct statistically. Subtle but important differences may fall below a statistically significant threshold. As a result, there may be limits to extracting process-related information solely from spatial and non-spatial patterns. This is particularly true if pattern is measured and evaluated using remote sensing techniques without field reconnaissance. Understanding the link between ecosystem processes and their related patterns as defined in a rigorous statistical and ecological sense will require the integration of both ecological and statistical analyses.

We provide an overview of the problems associated with inferring process from pattern and discuss concepts of the pattern–process duality leading to the relationship between ecological and statistical significance of landscape pattern. To illustrate some of these concepts for the Pacific Northwest (PNW), we designed a spatial simulation model to produce landscape patterns under varying sequences and intensities of natural disturbance. These simulations represent a pilot study to: (1) determine the statistical significance of landscape patterns, and (2) begin to unravel process attributes (e.g., type, intensity, sequence, size of disturbance) by examination of a given landscape pattern. Although the simulation study is directed at the landscape scale, it is amenable for application at larger scales as well.

Landscape Pattern–process Duality

During the past several years, the number of landscape- and regional-scale ecological studies has increased dramatically. In contrast to prior ecological analyses, the unit of study is now focused at the landscape level where the type, arrangement and function of its component parts are studied to understand the higher order spatial unit, the landscape. The exact dimensions of a landscape may vary according to a given ecosystem or biome. Obtaining a strict definition of a landscape is perhaps less important than understanding what the approach affords. Landscape ecology examines the interactions and relationships between individual biological and physical components which compose the system within the relevant temporal and spatial context (Forman and Godron, 1986). The underlying assumption is that these individual components (e.g., landscape subunits) are to varying degrees interdependent. Traditionally, the approach has been to view the system in terms of its pattern and processes. Implicit in this approach has been the underlying premise that pattern provides a window by which processes can be divined. While the pattern–process paradigm is well established, there is still a poor understanding of this relationship at the landscape scale, for several reasons.

A historical factor contributing to this pattern–process duality has its roots in the origin of landscape ecology itself. In Europe where many landscape ecology principles were first applied, the link between pattern and process is more obvious than in less developed landscapes of western North America. European landscapes have been organized in several iterations of human activities over the past centuries; Roman roads, Gaul settlements, and a further patchwork of private and state agricultural lands and forests. French farmland, English hedgerows, German farm and forest lands, and Dutch shelterbelts are landscape patterns as familiar as our own American lands (Forman and Godron, 1986). Even in the Alps, the signature of human activities is strongly etched in the landscape. The processes that are responsible for dominant structures forming these landscapes, namely, human activities, are readily observable within a lifetime. As a contrasting example, the conifer-dominated landscapes of the Pacific Northwest pose a different set of conditions. Here the sources and types of processes that form landscape patterns are less straightforward because of differences in the historical development of human activity in the landscape (Ripple *et al.*, 1991). Until only recently the signature of natural disturbance patterns has been masked by intense timber harvesting and road construction.

In addition, the accessibility of computer graphics technology, GIS, and spatial statistics has facilitated the first step in landscape analysis by allowing ecosystem scientists to identify and quantify landscape patterns. As a result, there is generally a better understanding of the range and types of landscape *pattern* in natural systems rather than the *processes* responsible for their generation. Thus, while we are much more cognizant of the types of patterns found in the Pacific Northwest forests (e.g., Ripple *et al.*, 1991, Spies *et al.*, in press), the identification of landscape-level processes is only beginning. Few studies exist which translate knowledge of the ecosystem to an explicit spatio-temporal context. In general, among empirically based work, there is a tendency to find highly technical studies on landscape pattern and more qualitative studies describing landscape processes.

Sufficient quantitative information relating disturbance processes to resultant landscape patterns (i.e., stand structure, location and extent) is lacking. While certain processes might be related to an observed pattern (e.g., forest stand complexity and condition), there is insufficient understanding regarding the pattern–process relationship to allow for accurate prediction of a disturbance event and its resultant pattern. For example, fire is a known disturbance agent in sculpting PNW conifer landscapes (Morrison and Swanson, 1990). Although the mechanisms of fire spread and initiation are generally well accepted, we are still far from being able to predict resulting landscape patterns accurately (i.e., the size, location and intensity of events).

Ecological Versus Statistical Significance of Pattern

Confounding the issue is the realization that there is not always a 1:1 mapping of a given pattern to a given process, that is, two independent ecological processes may produce the same pattern (Moloney *et al.,* 1991). A second factor leading to a non-unique process–pattern relationship is the intrinsic stochasticity of the system (e.g., random location of fire initiation). This seeming inconsistency leads one to enquire: what makes one pattern differ from another?

Two landscape patterns may vary in terms of their composition, spatial arrangement of units and variability through time. Thus, it is often insufficient to classify a given landscape with a single statistic. In fact, the identification and definition of components composing landscape pattern in itself will determine what form the pattern will take and therefore distinguish it from other patterns. For example, an aerial photograph or satellite image provides a snapshot of the landscape at a single moment. The process of pattern analysis can involve complex steps requiring subjective interpretation. Image classification is a subjective process in many ways; developing objective, repeatable algorithms requires both time and conceptual effort. Once an image is classified, the researcher is faced with the question of translating ecological units (e.g., a forest stand) to an equivalent unit in spectral space. The definition of homogeneous units will depend on many factors. To illustrate, consider the problem of defining a forest stand as a coherent mappable unit. Obtaining a consistent definition of a forest stand will depend on several factors each relating to various sources of variability (e.g., image resolution relative to landscape variability (heterogeneity), within-class variability and interclass variability), which in turn is a function of data resolution relative to the spatial autocorrelation character-izing landscape structure. Stand definition will also vary according to the ecological context in which it is being considered; what may be a homogeneous unit for a silviculturist may be interpreted differently by a wildlife biologist where considerations of landscape connectivity may play a role. To develop a consistent stand classification algorithm, meaningful definitions of ecological units and process must be established. Once ecological significance of pattern and its corresponding processes are established, we must relate statistical significance to its occurrence.

Statistical Detection of Ecological Patterns

The spatial patterns recorded by GIS and imagery represent the summation of a multitude of processes and patterns sustained by the landscape over time (i.e., a snapshot or a single pattern sample from a population of possible patterns). From this perspective it is perhaps suitable to view spatial pattern and temporal variability in a fashion analogous to the statistical mean and variance. Until recently, ecologists were chiefly concerned about the mean value of a phenomenon. With the advent of ecological concepts of heterogeneity, spatial correlation and patch dynamics, the variance has instead become the statistic of interest. Spatial pattern, like the mean, gives an average picture of the landscape; a sample averaging over time. However, similar to the mean, spatial pattern can mask the sets of processes and other patterns that have created it.

In determining statistical significance, it is important to stress that landscape pattern may be spatial or non-spatial in nature. Two landscapes may differ in pattern based simply on a non-spatial summary statistic like the mean or variance of total mean basal area (Ripple *et al.,* 1991). On the other hand, landscape pattern may comprise a difference in one or more elements composing spatial pattern (e.g., patch size, patch distribution, anisotropy, nested structure; Bradshaw, 1991). The existence of statistical significance is therefore a function of what aspect of pattern is being measured and compared. The pattern element of interest is most often designated by its ecological significance, bringing us full circle to the importance of establishing meaningful definitions of ecological patterns.

Pattern Generating Processes in the Western Cascades, Oregon

Simplistically, the state of the forest ecosystems can be described as two opposing forces: regeneration and growth, and disturbance and mortality. These processes are not decoupled but interact over space and time. Forest composition and structure reflect a number of different factors, including endemic genetics, site conditions, and the type, frequency and magnitude of disturbance events. A given region will be characterized by its own set of landscape patterns.

The Western Cascade forests of Oregon are patterned by an array of disturbance events such as drought, insect predation, fire, blowdown and root rot (Bradshaw and Spies, 1992). Each agent of disturbance can be characterized by a specific range in space and time. The nature of the disturbance will influence the type of landscape pattern observed. For example, drought may be regarded as a regional or even global-scale phenomenon resulting from a shift in the large-scale precipitation regime (e.g., ENSO events). Temporally, average precipitation may vary from annual to decadal or longer periods. Note that, while the domain over which drought varies is regional, the observable effects of the drought (i.e., pattern) follow the landscape (e.g., south-facing slopes may experience higher seedling mortality as opposed to north-facing slopes).

In contrast to drought, root-rot (*Phellinus*) is an individual-based disturbance agent attacking a single tree at a single point in time. The pathogen spreads radially from the source and may subsequently create patches of mortality within the stand. The difference in the two disturbance processes create parallel differences in the resultant landscape pattern; *Phellinus* pockets punctuate a stand while drought effects are pronounced both on the individual level and at the larger landscape or regional scales.

Two disturbance processes and hence their patterns may not always be mutually distinct and independent. *Phellinus* has been documented to attack both healthy and stressed trees by spreading along tree roots below ground. The rate and direction of spread depend on factors influenced by below-ground processes. In contrast, the Douglas fir bark beetle (*Dendroctonus pseudotsugae*) prefers downed logs and live but stressed large Douglas fir (*Pseudotsuga menziesii*) trees (Atkins and McMullen, 1958). Large concentrations of suitable host material, occurring from disturbances such as windthrow or high intensity wildfire, can cause epidemic outbreaks, resulting in infestation and subsequent death of live and healthy Douglas fir trees (Bedard, 1950). Bark beetle outbreaks have also been linked to areas that have sustained high moisture stress (i.e., loci where the effects of drought are pronounced). Thus, mortality patterns related to bark beetle infestation reflect an imprint of drought effects and pattern as well.

Until the last 100 years and in particular the last 30 years, the signature of natural disturbance in Northwest forests has been the predominant pattern. During the past century, the Northwest landscape has become dominated by timber harvesting and road construction; the range of cutting styles of private and public holdings has created a patchwork of young, even-aged stands with patches of older, intact stands, which has resulted in a gradual masking of the imprint of natural disturbance.

Pattern and Process Simulation of Forested Landscapes in the Western Cascades, Oregon

We developed a spatial simulation system capable of simulating fine-scale stand dynamics and multiscale disturbances to evaluate the relationship between pattern and process in the coniferous-dominant forests of the PNW region. The model was developed specifically for stand and disturbance conditions characterizing the H. J. Andrews Experimental Forest Long-Term Ecological Research (LTER) site, located in the central Western Cascades of Oregon.

The vegetation-dynamics model component of the system is an adaptation of the individual process-based stand-level gap model ZELIG.PNW.2.2 (Urban, 1993), which has been successfully parametrized and tested for the H. J. Andrews Experimental Forest (Garman *et al.,* 1992). Similar to the stand-level gap model, the present version uses a stochastic approach to simulate demographics of individual live trees, and snag and log dynamics in 0.1-ha cells. To accommodate simulation of stand dynamics across large spatial scales, computational speed of the stand model was substantially increased by replacing some of the detailed computation of tree growth with non-linear functions derived from simulation experiments using ZELIG.PNW.2.2. Despite this simplification, model output compares favourably between the two models. The state space of each cell in our version includes diameter at breast height (dbh) of live stems by species, and mass of slow (= Douglas fir) and fast (= all other species) decaying snags and logs. Each cell on the landscape can be initiated with specific composition and structure information or initiated from bare ground.

Of the numerous disturbances agents found in the Western Cascades, three types were considered for the simulation: Douglas fir bark beetle, drought and fire. Bark beetle infestation is simply modelled as a function of the amount of log mass of Douglas fir and the occurrence of live Douglas fir >60 cm dbh. A specified number of initial points of beetle outbreak are determined prior to a simulation. The initial location for each outbreak is randomly selected. For a cell to be infected, both it and at least three adjacent or diagonal neighbours must meet the log mass and tree size threshold. This neighbourhood approach was motivated by field observations of the tendency for bark beetle infestations to be locally aggregated. All cells satisfying the infestation criteria become infected. All neighbours of the initial location and of all newly infected cells are evaluated in a similar manner. Infestation will spread from the initial location across a landscape in a contiguous manner until all surrounding cells are below the criteria threshold. When infected, all Douglas fir >60 cm dbh within the cell are killed and transferred to the snag and log pool.

Drought is manifested as a lowering of the threshold requirements for the spread of bark beetle, that is, the amount of log mass required for beetle infestation decreases under increasing drought. This representation of drought was used to examine the effects of imposing a landscape-level disturbance on detection of pattern by globally altering the susceptibility threshold of the landscape to beetle outbreak. This simplistic approach was deemed adequate within the study context to represent the relative response of beetle outbreak to drought. Through sensitivity analysis, threshold values based on the mean and standard deviation of log mass of Douglas fir on the simulated landscape were derived to represent a range of drought conditions; ranging from no drought (= normal conditions) to severe drought. The mean minus 0.1 times the standard deviation was used to represent no drought; severe drought equalled the mean minus 0.5 times the standard deviation.

A modified version of a detailed wind-driven spatial fire model was used to emulate spread and intensity of wildfire (Garman, 1992). Frequency and size of each fire are specified prior to a simulation. The point of initiation and direction of maximum spread of each fire are randomly selected. The rate at which a fire spreads to neighbouring cells increases with log mass. In general, fires tend to be elliptical in shape, but deviation from this general form results where dead and downed fuels (= log mass) are much greater than that of neighbouring cells. The intensity of a fire in each cell was modelled as a linear function of log mass. Fire intensity increases with increasing log mass, resulting in the death of a greater proportion of the live basal area in a cell. In this version of the model, trees are stored in an array by decreasing dbh. Killing of trees begins at the bottom of this array and continues until the required basal area has been removed. Thus, a lower intensity fire kills the relatively smaller trees; larger trees are only killed in a high-intensity fire. Fire intensity was scaled so that 50 per cent of the basal area would be killed when the log mass of a cell was equal to the mean of the simulated landscape. Killed trees are divided equally between the snag and log pool.

Simulation Experiments

For all simulation experiments, we used a simulated 200-year-old landscape 100×100 cells in size (1000 ha), which was generated from bare ground using the vegetation dynamics model. Although the landscape was relatively even-aged, the stochasticity of the vegetation dynamics model ensured that log mass and overstory composition and structure of each cell varied in a manner similar to that expected on a real-world landscape recovering from a large disturbance such as catastrophic wildfire.

Using a variety of values for intensity and frequency, simulations were initially performed to evaluate the effects of beetle outbreak and wildfire separately (e.g., Figures 31.1a–d). For the purpose of illustration, we chose those conditions that would provide an adequate contrast. Two levels of initiation frequency for bark beetle outbreak (1 and 5), two levels of drought (severe and no drought) and two intensities of wildfire (one fire 1000 ha in size, and three fires each 3000 ha in size) were used. Temporal sequencing of the two levels of wildfire and beetle outbreak under the two levels of drought were varied. Initial locations of disturbances were held constant between paired simulation runs (i.e., runs differing only in the temporal order of disturbance). This ensured that any observed differences in temporal sequencing were not confounded by different locations of initiations. Disturbances were generated on the simulated 200-year-old landscape in selected order, and total basal area of each cell and the corresponding spatial co-ordinates (= row and column) were recorded. In this exercise, we did not vary the temporal dynamics of the landscape but merely

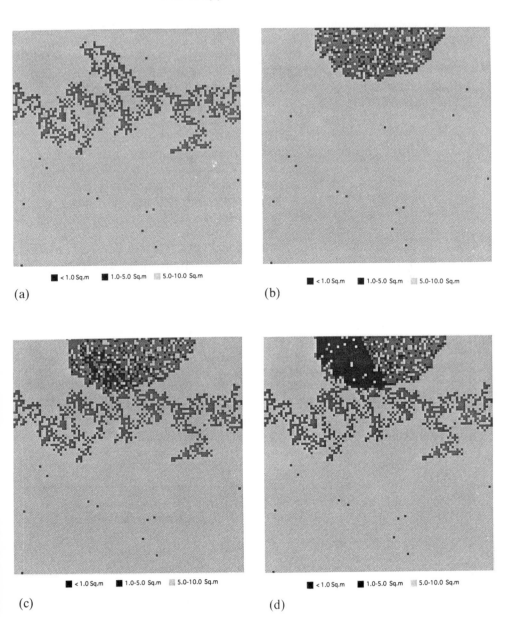

Figure 31.1 Examples of simulated patterning of total basal area resulting from different combinations of bark beetle infestation and wildfire.

Note: Initiation points for wildfire and for beetle outbreak were held constant among all runs: (a) bark beetle under the no drought condition and with one initiation point, (b) a 1000 ha wildfire, (c) bark beetle outbreak followed by wildfire, (d) wildfire followed by bark beetle outbreak.

imposed disturbances on the initial simulated landscape. Five replications of each scenario were performed, each initiated with a different random number seed to vary initial locations for each disturbance.

Spatial and non-spatial statistics were calculated for each simulation run and used to compare the response of landscape pattern to the temporal sequence of disturbances. Mean total basal area was used as the non-spatial metric. The continuous values of basal area were translated into three discrete classes (aggregation criteria of <1 m^2, 1–5 m^2 and >5 m^2) before calculating spatial metrics. These intervals were selected because they emphasized the effects of each disturbance. Because of the potential for biasing results, we varied our intervals by $\pm$ 30 per cent and compared results among several scenarios. This sensitivity analysis indicated little change in relative differences among temporal order of disturbance and among the different levels of disturbance, indicating that our initial interval values provided a relatively unbiased classification scheme. The landscape metrics program FRAGSTATS (McGarigal and Marks, 1993) was used to calculate a wide range of spatial metrics for each classified map. Mean nearest neighbour (summation of minimum distance between patches of similar type/total number of patches in the landscape) and total patch edge (summation of the amount of edge of each patch) metrics best represented the pattern of the classified maps, and were used for statistical comparisons.

Results

Visual examination of individual landscapes shows considerable variability of both spatial and non-spatial pattern. Three landscapes are included to illustrate a representative spectrum of pattern variability observed in the numerous simulation runs (Figures 31.1c, 31.3a and 31.3c). These landscapes were generated using the same parameters (low drought, and a single fire event following a single beetle outbreak); the sole factor that differed among the three was the random initiation point of disturbance. In contrast with significant differences in visual assessment of pattern, the variability within a given disturbance scenario as measured by the variance is relatively low (Figure 31.2a–c).

General trends in the three metrics used to analyse pattern under temporal sequencing of disturbance events were fairly consistent across all scenarios. When wildfire was implemented first, the landscape had lower total mean basal area (Figure 31.2a), higher mean nearest neighbour distances (Figure 31.2b), and lower total edge length (Figure 31.2c). Statistical comparison of metrics indicates a significant interaction between disturbance severity and temporal order. Under the single fire scenario, spatial and non-spatial metrics distinguished temporal order under severe drought regardless of the number of bark beetle initiations. This was also evident under the multiple fire–no drought scenario. Under the most severe level of disturbance simulated (multiple fires, severe drought), spatial metrics were significantly different ($p < 0.05$) between disturbance sequences, but mean total basal area was similar ($p > 0.05$). In

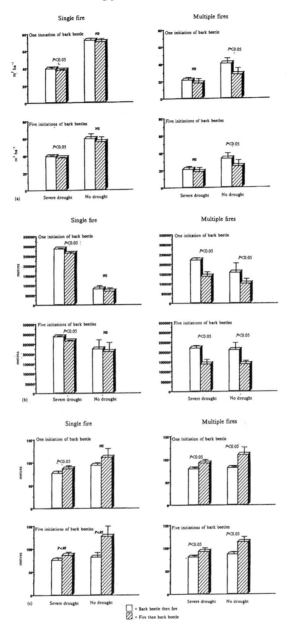

Figure 31.2 Non-spatial and spatial statistics of simulated landscapes under varying drought conditions and number of initiations of bark beetle and wildfire, and for different sequencing of disturbances.

Note: (a) mean total basal area, (b) mean minimum distance between patches of similar types, (c) mean total edge of patches. Means based on five replications. Error bar represents upper 95 per cent confidence interval. NS = not significantly different at the 0.05 level.

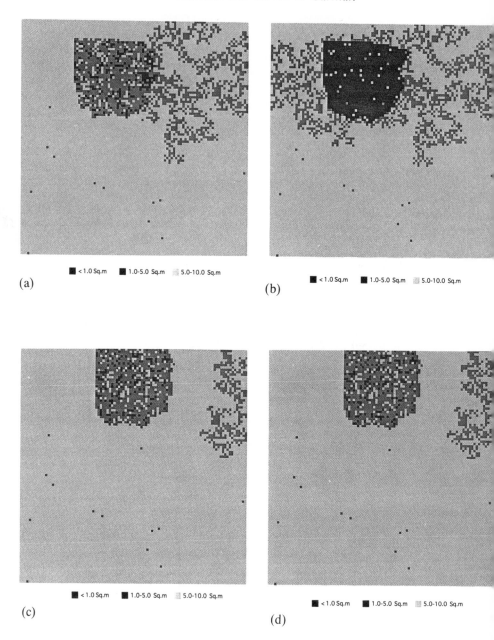

Figure 31.3 Examples of simulated patterning of total basal area under different sequences of a 1000 ha wildfire and bark beetle infestation under the no drought condition, showing importance of location of disturbance initiation.

Note: (a) bark beetle outbreak followed by wildfire, (b) reverse order of (a), (c) bark beetle outbreak followed by wildfire, (d) reverse order of (c).

contrast, metrics differed little between temporal order in the single fire–no drought scenario, which represents the least severe set of conditions evaluated.

In general, there is a similar trend observed when the number of disturbance events increases (i.e., fire events or beetle outbreaks) as when drought conditions are changed to severe (Figures 31.2 a–c). More specifically, it is not possible to say that spatial and non-spatial metrics are systematically significant. On closer scrutiny, there are some interesting deviations from this trend. For example, under conditions of severe drought and multiple fires, increasing the number of beetle outbreaks from one to five does not increase statistical significance for total mean basal area though there is a change in significance in the case of landscape metrics (Figures 31.2 a–c).

The effects of increasing beetle outbreaks is not symmetric with increasing the number of fire events, demonstrating that the individual process mechanism acts as a regulating factor in determining the type and spread of disturbance in the landscape. Wildfire tended to form large elliptical shapes comprising a mosaic of patches. This mosaic resulted from the heterogeneous distribution of log mass as well as stand conditions of the initial landscape (e.g., Figure 31.1b). When wildfire was the first disturbance event, the increased levels and the contiguous pattern of log mass promoted the spread of beetles throughout the burn, thus further reducing standing basal area. The initial mosaic produced by wildfire was reduced to a single large patch of low basal area, thus decreasing the overall number of patches on the landscape. This in turn accounts for the greater mean distance between patches and the increase in total edge of patches on the landscape. When wildfire occurred after beetle outbreak, the intensity of the fire would be greater where it overlapped with trees killed by beetles, resulting in small patches of low basal area. A mosaic of patches would tend to be produced, however, where the two disturbances did not overlap. Thus, the number of patches on the landscape would increase, which in turn would increase the total amount of edge and decrease the distance between similar patch types.

Discussion

The simulations presented here lack the complexity of natural landscape dynamics observed in the forests of the Pacific Northwest. This observation is not surprising as the central objective of the study has been to reduce the problem's dimensionality to isolate a single component of the dynamics; namely, the temporal order of disturbance events. Nonetheless, the simulations have provided some insight into the original inquiry relating pattern to process.

At the beginning of our discussion, two landscapes were considered similar when the ecological functions shared by the two landscapes were alike. In the examples presented here, the 'same' ecological processes were defined as those examples where the same disturbance algorithms with identical parameters and the same initial landscape conditions were used. Two cases were examined: (1) ecologically identical processes (i.e., pattern variability resulting from a difference in random initiation points), and (2) ecologically distinct processes (i.e., pattern variability resulting from a reversal in disturbance order (e.g., fire–beetle versus beetle–fire). These landscape patterns generated by the same ecological processes but differing in the disturbance initiation locations comprise a set of visually very different and distinct patterns (Figures 31.1c, 31.3 a and 31.3 c). In light of our definition, although these landscapes do not differ in terms of the ecological processes that created them, they are spatially and visually distinct (see Pfaltz and French, this volume, for what constitutes 'change').

In the second case, where landscape patterns were generated from ecologically distinct processes (i.e., reversal of temporal order of disturbance events, see figures 31.1d and 31.3 d), pattern variability as measured by spatial and non-spatial metrics was generally significant only under the restricted conditions of severe disturbance (severe drought or multiple large disturbance events; Figures 31.2 a–c). Under the simulation scenarios, the ord·r of the disturbance events was not a sufficient criterion to distinguish two landscape patterns statistically. While two landscapes may be generated by the same pattern-generating processes, their respective resultant patterns may differ. We would hypothesize that landscapes characterized by significantly different patterns would also be characterized by correspondingly different functional behaviour. These results suggest that a refinement of the original definition of 'ecological significance' of processes and patterns is appropriate.

Accurate evaluation of ecological processes may require considerations of several factors relating to disturbance (e.g., type, frequency and intensity) as well as the existing landscape pattern. When temporal order of disturbance events was reversed, the interaction of the process with pattern changed the final landscape pattern outcome. The first disturbance, be it fire or beetle infestation, acted to alter the landscape differently than did the second disturbance. Technically, it might be more suitable to modify our previous notion of 'sameness'. We may be best served to consider differences in landscape conditions both before and after the second disturbance rather than a single comparison of the final landscape pattern. The 'sameness' of a landscape would be defined based on a combination of disturbance attributes and the existing patterns. Essentially, the distinction between pattern and process becomes less distinct; in this approach, pattern is defined based on its potential response to various types and attributes of disturbance. In practice, this is usually not feasible; procuring pattern data before and after disturbance events is difficult because of the limited window over which imagery exists. However, interactive use of landscape simulations, GIS and field data promises to provide a better understanding of key factors involved.

Second, simulation results indicate that statistical significance does not correspond systematically to ecological significance. More specifically, while the spatial landscape metrics (total edge and nearest neighbour) distinguished the temporal order of disturbance on average, non-spatial statistical significance as measured by total mean basal area did not detect differences in temporal order of disturbance consistently. Deviations among the trends observed indicate that the interactions existing between disturbance intensity and type play an important role in pattern determination. This observation suggests the existence of an observable set of threshold conditions (e.g., patterns were distinguishable only under conditions of severe drought for a single fire event and no drought under multiple fire events; Figures 31.2 a–c). In general, statistical significance was observed above and below certain levels of disturbance severity. Defining numerical and spatial bounds of significance may provide a useful means to map regions of detectable ecological and statistical significance.

Implications for Management

In the simulations discussed here, the initial landscapes were uncorrelated, that is, the simulation was designed such that each cell was independent of the other with respect to stand growth and mortality. In reality, both natural landscape patterns and human-derived disturbances create a correlated landscape at several scales. Landscape-level features such as topography contribute to fire spread and initiation, creating a mosaic of successional stages and composition across the landscape. Timber harvesting has imposed a severe and high-frequency disturbance regime on the landscape. The PNW landscape is now dominated by spatial correlation at the scale of individual harvest units and their aggregate (Franklin and Forman, 1987). Our simulation results indicate considerable variability resulting from a change in disturbance initiation site alone across an uncorrelated landscape. The creation of highly correlated landscapes by intense and sustained clear-cutting in the PNW has dramatically altered the potential for sites to be disturbed. This superpositioning of patterns on the landscape may predetermine and restrict the natural disturbance regime, and prevent new ecosystem management practices from effectively mimicking natural disturbance conditions. In essence, we may have rendered the landscape into a disturbance regime well beyond the limits of natural variability.

It seems clear that intensive efforts by ecologists and statisticians need to be directed to reassessing the concepts of ecological and statistical significance of pattern. Efforts need to be directed in two main areas: (1) identification and definition of ecological processes and units at the landscape level, and (2) quantitative sensitivity analyses relating changes in process attributes (e.g., disturbance intensity) to changes in spatial and non-spatial pattern. In the course of these efforts, it is likely we will be able to develop a fuller understanding of what is really meant by the 'pattern–process' paradigm.

Acknowledgement

We would like to thank two anonymous reviewers for their comments and suggestions.

References

Atkins, M. D. and McMullen, L. H., 1958, Selection of host material by the Douglas fir beetle. Canadian Department of Agriculture, For. Biol. Div. Sci. Serv., *Bimonthly Progress Report,* **14**(3), 5–16.

Bedard, W. D., 1950, *The Douglas-fir Beetle,* US Department of Agriculture, Circular No. 918.

Bradshaw, G. A., 1991, 'Hierarchical analysis of pattern and process in Douglas fir forests using the wavelet transform', unpublished PhD dissertation, Oregon State University.

Bradshaw, G. A. and Spies, T. A., 1992, Characterizing canopy gap structure in forests using the wavelet analysis, *Journal of Ecology,* **80**, 205–15.

Cohen, W. B., Spies, T. A. and Bradshaw, G. A., 1990, Using semi-variograms of serial videography for analysis of conifer canopy structure, *Remote Sensing of the Environment,* **34**, 167–78.

Ford, E. D. and Renshaw, E., 1984, The interpretation of process from pattern using two-dimensional spectral analysis, *Vegetatio,* **56**, 113–23.

Forman, R. T. T. and Godron, M., 1986, *Landscape Ecology,* New York: Wiley.

Franklin, J. F. and Forman, R. T. T., 1987, Creating landscape patterns by cutting: ecological consequences and principles, *Landscape Ecology,* **1**, 5–18.

Garman, S. L., 1992, *The Acadia National Park Geographic-based Fire and Natural Resource Management System (AGEOFRSS),* Bar Harbor, Maine: US Department of the Interior, North Atlantic Region, National Park Service, Office of Scientific Studies, Acadia National Park.

Garman, S. L., Hansen, A. J., Urban, D. L. and Lee, P. F., 1992, Alternative silvicultural practices and diversity of animal habitat in Western Oregon: A computer simulation approach, in Luker, P. (Ed.) *Proceedings of the 1992 Summer Simulation Conference,* pp. 777–81, Reno, Nevada: Society for Computer Simulation.

Legendre, P. and Fortin, M. J., 1989, Spatial pattern and ecological analysis, *Vegetatio,* **80**, 108–38.

McGarigal, K. and Marks, B., 1993, 'FRAGSTATS User's Manual', unpublished, Department of Forest Science, Oregon State University, Corvallis.

Milne, B. T., 1988, Measuring the fractal geometry of landscapes, *Applied Mathematical Computing,* **27**, 67–79.

Moloney, K. A., Morin, A. and Levin, S. A., 1991, Interpreting ecological patterns generated through simple stochastic processes, *Landscape Ecology,* **5**(3), 163–74.

Morrison, P. H. and Swanson, P., 1990, *Fire History and Pattern in a Cascade Range Landscape,* Forest Service General Technical Report PNW-24, Portland, Oregon: USDA Forest Service, Pacific NW Station.

Ripple, W. J., Bradshaw, G. A. and Spies, T. A., 1991, Measuring landscape patterns in the Cascade Range of Oregon, USA, *Biological Conservation,* **57**, 73–88.

Spies, T. A., Ripple, W. J. and Bradshaw, G. A., in press, Dynamics and pattern of a managed coniferous forest landscape in Oregon, *Ecological Applications.*

Turner, S. J., O'Neill, R. V., Conley, W., Conley, M. and Humphries, H., 1991, Pattern and scale: statistics for landscape ecology, in Turner, M. G. and Gardner, R. H. (Eds) *Quantitative Methods and Landscape Ecology,* pp. 17–50, New York: Springer.

Urban, D. L., 1993, *A User's Guide to ZELIG Version 2.0,* Fort Collins, Colorado: Department of Forest Sciences, Colorado State University.

Subject index